D1234175

HIGHER CALCULUS

HIGHER CALCULUS

BY

FRANK BOWMAN

Formerly Head of the Department of Mathematics
Manchester College of Science and Technology

AND

F.A. GERARD

Professor of Engineering
Sir George Williams University, Montreal

CAMBRIDGE
AT THE UNIVERSITY PRESS
1967

Published by the Syndics of the Cambridge University Press
Bentley House, 200 Euston Road, London, N.W.1
American Branch: 32 East 57th Street, New York, N.Y.10022

© Cambridge University Press 1967

Library of Congress Catalogue Card Number: 66–11624

Printed in Great Britain
at the University Printing House, Cambridge
(Brooke Crutchley, University Printer)

PREFACE

The course of mathematical analysis presented in this book is intended to succeed an elementary course in which the reader will have become familiar with the leading ideas and applications of the differential and integral calculus. Accordingly, care has been taken to make the presentation sufficiently rigorous for most purposes, though it does not aim at the degree of rigour that might be expected in a modern advanced treatise.

The first four chapters are concerned with functions of a single variable, the next four with functions of two or more variables, and the last four with integration; but as this is not a first course, we have not refrained from assuming in the fourth chapter that the idea of integration of a continuous function will be well known to the reader. A feature of the fourth chapter is the prominence given at the beginning to the asymptotic property of the Taylor formula, so that practice may be gained in manipulating Taylor and Maclaurin expansions, especially those of the elementary functions, before embarking on a serious study of the convergence of infinite series. The rest of the book follows orthodox lines, after essential preliminaries have been reviewed and exemplified in the first chapter.

Examples for exercise have been collected at the ends of chapters so as not to interrupt the orderly progress of the text which will thereby, it is hoped, be found more easy to read.

The authors acknowledge their indebtedness to Mr F. L. Heywood, formerly senior mathematics master of Manchester Grammar School, and to the late Mr W. Hunter, formerly lecturer in the University and in the College of Technology, Manchester, both of whom made helpful comments on the early chapters. Thanks are also due to the advisers of the Cambridge University Press for many suggestions which have enabled us to add to any value that the original text may have possessed. Of the standard text-books to which we have frequently referred, we should at least mention those of Hardy and de la Vallée Poussin.

F. B.
F. A. G.

CONTENTS

Chapter 3. Successive differentiation

Chapter 4. Taylor's theorem

Chapter 5. Partial derivatives

Chapter 6. Implicit functions

Chapter 7. Successive partial differentiation

Chapter 8. Maxima and minima

Chapter 12. Double integrals

1

FUNCTIONS

1.1 Interval

If $a < b$, the values of a continuous variable x that satisfy the inequalities $a \leqslant x \leqslant b$ are said to form a *closed interval*. The interval may be denoted by $(a \leqslant x \leqslant b)$; it is said to be 'closed' because it includes the end values, or end 'points', a and b, as well as all the values, or 'points', between a and b.

An interval $(a < x < b)$ which does not include the end points is called an *open interval*.

The closed interval is often denoted by $[a, b]$, the open interval by (a, b). This notation will not necessarily imply that $a < b$; e.g. $[0, x]$ will not necessarily mean that x is positive.

An interval may be closed at one end and open at the other as, for example, $(a \leqslant x < b)$, $(a < x \leqslant b)$.

In this book, all numbers will be real unless otherwise indicated.

1.2 Neighbourhood

Let c be an interior point of the interval $(a \leqslant x \leqslant b)$, and let δ be a positive number such that $a \leqslant c - \delta < c + \delta \leqslant b$. Then the sub-interval $(c - \delta \leqslant x \leqslant c + \delta)$ or $[c - \delta, c + \delta]$ is called a *neighbourhood* of c. In a neighbourhood of c it is sometimes convenient to say briefly that x *is near* c.

At the end points a, b the corresponding neighbourhoods are $(a \leqslant x \leqslant a + \delta)$, $(b - \delta \leqslant x \leqslant b)$ or $[a, a + \delta]$, $[b - \delta, b]$, respectively.

The closed interval $(c - \delta \leqslant x \leqslant c + \delta)$ may also be denoted by

$$0 \leqslant |x - c| \leqslant \delta,$$

the open interval $(c - \delta < x < c + \delta)$ by $0 \leqslant |x - c| < \delta$.

1.3

Theorem. Let i be any irrational number between 0 and 1. Let p, q, n be any positive integers such that $p < q \leqslant n$ and n is fixed.

There exists a neighbourhood of i which has the property that no rational number of the form p/q belongs to it.

Proof. Let d be the least of the differences

$$|i - p/q|, \quad \text{for all } p, q \text{ such that } \quad p < q \leqslant n,$$

and let δ be chosen so that $0 < \delta < d$. Then $[i - \delta, \, i + \delta]$ is evidently a neighbourhood of i which has the property stated.

1.4   Function

A variable y is said to be a *single-valued function* of a variable x if, for some set of values of x, the value of y can be found uniquely for each value of x in the set.

To indicate that y is a function of x, notations such as

$$y = f(x), \quad y = g(x), \quad y = \phi(x), \quad y = y(x), \ldots$$

are used.

To say that y is a single-valued function of x implies the existence of a rule or rules by which the value of y is determined uniquely when the value of x is given. For any such value of x the function is said to be *defined*. A function is said to be *defined in an interval* if it is defined for every value of x in the interval (at every 'point' of the interval, or 'for all x' in the interval). If a function is defined in a closed interval, it is evidently defined in every subinterval, open or closed.

Examples

(1) The equations

$$y = x \quad (0 \leqslant x \leqslant 1), \quad y = 1 - x \quad (1 < x \leqslant 2),$$

define y for every value of x in the closed interval $[0, 2]$, but not for any other value of x.

Note that the function y is not 'continuous' at $x = 1$ (see §1.18).

(2) The equation $y = \sqrt{(1 - x^2)}$ defines y only in the interval $[-1, 1]$, since $\sqrt{(1 - x^2)}$ is real only for values of x in this interval.

(3) The equation $y = (\sin x)/x$ defines y for all values of x except the one value $x = 0$, for which $(\sin x)/x$ assumes the form $0/0$ which is meaningless.

(4) The equation $y = \log x$ defines y only when $x > 0.$†

(5) The equation $y = 1/(x - a)$ defines y for all x except $x = a$.

† In this book, $\log x$ will be used to denote $\log_e x$.

(6) If $f(x) = x/|x|$, then $f(x) = 1$ for all $x > 0$, $f(x) = -1$ for all $x < 0$; but $f(0)$ is not defined.

(7) The equation $y^3 = x$ defines y as a single-valued function of x in any interval of values of x, because, for every real value of x, the equation is satisfied by one and only one real value of y.

The equation $y^3 = x$ is said to define y *implicitly* as a function of x; in the preceding examples, y is said to be defined *explicitly*.

The function y defined by $y^3 = x$, or $x = y^3$, is called the *inverse* of the function defined by $y = x^3$ (see §1.25).

(8) The equation $x^2 + y^2 = a^2$ defines two single-valued functions of x in the interval $-a \leqslant x \leqslant a$. One is $y = \sqrt{(a^2 - x^2)}$, the other

$$y = -\sqrt{(a^2 - x^2)}.$$

Each is represented graphically by a semi-circle, the first above the axis of x, the second below it.

The equation $x^2 + y^2 = a^2$ is said to define y as a *two-valued* function of x (cf. §1.28 Example 3).

(9) The equation $\sin y = x$ defines y as an *infinitely many-valued* function of x; there is an infinity of single-valued functions of x in the interval $-1 \leqslant x \leqslant 1$ that satisfy the equation (see §1.29).

1.5 Limit of a function at a finite point

Let $f(x)$ be defined at every point of the interval $(c - \delta \leqslant x \leqslant c + \delta)$ with the possible exception of the point c itself.

We say that $f(x)$ tends to a left-hand limit l_1 at $x = c$ if, for every positive number ϵ, there exists a positive number h such that

$$|f(x) - l_1| < \epsilon \quad \text{for all } x \text{ such that} \quad c - h \leqslant x < c,$$

and we write

$$f(x) \to l_1 \quad \text{when} \quad x \to c - 0, \quad \text{or} \quad \lim_{x \to c - 0} f(x) = l_1.$$

Similarly, we say that $f(x)$ tends to a right-hand limit l_2 at $x = c$ if, for every positive number ϵ, there exists a positive number h such that

$$|f(x) - l_2| < \epsilon \quad \text{for all } x \text{ such that} \quad c < x \leqslant c + h,$$

and we write

$$f(x) \to l_2 \quad \text{when} \quad x \to c + 0, \quad \text{or} \quad \lim_{x \to c + 0} f(x) = l_2.$$

If both l_1 and l_2 exist and have the same value l, then for every positive ϵ, a positive number h exists such that $|f(x)-l| < \epsilon$ for all x such that $0 < |x-c| \leqslant h$, and we write simply

$$f(x) \to l \quad \text{when} \quad x \to c, \quad \text{or} \quad \lim_{x \to c} f(x) = l.$$

The left-hand limit l_1 is often denoted by $f(c-0)$ and the right-hand limit l_2 by $f(c+0)$, when they exist. We cannot replace l_1, l_2, or l by $f(c)$ because $f(c)$ denotes the value of the function at $x = c$, and this value may not be defined. Even if it is defined, it has no bearing on the existence or value of any of the limits.

Example

$$f(x) = \frac{\sin (x-c)}{x-c}.$$

Here $f(c-0) = 1$ and $f(c+0) = 1$, so that both left-hand and right-hand limits exist, but $f(c)$ is not defined.

It follows from the definition of a limit that, if $f(x)$ tends to the limit l when $x \to c$, we can put

$$f(c+h) = l+\eta, \tag{1}$$

where η is a function of h which tends to zero when $h \to 0$, that is, $\eta \to 0$ when $h \to 0$.

Note that here $h \to 0$ through positive or negative values, but that the value $h = 0$ is irrelevant since $f(c)$ is not necessarily defined. In what follows, the statement 'lim $f(x)$ exists when $x \to c$' will be assumed to imply, whether stated explicitly or not, that $f(x)$ is defined in the neighbourhood of $x = c$, though possibly not at $x = c$ itself.

1.6.   Limit of $f(x)$ when x tends to infinity

Let δ be any positive number, and let $f(x)$ be defined when $|x| > \delta$.

We say that $f(x)$ tends to a limit l_1 when $x \to +\infty$ if, for every positive ϵ, a positive number X exists such that

$$|f(x)-l_1| < \epsilon \quad \text{for all} \quad x > X$$

and we write

$$f(x) \to l_1 \quad \text{when} \quad x \to +\infty, \quad \text{or} \quad \lim_{x \to +\infty} f(x) = l_1.$$

Similarly, we say that $f(x)$ tends to a limit l_2 when $x \to -\infty$ if, for every positive ϵ, a positive number X exists such that

$$|f(x) - l_2| < \epsilon \quad \text{for all} \quad x < -X$$

and we write

$$f(x) \to l_2 \quad \text{when} \quad x \to -\infty, \quad \text{or} \quad \lim_{x \to -\infty} f(x) = l_2.$$

If both l_1 and l_2 exist and have the same value l, then, for every positive ϵ, a positive number X exists such that

$$|f(x) - l| < \epsilon \quad \text{for all } x \text{ such that} \quad |x| > X$$

and we write

$$f(x) \to l \quad \text{when} \quad x \to \infty, \quad \text{or} \quad \lim_{x \to \infty} f(x) = l.$$

Examples

(1) When $x \to +\infty$, $\tanh x \to 1$. For, since e^x cannot be zero, we can put

$$\tanh x = \frac{e^x - e^{-x}}{e^x + e^{-x}} = \frac{e^x(1 - e^{-2x})}{e^x(1 + e^{-2x})} = \frac{1 - e^{-2x}}{1 + e^{-2x}}$$

from which it follows that $\tanh x \to (1 - 0)/(1 + 0) = 1$ when $x \to +\infty$. In a similar way, we see that $\tanh x \to -1$ when $x \to -\infty$.

(2) If $x > 1$ or if $x < -\frac{1}{2}$, then $(2x + 1)/(x - 1)$ is positive and we can put

$$\log \frac{2x + 1}{x - 1} = \log \frac{2 + x^{-1}}{1 - x^{-1}}$$

from which it follows that $\log\{(2x + 1)/(x - 1)\}$ tends to the limit $\log 2$ when $x \to +\infty$ or when $x \to -\infty$.

Note that $\log\{(2x + 1)/(x - 1)\}$ is not defined if $-\frac{1}{2} \leqslant x \leqslant 1$.

1.7

We may here recall three elementary theorems on limits:

Let u, v be two functions of x, and let $\lim u$ and $\lim v$ exist when x tends to some finite value or to infinity; then

I. $\lim (u + v)$ exists and $\lim (u + v) = \lim u + \lim v$.

II. $\lim (uv)$ exists and $\lim (uv) = (\lim u)(\lim v)$.

III. $\lim (u/v)$ exists and $\lim (u/v) = (\lim u)/(\lim v)$, provided that $\lim v \neq 0$.

Similar theorems to I and II hold good for the sum and product of a finite number of functions. For instance, if $\lim u$, $\lim v$, and $\lim w$ exist, then $\lim (uvw)$ exists and $\lim (uvw) = (\lim u)(\lim v)(\lim w)$.

Proof. Since $\lim u$ and $\lim v$ exist, therefore, by II, $\lim (uv)$ exists. Hence, also by II, $\lim (uv.w)$ exists and

$$\lim (uv.w) = (\lim uv)(\lim w) = \{(\lim u)(\lim v)\}\lim w,$$

i.e. $\lim (uvw) = (\lim u)(\lim v)(\lim w)$.

1.8   Other ways in which a function may behave

It may happen that, as $x \to c - 0$, the value of $f(x)$ increases without limit, so that for every positive number N, there exists a positive number h such that

$$f(x) > N \quad \text{for all } x \text{ such that} \quad c - h \leqslant x < c;$$

we then say that $f(x) \to +\infty$ when $x \to c - 0$.

Similarly, we define the meaning of $f(x) \to -\infty$ when $x \to c - 0$, $f(x) \to +\infty$ when $x \to c + 0$, and $f(x) \to -\infty$ when $x \to c + 0$.

For example, $\tan x \to +\infty$ when $x \to \frac{1}{2}\pi - 0$, $\tan x \to -\infty$ when $x \to \frac{1}{2}\pi + 0$.

Again, it may happen that for every positive number N there exists a positive number X such that $f(X) > N$ for all $x > X$. We then say that $f(x) \to +\infty$ when $x \to +\infty$. Similarly, it may happen that $f(x) \to -\infty$ when $x \to +\infty$; or that $f(x) \to +\infty$ or $-\infty$ when $x \to -\infty$.

For example, $\log x \to +\infty$ when $x \to +\infty$; $x^2 - x^3 \to -\infty$ when $x \to +\infty$; $\sqrt{(-x)} \to +\infty$ when $x \to -\infty$; $\sinh x \to -\infty$ when $x \to -\infty$.

Further, it may happen that $f(x)$ *oscillates* as x approaches a finite value or as x tends to infinity.

Examples

(1) The function $\sin (1/x)$ neither approaches a limit nor tends to infinity as $x \to 0$, but oscillates between the finite values $+1$ and -1 infinitely often in any interval which includes the origin, since $\sin (1/x) = +1$ and -1 alternately when $x = 2/n\pi$ and n takes the values $\ldots -3, -1, 1, 3, 5, \ldots$.

(2) The function $\tan x$ oscillates between $+\infty$ and $-\infty$ as $x \to +\infty$ or $-\infty$.

1.9   Even functions and odd functions

We shall assume that $f(x)$ is defined for all the values of x concerned. If n is even, then $(-x)^n = x^n$. For this reason, any function $f(x)$

which has the property expressed by $f(-x) = f(x)$ is called an *even* function. Its graph is symmetrical about the axis of y. For example, $\cos x$ and $\cosh x$ are even functions.

If n is odd, then $(-x)^n = -x^n$: consequently, a function $f(x)$ is said to be an *odd* function if $f(-x) = -f(x)$. The graph of an odd function is its own image in the origin. For example, $\tan x$ and $\tanh x$ are odd functions.

Evidently, the product of two even functions or of two odd functions is an even function; the product of an even and an odd function is an odd function. For example, the function $(\sin x)/x$ or $(\sin x)(1/x)$ is even; the function $\sin x \cosh x$ is odd.

1.10   Periodic functions

A function $f(x)$ is said to be *periodic with period* T if it has the property expressed by $f(x+T) = f(x)$ for all values of x. The graph of such a function is a succession of repetitions of the portion from $x = 0$ to $x = T$. Evidently, if T is a period, then any whole multiple of T is a period. If no submultiple of T is a period, then T is called the *primitive* period.

Example

The function $\sin x$ is periodic with period 2π, since

$$\sin(x+2\pi) = \sin x;$$

$\sin \omega t$ is periodic with period $2\pi/\omega$ since $\sin \omega(t + 2\pi/\omega) = \sin \omega t$; $\sin \pi x$ is periodic with period 2, since $\sin \pi(x+2) = \sin \pi x$.

The function $\tan x$ has the period 2π; its primitive period is π.

1.11   Monotonic functions

Let $f(x)$ be defined in a closed or an open interval, of which x_1 and x_2 are two points.

We say that $f(x)$ is *monotonic increasing* in the interval if

$$f(x_1) \leqslant f(x_2) \quad \text{for all} \quad x_1 < x_2$$

and further, that $f(x)$ is *strictly monotonic increasing* if

$$f(x_1) < f(x_2) \quad \text{for all} \quad x_1 < x_2.$$

Monotonic decreasing functions are defined in a similar way.

1.12

Theorem. Let $f(x)$ be strictly monotonic increasing in an interval of which x_1 and x_2 are two points. If $f(x_1) < f(x_2)$, then $x_1 < x_2$.

Proof. The only alternatives are $x_1 = x_2$ and $x_2 < x_1$, both of which are impossible, since from $x_1 = x_2$ would follow $f(x_1) = f(x_2)$, and from $x_2 < x_1$ would follow $f(x_2) < f(x_1)$, whereas $f(x_1) < f(x_2)$ by hypothesis; therefore, $x_1 < x_2$.

1.13   Bounded functions

A function $f(x)$ is said to be *bounded* in an interval if it is defined in the interval and if there exist numbers H, K, independent of x, such that $H < f(x) < K$ for all x in the interval.

If H exists, $f(x)$ is said to be *bounded below*; if K exists, $f(x)$ is said to be *bounded above*.

A function may be defined in an interval and yet not be bounded (see the first two examples of §1.24).

1.14   Limits of a bounded monotonic function

In this book we shall assume that, if $f(x)$ is bounded and monotonic in a finite interval whose end points are a, $b(a < b)$, then the end limits $f(a+0), f(b-0)$ exist.†

It will follow that, at every interior point c, the limits $f(c-0)$, $f(c+0)$ always exist for a bounded monotonic function.

We shall also assume that, if $f(x)$ is bounded and monotonic for all sufficiently large positive values of x, then $f(x)$ tends to a limit when $x \to +\infty$; and that, if $f(x)$ is bounded and monotonic for all sufficiently large negative values of x, then $f(x)$ tends to a limit when $x \to -\infty$.

If $f(x)$ is unbounded and monotonic, then, according as $f(x)$ is increasing or decreasing: when $x \to +\infty$, $f(x) \to +\infty$ or $-\infty$; and when $x \to -\infty$, $f(x) \to -\infty$ or $+\infty$, respectively.

We note particularly that, when x tends to $+\infty$ or to $-\infty$, or when x tends to a finite value, a monotonic function of x must either tend to a limit or to plus or minus infinity (it cannot oscillate).

An important corollary is that, if there is a number X such that $f(x)$ is monotonic increasing for all $x > X$, and if a number K, independent of x, exists such that $f(x) < K$ for all $x > X$, then $f(x)$ tends to a limit not greater than K when $x \to +\infty$.

† G. H. Hardy, *A Course of Pure Mathematics*, Cambridge University Press, §§ 69, 92, 95.

Little modification in the wording is required to extend the corollary to the cases in which $f(x)$ is monotonic decreasing, or $x \to -\infty$, and also to cases in which x tends to a finite value.

1.15   Closest bounds of a bounded function

If $f(x)$ is bounded in an interval, open or closed, there exist† two numbers M and m called the least upper bound (l.u.b.) and the greatest lower bound (g.l.b.) of $f(x)$ in the interval. They are defined as follows:

Definition. The *least upper bound M* is the least number of which it is true that $f(x) \leqslant M$ for all x in the interval; the *greatest lower bound m* is the greatest number of which it is true that $f(x) \geqslant m$ for all x in the interval.

The bounds M and m may or may not be values of the function. For example, if $f(x)$ is defined in the interval $[-\pi, \pi]$ by

$$f(x) = (\sin x)/x \quad (x \neq 0); \quad f(0) = \tfrac{1}{2},$$

then $M = 1$ and $m = 0$. Here m is a value of $f(x)$, but M is not.

1.16

It follows from the definition of M that, if ϵ is any positive number, there exists at least one value of x in the interval for which $f(x) > M - \epsilon$.

For if this were not true, the number $M - \epsilon$ would have the property expressed by $f(x) \leqslant M - \epsilon$ for all values of x in the interval, contrary to the definition of M as the least number which has this property.

Similarly, there must be at least one value of x in the interval for which $f(x) < m + \epsilon$.

The bounds M and m are sometimes called 'the closest bounds' of $f(x)$ in the interval.

1.17   Spread of a bounded function

Let M and m be the closest bounds of $f(x)$ in the interval $[a, b]$. The difference $M - m$ will be called the *spread*‡ of $f(x)$ in the interval. Evidently $M - m \geqslant 0$.

Let $M(h)$ and $m(h)$ denote the closest bounds of $f(x)$ in the sub-interval $[c-h, c+h]$, where c is an interior point. Then as h decreases, $M(h)$ is monotonic decreasing and $m(h)$ is monotonic increasing, and consequently the spread $M(h) - m(h)$ is monotonic decreasing. Also,

---

† Hardy, loc. cit. §§ 80, 81, 103.
‡ Oscillation (Hardy, Goursat).

$M(h) - m(h) \geqslant 0$. It follows (end of §1.14) that $M(h) - m(h)$ tends to a limit when $h \to 0$ and that $\lim \{M(h) - m(h)\} \geqslant 0$.

Similarly, the spread of $f(x)$ in each of the subintervals $[a, a+h]$, $[b-h, b]$ tends to a limit $\geqslant 0$ when $h \to 0$.

1.18   Continuity of a function at a point

Three equivalent definitions of the *continuity of a function at a point* will be given.

Let $f(x)$ be defined in the interval $[a, b]$, $a < b$, and let c be an interior point of the interval.

I.   The function $f(x)$ is continuous at $x = c$ if the left-hand and right-hand limits $f(c-0)$ and $f(c+0)$ both exist and if each is equal to $f(c)$.

II.   The function $f(x)$ is continuous at $x = c$ if, for every positive ϵ, a number h exists such that

$$|f(x) - f(c)| < \epsilon, \quad \text{for all } x \text{ such that} \quad |x - c| \leqslant h.$$

III.   The function $f(x)$ is continuous at $x = c$ if the spread of $f(x)$ in the subinterval $[c-h, c+h]$ tends to zero when $h \to 0$.

At the end point $x = a$, $f(x)$ is said to be continuous:

I, if the limit $f(a+0)$ exists and is equal to $f(a)$; or

II, if, for every positive ϵ, a number h exists such that

$$|f(x) - f(a)| < \epsilon, \quad \text{for all } x \text{ such that} \quad 0 \leqslant x - a \leqslant h; \text{ or}$$

III, if the spread of $f(x)$ in the subinterval $[a, a+h]$ tends to zero when $h \to 0$.

Appropriate modifications must be made at the end point $x = b$.

1.19

That I, II, III are equivalent follows from the previous definitions. For instance, that I and II are equivalent, i.e. that each is a necessary consequence of the other, follows from the definition of a limit and the fact that $f(c)$ exists.†

1.20   Continuity of a function in an interval

Two equivalent definitions of the *continuity of a function in an interval* are:

I.   A function is continuous in a closed interval if it is continuous at every point of the interval (§1.18).

II.   A function is continuous in a closed interval if, for every positive

---

† On *continuous functions* see Hardy, loc. cit. §§99–103.

ϵ, a number δ exists such that the spread of the function is less than ϵ in every subinterval of length δ.

When it is necessary to allude specially to the second definition, we often say that continuity in an interval is *uniform*, or refer to *the uniformity of continuity*.

1.21

To prove that I and II are equivalent, we have to show that each necessarily follows from the other. It is easy to see that I follows from II: we shall prove that II follows from I.

Proof. In accordance with I let $f(x)$ be continuous at every point of an interval $[a,b]$, $a < b$, and assume that, for some positive ϵ, the number δ does not exist.

Let 'spr$f(x)$' denote 'the spread of $f(x)$'. Bisect $[a,b]$ into two halves: then, since δ does not exist, spr$f(x) \geqslant \epsilon$ in one half or in both. Let $[a_1,b_1]$ denote the half in which spr$f(x) \geqslant \epsilon$ or the left-hand half in case spr$f(x) \geqslant \epsilon$ in both halves. Next, let $[a_2,b_2]$ be the half of $[a_1,b_1]$ obtained from $[a_1,b_1]$ in the same way as $[a_1,b_1]$ was obtained from $[a,b]$. Then let $[a_3,b_3]$ be similarly obtained from $[a_2,b_2]$, and so on.

Now the sequence $a, a_1, a_2,\ldots$ is monotonic increasing and is bounded above, since $a_n < b$; while $b, b_1, b_2, \ldots$ is monotonic decreasing and is bounded below, since $b_n > a$. Hence both sequences tend to limits (§1.14). Moreover, since $b_n - a_n = (b-a)/2^n$, which tends to zero as $n \to \infty$, both sequences have the same limit; let this limit be c.

If c is an interior point of $[a,b]$ it follows that spr$f(x) \geqslant \epsilon$ in the interval $(c-h, c+h)$ however small h may be; hence $f(x)$ cannot be continuous at c. Similarly, if c is an end point, $f(x)$ cannot be continuous at c. Thus there is at least one point at which $f(x)$ is not continuous, contrary to the hypothesis that $f(x)$ is continuous at every point. Thus, the assumption that δ does not exist leads to a contradiction; therefore, δ exists.

1.22 Properties of continuous functions

I. If $f(x)$ is continuous at $x = c$, it follows from (1) and §1.18, I, that we can put

$$f(x) = f(c) + \eta, \tag{2}$$

where η is a function of x such that $\eta \to 0$ when $x \to c$ and $\eta = 0$ when $x = c$; or

$$f(c+h) = f(c) + \eta, \tag{3}$$

where η is a function of h such that $\eta \to 0$ when $h \to 0$ and $\eta = 0$ when $h = 0$.

Note that η is defined at $x = c$ or at $h = 0$, as the case may be (cf. §1.5).

Conversely, if $f(x)$ can be expressed in the form $f(x) = C + \eta$, where η is a function of x which tends to zero when $x \to c$, $\eta = 0$ when $x = c$, and C is independent of x, then $f(x)$ is continuous at $x = c$ and $f(c) = C$.

Or, if $f(c + h)$ can be expressed in the form $f(c + h) = C + \eta$, where η is a function of h which tends to zero when $h \to 0$, $\eta = 0$ when $h = 0$, and C is independent of h, then $f(x)$ is continuous at $x = c$ and $f(c) = C$.

II. If $f(x)$ is continuous at $x = c$ and if $f(c) \neq 0$, there is a neighbourhood $[c - h, c + h]$ of c in which $f(x)$ has the same sign as $f(c)$.

Proof. Since $\eta \to 0$ in (2), given any $\epsilon > 0$ there is an interval $[c - h, c + h]$ in which $|\eta| < \epsilon$.

Since $f(c) \neq 0$, we can put $\epsilon = \frac{1}{2}|f(c)|$. It follows from (3) that in this interval $f(x)$ lies between $f(c) - \frac{1}{2}f(c)$ and $f(c) + \frac{1}{2}f(c)$, both of which have the same sign as $f(c)$. This interval is therefore a neighbourhood of c in which $f(x)$ has at every point the same sign as $f(c)$.

III. The following statements may be easily proved with the aid of (2) and (3).

The sum and product of two, and hence of a finite number of, continuous functions define continuous functions.

The quotient of two continuous functions defines a continuous function except at points where the denominator has the value zero; in particular, the reciprocal of a continuous function is continuous except at a point or points where the function has the value zero.

IV. A function which is continuous in a closed interval is bounded in the interval; briefly, a continuous function is bounded.

Proof. Let $f(x)$ be continuous in the interval $[a, b]$. It follows from §1.20, II, that for any given $\epsilon > 0$ (e.g. $\epsilon = 1$) there exists an integer n such that, when the interval is divided into n equal subintervals, $\mathrm{spr} f(x) < \epsilon$ in each subinterval. Consequently, the spread of $f(x)$ in the whole interval is less than $n\epsilon$, and therefore for every value of x in the interval, $f(x)$ lies between the two numbers $f(a) - n\epsilon$ and $f(a) + n\epsilon$, both of which are independent of x. Hence $f(x)$ is bounded (§1.13).

V. The closest bounds of a continuous function are values of the function: briefly, a continuous function attains its closest bounds.

Let M be the l.u.b. of $f(x)$, continuous in the closed interval $[a, b]$, $a < b$. We shall prove that at least one number ξ exists ($a \leqslant \xi \leqslant b$) such that $f(\xi) = M$.

Proof. Assume that no such number ξ exists. Then, first, the function

$M - f(x)$ is continuous and does not take the value zero in the interval. Hence its reciprocal is continuous, by III, and is bounded, by IV.

Secondly, let K be any positive number. Then, by §1.16, there is at least one value of x for which $f(x) > M - 1/K$, and hence for which $1/\{M - f(x)\} > K$; consequently, the reciprocal of $M - f(x)$ is not bounded.

Thus, we have first shown that $1/\{M - f(x)\}$ is bounded, and secondly that it is not bounded. Consequently, the assumption that ξ does not exist leads to a contradiction; therefore, ξ exists.

A similar proof applies to the g.l.b.

VI. If $f(x)$ is continuous in $[a, b]$ and if $f(a), f(b)$ have opposite signs, then $f(x) = 0$ for at least one value of x between a and b.

Suppose that $f(a) < 0$ and $f(b) > 0$, and $a < b$. We shall prove that at least one number ξ exists ($a < \xi < b$) for which $f(\xi) = 0$.

Proof. Assume that no such number ξ exists. Then, first, $f(x)$ is continuous and is not zero at any point in the interval. Hence the reciprocal $1/f(x)$ is continuous, by III, and bounded, by IV.

Secondly, let K be any positive number. Then by §1.20, II, the interval $[a, b]$ can be divided into subintervals in each of which $\mathrm{spr}\, f(x) < 1/K$. At every point of division, $f(x)$ is either positive or negative, and has not the same sign at all points of division; hence there is at least one subinterval with end points α, β ($\alpha < \beta$) such that $f(\alpha) < 0$ and $f(\beta) > 0$. Since $\mathrm{spr}\, f(x) < 1/K$ in this subinterval, it follows that $f(\beta) - f(\alpha) < 1/K$ and therefore $f(\beta) < 1/K$ and $1/f(\beta) > K$. Hence $1/f(x)$ is not bounded.

Thus, we have first shown that $1/f(x)$ is bounded, and secondly that it is not bounded. Consequently, the assumption that ξ does not exist leads to a contradiction; therefore, ξ exists.

Corollary. If $f(x)$ is a continuous strictly monotonic function, there is one and only one such value of ξ.

Proof. If there were two values ξ_1, ξ_2 then $f(\xi_1) = f(\xi_2)$, ($\xi_1 \neq \xi_2$), which is not consistent with $f(x)$ being strictly monotonic (§1.11).

VII. A function $f(x)$, continuous in the interval $[a, b]$, takes every value between $f(a)$ and $f(b)$ for at least one value of x in the interval.

We shall prove that if $f(a) \neq f(b)$, $a < b$, and if C is any value between $f(a)$ and $f(b)$, then at least one number ξ exists ($a < \xi < b$) for which $f(\xi) = C$.

Proof. Consider the function $f(x) - C$. It is continuous in the interval and has opposite signs at the end points a, b. It follows from VI that at least one number ξ exists $(a < \xi < b)$ such that $f(\xi) - C = 0$, that is, such that $f(\xi) = C$.

Corollary. If $f(x)$ is continuous and strictly monotonic in the interval, there is one and only one such value ξ, by VI, Corollary.

1.23   Discontinuities

A function is said to be discontinuous at a point where it is not continuous.

I. A function $f(x)$ is said to have a *simple discontinuity* at $x = c$ if (i) $f(c-0)$, $f(c+0)$, $f(c)$ all exist but are not all equal; or if (ii) $f(c-0)$, $f(c+0)$ both exist but $f(c)$ is not defined.

In case (ii) if $f(c-0) = f(c+0)$, the discontinuity is sometimes called a *removable discontinuity*, because we need only define $f(c)$ by

$$f(c) = f(c-0) = f(c+0)$$

in order to make $f(x)$ continuous at $x = c$.

Examples

(1) Heaviside's unit function $1(t)$ may be defined by

$$1(t) = 0 \quad (t < 0), \quad 1(0) = \tfrac{1}{2}, \quad 1(t) = 1 \quad (t > 0).$$

It has a simple discontinuity at $t = 0$, since the left-hand and right-hand limits (0 and 1, respectively) exist but are unequal, and neither is equal to $1(0)$.

(2) The function $f(x) = (\sin x)/x$ has a removable discontinuity at $x = 0$, since the left-hand limit $f(0-0)$ and the right-hand limit $f(0+0)$ both exist and are equal (each equal to 1); so if we define $f(0)$ to be 1, then $f(x)$ becomes continuous at $x = 0$. (It is evidently continuous for all other values of x, since the denominator is not zero if $x \neq 0$).

(3) The function y defined by

$$y = 3 \quad (0 < x \leqslant 1), \quad y = 4 \quad (1 < x \leqslant 2), \quad y = 6 \quad (2 < x \leqslant 4)$$

is defined at every point of the interval $(0 < x \leqslant 4)$. It has a simple discontinuity at $x = 1$ and at $x = 2$.

II. A function $f(x)$ is said to have an *infinity* at $x = c$ if $f(c)$ is not defined and $f(x) \to +\infty$ or $-\infty$ when $x \to c$.

Examples

(1) The function $1/(x-a)$ has a *simple infinity* or a *simple pole* at $x = a$.

(2) The function $1/(x-a)^2$ has a *double infinity*, or a *double pole*, or a *pole of order* 2 at $x = a$.

(3) The function $\log (\sin x) \to -\infty$ when $x \to 0+$. It is not defined for $x \leqslant 0$ near $x = 0$. It is said to have a *logarithmic infinity* at $x = 0$ because it behaves like $\log x$ near $x = 0$.

(4) The function $(\cos x)/x^2$ behaves like $1/x^2$ near $x = 0$, since $\cos x \to 1$ as $x \to 0$; in other words, $(\cos x)/x^2$ has a *double pole* at $x = 0$.

III. Such a function as $\sin (1/x)$ is said to have an *oscillating discontinuity* at $x = 0$ (see § 1.8).

1.24 Further examples

(1) $$f(x) = 1/x \quad (x \neq 0), \quad f(0) = 0.$$

Here $f(x)$ is defined for all values of x, but $f(x)$ is not bounded ($1/x \to +\infty$ when $x \to 0+$; $1/x \to -\infty$ when $x \to 0-$).

(2) $$y = \lim_{a \to 0} \frac{\sin x \cos x}{\sin^2 x + a \cos^2 x}.$$

Here $y = 0$ when $x = \frac{1}{2}k\pi$ (k any integer), $y = \cot x$ for all other values of x.

The function y is defined for all values of x but is not bounded.

(3) $$f(x) = (\sin x)/x \quad (x \neq 0); \quad f(0) = 0.$$

There is a simple discontinuity at $x = 0$, where the left-hand and right-hand limits both exist and each is equal to 1, but this is not equal to $f(0)$.

(4) $y = 1$ when x is rational, $y = 0$ when x is irrational.

Here $\operatorname{spr} y = 1$ in any neighbourhood of any value of x; so that y is discontinuous for every value of x.

(5) Let p/q be a proper rational fraction in its lowest terms. In the interval $(0, 1)$ let $f(x)$ be defined by $f(x) = 1/q$ at every rational point $x = p/q$, but by $f(x) = 0$ at every irrational point. Then $f(x)$ is continuous at every irrational point, but discontinuous at every rational point.

Proof. Let $x = i$ be any irrational point between 0 and 1; then $f(i) = 0$. Given $\epsilon > 0$, choose n so large that $1/n < \epsilon$. By § 1.3 there exists a neighbourhood N of i such that every rational number belonging to N is of the form p/q ($q > n$). Consequently $f(x) = 1/q < \epsilon$

at every rational point in N. Also, $f(x) = 0$ at every irrational point. Thus N is a neighbourhood of i at every point of which $|f(x) - f(i)| < \epsilon$. Hence $f(x)$ is continuous at $x = i$.

Next, let $x = p/q$ be any rational point $(0 < p/q < 1)$. Then $f(p/q) = 1/q$. Now, every neighbourhood of p/q contains irrational points, and at each of them $f(x) = 0$. Consequently, there exists no neighbourhood of $x = p/q$ in which $\mathrm{spr} f(x) < 1/q$, and therefore $f(x)$ is discontinuous at $x = p/q$.

This example warns us that intuitive ideas of continuity are not to be relied upon. It shows, for instance, that a function which is continuous at a point is not necessarily continuous in the neighbourhood of that point.

1.25   Inverse functions

Let $y = f(x)$ be a continuous function of x, strictly increasing from α to β when x increases from a to b.

Then by §1.22, VII, Corollary, for each value of y between α and β there exists one and only one value of x between a and b. Interchanging x and y, we see that the equation $x = f(y)$, together with the condition $a \leqslant y \leqslant b$, defines y as a single-valued function of x for values of x in the interval $\alpha \leqslant x \leqslant \beta$.

Denote this function by $\phi(x)$. Then $\phi(x)$ is called the *inverse function* of the function $f(x)$, or a *branch of the inverse function* in case the equation $x = f(y)$ is satisfied by other values of y besides the one lying in the interval $a \leqslant y \leqslant b$.

1.26

Theorem. The inverse function $y = \phi(x)$, or the branch $y = \phi(x)$ of the inverse function if it is only a branch, is strictly increasing and continuous for values of x in the interval $\alpha \leqslant x \leqslant \beta$.

Proof. Let x_0, x be such that $\alpha \leqslant x_0 < x \leqslant \beta$, and let $y_0 = \phi(x_0)$, $y = \phi(x)$.

Then since $x_0 = f(y_0)$ and $x = f(y)$ and $x_0 < x$, it follows from §1.12 that $y_0 < y$. Hence the inverse function $y = \phi(x)$ is strictly increasing.

Next, let x_0 be fixed and let $x \to x_0$ monotonically. Since x is now decreasing, y will decrease monotonically, and being bounded below (since $y_0 < y$), y must tend to a limit y_1 such that $y_0 \leqslant y_1$. Since $f(y)$ is continuous, it follows that $x_0 = f(y_1)$. But $x_0 = f(y_0)$, and therefore $f(y_1) = f(y_0)$ and hence $y_1 = y_0$, the alternative $y_1 > y_0$ being excluded by the strictly monotonic property of $f(y)$.

We have thus proved that $y \to y_0$ when $x \to x_0$ from the right. Similarly, $y \to y_0$ when $x \to x_0$ from the left. Hence $y = \phi(x)$ is continuous at $x = x_0$.

Appropriate modifications must be made when x_0 is an end point, or when $f(x)$ is strictly decreasing.

1.27

Graphically, since the inverse of the function $y = f(x)$, or any branch of the inverse in its corresponding interval, is defined by $x = f(y)$, that is, by interchanging x and y, the graphs of the two functions, each the inverse of the other, have the same shape and are the reflections of each other in the line $y = x$, the scales on the coordinate axes being assumed to be the same.

1.28

Examples

(1) The function $y = x^3$ is continuous and strictly increasing from $-h^3$ to $+h^3$ in any interval $-h \leqslant x \leqslant h$.

The inverse function given by $x = y^3$ has only one real branch $y = x^{\frac{1}{3}}$ which is continuous and strictly increasing from $-k^{\frac{1}{3}}$ to $+k^{\frac{1}{3}}$ in any interval $-k \leqslant x \leqslant k$.

(2) The exponential function $y = e^x$ ($\exp x$) is continuous and strictly increasing for $-\infty < x < +\infty$. The inverse function given by $x = e^y$ has only one real branch $y = \log x$ which is continuous and strictly increasing in the interval $0 < x < +\infty$.

The graphs of $y = e^x = \exp x$ and $y = \log x$ are the reflections of each other in the line $y = x$.

Note. Since $\log x$ is monotonic, the equation $\log x = \log a (0 < a)$ has only the one real solution, $x = a$.

The equation $e^x = e^a$ has only the one solution, $x = a$, for any a.

(3) The function $y = x^2$ is continuous and strictly increasing from 0 to h^2 in any interval $0 \leqslant x \leqslant h$, but strictly decreasing from h^2 to 0 in any interval $-h \leqslant x \leqslant 0$.

The inverse function, given by $x = y^2$, has two branches: one branch $y = \sqrt{x} = x^{\frac{1}{2}}$ strictly increases from 0 to $\sqrt{k}$ in any interval $0 \leqslant x \leqslant k$, the other $y = -\sqrt{x} = -x^{\frac{1}{2}}$ strictly decreases from 0 to $-\sqrt{k}$.

The branch $y = \sqrt{x} = x^{\frac{1}{2}}$, on which $y > 0$, is called the *principal branch*, and for any given value of x the corresponding value of y on the principal branch is called the *principal value* of y.

1.29   The inverse circular (trigonometric) functions

The sine function $y = \sin x$ is continuous for all values of x. The *inverse sine* function is determined by the equation $x = \sin y$. When $-1 < x < 1$, this equation is satisfied by an infinity of values of y, which lie in the intervals

$$\ldots -\tfrac{3}{2}\pi < y < -\tfrac{1}{2}\pi, \quad -\tfrac{1}{2}\pi < y < \tfrac{1}{2}\pi, \quad \tfrac{1}{2}\pi < y < \tfrac{3}{2}\pi, \ldots.$$

These values of y vary continuously and strictly increase or decrease while x increases from -1 to $+1$. The inverse sine function therefore consists of an infinity of branches, each of which is continuous and strictly monotonic in the interval $-1 \leqslant x \leqslant 1$.

The branch determined by $-\tfrac{1}{2}\pi \leqslant y \leqslant \tfrac{1}{2}\pi$ is called the *principal branch*. The principal branch is usually denoted by $\sin^{-1} x$ (or, arc $\sin x$). The branch determined by $\tfrac{1}{2}\pi \leqslant y \leqslant \tfrac{3}{2}\pi$ is then $\pi - \sin^{-1} x$, and the other branches can also be expressed in terms of $\sin^{-1} x$.

Occasionally, it is convenient to use $\sin^{-1} x$ (or, arc $\sin x$) to convey the idea of the many-valued inverse sine function.

1.30

The remaining inverse circular functions are defined similarly. They are all infinitely many-valued. Their principal values are usually denoted by $\cos^{-1} x$ (or, arc $\cos x$), $\tan^{-1} x$ (or, arc $\tan x$), and so on, and these are determined respectively by the intervals:

$$0 \leqslant \cos^{-1} x \leqslant \pi \qquad (-1 \leqslant x \leqslant 1),$$

$$-\tfrac{1}{2}\pi < \tan^{-1} x < \tfrac{1}{2}\pi \qquad (-\infty < x < +\infty),$$

$$0 < \cot^{-1} x < \pi \qquad (-\infty < x < +\infty),$$

$$0 \leqslant \sec^{-1} x \leqslant \pi \qquad (|x| \geqslant 1),$$

$$-\tfrac{1}{2}\pi \leqslant \operatorname{cosec}^{-1} x \leqslant \tfrac{1}{2}\pi \qquad (|x| \geqslant 1).$$

1.31   The hyperbolic and inverse hyperbolic functions

The functions defined by

$$\sinh x = \frac{e^x - e^{-x}}{2}, \quad \cosh x = \frac{e^x + e^{-x}}{2}, \quad \tanh x = \frac{\sinh x}{\cosh x} = \frac{e^x - e^{-x}}{e^x + e^{-x}},$$

$\operatorname{cosech} x = 1/\sinh x$, $\operatorname{sech} x = 1/\cosh x$, $\coth x = 1/\tanh x$, are called the *hyperbolic* functions because, if we put $X = \cosh t$, $Y = \sinh t$, then $X^2 - Y^2 = 1$, and it follows that, as t increases from $-\infty$ to $+\infty$, the point (X, Y) describes the half of the hyperbola $X^2 - Y^2 = 1$ on which $X > 0$.

It will be assumed that the reader is familiar with these functions and their inverses. For the sake of reference we recall that the inverses of $\sinh x$, $\tanh x$, $\coth x$, $\operatorname{cosech} x$, are all single-valued functions denoted respectively by $\sinh^{-1} x$, $\tanh^{-1} x$, $\coth^{-1} x$, $\operatorname{cosech}^{-1} x$ and such that

$$-\infty < \sinh^{-1} x = \log\{x + \sqrt{(1 + x^2)}\} < +\infty \qquad (-\infty < x < +\infty),$$

$$-\infty < \tanh^{-1} x = \tfrac{1}{2}\log\frac{1 + x}{1 - x} < +\infty \qquad (-1 < x < 1),$$

$$-\infty < \coth^{-1} x = \tfrac{1}{2}\log\frac{x + 1}{x - 1} < +\infty \qquad (|x| > 1),$$

$$-\infty < \operatorname{cosech}^{-1} x = \log\left\{\frac{1}{x} + \sqrt{\left(\frac{1}{x^2} + 1\right)}\right\} < +\infty \quad (|x| > 0).$$

The inverses of $\cosh x$ and $\operatorname{sech} x$ have two real branches. The two branches of the inverse of $\cosh x$ are denoted by $\pm \cosh^{-1} x$, the positive branch being implied by $\cosh^{-1} x$ and being such that

$$0 \leqslant \cosh^{-1} x = \log\{x + \sqrt{(x^2 - 1)}\} < +\infty \quad (1 \leqslant x < \infty).$$

The two branches of the inverse of $\operatorname{sech} x$ are denoted by $\pm \operatorname{sech}^{-1} x$, where $\operatorname{sech}^{-1} x$ denotes the positive branch, and

$$0 \leqslant \operatorname{sech}^{-1} x = \log\frac{1 + \sqrt{(1 - x^2)}}{x} < +\infty \quad (0 < x \leqslant 1).$$

1.32 Necessary and sufficient conditions

Given two statements A and B, so related that B is true when A is true, and hence that A is false when B is false, we say that A is a *sufficient condition* for B and that B is a *necessary condition* for A.

If B is true when and only when A is true, we say that A is a necessary and sufficient condition for B.

Example

That $\lim u$ and $\lim v$ should exist are *sufficient* conditions (§1.7) for the existence of $\lim (u + v)$ or $\lim (uv)$. They are not *necessary* conditions. Thus, if $u = x^2 + 1/x$ and $v = 3 - x^2$, then $\lim (u + v)$ exists when $x \to \infty$, though $\lim u$ and $\lim v$ do not exist. Again, $\lim (u \times 1/v)$ exists, though $\lim u$ does not exist, while $\lim (1/v)$ does.

Note that necessary and sufficient conditions (n.s.c.) are not unique: e.g. if $a \neq 0$ then n.s.c. in order that $ax^2 + 2bx + c$ may be positive for all x are

$$a > 0, \quad ac - b^2 > 0,$$

but we could equally well say that n.s.c. are

$$a > 0, \quad ac - b^2 > 0, \quad c > 0,$$

or $\qquad\qquad a > 0, \quad ac - b^2 > 0, \quad c > 0, \quad a + 2b + c > 0,$

and so on. We usually try to reduce the number of n.s.c. to a minimum.

Examples 1

It is assumed in some of the following examples that the reader will be familiar with the rules of differentiation and the geometrical meaning of the derivative.

(1) Show in diagrams the intervals to which the 'point' x belongs when x satisfies the inequalities:

(i) $x^2 < 4$,   (ii) $-1 \leqslant x \leqslant 1$,

(iii) $(x-1)^2 \leqslant 3$,   (iv) $|x| < 1$,

(v) $|x-2| > 1$,   (vi) $|2x+3| < 4$,

(vii) $x(2-x) > 0$,   (viii) $(x+1)\,x(x-3) > 0$,

(ix) $(x+1)/(x-1) > 2$,   (x) $(5-2x)^2 < 10$,

(xi) $x(1-x)^3\,(x+4) > 0$,   (xii) $\dfrac{|x+1|}{|x-1|} > 2$,

(xiii) $\dfrac{x-1}{x+2} > \dfrac{x+2}{x-1}$,   (xiv) $\dfrac{x(x-2)}{x-1} + \tfrac{3}{2} > 0$,

(xv) $\dfrac{(x-1)(x-2)}{(x-3)(x-4)} > 6$,   (xvi) $\dfrac{1}{x} + \dfrac{1}{x+1} > \tfrac{1}{2}$,

(xvii) $\dfrac{(x+1)\,(x+2)}{x(1-x)} > 1$,   (xviii) $\dfrac{(x+1)\,(x+2)}{x(x-1)} > 1$.

Note: When fractions occur in an inequality, it is often helpful to apply the rule: multiply throughout by the *square* of the L.C.M. of the denominators (the square of a real number being positive).

(2) Sketch the graphs of the functions given by

(i) $f(x) = -1$ $(x < 0)$, $f(0) = 0$, $f(x) = 1$ $(x > 0)$,

(ii) $f(x) = \exp(1/x)$ $(x \neq 0)$, $f(0) = 0$,

(iii) $f(x) = \tanh(1/x)$ $(x \neq 0)$, $f(0) = 0$,

(iv) $f(x) = \tan^{-1}(1/x)$ $(x \neq 0)$, $f(0) = 0$, $-\tfrac{1}{2}\pi < f(x) < \tfrac{1}{2}\pi$,

(v) $f(x) = \lim_{a \to 0} \dfrac{2a+x}{a+x}$, (vi) $f(x) = \lim_{y \to 0} \dfrac{3y+\sin x}{y+x}$,

(vii) $f(x) = \lim_{n \to \infty} \dfrac{nx}{1+n^2x^2}$, (viii) $f(x) = \lim_{n \to \infty} \dfrac{x}{1+n\sin x}$,

(ix) $f(x) = \lim_{n \to \infty} x^n \quad (0 \leqslant x \leqslant 1)$, (x) $f(x) = \lim_{n \to \infty} x^{1/n} \quad (x \geqslant 0)$,

(xi) $f(x) = \lim_{n \to \infty} (\sin^2 x)^n$, (xii) $f(x) = \lim_{n \to \infty} (\sin^2 x)^{1/n}$.

(3) Sketch the limiting forms of the curves

(i) $x^{2n}+y^{2n} = a^{2n}$, (ii) $x^{2n}-y^{2n} = a^{2n}$, (iii) $x^{2n+1}+y^{2n+1} = a^{2n+1}$,

when $n \to \infty$ through positive integral values.

(4) If n is a positive integer, state the least period of each of the periodic functions

(i) $\sin^n \pi x$, (ii) $\tan^n \pi x$, (iii) $\sin^n x \cos^n x$.

(5) Show that each of the functions $x|x|$, $x/|x|$ is odd, and sketch their graphs. Sketch the graph of $\frac{1}{2}(1+x/|x|)$.

(6) Prove that the function $\frac{1}{2} + 1/(e^x - 1)$ is odd. Sketch its graph.

(7) From the identity

$$f(x) \equiv \tfrac{1}{2}\{f(x)+f(-x)\} + \tfrac{1}{2}\{f(x)-f(-x)\}$$

prove that any function can be expressed as the sum of an even and an odd function. Express e^x and e^{-x} in this way.

(8) Sketch the graphs of

(i) $y = |x|$, (ii) $y = |x|^{\frac{1}{2}}$, (iii) $y = |x|^{-\frac{1}{2}}$.

(9) Sketch the graphs of

(i) $\log|x|$, (ii) $\log(1+x^2)$, (iii) $\log|\sin x|$,

(iv) $\log\sin x$, (v) $\log\tan x$, (vi) $\log\cosh x$,

(vii) $\log\{(x+1)(x-1)\}$, (viii) $\log(x+1)+\log(x-1)$,

(ix) $\log\{(x+1)/(x-1)\}$, (x) $\log(x+1)-\log(x-1)$,

(xi) $2\tan^{-1}x$ and $\tan^{-1}\{2x/(1-x^2)\}$.

(10) Describe the behaviour near $x = 0$ of the functions:

(i) $x^{\frac{5}{3}}$, (ii) $x^{\frac{2}{3}}$, (iii) $x^{-\frac{2}{3}}$,

(iv) $x^{\frac{1}{2}}\sin x$, (v) $x\cot x$, (vi) $x\sin(1/x)$,

(vii) $x^2\sin^2(1/x)$, (viii) $1/\log x$, (ix) $\log(1/x)$,

(x) $1/(1+e^{-1/x})$, (xi) $1/\log(x^2)$, (xii) $1/(\log x)^2$.

(11) Show that the function $(x-1)/(6-x)$ is monotonic increasing in the interval $1 \leqslant x \leqslant 5$ but not in the interval $1 \leqslant x \leqslant 7$.

(12) Show that if $a^2 > b$, the function y given by the equation

$$xy + a(x+y) + b = 0$$

decreases as x increases in any interval not containing the point $x = -a$. Also show that the function is its own inverse. In particular, consider the case $a = 0$.

(13) If $\sinh u = 1$ and $\cosh v = 3$, prove that $v = \pm 2u$.

(14) If $e^x = \coth u$, prove that $e^{2u} = \coth \frac{1}{2}x$.

(15) Sketch the graphs of $\sinh x$ and $\tanh 2x$ in one diagram.

Show that besides intersecting at the origin, the graphs intersect at two points at each of which $\cosh x = \frac{1}{2}(1 + \sqrt{3})$.

Calculate the coordinates of these two points.

(16) Show that the curves $y = \sinh 2x, y = a \cosh x$ intersect in one point. In particular, show that the curves $y = \sinh 2x, y = \frac{3}{2} \cosh x$ intersect at the point $(\log 2, 15/8)$ and that the angle of intersection is $\tan^{-1} (20/37)$.

(17) If $f(x)$, $g(x)$ are continuous at $x = c$, prove that $f(x) + g(x)$, $f(x) - g(x)$, $f(x) \cdot g(x)$ are continuous at $x = c$. Also, if $f(c) \neq 0$, prove that $1/f(x)$ is continuous at $x = c$. (See §1.18. Use §1.22, I.)

(18) If $f(x)$ is continuous at $x = c$, prove that $Af(x)$ is continuous at $x = c$, where A is any constant.

(19) (i) If n is a positive integer, prove that x^n is continuous at every value of x.

    (ii) Prove that a polynomial in x is continuous at every value of x.

(20) (i) If $f(a) = b$, and if $f(x)$ is continuous at $x = a$, and $g(y)$ is continuous at $y = b$, prove that $g\{f(x)\}$ is continuous at $x = a$. (Use §1.22, I.)

    (ii) If $f(x) \to b$ when $x \to a$, and if $g(y)$ is continuous at $y = b$ and $g(b) = c$, prove that $g\{f(x)\} \to c$ when $x \to a$.

    (iii) If $f(x) \to b$ when $x \to a$, and $g(y) \to c$ when $y \to b$, show that it is not necessarily true that $g\{f(x)\} \to c$ when $x \to a$.

[Note that, in (i) $g(b)$ and $f(a)$ are both defined; in (ii) $g(b)$ is defined but $f(a)$ is not; in (iii) neither $g(b)$ nor $f(a)$ is defined, and it may happen that there are values of x, as close to a as we please, at which $f(x) = b$. In that case $g\{f(x)\}$ is not defined in the neighbourhood of $x = a$.

    Example:

$$g(y) = y \sin (1/y) \quad (y \neq 0); \qquad f(x) = x \sin (1/x) \quad (x \neq 0).$$

Here, there are values of $x(x = 1/k\pi$, where k is an integer) as close to $x = 0$ as we please, at which $f(x) = 0$. But $g(0)$ is not defined, so that $g\{f(x)\}$ is not defined in the neighbourhood of $x = 0$.]

2

DIFFERENTIATION

2.1 Derivative at a point

Definition. The *derivative* of a function $f(x)$ at $x = x_0$ is denoted by $f'(x_0)$ and is defined by

$$f'(x_0) = \lim_{h \to 0} \frac{f(x_0 + h) - f(x_0)}{h} \tag{1}$$

provided that this limit exists.

The definition assumes that $f(x)$ is defined in the neighbourhood of $x = x_0$, including the point $x = x_0$.

If x_0 is an interior point of an interval, the existence of the limit implies that the left-hand and right-hand limits both exist and are equal (§1.5), and it is only then that $f(x)$ has a unique derivative at x_0. When one of these limits exists but not the other, $f(x)$ is said to have a left-hand or right-hand derivative at $x = x_0$, as the case may be. When both exist but are not equal, $f(x)$ has both a left-hand and a right-hand derivative, but not a unique derivative at $x = x_0$.

2.2 Differentiability at a point

Definition. The function $f(x)$ is said to be *differentiable* at $x = x_0$ if the derivative $f'(x_0)$ exists.

The existence of $f'(x_0)$ implies, by (1), that

$$\frac{f(x_0 + h) - f(x_0)}{h} = f'(x_0) + \eta,$$

where $h \neq 0$ and η is a function of h that tends to zero when $h \to 0$. It follows that, if $f(x)$ is differentiable at $x = x_0$, then we can put

$$f(x_0 + h) - f(x_0) = \{f'(x_0) + \eta\} h, \tag{2}$$

where $h \neq 0$ and $\eta \to 0$ when $h \to 0$.

Conversely, if $f(x_0 + h)$ can be expressed in the form

$$f(x_0 + h) = f(x_0) + (p + \eta) h, \tag{3}$$

where p is independent of h and $\eta \to 0$ when $h \to 0$, then $f(x)$ is differentiable at $x = x_0$ and $f'(x_0) = p$. For, from (3), after subtracting $f(x_0)$ from both sides, dividing by $h(h \neq 0)$ and letting $h \to 0$, we see that $f'(x_0)$ exists and is equal to p.

2.3

Note that a *necessary* condition for $f(x)$ to be differentiable at $x = x_0$ is that $f(x)$ should be continuous at $x = x_0$. For $f(x_0)$ is defined in (1), and (2) shows that $f(x_0 + h) \to f(x_0)$ when $h \to 0$.

That this condition is not *sufficient* an example will prove. Thus, the function $f(x) = |x|$ is continuous but has no unique derivative at $x = 0$. The left-hand and right-hand derivatives both exist but are not equal, being -1 and $+1$ respectively.

Weierstrass showed that it is possible to construct functions that are continuous at every point of an interval but not differentiable at any point.†

2.4

All that has been said in §§ 2.1, 2.2 holds good whether η is defined at $h = 0$ or not. We may therefore *define* the value of η arbitrarily at $h = 0$ if it is convenient for any purpose to do so: for instance, in order that we may put $h = 0$ in (2). In particular, we may define η to be zero at $h = 0$, and this has the advantage of making η continuous at $h = 0$. Consequently, if $f(x)$ is differentiable at $x = x_0$ we may put

$$f(x_0 + h) = f(x_0) + \{f'(x_0) + \eta\} h, \tag{4}$$

where $\eta \to 0$ when $h \to 0$ and $\eta = 0$ at $h = 0$.

Example

$$f(x) = x^2 \sin (1/x) \quad (x \neq 0), \quad f(0) = 0.$$

For this function, with $x_0 = 0$, we can put, if $h \neq 0$,

$$f(0 + h) = f(0) + (0 + \eta) h$$

where $\eta = h \sin (1/h)$. Comparing with (3), we see that $f'(0)$ exists and is equal to 0. Without affecting this conclusion, we may define η to be zero at $h = 0$. Any other value besides zero would do so far as differentiability is concerned, but $\eta = 0$ has the property of making η continuous at $h = 0$.

2.5 Differentiation

The process of evaluating the limit in (1) in order to obtain the derivative is called *differentiating from first principles*. We shall suppose the reader to be familiar with this process and with its applications to the elementary functions.

† See, e.g., Titchmarsh, *Theory of Functions*, § 11.22.

2.6. Rules of differentiation

It will be convenient for the next few sections to drop the zero suffix in x_0 and refer to 'the derivative at x', keeping in mind that the reference will still be to the derivative or differentiability *at a point*.

The following rules will be well known to the reader who has had an introduction to the differential calculus. It is assumed that $u(x)$, $v(x)$, $w(x)$ are functions that are differentiable at x.

I. *Sum or difference.*

$$f(x) = u(x) \pm v(x),$$

$$f'(x) = u'(x) \pm v'(x). \tag{5}$$

II. *Product.* $\quad f(x) = u(x) v(x),$

$$f'(x) = u'(x) v(x) + u(x) v'(x). \tag{6}$$

Corollary 1. If $f(x) = cv(x)$ where c is a constant, then

$$f'(x) = cv'(x). \tag{7}$$

Proof. Put $u(x) = c$, $u'(x) = 0$, in (6).

Corollary 2. If $f(x) = u(x) v(x) w(x)$, then

$$f'(x) = u'(x) v(x) w(x) + u(x) v'(x) w(x) + u(x) v(x) w'(x). \tag{8}$$

Proof. In (6), replace $v(x)$ by $v(x) w(x)$, and $v'(x)$ by

$$v'(x) w(x) + v(x) w'(x).$$

III. *Quotient.* Provided that $v(x) \neq 0$, if

$$f(x) = \frac{u(x)}{v(x)}, \quad \text{then} \quad f'(x) = \frac{u'(x) v(x) - u(x) v'(x)}{\{v(x)\}^2}; \tag{9}$$

in particular, if

$$f(x) = \frac{1}{v(x)}, \quad \text{then} \quad f'(x) = -\frac{v'(x)}{\{v(x)\}^2}. \tag{10}$$

IV. *Complex function.* If $f(x) = u(x) + iv(x)$, where $i = \sqrt{(-1)}$ and $u(x)$, $v(x)$ are real functions, both differentiable at x, then the derivative of $f(x)$ at x is defined by

$$f'(x) = u'(x) + iv'(x). \tag{11}$$

It may be verified that the above rules of differentiation hold good when we replace $u(x)$, $v(x)$, $w(x)$ by complex functions.

V. It is not necessary for both u' and v' to exist in order that the product uv, for example, should be differentiable, though, of course, if they do not both exist, the derivative of uv cannot be found by the product rule.

Example

Let $\phi(x)$ be a function which is continuous at every point, but has no derivative at any point (end of §2.3).

Let the function $f(x)$ be then defined by

$$f(x) = x\phi(x).$$

We shall prove that $f'(0)$ exists and that $f'(0) = \phi(0)$.

Proof. Since $\phi(x)$ is continuous at $x = 0$, therefore $\phi(0)$ exists and we can put

$$\phi(x) = \phi(0) + \eta, \quad f(x) = x\{\phi(0) + \eta\},$$

where η is a function of x which tends to zero when $x \to 0$, by §1.22, I. Consequently $f(0) = 0$ and

$$f(0 + h) - f(0) = f(h) = h\{\phi(0) + \eta\},$$

where η is now a function of h which tends to zero when $h \to 0$. By dividing by h, we get

$$\{f(0 + h) - f(0)\}/h = \phi(0) + \eta$$

from which, by letting $h \to 0$, follows $f'(0) = \phi(0)$.

Note 1. The derivative $f'(0)$ of the product $f(x) = x\phi(x)$ cannot be found by applying the product rule, because $\phi(x)$ has no derivative at $x = 0$ (nor indeed at any point).

Note 2. It may easily be proved, by using first principles, that the function $x\phi(x)$ has no derivative at any point except $x = 0$. Thus, it is possible for a function to have a derivative at one and only one point.

2.7   Differentials

Changing the notation, if we put $y = f(x)$,

$$\delta x = h = (x + h) - x, \quad \delta y = f(x + h) - f(x), \tag{12}$$

then δx, whether positive or negative, is called an *increment* in x, and δy, positive or negative, is called the *corresponding increment* in y; δx may be arbitrary, but δy depends on δx.

In the new notation, (2) reads

$$\delta y = f'(x)\,\delta x + \eta\,\delta x, \tag{13}$$

where η is now a function of δx which tends to zero when $\delta x \to 0$.

The first term on the right is called the *differential* of y with respect to x, and is denoted by dy, so that

$$dy = df(x) = f'(x)\, \delta x. \tag{14}$$

In particular, if $f(x) = x$, then $f'(x) = 1$ and hence $dx = \delta x$, so that the differential of x with respect to x is identical with an arbitrary increment in x. Hence (14) may be written

$$dy = df(x) = f'(x)\, dx; \tag{15}$$

thus the differential $df(x)$ is the product of the derivative $f'(x)$ and the differential dx. For this reason, the derivative $f'(x)$ is also called the *differential coefficient* of $f(x)$ with respect to x.

Again, from (15), by dividing by dx, we obtain

$$\frac{dy}{dx} = f'(x); \tag{16}$$

consequently, dy/dx is another way of expressing the derivative $f'(x)$.

Further, from (13) and (14) we have

$$\delta y = dy + \eta\, \delta x$$

and hence

$$\frac{\delta y - dy}{\delta x} = \eta,$$

from which it follows, since $\eta \to 0$ when $\delta x \to 0$, that the differential dy is an *approximation* to the increment δy in the sense that their difference is as small as we please compared with δx when δx is small enough. In this sense we can put

$$\delta y \doteq dy = f'(x)\, \delta x. \tag{17}$$

2.8   Graphical meanings

Let $y = f(x)$ be the equation of the curve indicated in Fig. 1. Let P be the point (x, y) on the curve and Q the point $(x + \delta x, y + \delta y)$.

We give below the usual graphical meanings associated with the derivative $f'(x)$ etc. without going into any discussion of the limiting process involved in defining the tangent at P as the limiting position of the chord PQ.

Let PT be the tangent at P, making an angle ψ with the directed axis of x. Then $\tan \psi$ is the gradient of the tangent at P. The *gradient of the curve* at P is defined to be that of the tangent at P, and is therefore also equal to $\tan \psi$.

Draw PK parallel to the axis of x and KQ parallel to the axis of y, and let the tangent PT meet KQ in T. Then

$$\delta x = PK = \text{increment in } x,$$

$$\delta y = KQ = \text{increment in } y,$$

$$dy = df(x) = KT = \text{differential of } f(x),$$

$$\frac{\delta y}{\delta x} = \frac{KQ}{PK} = \text{gradient of chord } PQ,$$

$$\frac{dy}{dx} = f'(x) = \frac{KT}{PK} = \tan \psi = \text{gradient of curve at } P.$$

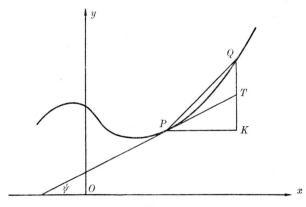

Fig. 1

2.9   Rate of change of a function at a point

We now go back to the notation $x = x_0$ to indicate a fixed value of x.

Let $\delta y = f(x_0 + \delta x) - f(x_0)$. Then $\delta y/\delta x$ is called the *average rate of change* of the function $f(x)$ with respect to x in the interval from x_0 to $x_0 + \delta x$. The limit of this average rate of change when $\delta x \to 0$ is called the rate of change of $f(x)$ *at the point* x_0, provided that the limit exists. But in that case the limit is the derivative $f'(x_0)$. Consequently, the derivative $f'(x_0)$ is said to measure the rate at which $f(x)$ is changing with respect to x *at the point* x_0.

2.10   Function strictly increasing or decreasing at a point

Let $f(x)$ be defined in $[a, b]$ and let $x = x_0$ be an interior point. We say that $f(x)$ is *strictly increasing* at $x = x_0$ if an interval

$$[x_0 - \delta, x_0 + \delta]$$

exists within which

$$f(x) < f(x_0) \quad \text{for} \quad x < x_0, \quad f(x_0) < f(x) \quad \text{for} \quad x_0 < x.$$

We say that $f(x)$ is *strictly decreasing* at $x = x_0$ if $-f(x)$ is strictly increasing at $x = x_0$.

The appropriate modifications at the end points a and b suggest themselves.

2.11 Properties of the derivative at a point

I. If $f'(x_0)$ exists, there is an interval $[x_0 - \delta, x_0 + \delta]$ in which we can put
$$f(x) = f(x_0) + (x - x_0)\{f'(x_0) + \eta\}, \tag{18}$$
where η is a function of x which tends to zero when $x \to x_0$, and $\eta = 0$ when $x = x_0$ (see §2.4).

II. If $f'(x_0)$ exists, there is an interval $[x_0 - \delta, x_0 + \delta]$ in which $f(x)$ is defined (§2.1), and $f(x)$ is continuous at x_0 (§2.3).

III. (i) If $f'(x_0) > 0$, then $f(x)$ is strictly increasing at $x = x_0$.

(ii) If $f'(x_0) < 0$, then $f(x)$ is strictly decreasing at $x = x_0$.

Proof of (i). Since $\eta \to 0$ when $x \to x_0$ and since $f'(x_0) > 0$, there is an interval $[x_0 - \delta, x_0 + \delta]$ in which $|\eta| < f'(x_0)$. It follows from (18) that in this interval

$$f(x) < f(x_0) \quad \text{for} \quad x < x_0, \quad f(x_0) < f(x) \quad \text{for} \quad x_0 < x,$$

and therefore, by §2.10, that $f(x)$ is strictly increasing at x_0.

The proof of (ii) is similar.

Corollary 1. A sufficient condition that $f(x)$ should be strictly increasing or decreasing at x_0 is $f'(x_0) \neq 0$.

This condition is sufficient but not necessary; for example, $f(x) = x^3$ is strictly increasing at $x = 0$ according to the definition (§2.10), although $f'(0) = 0$.

Corollary 2. If $f'(x_0) \neq 0$, there is an interval $[x_0 - \delta, x_0 + \delta]$ in which $f(x) \neq f(x_0), x \neq x_0$.

Corollary 3. If $f(x_0) = 0$ and $f'(x_0) \neq 0$, there is an interval
$$[x_0 - \delta, x_0 + \delta]$$
in which $f(x) \neq 0$, $x \neq x_0$.

This is a particular case of Corollary 2, with $f(x_0) = 0$.

Corollary 4. If $f(x)$ is strictly increasing at x_0 and if $f'(x_0)$ exists, then $f'(x_0) \geq 0$.

For if $f'(x_0) < 0$, then $f(x)$ would be strictly decreasing at x_0, by (ii) above.

Note. If $f'(x_0) > 0$, this is not a sufficient reason why there should exist any interval $[x_0 - \delta, x_0 + \delta]$ throughout which $f(x)$ is strictly monotonic increasing; $f(x)$ may, in fact, be strictly decreasing at points as close to x_0 as we please. This will be evident from the next example and the adjoining figure, which exhibits the graph of the function $x^2 \sin(1/x)$ between $x = 0$ and $x = 0{\cdot}1$ on a large scale (see Fig. 2).

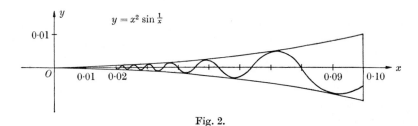

Fig. 2.

Example

$$f(x) = \tfrac{1}{2}x + x^2 \sin\frac{1}{x} \quad (x \neq 0), \quad f(0) = 0.$$

For this function (see §2.4, Example) $f'(0) = \tfrac{1}{2}$, and it may be easily shown by the rules of differentiation that $f'(x_0) = -\tfrac{1}{2}$ at $x_0 = 1/(2n\pi)$ where n is any integer. Now if n is large enough, x_0 is as small as we please. Consequently, by III, (i) and (ii), $f(x)$ is strictly increasing at $x = 0$ but is strictly decreasing at points as close to $x = 0$ as we please.

IV. Let $F(x)$ and $G(x)$ be differentiable at $x = x_0$.

Let $F(x_0) = 0, G(x_0) = 0$, and $G'(x_0) \neq 0$. Then, in the neighbourhood of $x = x_0$,

$$\frac{F(x)}{G(x)} = \frac{F'(x_0)}{G'(x_0)} + \eta, \tag{19}$$

where $\eta \to 0$ when $x \to x_0$.

Proof. By (18) and since $F(x_0) = G(x_0) = 0$,

$$F(x) = (x - x_0)\{F'(x_0) + \alpha\}, \quad G(x) = (x - x_0)\{G'(x_0) + \beta\},$$ where $\alpha \to 0$ and $\beta \to 0$ when $x \to x_0$. Also, by III, Corollary 3, since $G'(x_0) \neq 0$, there is an interval in which $G(x) \neq 0, x \neq x_0$; and hence in which, if $x \neq x_0$,

$$\frac{F(x)}{G(x)} = \frac{F'(x_0) + \alpha}{G'(x_0) + \beta};$$

consequently, $F(x)/G(x)$ tends to the limit $F'(x_0)/G'(x_0)$ when $x \to x_0$, and (19) follows.

Corollary. If $f(x)$ and $g(x)$ are differentiable at $x = x_0$, and if $g'(x_0) \neq 0$, then, in the neighbourhood of x_0,

$$\frac{f(x) - f(x_0)}{g(x) - g(x_0)} = \frac{f'(x_0)}{g'(x_0)} + \eta, \tag{20}$$

where $\eta \to 0$ when $x \to x_0$.

Proof. In (19) put $F(x) = f(x) - f(x_0)$, $G(x) = g(x) - g(x_0)$.

2.12   Differentiability in an interval. Derived function

A function is said to be *differentiable in an interval* if it is differentiable at every point of the interval.

Let $y = f(x)$ be differentiable in an interval. Then the function whose value at every point is the derivative of $f(x)$ at that point is called *the derived function* of $f(x)$. The derived function is denoted by $f'(x)$, y', y_1, dy/dx, Dy, $D_x y$, ... as convenient.

Note. The existence of the derivative at a particular point is not a sufficient reason why there should be any interval in which the derived function exists (see §2.6, V, Example).

In the following six sections will be given theorems on properties of the derived function.

2.13   Rolle's theorem

Theorem. If a function $\phi(x)$ is continuous in the closed interval $[a, b]$ and differentiable in the open interval (a, b), and if

$$\phi(a) = \phi(b) = 0,$$

then the derived function $\phi'(x)$ vanishes at one or more points in the open interval; that is to say, at least one number ξ exists, between a and b, such that $\phi'(\xi) = 0$; or, the equation $\phi'(x) = 0$ has at least one root $x = \xi$ in the open interval (a, b).

Proof. The proof depends upon the property of a function continuous in a closed interval, that it attains its closest bounds (§1.22, V).

Suppose $a < b$.

One possibility is $\phi(x) \equiv 0$. In this case $\phi'(x) \equiv 0$, so that $\phi'(\xi) = 0$ for every ξ in the interval.

In all other cases $\phi(x)$ must have positive values or negative values or both. Therefore its least upper bound M must be positive, or its greatest lower bound m negative, or both.

Suppose that $M > 0$. Then at least one number ξ exists at which $\phi(\xi) = M$, since $\phi(x)$ is continuous. This number ξ cannot be a or b because $\phi(a) = \phi(b) = 0$, whereas $M > 0$; therefore $a < \xi < b$.

Now $\phi'(\xi)$, which exists by hypothesis, can neither be positive nor negative. For if it were either, there would be points near ξ where $\phi(x) > \phi(\xi)$, by §2.11, III, (i) and (ii); that is, there would be points where $\phi(x) > M$, contrary to the definition of M; it follows that $\phi'(\xi) = 0$.

Thus, if $M > 0$ at least one number ξ exists. Similarly, we could prove that ξ exists if $m < 0$, or if $a > b$. The theorem follows.

Corollary. If $\phi(a) = 0$, and $\phi'(x) \neq 0$ at any point in $(a < x < b)$, then $\phi(x) \neq 0$ at any point in $(a < x \leqslant b)$.

Note 1. The conditions $\phi(a) = \phi(b) = 0$ of the theorem could be replaced by the single condition $\phi(a) = \phi(b)$.

Note 2. The conditions imposed upon $\phi(x)$ and $\phi'(x)$ are *sufficient* to ensure the existence of the number ξ: they are evidently not *necessary*.

Examples

(1) If $\phi(x)$ is a polynomial in x, then between any two real roots of the equation $\phi(x) = 0$ there lies at least one real root of the equation $\phi'(x) = 0$.

For, let α and β be any two real roots of the equation $\phi(x) = 0$. Then $\phi(\alpha) = 0$, $\phi(\beta) = 0$. Also $\phi(x)$ is continuous in $[\alpha, \beta]$, since a polynomial in x is continuous for all values of x; and $\phi'(x)$ exists in (α, β), since a polynomial in x is differentiable for all values of x. It follows from the theorem that the equation $\phi'(x) = 0$ has at least one real root between α and β.

(2) If $\phi(x) = \sqrt[3]{(3x - x^3)}$, the equation $\phi'(x) = 0$ has at least one root between 0 and $\sqrt{3}$.

Here $a = 0$, $b = \sqrt{3}$. Also $\phi(x)$ is continuous in $[0, \sqrt{3}]$; while $\phi'(x)$ exists in $(0, \sqrt{3})$, tending to $+\infty$ as $x \to 0$ and to $-\infty$ as $x \to \sqrt{3}$. By the theorem, the equation $\phi'(x) = 0$ must have at least one root $x = \xi$ between 0 and $\sqrt{3}$. It is easy to show that, in this case, there is only one value of ξ and that this is $\xi = 1$.

2.14 The mean value theorem

Theorem. If $f(x)$ is continuous in the closed interval $[a, b]$ and differentiable in the open interval (a, b), then at least one number ξ exists, between a and b, such that

$$f(b) - f(a) = (b - a)f'(\xi). \tag{21}$$

Proof. Suppose $a < b$. Put

$$\phi(x) \equiv (x - a)\{f(b) - f(a)\} - (b - a)\{f(x) - f(a)\}.$$

Then $\phi(x)$ is continuous in the closed interval and differentiable in the open interval. Also $\phi(a) = \phi(b) = 0$. Hence, by Rolle's theorem, at least one number ξ exists, between a and b, such that $\phi'(\xi) = 0$, that is, such that

$$f(b) - f(a) - (b - a)f'(\xi) = 0,$$

from which (21) follows.

Corollary 1. If $f(x)$ is continuous in the closed interval $[a, a+h]$ and differentiable in the open interval $(a, a+h)$, then at least one number θ exists $(0 < \theta < 1)$ such that

$$f(a+h) - f(a) = hf'(a + \theta h) \tag{22}$$

or
$$f(a+h) = f(a) + hf'(a + \theta h). \tag{23}$$

This follows from (21) by putting $b = a + h$, $\xi = a + \theta h$.

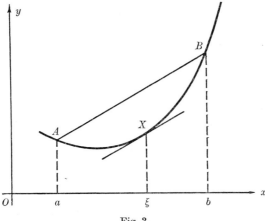

Fig. 3.

Corollary 2. If $f(x)$ is continuous at every point of an interval which includes $x = 0$, and is differentiable at every point of the interval, with the possible exception of $x = 0$, then at least one number θ exists $(0 < \theta < 1)$ such that

$$f(x) = f(0) + xf'(\theta x). \tag{24}$$

The mean value theorem has a simple graphical meaning. On the graph of $y = f(x)$ let A, B be the points where $x = a$, $x = b$, respectively (see Fig. 3). Then $\{f(b) - f(a)\}/(b - a)$ is the gradient of the chord AB.

Now after division by $b - a$, equation (21) takes the form

$$\frac{f(b) - f(a)}{b - a} = f'(\xi) \tag{25}$$

which indicates that there is a point X between A and B at which the gradient of the curve is equal to that of the chord AB, that is, at which the tangent to the curve is parallel to the chord AB.

2.15

Let $f(x)$ be continuous in $(a \leqslant x \leqslant b)$ and differentiable in $(a < x < b)$.

Theorem. (i) If $f'(x) > 0$ for all x in $(a < x < b)$, then $f(b) > f(a)$;

(ii) If $f'(x) < 0$ for all x in $(a < x < b)$, then $f(b) < f(a)$.

Proof. By (21), $f(b) - f(a)$ has the same sign as $f'(\xi)$, since $b - a > 0$. Hence follow (i) and (ii).

Corollary 1. If $f(a) = 0$ and $f'(x) > 0$ for all x in $(a < x < b)$, then $f(x) > 0$ for all x in $(a < x \leqslant b)$.

If $f(a) = 0$ and $f'(x) < 0$ for all x in $(a < x < b)$, then $f(x) < 0$ for all x in $(a < x \leqslant b)$.

Example

If $-\frac{1}{2} \leqslant x \leqslant \frac{1}{2}$, prove that

$$x - x^2 \leqslant \log(1+x) \leqslant x - \tfrac{1}{3}x^2.$$

Proof. First, put $f(x) = x - x^2 - \log(1+x)$. Then we find

$$f'(x) = -x(2x+1)/(1+x)$$

and therefore

$$f'(x) > 0 \quad \text{if} \quad -\tfrac{1}{2} < x < 0, \quad f'(x) < 0 \quad \text{if} \quad 0 < x.$$

Since $f(0) = 0$, it follows that $f(x) < 0$ if $-\frac{1}{2} \leqslant x < 0$, or if $0 < x$, and hence that $f(x) \leqslant 0$ if $-\frac{1}{2} \leqslant x$.

Next, put $g(x) = \log(1+x) - (x - \tfrac{1}{3}x^2)$. Then we find

$$g'(x) = \tfrac{1}{3}x(2x-1)/(1+x)$$

and therefore

$$g'(x) > 0 \quad \text{if} \quad -1 < x < 0, \quad g'(x) < 0 \quad \text{if} \quad 0 < x < \tfrac{1}{2}.$$

Since $g(0) = 0$, it follows that $g(x) < 0$ if $-1 < x < 0$, or if $0 < x \leqslant \frac{1}{2}$, and hence that $g(x) \leqslant 0$ if $-1 < x \leqslant \frac{1}{2}$.

Consequently, if $-\frac{1}{2} \leqslant x \leqslant \frac{1}{2}$, both $f(x) \leqslant 0$ and $g(x) \leqslant 0$. The required inequalities follow at once.

Corollary 2. If $f'(x) > 0$ for all x in $(a < x < b)$, then $f(x)$ is strictly monotonic increasing in $(a \leqslant x \leqslant b)$.

For if $a \leqslant x_1 < x_2 \leqslant b$, then $f(x)$ is continuous in $(x_1 \leqslant x \leqslant x_2)$ and $f'(x) > 0$ in $(x_1 < x < x_2)$. Hence follows $f(x_2) > f(x_1)$, by the theorem.

2.16

Theorem. If $f(x)$ is continuous in $[a,b]$ and if $f'(x) \equiv 0$ in (a,b), then $f(x)$ is constant in $[a,b]$.

Proof. Suppose $a < x \leqslant b$. Then, by the mean value theorem, at least one number ξ exists $(a < \xi < x)$ such that

$$f(x) = f(a) + (x-a)f'(\xi).$$

But $f'(\xi) = 0$ by hypothesis; therefore $f(x) = f(a) = $ const.

Corollary. If $f(x)$, $g(x)$ are continuous in $[a,b]$, and if $f'(x) \equiv g'(x)$ in (a,b), then the difference $f(x) - g(x)$ is constant in $[a,b]$.

For if $\phi(x) = f(x) - g(x)$, then $\phi'(x) = f'(x) - g'(x) \equiv 0$, and by the theorem $\phi(x) = \phi(a) = $ constant, that is,

$$f(x) - g(x) = f(a) - g(a) = \text{constant}.$$

2.17

Theorem. Let $f(x)$ be continuous at $x = x_0$. Let $\lim f'(x) \, (x \to x_0)$ exist. Then $f'(x)$ must be continuous at $x = x_0$. In other words, $f'(x_0)$ must exist and be equal to $\lim f'(x) \, (x \to x_0)$.

Proof. Since $\lim f'(x) \, (x \to x_0)$ exists, therefore $f'(x)$ must exist near x_0 and hence $f(x)$ is continuous near x_0, with the possible exception of x_0 itself. But it is given that $f(x)$ is continuous at x_0. Hence, by the mean value theorem, there is a neighbourhood of x_0 within which, for every $h \neq 0$, at least one number θ exists $(0 < \theta < 1)$ such that

$$\frac{f(x_0 + h) - f(x_0)}{h} = f'(x_0 + \theta h).$$

It is also given that the limit of the R.H.S. exists when $h \to 0$. Therefore the limit of the L.H.S. exists, i.e. $f'(x_0)$ exists, and the result of making $h \to 0$ is equivalent to

$$f'(x_0) = \lim f'(x) \quad (x \to x_0),$$

from which it follows that $f'(x)$ is continuous at $x = x_0$.

Corollary. If, in addition, $f''(x)$ tends to a limit when $x \to x_0$, then $f''(x_0)$ exists and

$$f''(x_0) = \lim f''(x) \quad (x \to x_0);$$

similarly for $f'''(x_0)$, ...

Note 1. It is not *necessary* for the existence of $f'(x_0)$ that

$$\lim f'(x) \quad (x \to x_0)$$

should exist, or even that $f'(x) \, (x \neq x_0)$ should exist. (See Example 2 below, and §2.6, v, Example.)

Note 2. It can be proved in the same way that, if $f(x)$ is continuous at x_0 and if the limits $f'(x_0 - 0)$ and $f'(x_0 + 0)$ exist, then the left-hand and right-hand derivatives of $f(x)$ at $x = x_0$ must exist and be equal to these limits respectively, whether they are themselves equal or not.

Examples

(1) $$f(x) = x^3 \sin \frac{1}{x} \quad (x \neq 0), \quad f(0) = 0.$$

If $x \neq 0$, by the product rule,

$$f'(x) = 3x^2 \sin (1/x) - x \cos (1/x).$$

It follows that $f'(x) \to 0$ when $x \to 0$. Also $f(x)$ is continuous at $x = 0$. Consequently, by the theorem, $f'(0)$ exists and is equal to $\lim f'(x)\,(x \to 0)$, i.e. $f'(0) = 0$. This can, of course, be proved easily from first principles.

(2) $$f(x) = x^2 \sin \frac{1}{x} \quad (x \neq 0), \quad f(0) = 0.$$

If $x \neq 0$, by the product rule,

$$f'(x) = 2x \sin (1/x) - \cos (1/x)$$

from which we see that when $x \to 0$, $f'(x)$ does not tend to a limit. This shows that $f'(x)$ cannot be continuous at $x = 0$, but does not indicate whether $f'(0)$ exists or not. In fact, we have seen (§2.4, Example) that $f'(0)$ exists and that $f'(0) = 0$.

(3) $$f(x) = x \sin \frac{1}{x} \quad (x \neq 0), \quad f(0) = 0.$$

In this case, neither $f'(0)$ nor $\lim f'(x)\,(x \to 0)$ exists.

2.18   Cauchy's mean value theorem

Let $F(x)$, $G(x)$ be continuous in the closed interval $[a, b]$ and differentiable in the open interval (a, b). Let $F(a) = G(a) = 0$.

Theorem 1. If (i) $G(b) \neq 0$, (ii) $F'(x)$, $G'(x)$ do not vanish together, then at least one number ξ exists, between a and b, such that

$$\frac{F(b)}{G(b)} = \frac{F'(\xi)}{G'(\xi)}. \tag{26}$$

That is, if $F(x)$, $G(x)$ both vanish at one end of a closed interval, then, under the given conditions, the ratio of their values at the other end is equal to the ratio of their derivatives at one interior point at least.

Proof. Put $\qquad \phi(x) = F(b)\,G(x) - F(x)\,G(b).$

Then $\phi(a) = 0, \phi(b) = 0$, and $\phi(x)$ is continuous in the closed interval and differentiable in the open interval. Hence, by Rolle's theorem, at least one number ξ exists in the open interval such that $\phi'(\xi) = 0$, that is, such that $\qquad F(b)\,G'(\xi) - F'(\xi)\,G(b) = 0.$

Now $G'(\xi) \neq 0$, for $G'(\xi) = 0$ would require either $G(b) = 0$ contrary to (i), or $F'(\xi) = 0$ contrary to (ii). Hence (26) follows on dividing by $G(b)\,G'(\xi)$.

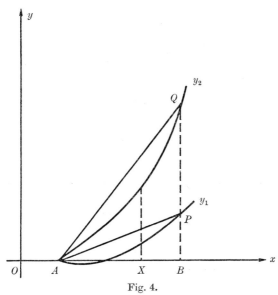

Fig. 4.

Graphical illustration. Two ways of illustrating (26) graphically will be indicated.

In Fig. 4 let the two curves be $y_1 = F(x)$, and $y_2 = G(x)$.

First, equation (26) shows that the ratio BP/BQ of the ordinates of the two curves at $x = b$ is equal to the ratio of their gradients for at least one value ξ, of x, between a and b, provided that conditions (i) and (ii) of the theorem, or other sufficient conditions, are satisfied.

Secondly, equation (26) can be expressed in the form

$$\frac{F(b)/(b-a)}{G(b)/(b-a)} = \frac{F'(\xi)}{G'(\xi)} \tag{27}$$

which shows that the ratio of the average gradients of the curves between $x = a$ and $x = b$ is equal to the ratio of their gradients for at least one value ξ between a and b; in other words, the ratio of the gradients of the chords AP, AQ is equal to the ratio of the gradients of the tangents to the curves for at least one value of x between a and b.

Note. Conditions (i), (ii) in the statement of the above theorem are *sufficient* conditions for the existence of at least one ξ in (26).

Corollary. A sufficient single condition is $G'(x) \neq 0$ throughout (a, b). For conditions (i) and (ii) are then both satisfied, the former in virtue of §2.13, Corollary.

The above theorem may be put in the alternative form:

Theorem 2. Let $f(x)$, $g(x)$ be continuous in the closed interval $[a, b]$ and differentiable in the open interval (a, b).

Then (i) $g(b) \neq g(a)$, (ii) $f'(x), g'(x)$ do not vanish together in (a, b), are sufficient conditions for at least one number ξ to exist between a and b such that

$$\frac{f(b) - f(a)}{g(b) - g(a)} = \frac{f'(\xi)}{g'(\xi)}. \tag{28}$$

This follows from (26) by putting

$$F(x) = f(x) - f(a), \quad G(x) = g(x) - g(a).$$

Corollary. A sufficient single condition for (28) to be true is $g'(x) \neq 0$ throughout (a, b).

Equation (28), due to Cauchy, is a generalization of the simple mean value theorem (a special case got by putting $g(x) = x$). Accordingly, Theorem 2 is called *Cauchy's mean value theorem.*

The reader may test this theorem, e.g. by considering the following cases:

(1) $f(x) = x^3$, $g(x) = x^2$; $a = 0$, $b = 3$.

In this case, $g'(x) \neq 0$ for any x in the open interval $(0, 3)$. This is sufficient to ensure that at least one number ξ exists $(0 < \xi < 3)$. We find $\xi = 2$.

(2) $f(x) = (x+1)^3, g(x) = x^2; a = -2, b = 1$.

There is no value of x in $(-2, 1)$ at which both $f'(x) = 0$ and $g'(x) = 0$. This, together with $g(b) = g(1) = 1 \neq g(a) = g(-2) = 4$, is sufficient to ensure that at least one number ξ exists $(-2 < \xi < 1)$. We find $\xi = -2 + \sqrt{3}$.

(3) $f(x) = (x-1)^3$, $g(x) = (x-1)^2$; $a = -1, b = 4$.

Both $f'(x)$ and $g'(x)$ vanish at $x = 1$ and no value of ξ exists $(-1 < \xi < 4)$ for which (28) is true.

(4) $f(x) = x^3, g(x) = x^2; a = -1, b = 3$.

Here $f'(x)$ and $g'(x)$ both vanish at $x = 0$. Nevertheless, a number ξ exists $(-1 < \xi < 3)$, for we find $\xi = \frac{7}{3}$, showing that conditions (i), (ii) of the theorem, though sufficient, are not necessary.

2.19 Calculus methods of finding limits. L'Hospital's rules

The name of l'Hospital is associated with certain rules for finding the limit, when it exists, of a fraction $F(x)/G(x)$ when for some value of x it assumes the indeterminate form $0/0$ or ∞/∞.

Rule 1. *Form* $0/0$. Let $F(a) = G(a) = 0$, and let $F'(a)$, $G'(a)$ exist, $G'(a) \neq 0$; then

$$\lim_{x \to a} \frac{F(x)}{G(x)} = \frac{F'(a)}{G'(a)}. \tag{29}$$

Proof. This rule follows at once from (19).

Note that $F(x)$, $G(x)$ are defined at $x = a$.

Rule 2. *Form* $0/0$. Let $F(a) = G(a) = 0$. Let $F(x)$, $G(x)$ be continuous at a and differentiable in the neighbourhood of a, with the possible exception of the point $x = a$ itself; then

$$\lim_{x \to a} \frac{F(x)}{G(x)} = \lim_{x \to a} \frac{F'(x)}{G'(x)}, \tag{30}$$

provided that the limit on the R.H.S. exists.

Proof. Since the limit on the R.H.S. exists, $F'(x)/G'(x)$ must be defined near $x = a$, and hence there is a neighbourhood of a in which

$$G'(x) \neq 0 \quad (x \neq a).$$

It follows, by §2.18, Theorem 1, Corollary, that a number ξ exists between a and x such that

$$\frac{F(x)}{G(x)} = \frac{F'(\xi)}{G'(\xi)}.$$

Since $\xi \to a$ when $x \to a$, the rule follows.

Examples

(1) $$\lim_{x \to 0} \frac{x - \sin x}{x^3} = \lim \frac{1 - \cos x}{3x^2} = \lim \frac{\sin x}{6x} = \frac{1}{6}.$$

Here we apply Rule 2 twice and then either Rule 1, or Rule 2 a third time.

(2) $$\lim_{x \to 0+} \frac{x \sin(1/x)}{\sqrt{(\sin x)}} = \lim \frac{\sqrt{x} \sin(1/x)}{\sqrt{\{(\sin x)/x\}}} = \frac{0}{1} = 0.$$

The result follows from known limits after dividing the numerator and denominator of the first fraction by $\sqrt{x}$, without applying either rule. In fact, Rule 1 will not apply, since $F'(0)$ and $G'(0)$ do not exist. Rule 2 will not apply, since $\lim F'(x)/G'(x)$, $x \to 0$, does not exist.

$$
(3) \qquad \lim_{x \to 0} \frac{x + x^2 \sin(1/x)}{\tan x} = 1.
$$

The result is obvious after dividing numerator and denominator by x.

Rule 1 will apply if we first define the value of the numerator at $x = 0$ to be zero; then $F'(0)$ and $G'(0)$ both exist (see §2.4, Example).

Rule 2 will not apply, because $\lim F'(x)/G'(x)$, $x \to 0$, does not exist.

Rule 3. *Form* ∞/∞. At every point of the interval (a, b): (i) let $f(x)$ and $g(x)$ be continuous, (ii) let $g'(x)$ exist and not be zero. Suppose $a < b$.

Further, when $x \to b$, (iii) let $f(x) \to +\infty$ and $g(x) \to +\infty$; (iv) let $\lim f'(x)/g'(x)$ exist and have the value l.

Then $\lim f(x)/g(x)$ also exists and has the value l, i.e.

$$
\lim_{x \to b} \frac{f(x)}{g(x)} = \lim_{x \to b} \frac{f'(x)}{g'(x)} = l. \qquad (31)
$$

[In this statement of Rule 3 we are concerned with left-hand limits. See *Note* below.]

Proof. Let ϵ_1 be any positive number. Then, by (iii) and (iv), there exists an interval $(a < c \leqslant x < b)$ in which

(v) $f(x) > f(c) > 0$, $g(x) > g(c) > 0$, and in which

(vi) $f'(x)/g'(x) = l + \alpha$, where $|\alpha| < \epsilon_1$.

Also, by (i) and (ii), if $x > c$ we can apply Cauchy's mean value theorem and put

$$
\frac{f(x) - f(c)}{g(x) - g(c)} = \frac{f'(\xi)}{g'(\xi)}, \qquad (32)
$$

where $c < \xi < x$. Now by (v), $f(x) \neq 0$ and $g(x) \neq 0$, so (32) can be re-arranged as

$$
\frac{f(x)}{g(x)} = \frac{f'(\xi)}{g'(\xi)} \frac{1 - g(c)/g(x)}{1 - f(c)/f(x)}. \qquad (33)
$$

Referring to (iii) and keeping c fixed, we can find a number $X > c$ such that $g(c)/g(x)$ and $f(c)/f(x)$ are as small as we please for all x in $(X < x < b)$. Consequently, remembering (vi), we can put (33) in the form

$$
f(x)/g(x) = (l + \alpha)(1 + \beta),
$$

where $|\alpha| < \epsilon_1$ and $|\beta| < \epsilon_1$. Hence follows

$$
|f(x)/g(x) - l| = |l\beta + \alpha + \alpha\beta| < l\epsilon_1 + \epsilon_1 + \epsilon_1^2.
$$

Now given any positive ϵ, we can choose $\epsilon_1 > 0$ so that

$$l\epsilon_1 + \epsilon_1 + \epsilon_1^2 < \epsilon.$$

It follows that $f(x)/g(x) \to l$ when $x \to b$.

Note. Rule 3 holds good also when $f(x) \to +\infty$ or $-\infty$ and $g(x) \to +\infty$ or $-\infty$. It also holds good if b is replaced by $+\infty$ or $-\infty$; and if b is replaced by a, in which case we should be concerned with right-hand limits.

The proof will require minor modifications to meet these cases.

Other indeterminate forms. A function $f(x)$ which is undefined at $x = a$, say, may assume other indeterminate forms besides $0/0$ or ∞/∞; for example, $\infty - \infty$, $0 \times \infty$, ∞^0, 1^∞. These forms can often be examined by reducing them to the form $0/0$ or ∞/∞ by an algebraic operation. Others can only be studied from first principles.

Examples

(1) $$\lim_{x \to +\infty} \left(\frac{\log x}{x^n} \right) = 0 \quad (n > 0). \tag{34}$$

The indeterminate form as $x \to +\infty$ is ∞/∞. By applying Rule 3 we have

$$\lim_{x \to +\infty} \frac{\log x}{x^n} = \lim \frac{x^{-1}}{nx^{n-1}} = \lim \frac{1}{nx^n} = 0, \quad \text{since} \quad n > 0.$$

In words, we say that when $x \to +\infty$, $\log x \to +\infty$ more slowly than x^n, however small the (positive) index n.

(2) $$\lim_{x \to 0+} (x^n \log x) = 0 \quad (n > 0). \tag{35}$$

Here the indeterminate form is $0 \times (-\infty)$. But after putting

$$x^n \log x = (\log x)/x^{-n},$$

of which the indeterminate form is $-\infty/\infty$, we can apply Rule 3, thus

$$\lim_{x \to 0+} (x^n \log x) = \lim \frac{\log x}{x^{-n}} = \lim \frac{x^{-1}}{-nx^{-n-1}} = \lim \frac{x^n}{-n} = 0.$$

(3) $$\lim_{x \to +\infty} (x^n e^{-x}) = 0 \quad (\text{all } n). \tag{36}$$

If $n < 0$, when $x \to +\infty$, $x^n e^{-x} \to 0 \times 0 = 0$.

If $n = 0$, $x^n e^{-x} = e^{-x} \to 0$.

If $n > 0$, $x^n e^{-x}$ assumes the indeterminate form $\infty \times 0$. But

$$x^n e^{-x} = x^n / e^x$$

of which the indeterminate form is ∞/∞. Supposing, first, that n is an integer we see, by n applications of Rule 3, that

$$\lim\,(x^n/e^x) = \lim\,(n!/e^x) = 0.$$

Secondly, supposing that $n = m - \theta$, where m is a positive integer and $0 < \theta < 1$, we have

$$\lim_{x \to +\infty} \frac{x^n}{e^x} = \lim \frac{x^m}{e^x} \frac{1}{x^\theta} = 0 \times 0 = 0.$$

Corollary.　　　$\displaystyle\lim_{x \to +\infty}\,(x^n e^{-px}) = 0$　　(all $n, 0 < p$).　　　(37)

This follows from (36) after putting $y = px$, assuming that p is independent of x and of n.

2.20　Derivative of a function of a function

Let $u = f(x)$ be defined near $x = x_0$ and let $y = g(u)$ be defined for the corresponding values of u; then the equations

$$y = g(u), \quad u = f(x)$$

define y as a function of x through u, say $y = \phi(x)$, near $x = x_0$.

Theorem. If $u_0 = f(x_0)$, and if $f'(x_0)$, $g'(u_0)$ exist, then $\phi'(x_0)$ exists and is given by the formula

$$\phi'(x_0) = g'(u_0)f'(x_0).\tag{38}$$

Further, it will follow that, when the derived functions $f'(x)$, $g'(u)$ exist over corresponding intervals of x, u respectively, then

$$\phi'(x) = g'(u)f'(x), \quad \text{or} \quad \frac{dy}{dx} = \frac{dy}{du}\frac{du}{dx}.\tag{39}$$

Proof. Let $x_0 + \delta x, u_0 + \delta u, y_0 + \delta y$ be corresponding values of x, u, y respectively. Then, by (13) since y is differentiable at u_0,

$$\delta y = \{g'(u_0) + \eta\}\,\delta u$$

where $\eta \to 0$ when $\delta u \to 0$. Dividing by δx, we have, if $\delta x \neq 0$,

$$\frac{\delta y}{\delta x} = \{g'(u_0) + \eta\}\frac{\delta u}{\delta x}$$

from which (38) follows when $\delta x \to 0$, since $\delta u \to 0$ and $\delta u/\delta x \to f'(x_0)$.

Extension. Let $u = f(x)$, $v = g(u)$, $y = h(v)$, and let the derived functions $f'(x)$, $g'(u)$, $h'(v)$ or du/dx, dv/du, dy/dv exist over corresponding intervals of x, u, v respectively. Then dy/dx exists and is given by

$$\frac{dy}{dx} = \frac{dy}{dv}\frac{dv}{du}\frac{du}{dx} = h'(v)g'(u)f'(x). \tag{40}$$

For, by using (39) twice,

$$\frac{dy}{dx} = \frac{dy}{dv}\frac{dv}{dx} = \frac{dy}{dv}\frac{dv}{du}\frac{du}{dx}. \tag{41}$$

The result can be extended to any finite number of functions connected in the same kind of way.

2.21   Derivative of an inverse function

In §1.25 were considered conditions under which the equation $x = f(y)$ defines a function $y = \phi(x)$, called the inverse of the function $y = f(x)$. We now note certain sufficient conditions for the derivative of the inverse function to be given by the usual formula

$$dy/dx = (dx/dy)^{-1}.$$

Theorem. Let $x = a$, $y = A$ satisfy the equation $y = f(x)$, so that $A = f(a)$. Then, sufficient conditions in order that the equation

$$f(y) = x, \quad \text{with} \quad y = a \quad \text{at} \quad x = A,$$

should define y as the inverse function $y = \phi(x)$, or a branch of the inverse function, in the neighbourhood of $x = A$, having at $x = A$ the value a and a derivative $\phi'(A)$ given by

$$\phi'(A) = 1/f'(a) \tag{42}$$

are that $f'(x)$ should be continuous at $x = a$ and $f'(a) \neq 0$.

Proof. Since $f'(x)$ is continuous at $x = a$ and $f'(a) \neq 0$, it follows that there is a neighbourhood of $x = a$ in which $f'(x)$ is one-signed (§1.22, II), and hence in which $f(x)$ is strictly monotonic and continuous; consequently, by §1.26 the equation $x = f(y)$ defines y inversely as a strictly monotonic continuous function $y = \phi(x)$ near $x = A$.

Now consider $y = \phi(x)$. While x increases from A to $A + \delta x$, let y increase from a to $a + \delta y$; then

$$\frac{\phi(A + \delta x) - \phi(A)}{(A + \delta x) - A} = \frac{(a + \delta y) - a}{f(a + \delta y) - f(a)}.$$

Since $\phi(x)$ is strictly monotonic, $\delta y \neq 0$ if $\delta x \neq 0$, and therefore

$$\frac{\phi(A + \delta x) - \phi(A)}{\delta x} = \frac{1}{\{f(a + \delta y) - f(a)\}/\delta y}.$$

Now let $\delta x \to 0$, then $\delta y \to 0$ since $\phi(x)$ is continuous, and since $f'(a) \neq 0$ exists the R.H.S. has the limit $1/f'(a)$. It follows that the limit of the L.H.S. exists and is equal to $1/f'(a)$, which proves (42).

Corollary. Let $f'(x)$ be continuous and positive in an interval $(a < x < b)$. In this interval let $f(x)$ increase from $A = f(a)$ to $B = f(b)$, taking the value $C = f(c)$ at some interior point $x = c$. Then the equation

$$f(y) = x, \quad \text{with} \quad y = c \quad \text{at} \quad x = C,$$

defines the inverse function $y = \phi(x)$, or a branch of the inverse function, having at $x = C$ the value c and having throughout the interval $(A < x < B)$ a derived function $\phi'(x)$ given by

$$\phi'(x) = 1/f'(y), \quad \text{or} \quad dy/dx = \frac{1}{dx/dy}. \tag{43}$$

If $f'(x)$ is negative instead of positive, $f(x)$ will decrease from A to B and formula (43) will hold good over the interval $(B < x < A)$.

Example

Consider the function $f(x) = \sin x$, $f'(x) = \cos x$.

Here $f'(x)$ is continuous for all x. In addition, over the interval $(-\tfrac{1}{2}\pi < x < \tfrac{1}{2}\pi)$, $f'(x) = \cos x > 0$ and $f(x) = \sin x$ steadily increases from -1 to $+1$ taking the value $\tfrac{1}{2}$ at $x = \tfrac{1}{6}\pi$. Consequently, the equation

$$\sin y = x, \quad \text{with} \quad y = \tfrac{1}{6}\pi \quad \text{at} \quad x = \tfrac{1}{2}, \tag{44}$$

defines a branch of the inverse function $y = \phi(x)$, having the value $\tfrac{1}{6}\pi$ at $x = \tfrac{1}{2}$ and having a derived function $\phi'(x)$ in the interval $(-1 < x < 1)$ given by

$$\phi'(x) = \frac{1}{f'(y)} = \frac{1}{\cos y}.$$

This branch is called the principal branch of the inverse sine function and is denoted by $\sin^{-1} x$ (see §1.29). Since $-\tfrac{1}{2}\pi < y < \tfrac{1}{2}\pi$, therefore $\cos y > 0$. Also, $\cos^2 y = 1 - \sin^2 y = 1 - x^2$; hence

$$\cos y = \sqrt{(1 - x^2)}$$

and $$\phi'(x) = \frac{d}{dx}(\sin^{-1} x) = \frac{1}{\sqrt{(1 - x^2)}}. \tag{45}$$

Note. The condition 'with $y = \tfrac{1}{6}\pi$ at $x = \tfrac{1}{2}$' in (44) could be replaced by 'with $y = 0$ at $x = 0$', or by any convenient pair of values of x and y in which $-\tfrac{1}{2}\pi < y < \tfrac{1}{2}\pi$.

Next, consider the interval $(\tfrac{1}{2}\pi < x < \tfrac{3}{2}\pi)$. In this interval $\cos x < 0$ and $\sin x$ steadily decreases from 1 to -1, taking the value $\tfrac{1}{2}$ at $x = \tfrac{5}{6}\pi$.

Consequently, the equation

$$\sin y = x, \quad \text{with} \quad y = \tfrac{5}{6}\pi \quad \text{at} \quad x = \tfrac{1}{2}, \tag{46}$$

defines another branch of the inverse sine function, say $y = \phi_1(x)$, having the value $\tfrac{5}{6}\pi$ at $x = \tfrac{1}{2}$ and having a derived function $\phi_1'(x)$ given in the interval $-1 < x < 1$ by

$$\phi_1'(x) = \frac{1}{f'(y)} = \frac{1}{\cos y} = -\frac{1}{\sqrt{(1-x^2)}} \tag{47}$$

since $\cos y < 0$ over the interval $\tfrac{1}{2}\pi < y < \tfrac{3}{2}\pi$ (see Fig. 5).

The condition 'with $y = \tfrac{5}{6}\pi$ at $x = \tfrac{1}{2}$' could be replaced by, e.g. 'with $y = \pi$ at $x = 0$' (see above *Note*).

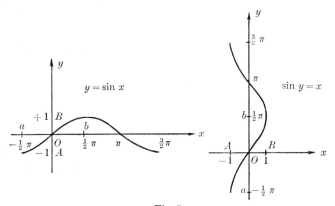

Fig. 5.

2.22   The derivative dy/dx when x and y are given parametrically

Theorem. Let x and y be defined as functions of t by the equations

$$x = f(t), \quad y = g(t) \tag{48}$$

and let there be an interval of values of t in which $f'(t)$ is continuous and non-zero and $g'(t)$ exists. Then equations (48) define y as a function of x whose derivative is given by

$$\frac{dy}{dx} = \frac{dy/dt}{dx/dt} = \frac{g'(t)}{f'(t)}. \tag{49}$$

Proof. By §2.21, the equation $x = f(t)$ defines t inversely as a function of x whose derivative is given by $dt/dx = (dx/dt)^{-1}$. Hence y can be regarded as a function of x through t and, by (39),

$$\frac{dy}{dx} = \frac{dy}{dt}\frac{dt}{dx} = \frac{dy/dt}{dx/dt}. \tag{50}$$

2.23   Simple case of an implicit function

Theorem. Let $x = x_0$, $y = y_0$ satisfy the equation $g(y) = f(x)$, so that $g(y_0) = f(x_0)$. Then, sufficient conditions that the equation

$$g(y) = f(x), \quad \text{with} \quad y = y_0 \quad \text{at} \quad x = x_0, \tag{51}$$

should define y as a function of x, say $y = \psi(x)$, in the neighbourhood of $x = x_0$, having at $x = x_0$ the value $y_0 = \psi(x_0)$ and a derivative $\psi'(x_0)$ given by

$$\psi'(x_0) = f'(x_0)/g'(y_0) \tag{52}$$

are that $g'(y)$ should be continuous at $y = y_0$, with $g'(y_0) \neq 0$, and that $f'(x_0)$ should exist.

Equation (51) is said to define the function $y = \psi(x)$ implicitly (cf. §1.4, examples 7–9).

Proof. Introduce an auxiliary variable t defined by $t = f(x)$ and put $t_0 = f(x_0)$, so that $g(y_0) = f(x_0) = t_0$.

Then, since $g'(y)$ is continuous at $y = y_0$, with $g'(y_0) \neq 0$, it follows from §2.21 that the equation

$$g(y) = t, \quad \text{with} \quad y = y_0 \quad \text{at} \quad t = t_0,$$

defines y inversely as a function of t, say $y = \phi(t)$, in the neighbourhood of $t = t_0$, having at $t = t_0$ the value $y_0 = \phi(t_0)$ and a derivative

$$\phi'(t_0) = 1/g'(y_0).$$

Also, since $f'(x_0)$ exists, the function $t = f(x)$ is necessarily defined in the neighbourhood of $x = x_0$.

It follows from §2.20 that the equations

$$y = \phi(t), \quad t = f(x)$$

define y as a function of x through t, say $y = \psi(x)$, and that, since $\phi'(t_0), f'(x_0)$ exist, the derivative of $\psi(x)$ at $x = x_0$ is given by

$$\psi'(x_0) = \phi'(t_0)f'(x_0),$$

that is, by $\qquad \psi'(x_0) = f'(x_0)/g'(y_0).$

Corollary. It will follow that, if the same conditions hold good over an interval of values of x and over the corresponding interval of values of y, then the function $\psi(x)$ will be defined and its derived function will be given by

$$\frac{dy}{dx} = \psi'(x) = \frac{f'(x)}{g'(y)}, \quad \text{or by} \quad g'(y)\frac{dy}{dx} = f'(x). \tag{53}$$

Note. In practice we obtain this result by the Rule: 'Differentiate both sides of the equation $g(y) = f(x)$ with respect to x, regarding y as a function of x', but this rule assumes that the equation defines y and that y is differentiable.

Examples

(1) If x and y are real, the rule cannot be applied to the equation $y^2 = -1 - x^4$ to give $2y\,dy/dx = -4x^3$, because the equation does not define y as a function of x, there being no real pair of values of x, y that satisfy it.

(2) The equation $y^2 = 1 + x^4$ defines two functions of x near $x = 0$. One is $y = \sqrt{(1 + x^4)}$ defined by

$$y^2 = 1 + x^4, \quad y = 1 \quad \text{at} \quad x = 0,$$

the other is $y = -\sqrt{(1 + x^4)}$ defined by

$$y^2 = 1 + x^4, \quad y = -1 \quad \text{at} \quad x = 0.$$

In each case dy/dx is given by $2y\,dy/dx = 4x^3$ for all values of x.

2.24   Logarithmic differentiation

Theorem. If $\log y = f(x)$ and $f(x)$ is differentiable, then

$$\frac{1}{y}\frac{dy}{dx} = f'(x) \quad (0 < y). \tag{54}$$

Proof. Put $g(y) = \log y$ in (53) and (54) follows.

This theorem is useful when $\log y$ is easier to differentiate than y itself. When this is so, 'take logs' before differentiating.

Example

If $y = u_1^m u_2^n/(v_1^p v_2^q)$, by taking logarithms we have

$$\log y = m \log u_1 + n \log u_2 - p \log v_1 - q \log v_2.$$

Hence, if u_1, u_2, v_1, v_2 are positive and differentiable, and if m, n, p, q are constants, dy/dx is given by

$$\frac{1}{y}\frac{dy}{dx} = \frac{m}{u_1}\frac{du_1}{dx} + \frac{n}{u_2}\frac{du_2}{dx} - \frac{p}{v_1}\frac{dv_1}{dx} - \frac{q}{v_2}\frac{dv_2}{dx}.$$

Examples 2 A

It will be assumed below, when reference is made to d^2y/dx^2 or $f''(x)$, that the reader is familiar with the idea and notation of successive differentiation.

(1) $$f(x) = |\sin x|.$$

Show that $f(x)$ is continuous for all x. Find the left-hand and right-hand derivatives at the points where there is no unique derivative.

(2) If $f(x) = x\,(0 \leqslant x \leqslant 1)$, and $f(x)$ is even and periodic with period 2, show that $f(x)$ is continuous for all x. Find the left-hand and right-hand derivatives at the points where no unique derivative exists.

(3) $1(t) = 0 \quad (t < 0), \quad 1(0) = \frac{1}{2}, \quad 1(t) = 1 \quad (0 < t).$
Show that the left-hand and right-hand derivatives of $1(t)$ do not exist at $t = 0$.
Show that

    (i) $\;1(t) + 1(-t) = 1;$ (ii) $\;1(at) = 1(t) \quad (0 < a);$

    (iii) $\;\{1(t)\}^n = 1(t) \quad (t \neq 0, n > 0);$

    (iv) $\;1(t-a)\,1(t-b) = 1(t-b) \quad (a < b).$

[See Example (1), p. 14. The value of $1(0)$ is of no importance in applications.]

(4) $f(x) = x\tan^{-1}(1/x), \quad x \neq 0, \quad -\frac{1}{2}\pi < \tan^{-1}(1/x) < \frac{1}{2}\pi; \quad f(0) = 0.$
Find the left-hand and right-hand derivatives at $x = 0$.

(5) $$f(x) = -x^3 + (6ab - 3a^2)\,x \quad (0 < x \leqslant a),$$
$$f(x) = -a(3x^2 - 6bx + a^2) \quad (a < x).$$

Show that $f(x), f'(x), f''(x)$ are all continuous at $x = a$, but that $f'''(x)$ is discontinuous.

(6) $$f(x) = 1 - \frac{|x|}{\sqrt{(a^2 + x^2)}}.$$

Sketch the graph of $f(x)$ and find the left-hand and right-hand derivatives at $x = 0$.

[Here $f(x)$ is proportional to the gravitational potential at a point on the axis of x, due to a uniform circular disc of radius a with its centre at the origin and its plane perpendicular to the axis of x.]

(7) $$f(x) = \frac{3a^2 - x^2}{2a^3} \quad (0 \leqslant x \leqslant a), \quad f(x) = \frac{1}{x} \quad (a < x).$$

Sketch the graph of $f(x)\,(0 \leqslant x)$. Show that $f(x)$ and $f'(x)$ are continuous at $x = a$, but that $f''(x)$ is discontinuous.

[Here $f(x)$ is proportional to the gravitational potential due to a uniform solid sphere of radius a, at a point at distance x from the centre.]

Examples 2B

(1) $$\phi(x) = \sqrt{(x-a)}\sqrt{(b-x)} \quad (a \leqslant x \leqslant b).$$

Show that the conditions of Rolle's theorem are satisfied by $\phi(x)$ and find ξ. [Note that $\phi'(a)$, $\phi'(b)$ do not exist finitely.]

(2) $$\phi(x) = \frac{b-x}{x-a} \quad (a < x \leqslant b), \quad \phi(a) = 0.$$

Show that the conditions of Rolle's theorem are not all satisfied by $\phi(x)$, and that no number ξ exists.

(3) $$\phi(x) = \sin x - \frac{2x}{\pi}.$$

Show that $\phi(0) = 0$, $\phi(\frac{1}{2}\pi) = 0$, and find ξ between 0 and $\frac{1}{2}\pi$ such that $\phi'(\xi) = 0$.

(4) $$\phi(x) = x\sin\frac{1}{x} \quad \left(0 < x \leqslant \frac{1}{\pi}\right), \quad \phi(0) = 0.$$

Show that the conditions of Rolle's theorem are satisfied by $\phi(x)$, and that an infinity of values of ξ exist.
[Note that $\phi'(0)$ does not exist.]

(5) $$\phi(x) = 2x-1 \quad (x \leqslant 1), \quad \phi(x) = (x-2)^2 \quad (1 < x).$$

Show that Rolle's theorem does not apply to $\phi(x)$ in the interval $\frac{1}{2} \leqslant x \leqslant 2$, and explain why.

(6) Apply (24) to the functions (i) $f(x) = x^{\frac{1}{2}}$, (ii) $f(x) = x^{\frac{3}{2}}$, where $0 \leqslant x$, and find the value of θ in each case.

(7) By applying (24) to the cubic function $f(x) = x^3 + ax^2 + bx + c$ show that the quadratic equation in θ,

$$3x\theta^2 + 2a\theta - x - a = 0 \quad (x \neq 0),$$

must have a root between 0 and 1, whatever the (real) values of a and x. Show that in fact there is a root between 0 and $\frac{2}{3}$ if $x + a \neq 0$, and that $\theta = \frac{2}{3}$ is a root if $x + a = 0$.

(8) If $f(x)$ is continuous in the closed interval $n \leqslant x \leqslant n+1$, and differentiable in the open interval $n < x < n+1$, show that at least one number θ exists $(0 < \theta < 1)$ such that
$$f(n+1) - f(n) = f'(n+\theta).$$

Show that θ exists $(0 < \theta < 1)$ such that

(i) $$\frac{1}{n} - \frac{1}{n+1} = \frac{1}{(n+\theta)^2} \quad (0 < n);$$

(ii) $\log(n+1) - \log n = 1/(n+\theta) \quad (0 < n)$.

(9) Given that $f(a) = g(a)$, show that:
   (i) if $f'(x) > g'(x) \quad (a \leqslant x)$, then $f(x) > g(x) \quad (a < x)$;
   (ii) if $f'(x) < g'(x) \quad (x \leqslant a)$, then $f(x) > g(x) \quad (x < a)$.

4

(10) Prove that:

(i) $e^x \geqslant 1+x$,   (ii) $e^{-x} \geqslant 1-x$,

(iii) $1/(1-x) \geqslant e^x \geqslant 1+x$ $(x < 1)$,

(iv) $x \geqslant \log(1+x) \geqslant x/(1+x)$ $(-1 < x)$,

(v) $\log(1+x) > x/(1+\frac{1}{2}x)$ $(0 < x)$.

(11) (i) If $0 < x < \frac{1}{2}\pi$, show that $\tan x > x$.

(ii) If $0 < |x| < \frac{1}{2}\pi$, show that $\log \sec x > \frac{1}{2}x^2$.

(12) (i) Show that the function $x\sin(a/2x)$ is strictly increasing if $x > a/\pi > 0$.

(ii) Show that the function

$$\frac{1}{x} + \frac{1}{x+1} - 2\log\frac{x+1}{x}$$

strictly decreases from $+\infty$ to 0 as x increases from 0 to $+\infty$.

(13) If $0 < a < 1$, show that $(\cosh ax)/\sinh x$ steadily decreases from $+\infty$ to 0 as x increases from 0 to $+\infty$.

(14) Show that, if $x > 0$:

(i) $\sin x < x$,   (ii) $\cos x > 1 - \frac{1}{2}x^2$,

(iii) $\sin x > x - x^3/3!$,   (iv) $\cos x < 1 - x^2/2! + x^4/4!$

(15) Show that:

(i) $e^x > 1+x$ $(x \neq 0)$;

(ii) $e^x >$ or $< 1+x+\frac{1}{2}x^2$, according as $x > 0$ or $x < 0$;

(iii) $e^x > 1+x+x^2/2!+x^3/3!$ $(x \neq 0)$;

(iv) $e^x >$ or $< 1+x+x^2/2!+x^3/3!+x^4/4!$ according as $x > 0$ or $x < 0$.

(16) If $F_n(x) = \log(1+x) - S_n(x)$,

where $S_n(x) = x/1 - x^2/2 + x^3/3 - \dots$ to n terms,

prove that $F_n'(x) = (-x)^n/(1+x)$. Hence show that $\log(1+x)$ lies between $S_n(x)$ and $S_{n+1}(x)$ if $0 < x$. By putting $x = 1$ deduce that

$$\log 2 > S_{2n}(1) > \log 2 - \frac{1}{2n+1}$$

and hence that $\log 2 = 1 - \frac{1}{2} + \frac{1}{3} - \frac{1}{4} + \dots$ *ad inf.*

(17) If $p > q > 0$ and $x > 0$, prove that

$$\frac{x^p - 1}{p} \geqslant \frac{x^q - 1}{q},$$

with equality only when $x = 1$.

(18) $f(x) = \left(x\sin\frac{1}{x}\right)^2$ $(x \neq 0)$, $f(0) = 0$.

Show that $f'(0) = 0$, but that $f'(0)$ cannot be obtained by first finding $f'(x)$, $x \neq 0$, and then putting $x = 0$.

(19) Consider Cauchy's mean value theorem in the following cases; in each case determine whether or not the number ξ exists:

(i) $f(x) = \sin \pi x$, $\quad g(x) = x^2$; $\quad a = 0, b = \frac{1}{2}$;

(ii) $f(x) = \cos \pi x$, $\quad g(x) = x^2$; $\quad a = 0, b = \frac{1}{2}$;

(iii) $f(x) = \cos \pi x$, $\quad g(x) = x^2$; $\quad a = -\frac{1}{3}, b = \frac{1}{2}$;

(iv) $f(x) = \cos \pi x$, $\quad g(x) = x^2$; $\quad a = -1, b = \frac{1}{2}$.

(20) Consider the equation $x^4 - y^4 - 2xy(1 - x^2 y^2) = 0$.

Regarding x as the unknown, show that if $y(1 - y^2) \neq 0$ the equation has a root between y and 1. Regarding y as the unknown, show that if $0 < x^2 < 1$ the equation has a root between 0 and x.

(21) Let $f(x)$ be a polynomial with real coefficients. If the equation $f'(x) = 0$ has not more than r real roots, show that the equation $f(x) = 0$ cannot have more than $r + 1$ real roots.

If $n > 2$, and a, b, c are real, show that the equation

$$x^n + ax^2 + bx + c = 0$$

(i) has either 1 or 3 real roots if n is odd,

(ii) has 0, 2 or 4 real roots if n is even.

(22) Let $f(x)$ be a polynomial, and let λ, μ be any two real numbers. If all the roots of the equation $f(x) = 0$ are real and distinct, show that the same is true of the equation $f(x) - \lambda f'(x) = 0$, and also of the equation

$$f(x) - (\lambda + \mu)f'(x) + \lambda \mu f''(x) = 0.$$

Generalize this.

If $b^2 \geqslant 4ac$ and if the roots of $f(x) = 0$ are all real and distinct, show that the same is true of the equation $af(x) + bf'(x) + cf''(x) = 0$.

Examples 2C

(1) Evaluate the following limits:

(i) $\displaystyle\lim_{x \to 0} \frac{e^{ax} - 1}{x}$,

(ii) $\displaystyle\lim_{x \to 0} \frac{a^x - 1}{x}$ $\quad (0 < a)$,

(iii) $\displaystyle\lim_{x \to 1} \frac{x^p - x^q}{x^m - x^n}$,

(iv) $\displaystyle\lim_{x \to +\infty} x\{x - \sqrt{(x^2 - a^2)}\}$,

(v) $\displaystyle\lim_{x \to 0+} \frac{x + \exp(-1/x)}{\sin \pi x}$,

(vi) $\displaystyle\lim_{x \to 0+} \frac{1}{x^n} \operatorname{sech} \frac{1}{x}$ $\quad (0 < n)$,

(vii) $\displaystyle\lim_{x \to -\infty} \frac{\cosh(a + x)}{\sinh x}$,

(viii) $\displaystyle\lim_{x \to \frac{1}{4}} \frac{1 - \sin \pi x}{2x + \cos 2\pi x}$,

(ix) $\displaystyle\lim_{x \to 0+} (1 + x)^{\log x}$,

(x) $\displaystyle\lim_{x \to \infty} x(a^{1/x} - 1)$ $\quad (0 < a)$,

(xi) $\displaystyle\lim_{x \to \frac{1}{2}\pi} (\pi - 2x) \tan x$,

(xii) $\displaystyle\lim_{x \to +\infty} x^n(1 - \tanh x)$,

(xiii) $\displaystyle\lim_{\theta \to \alpha} \frac{\cos(\pi\theta/2\alpha)}{\cos \theta - \cos \alpha}$ $\quad (\alpha \neq n\pi)$,

(xiv) $\displaystyle\lim_{x \to \pm\infty} \log \frac{ax+b}{px+q}$ $(0 < ap)$,

(xv) $\displaystyle\lim_{x \to \infty} (x \tanh x - \log \cosh x)$,

(xvi) $\displaystyle\lim_{x \to 0+} \frac{\log(ax^m)}{\log(bx^n)}$ $(a, b, m, n \text{ all positive})$,

(xvii) $\displaystyle\lim_{x \to \frac{1}{2}\pi} \frac{2 \tan x - \tan(\frac{1}{4}\pi + \frac{1}{2}x)}{\pi - 2x}$,

(xviii) $\displaystyle\lim_{x \to +\infty} \{\sinh^{-1}(ax+b) - \sinh^{-1}(cx+d)\}$ $(0 < ac)$.

(2) $\qquad\qquad f(x) = x^n \log x \quad (0 < x), \quad f(0) = 0.$

Show that $f'(0) = 0$ if $1 < n$, and that

$$f'(x) \to -\infty, \quad x \to 0+, \quad \text{if} \quad 0 < n \leqslant 1.$$

(3) $\qquad\qquad f(x) = \dfrac{x^2}{1 + \exp(-1/x)} \quad (x \neq 0), \quad f(0) = 0.$

Show that $f'(x)$ is continuous at $x = 0$, but that $f''(x)$ is discontinuous.

(4) $\qquad\qquad f(x) = \exp(-1/x)\{\sin\exp(1/x)\} \quad (0 < x), \quad f(0) = 0.$

Find $f'(x)$ and $f'(0+)$ and show that $f'(x)$ is discontinuous at $x = 0$.

(5) $\qquad f(x) = \dfrac{a + b\exp(1/x)}{c + d\exp(1/x)} \quad (x \neq 0), \quad (c \neq 0, d \neq 0, ad - bc \neq 0).$

Show that the left-hand and right-hand limits of $f(x)$ and of $f'(x)$ exist at $x = 0$.

[The left-hand and right-hand derivatives at $x = 0$ do not exist, as $f(0)$ is not defined.]

Sketch the graph of $f(x)$ near $x = 0$ for $a = 2, b = 0, c = 1, d = -1$.

(6) If $f'(x)$ tends to a limit when $x \to \infty$, and if the limit is denoted by $f'(\infty)$, show that

$$\lim_{x \to \infty} \frac{f(x)}{x} = \lim_{x \to \infty}\{f(x+1) - f(x)\} = f'(\infty).$$

(7) $\qquad\qquad f(x) = \left(1 + \dfrac{1}{x}\right)^x.$

(i) Show that $f(x)$ is not defined in $(-1 \leqslant x \leqslant 0)$.

(ii) Find $\lim f(x)$ when $x \to 0+$.

(iii) Find $\lim f(x)$ when $x \to +\infty$ and when $x \to -\infty$.

(iv) Show that $f'(x) > 0$ for $x < -1$ and for $0 < x$.

(v) Sketch the graph of $f(x)$ for $x < -1$ and for $0 < x$.

(8) In the following cases examine the behaviour of $f(x)$ and $f'(x)$ near $x = 0$, and the sign of $f'(x)$ for all values of x, and sketch the graph of $f(x)$:

(i) $f(x) = x^x \quad (0 < x)$, $\qquad\qquad$ (ii) $f(x) = x^2 \log(x^2)$,

(iii) $f(x) = \text{sech}(1/x)$, $\qquad\qquad$ (iv) $f(x) = \tanh(1/x)$,

(v) $f(x) = \exp(-1/x^2)$, $\qquad\qquad$ (vi) $f(x) = (1+x)^{1/x} \quad (-1 < x)$.

(9) (a) Let $f(x)$ be defined by

$$f(x) = \frac{\sin x}{x} \quad (x \neq 0), \quad f(0) = 1.$$

Use §2.17 to find $f'(0), f''(0), \ldots$.

(b) Repeat for the function defined by

$$f(x) = \frac{\log(1+x)}{x} \quad (x \neq 0), \quad f(0) = 1.$$

(10) If $F'(a), G'(a)$ both exist, with $G(a) = 0$, $G'(a) \neq 0$, prove that

$$\lim_{x \to a} \frac{(x-a) F(x)}{G(x)} = \frac{F(a)}{G'(a)}.$$

Examples 2D

Examples in formal differentiation:

(1) If $y = ax/(x+b)$, prove that

$$\frac{d^2y}{dx^2} = \frac{2}{y} \left(\frac{dy}{dx}\right)^2 - \frac{2}{x} \frac{dy}{dx}.$$

(2) If $y = e^{-kt}(A \sin \mu t + B \cos \mu t)$, prove that

$$\frac{d^2y}{dt^2} + 2k \frac{dy}{dt} + (k^2 + \mu^2) y = 0.$$

(3) Show that $y = Kx^n$ satisfies the equation

$$x \frac{d^2y}{dx^2} + \frac{dy}{dx} = \frac{a}{\sqrt{y}}$$

provided that the constants K and n have certain values.

(4) Verify that, if κ and A are constants,

$$v = \left(\frac{3}{r^2} - \frac{1}{\kappa^2}\right) \sin \left(\frac{r}{\kappa} + A\right) - \frac{3}{\kappa r} \cos \left(\frac{r}{\kappa} + A\right)$$

satisfies the equation $\qquad \dfrac{d^2v}{dr^2} + \left(\dfrac{1}{\kappa^2} - \dfrac{6}{r^2}\right) v = 0.$

(5) If $y = A \cos \omega t + B \sin \omega t$, show that $(dy/dt)^2 = \omega^2(A^2 + B^2 - y^2)$, where A, B, ω are constants. Find similar results for

(i) $y = A \cosh mt + B \sinh mt$,   (ii) $y = A e^{nt} + B e^{-nt}$.

(6) Show that $y = A e^{x^2} + B e^{\frac{1}{2}x^2}$, where A, B are constants, satisfies the equation

$$\frac{d^2y}{dx^2} - \left(3x + \frac{1}{x}\right) \frac{dy}{dx} + 2x^2 y = 0.$$

(7) If $s = \sin\theta$, $c = \cos\theta$, show that

(i) $\dfrac{d}{d\theta}(sc^n) = (n+1)c^{n+1} - nc^{n-1}$,

(ii) $\dfrac{d}{d\theta}(s^m c) = ms^{m-1} - (m+1)s^{m+1}$,

(iii) $\dfrac{d}{d\theta}(s^m c^n) = ms^{m-1}c^{n-1} - (m+n)s^{m+1}c^{n-1}$

$\qquad\qquad = (m+n)s^{m-1}c^{n+1} - ns^{m-1}c^{n-1}$.

(8) If $y = \sin^n\theta$ show that

$$d^2y/d\theta^2 = n(n-1)\sin^{n-2}\theta - n^2\sin^n\theta.$$

If n is a positive integer, show that the nth differential coefficient of $\sin^n\theta$ at $\theta = 0$ is $n!$

(9) If $w = (z^2 - z + 1)^3/(z^2 - z)^2$, show that

$$dw/dz = (z^2 - z + 1)^2(2z - 1)(z - 2)(z + 1)/(z^2 - z)^3.$$

(10) Differentiate the following functions:

    (i) $\log\{\log(\log x)\}$,   (ii) $\log\sin(\log\sin x)$,

    (iii) $x^{\log x}$,   (iv) $x^{\log\log x}$,

    (v) $(\log x)^{\log x}$,   (vi) a^{x^x},

    (vii) $\log_x a$,   (viii) $(\log_x a)^x$,

    (ix) $\sin^{-1}\{2x/(1+x^2)\}$,   (x) $\sin^{-1}\{2x\sqrt{(1-x^2)}\}$,

    (xi) $\tan^{-1}\dfrac{\sqrt{(x^2-1)}}{1+x}$,   (xii) $\tan^{-1}\dfrac{x\sin\alpha}{1-x\cos\alpha}$,

    (xiii) $\sin^{-1}\dfrac{2\sqrt{(ax)}}{a+x}$,   (xiv) $\tanh^{-1}\dfrac{4x(1+x^2)}{1+6x^2+x^4}$,

    (xv) $\sin^{-1}\dfrac{\sqrt{(\cos 2\theta)}}{\sin\theta}$,   (xvi) $\tan^{-1}\left\{\dfrac{a+x}{a-x}\tan\dfrac{\alpha}{2}\right\}$,

    (xvii) $\tan^{-1}\dfrac{x}{1+\sqrt{(1+x^2)}}$,   (xviii) $\cos^{-1}\dfrac{a+b\cos x}{b+a\cos x}$ $(a < b)$,

    (xix) $\cosh^{-1}\dfrac{\sqrt{(x+1)}\sqrt{(x+4)}}{x+2}$,   (xx) $\log\dfrac{3\sqrt{(1+x^3)} - (1+x)^2}{3\sqrt{(1+x^3)} + (1+x)^2}$,

    (xxi) $\log\dfrac{\sqrt{(1-k^2x)} + (1-k')\sqrt{(x-x^2)}}{1-(1-k')x}$ $(k^2 + k'^2 = 1)$.

Examples 2 E

(1) Four inextensible jointed rods form a quadrilateral. If p, q are the lengths of the lines joining the mid-points of opposite sides, prove that, in a small deformation of the quadrilateral, $p\,dp = q\,dq$.

(2) Four inextensible jointed rods form a quadrilateral $ABCD$. If $x = AC$, $y = BD$, prove that in a small deformation of the quadrilateral

$$\frac{dA}{\Delta BCD} = -\frac{dB}{\Delta CDA} = \frac{dC}{\Delta DAB} = -\frac{dD}{\Delta ABC}$$

$$= -\frac{x\,dx}{2\Delta ABC.\Delta CDA} = \frac{y\,dy}{2\Delta DAB.\Delta BCD}.$$

(3) If $x = r\cos\theta$, $y = r\sin\theta$, $z = x + iy = r\,e^{i\theta}$, and if r, θ are functions of t, verify that (see §2.6, IV)

(i) $\dfrac{dz}{dt} = \dfrac{dx}{dt} + i\dfrac{dy}{dt} = \left(\dfrac{dr}{dt} + ir\dfrac{d\theta}{dt}\right)e^{i\theta},$

(ii) $\dfrac{d^2z}{dt^2} = \dfrac{d^2x}{dt^2} + i\dfrac{d^2y}{dt^2} = \left\{\dfrac{d^2r}{dt^2} - r\left(\dfrac{d\theta}{dt}\right)^2 + \dfrac{i}{r}\dfrac{d}{dt}\left(r^2\dfrac{d\theta}{dt}\right)\right\}e^{i\theta}.$

(4) *Derivative of a determinant.* If $u_1, v_1, w_1, \ldots,$ are differentiable functions of x, show, by expanding the determinant $|u_1\,v_2\,w_3|$ and using the product rule, that the derivative of the determinant can be expressed by

$$\begin{vmatrix} u_1' & v_1 & w_1 \\ u_2' & v_2 & w_2 \\ u_3' & v_3 & w_3 \end{vmatrix} + \begin{vmatrix} u_1 & v_1' & w_1 \\ u_2 & v_2' & w_2 \\ u_3 & v_3' & w_3 \end{vmatrix} + \begin{vmatrix} u_1 & v_1 & w_1' \\ u_2 & v_2 & w_2' \\ u_3 & v_3 & w_3' \end{vmatrix},$$

or by a similar expression in which rows are differentiated in turn, instead of columns.

Generalize for a determinant of order n.

(5) Let $y = (1 + 1/x)^{x+a}$. Prove that, as x increases from 0 to ∞,

   (i) if $a = 0$, then y increases monotonically from 1 to e,

   (ii) if $a < 0$, then y increases monotonically from 0 to e,

   (iii) if $a \geqslant \frac{1}{2}$, then y decreases monotonically from ∞ to e.

Also, examine how y varies when $0 < a < \frac{1}{2}$.

(6) Prove that the function

$$\frac{x+1}{x+\frac{1}{2}}\left(1+\frac{1}{x}\right)^x$$

increases monotonically from 2 to e while x increases from 0 to ∞.

(7) If $\phi(x)$ is defined as in §2.6, v, Example, show that $\phi(x)\sin x$ is differentiable at $x = 0$, $\pm\pi$, $\pm 2\pi$, ... but at no other point.

3

SUCCESSIVE DIFFERENTIATION

3.1 Notation

The derived function of $f'(x)$, if it exists, is called the *second derived function* of $f(x)$ and, if $y = f(x)$, is denoted as convenient by

$$f''(x), \quad y'', \quad y_2, \quad d^2y/dx^2, \quad (d/dx)^2 y, \quad D_x^2 y, \quad D^2 y.$$

The third, fourth, ..., nth derived functions of $f(x)$ are successively defined, each as the derived function of its predecessor. The nth derived function is denoted by

$$f^{(n)}(x), \quad y^{(n)}, \quad y_n, \quad d^n y/dx^n, \quad (d/dx)^n y, \quad D_x^n y \quad \text{or} \quad D^n y.$$

The existence of $f^{(n)}(x)$ implies the existence and continuity of $f^{(n-1)}(x), ..., f'(x), f(x)$ (see §2.11, II).

If $f^{(n)}(a)$ exists, $f(x)$ is said to be n-times differentiable at $x = a$; if $f^{(n)}(x)$ exists throughout an interval, $f(x)$ is said to be n-times differentiable in the interval.

Note. If u is a function of x, $f^{(n)}(u)$ means $(d/du)^n f(u)$, not $(d/dx)^n f(u)$. Thus, if $f(3x+5) = \sin(3x+5)$, then $f'(3x+5)$ means $\cos(3x+5)$, not $3\cos(3x+5)$.

3.2 The nth derivatives of some elementary functions

The nth derived functions of some of the elementary functions of x can be simply expressed in terms of x and n: thus, if a, b, c, m are constants,

$$D^n(x-a)^m = m(m-1)...(m-n+1)(x-a)^{m-n}, \tag{1}$$

$$D^n \frac{1}{(x-a)^m} = (-)^n \frac{m(m+1)...(m+n-1)}{(x-a)^{m+n}}, \tag{2}$$

$$D^n \log x = (-)^{n-1} \frac{(n-1)!}{x^n}, \tag{3}$$

$$D^n e^{cx} = c^n e^{cx}. \tag{4}$$

By putting $c = a+ib$ in (4), we obtain

$$D^n e^{(a+ib)x} = (a+ib)^n e^{(a+ib)x} = (re^{i\theta})^n e^{(a+ib)x}$$

and hence, by equating real and imaginary parts,

$$D^n e^{ax} \cos bx = r^n e^{ax} \cos(bx+n\theta), \tag{5}$$

$$D^n e^{ax} \sin bx = r^n e^{ax} \sin(bx+n\theta), \tag{6}$$

where
$$r e^{i\theta} = a+ib, \quad r = |a+ib| = \sqrt{(a^2+b^2)}, \quad \theta = \arg(a+ib).$$

In particular, when $a = 0$,

$$D^n \cos bx = b^n \cos(bx + \tfrac{1}{2}n\pi), \tag{7}$$

$$D^n \sin bx = b^n \sin(bx + \tfrac{1}{2}n\pi), \tag{8}$$

and when $b = 1$, $\qquad D^n \cos x = \cos(x + \tfrac{1}{2}n\pi), \tag{9}$

$$D^n \sin x = \sin(x + \tfrac{1}{2}n\pi). \tag{10}$$

3.3

By means of (1) and (2) the nth derivative of any *rational fraction* can be found. If the fraction is a proper fraction, i.e. if the degree of the numerator is less than that of the denominator, the denominator must first be factorized into its linear factors, real or complex, and the fraction then expressed as the sum of partial fractions each with a denominator which is a linear factor or a power of a linear factor. If the fraction is an improper fraction, the numerator must first be divided by the denominator so as to express the fraction as the sum of a polynomial and a proper fraction, and the proper fraction must then be expressed in partial fractions, as just explained. (See §11.9.)

Examples

(1) $$y = \frac{2x^3}{x^2-1} = 2x + \frac{1}{x-1} + \frac{1}{x+1},$$

$$y_n = (-)^n n! \left\{ \frac{1}{(x-1)^{n+1}} + \frac{1}{(x+1)^{n+1}} \right\} \quad (n > 1).$$

(2) $y = \tan^{-1} x$,

$$y_1 = \frac{1}{1+x^2} = \frac{1}{2}\left(\frac{1}{1-ix} + \frac{1}{1+ix}\right),$$

$$y_n = \frac{(n-1)!}{2}\left\{ \frac{i^{n-1}}{(1-ix)^n} + \frac{(-i)^{n-1}}{(1+ix)^n} \right\}.$$

Since $x = \tan y$ and $i = e^{\frac{1}{2}\pi i}$, $-i = e^{-\frac{1}{2}\pi i}$, we can express y_n neatly in terms of y, thus

$$y_n = \frac{(n-1)! \cos^n y}{2i} \left\{ e^{in(\frac{1}{2}\pi + y)} - e^{-in(\frac{1}{2}\pi + y)} \right\}$$

$$= (n-1)! \cos^n y \sin n(\tfrac{1}{2}\pi + y).$$

Note that $-\tfrac{1}{2}\pi < y < \tfrac{1}{2}\pi$, so that $\cos y > 0$.

3.4   The nth derivative of a product. Leibniz's rule

Theorem. If u, v are functions of x and if u_n, v_n exist, then

$$(uv)_n = u_n v + {}^nC_1 u_{n-1} v_1 + {}^nC_2 u_{n-2} v_2 + \ldots + uv_n, \qquad (11)$$

the coefficients being those of the binomial theorem.

Proof. Assume that, for any positive integer m,

$$(uv)_m = u_m v + {}^mC_1 u_{m-1} v_1 + {}^mC_2 u_{m-2} v_2 + \ldots. \qquad (12)$$

Also, assume that u_{m+1} and v_{m+1} exist. Then we can differentiate (12), obtaining

$$(uv)_{m+1} = u_{m+1} v + (1 + {}^mC_1) u_m v_1 + ({}^mC_1 + {}^mC_2) u_{m-1} v_2 + \ldots$$

or, remembering that ${}^mC_r + {}^mC_{r+1} = {}^{m+1}C_{r+1}$,

$$(uv)_{m+1} = u_{m+1} v + {}^{m+1}C_1 u_m v_1 + {}^{m+1}C_2 u_{m-1} v_2 + \ldots. \qquad (13)$$

But (13) is merely (12) with $m+1$ in place of m. Consequently, if (12) is true for any positive integer m and if u_{m+1} and v_{m+1} exist, then (12) is true for the integer $m+1$. But (12) is true for $m = 1$ if u_1 and v_1 exist; it is therefore true for $m = 2$ if u_2 and v_2 exist, and therefore for $m = 3$ if u_3 and v_3 exist, and by induction for any positive integer n if u_n and v_n exist.

Corollary. When $v = e^{ax}$, Leibniz's rule gives

$$(ue^{ax})_n = u_n e^{ax} + {}^nC_1 u_{n-1} a e^{ax} + {}^nC_2 u_{n-2} a^2 e^{ax} + \ldots$$

or $\qquad D^n(ue^{ax}) = e^{ax}(D^n u + {}^nC_1 a D^{n-1}u + {}^nC_2 a^2 D^{n-2}u + \ldots)$

which is symbolically expressed by

$$D^n(ue^{ax}) = e^{ax}(D+a)^n u. \qquad (14)$$

By replacing a by $a+ib$ and equating real and imaginary parts, we could deduce formulae for

$$D^n(ue^{ax}\cos bx), \quad D^n(ue^{ax}\sin bx),$$

where u denotes a real function of x.

3.5   The nth derivative of a quotient

Theorem. If $y = u/v$ and if u_n, v_n exist and $v \neq 0$, then y_n is given by

$$v^{n+1} y_n = \begin{vmatrix} v & 0 & 0 & \ldots & u \\ v_1 & v & 0 & \ldots & u_1 \\ v_2 & 2v_1 & v & \ldots & u_2 \\ \ldots\ldots\ldots\ldots\ldots\ldots\ldots\ldots\ldots\ldots \\ v_n & {}^nC_1 v_{n-1} & {}^nC_2 v_{n-2} & \ldots & u_n \end{vmatrix}. \qquad (15)$$

Proof. Evidently y_n exists, for it could be found by n successive applications of the quotient rule. However, this would not lead to a formula as simple as (15). To obtain (15) we multiply $y = u/v$ by v and then differentiate the equation $vy = u$ successively n times, using Leibniz's rule to differentiate the product vy. We then have the following equations:

$$vy = u,$$
$$v_1 y + v y_1 = u_1,$$
$$v_2 y + 2 v_1 y_1 + v y_2 = u_2,$$
$$\dots\dots\dots\dots\dots\dots\dots\dots\dots\dots\dots\dots\dots\dots\dots\dots\dots\dots .$$
$$v_n y + {}^n C_1 v_{n-1} y_1 + {}^n C_2 v_{n-2} y_2 + \dots + v y_n = u_n.$$

These $n+1$ equations are linear in $y, y_1, y_2, \dots, y_n$ which may be regarded as $n+1$ unknowns. When the equations are solved by the method of determinants, y_n is given by $y_n = \Delta_{n+1}/\Delta$, provided that $\Delta \neq 0$; where Δ is the determinant of the coefficients and Δ_{n+1} is the determinant obtained by replacing the last column of Δ by the column of u's.

We see at once that $\Delta = v^{n+1}$ and that Δ_{n+1} is the determinant in (15). Because $v \neq 0$, therefore $\Delta \neq 0$, and (15) follows.

3.6 The nth derivative of a function of a function

Let $y = f(u)$, $u = g(x)$, so that y is a function of x through u. Let $y = \phi(x)$. Then by §2.20 and the product rule, provided that

$$f'(u), \quad g'(x), \quad f''(u), \quad g''(x)$$

exist, $\quad \phi'(x) = f'(u) g'(x), \quad \phi''(x) = f''(u)\{g'(x)\}^2 + f'(u) g''(x)$

and it is evident that repeated differentiation would show that $\phi^{(n)}(x)$ exists if $f^{(n)}(u)$, $g^{(n)}(x)$ exist, i.e. $d^n y/dx^n$ exists if $d^n y/du^n$ and $d^n u/dx^n$ exist.

3.7

We shall now assume the validity of the following formula, known as Taylor's formula, to be proved later (§4.4): If $\phi(x)$ is a function of x and if $\phi^{(n)}(x)$ exists, then $\phi(x+h)$ can be expressed in the form

$$\phi(x+h) = \phi_0 + h\phi_1 + h^2\phi_2/2! + \dots + h^n\phi_n/n! + \eta h^n, \tag{16}$$

where $\phi_0 = \phi(x)$, $\phi_1 = \phi'(x)$, $\phi_2 = \phi''(x)$, $\dots$ and η is a function of x and h which, with x fixed, tends to zero when $h \to 0$.

Moreover, Taylor's formula is the one and only formula that exists of the type
$$\phi(x+h) = c_0 + c_1 h + c_2 h^2 + \ldots + c_n h^n + \eta h^n \qquad (17)$$
in which the coefficients c_s $(0 \leqslant s \leqslant n)$ are independent of h and $\eta \to 0$ when $h \to 0$ (see §4.6). It follows that if ϕ_n is known to exist, and if a formula of the type (17) can be found by a method which does not depend upon a prior knowledge of ϕ_n, then this formula will determine ϕ_n, since $\phi_n/n! = c_n$.

3.8

It will now be shown how such a method can be applied to determine the nth derivative of a function of a function (see also §4.7, III and Examples 3B, 40).

As in §3.6, let $y = f(u)$, $u = g(x)$, and put $y = \phi(x)$, so that
$$y = f\{g(x)\} = \phi(x).$$
For brevity put $g_s = g^{(s)}(x)$, $\phi_s = \phi^{(s)}(x)$, $f_s = f^{(s)}(u)$ $(s \leqslant n)$.
Let g_n exist; then, by Taylor's formula,
$$\phi(x+h) = f\{g(x+h)\}$$
$$= f\{g_0 + g_1 h + g_2 h^2/2! + \ldots + g_n h^n/n! + \alpha h^n\},$$
where $\alpha \to 0$ when $h \to 0$; and therefore, since $g_0 = g(x) = u$,
$$\phi(x+h) = f(u+k), \qquad (18)$$
where
$$k = g_1 h + g_2 h^2/2! + \ldots + g_n h^n/n! + \alpha h^n. \qquad (19)$$

Let f_n exist; then, again by Taylor's formula,
$$\phi(x+h) = f_0 + f_1 k + f_2 k^2/2! + \ldots + f_n k^n/n! + \beta k^n, \qquad (20)$$
where $\beta \to 0$ when $k \to 0$.

Now $k \to 0$ when $h \to 0$ from (19), and therefore $\beta \to 0$ when $h \to 0$. Consequently, if we substitute for k from (19) in (20) and re-arrange in ascending powers of h, we shall obtain a formula of type (17). But ϕ_n exists, since g_n and f_n exist (§3.6), and so this formula will be the Taylor formula for $\phi(x+h)$ and in it the coefficient c_n of h^n will determine ϕ_n since $c_n = \phi_n/n!$

Examples

(1) Prove that
$$D^n f(x^2) = f_n(2x)^n + \frac{n(n-1)}{1!} f_{n-1}(2x)^{n-2}$$
$$+ \frac{n(n-1)(n-2)(n-3)}{2!} f_{n-2}(2x)^{n-4} + \ldots \qquad (21)$$
where $f_s = f^{(s)}(u)$, $u = x^2$, and we assume that $f^{(n)}(u)$ exists.

Proof. Put $\phi(x) = f(x^2) = f(u)$. Then

$$\phi(x+h) = f\{(x+h)^2\} = f(u+k),$$

where $\qquad u = x^2, \quad k = 2hx + h^2 = h(2x+h);$

and hence

$$\phi(x+h) = f_0 + f_1 k + f_2 k^2/2! + \dots + f_n k^n/n! + \beta k^n,$$

where $\beta \to 0$ when $k \to 0$ and therefore when $h \to 0$.

After substituting for k and re-arranging in ascending powers of h, we obtain a formula of the type

$$\phi(x+h) = c_0 + c_1 h + c_2 h^2 + \dots + c_n h^n + \eta h^n,$$

where the coefficients c_s are independent of h, and $\eta \to 0$ when $h \to 0$. In this formula c_n is the result of finding the coefficient of h^n in

$$f_n k^n/n! + f_{n-1} k^{n-1}/(n-1)! + f_{n-2} k^{n-2}/(n-2)! + \dots$$

that is, in

$$f_n h^n (2x+h)^n/n! + f_{n-1} h^{n-1}(2x+h)^{n-1}/(n-1)! + \dots$$

But $c_n = \phi_n(x)/n!$. Hence, after finding c_n and multiplying by $n!$ we obtain (21).

(2) Verify (21) when $f(u) = \log u$, and therefore when

$$\phi(x) = \log x^2 = 2 \log x \quad (0 < x).$$

By (3), $\qquad\qquad \phi_n(x) = 2(-)^{n-1}(n-1)!/x^n. \qquad\qquad (22)$

Also by (3), $f_s(u) = (-)^{s-1}(s-1)!/u^s$ and hence (21) gives

$$\phi_n(x) = (-)^{n-1}\left\{\frac{(n-1)!}{u^n}(2x)^n - \frac{n(n-1)}{1!}\frac{(n-2)!}{u^{n-1}}(2x)^{n-2} + \dots\right\}$$

$$= \frac{(-)^{n-1}(n-1)!}{x^n}\left\{2^n - \frac{n}{1!}2^{n-2} + \frac{n(n-3)}{2!}2^{n-4} - \dots\right\}.$$

This agrees with (22) since the series in brackets has the sum 2 by a known result, which can be proved by equating coefficients of x^n in the identity $\quad 2\log(1-x) = \log(1 - 2x + x^2)$

after expanding both sides in ascending powers of x, the R.H.S. being first expanded in ascending powers of $2x - x^2$.

(3) Prove that $D^n f(\log x)$ is given symbolically by

$$x^n D^n f(\log x) = \vartheta(\vartheta - 1)(\vartheta - 2)\dots(\vartheta - n + 1)f(u),$$

where $u = \log x$, $D = d/dx$, $\vartheta = d/du = xd/dx = xD$.

Let $f(\log x) = \phi(x)$. Now

$$xD\{x^{n-1}D^{n-1}\phi(x)\} = x^n D^n \phi(x) + (n-1)x^{n-1}D^{n-1}\phi(x)$$

and hence, after putting $F_n = x^n D^n \phi(x)$ and transposing,

$$F_n = \vartheta F_{n-1} - (n-1)F_{n-1} = (\vartheta - n + 1)F_{n-1}.$$

By applying this formula repeatedly to the R.H.S. we get

$$F_n = (\vartheta - n + 1)(\vartheta - n + 2)...(\vartheta - 1)\vartheta f(u),$$

as required, since $\phi(x) = f(u)$.

(4) Verify the formula of Example 3 when $f(\log x) = (\log x)^2$. (See Examples 3B, 14, iv.)

3.9

Various details connected with repeated differentiation will be illustrated in the following two examples.

Example

(1)
$$y = \frac{1}{\sqrt{(1+x^2)}}.$$
(23)

I. By squaring and multiplying by $1 + x^2$,

$$(1+x^2)y^2 = 1$$

and hence, after differentiating and dividing by $2y$,

$$(1+x^2)y_1 + xy = 0.$$
(24)

Now differentiate $n - 1$ times, using Leibniz's rule; we get, after a little reduction,

$$(1+x^2)y_n + (2n-1)xy_{n-1} + (n-1)^2 y_{n-2} = 0,$$
(25)

where y_0 will mean y when we put $n = 2$.

This is a *recurrence formula* for y_n. By means of it and (23), (24), we could successively find $y_2, y_3, ..., y_n$ in terms of x without further differentiation.

II. Putting $n = 2, 3, ..., n-1$ in (25), we should obtain, including (25) itself, $n - 1$ algebraic equations linear in $y_2, y_3, ..., y_n$. By solving these equations by the method of determinants we could express y_n in determinantal form.

III. By replacing n by $n+2$ in (25), we see that y_n satisfies the second-order linear differential equation

$$(1+x^2)y_n'' + (2n+3)xy_n' + (n+1)^2 y_n = 0.$$
(26)

IV. An explicit expression for y_n is found by putting

$$f(x^2) = 1/(1+x^2)^{\frac{1}{2}}$$

in (21). This shows that y_n can be put in the form

$$y_n = \phi_n(x)/(1+x^2)^{n+\frac{1}{2}} \tag{27}$$

where $\phi_n(x)$ denotes a polynomial of the form

$$\phi_n(x) = c_0 x^n + c_2 x^{n-2} + c_4 x^{n-4} + \dots \tag{28}$$

V. By substituting (27) in (25), we find that the polynomial $\phi_n(x)$ satisfies the recurrence relation

$$\phi_n + (2n-1)x\phi_{n-1} + (n-1)^2(1+x^2)\phi_{n-2} = 0. \tag{29}$$

VI. By substituting (27) in (26), we find that $\phi_n(x)$ satisfies the differential equation of the second order

$$(1+x^2)\phi_n'' - (2n-1)x\phi_n' + n^2\phi_n = 0. \tag{30}$$

VII. The polynomial $\phi_n(x)$ can be found by substituting (28) in the differential equation (30). First write (30) in the form

$$x^2\phi_n'' - (2n-1)x\phi_n' + n^2\phi_n + \phi_n'' = 0$$

and then make the substitution (28). The coefficients of the several powers of x must then vanish. Equating to zero the coefficient of x^{n-2s}, we find

$$4s^2 c_{2s} + (n-2s+2)(n-2s+1)c_{2s-2} = 0.$$

Putting $s = 1, 2, 3, \dots$ in turn, we find successively the values of $c_2, c_4, c_6, \dots$ in terms of c_0 and hence

$$\phi_n(x) = c_0\left\{x^n - \frac{n(n-1)}{4 \cdot 1^2}x^{n-2} + \frac{n(n-1)(n-2)(n-3)}{4^2 \cdot (2!)^2}x^{n-4} - \dots\right\},$$

where c_0 remains to be found. To find c_0 we consider the behaviour of y when x is large. When x is large and positive we have, from (23),

$$y = \frac{1}{x} + \dots, \quad y_n = \frac{(-)^n n!}{x^{n+1}} + \dots;$$

but from (27) and (28)

$$y_n = \frac{c_0 x^n}{x^{2n+1}} + \dots = \frac{c_0}{x^{n+1}} + \dots,$$

and hence $c_0 = (-)^n n!$

VIII. The values of the successive derivatives of y when $x = 0$, viz.

$(y_1)_0$, $(y_2)_0$, $(y_3)_0$, ... can readily be found; for when $x = 0$ the recurrence formula (25) reduces to

$$(y_n)_0 = -(n-1)^2 (y_{n-2})_0$$

from which, since $(y_1)_0 = 0$, follows $(y_3)_0 = (y_5)_0 = ... = 0$; and since $y = 1$ when $x = 0$,

$$(y_2)_0 = -1^2,$$

$$(y_4)_0 = -3^2(y_2)_0 = 3^2 \cdot 1^2,$$

$$(y_6)_0 = -5^2(y_4)_0 = -5^2 \cdot 3^2 \cdot 1^2, ...$$

IX. The values of $(y_n)_0$ can also be found by using Maclaurin's formula. For the expansion of $1/\sqrt{(1+x^2)}$ is (cf. §4.4 (16))

$$\frac{1}{\sqrt{(1+x^2)}} = 1 - \frac{1}{2}\frac{x^2}{1!} + \frac{1}{2}\frac{3}{2}\frac{x^4}{2!} - \frac{1}{2}\frac{3}{2}\frac{5}{2}\frac{x^6}{3!} + \cdots$$

in which the coefficient of x^n is known to be $(y_n)_0/n!$ Hence we find again the values of $(y_n)_0$ given in VIII.

X. The equation $\phi_n(x) = 0$ has n distinct real roots; that is, all the n roots of the polynomial equation $\phi_n(x) = 0$ are real, and no two are equal.

Proof. Put $f(x) = 1/(1+x^2)^{\frac{1}{2}}$. Then, by (27), the polynomials $\phi_1(x)$, $\phi_2(x)$, ... are determined by

$$f'(x) = \phi_1(x)/(1+x^2)^{\frac{3}{2}}, \quad f''(x) = \phi_2(x)/(1+x^2)^{\frac{5}{2}}, ...$$

Assume that the equation $\phi_s(x) = 0$, for some positive integer s, has s distinct real roots $\alpha_1, \alpha_2, ..., \alpha_s$. These are also roots of the equation $f^{(s)}(x) = 0$.

Now the function $f^{(s)}(x)$ is continuous and its derivative $f^{(s+1)}(x)$ exists for all x; consequently, by Rolle's theorem, between every pair of real roots of $f^{(s)}(x) = 0$ there lies at least one real root of $f^{(s+1)}(x) = 0$. The equation $f^{(s+1)}(x) = 0$ has therefore at least $s-1$ real roots $\beta_2, \beta_3, ..., \beta_s$ such that

$$\alpha_1 < \beta_2 < \alpha_2 < \beta_3 < \alpha_3 < ... < \beta_s < \alpha_s.$$

Further, the polynomial equation $\phi_s(x) = 0$ cannot have more than the s roots $\alpha_1, \alpha_2, ..., \alpha_s$, so it has no root in the interval $\alpha_s < x < +\infty$. The same is therefore true of the equation $f^{(s)}(x) = 0$. The function $f^{(s)}(x)$, being continuous, must therefore be one-signed in this interval. But $f^{(s)}(x) \to 0$ when $x \to +\infty$ and, being continuous, must have either a positive l.u.b. or a negative g.l.b. in this interval. Also, $f^{(s)}(x)$ being differentiable for all x, the derivative $f^{(s+1)}(x)$ must be zero at this bound (as in the proof of Rolle's theorem). Consequently, the equation $f^{(s+1)}(x) = 0$ has a real root β_{s+1} in the interval $\alpha_s < x < +\infty$. Similarly, it has a real root β_1 in the interval $-\infty < x < \alpha_1$. The equation $f^{(s+1)}(x) = 0$ therefore has $s+1$ real distinct roots, and these must be roots of the equation $\phi_{s+1}(x) = 0$.

Thus, on the assumption that the polynomial equation $\phi_s(x) = 0$ has all its roots real and distinct, it follows that the same is true of the equation $\phi_{s+1}(x) = 0$.

Now $\phi_1(x) = x$. Therefore the equation $\phi_1(x) = 0$ has the one real root $x = 0$. It follows that the equation $\phi_2(x) = 0$ has two distinct real roots; and hence that $\phi_3(x) = 0$ has three distinct real roots; and by induction that $\phi_n(x) = 0$ has n distinct real roots.

Example

(2) $$y = \tan x.$$

I. Writing $y = \sin x/\cos x$, we could express y_n as a determinant (§3.5).

II. Differentiating $y = \tan x$ twice gives

$$y_1 = \sec^2 x = 1+y^2, \quad y_2 = 2yy_1.$$

After differentiating $n-2$ more times by Leibniz's rule,

$$y_n = 2(yy_{n-1} + {}^{n-2}C_1 y_1 y_{n-2} + {}^{n-2}C_2 y_2 y_{n-3} + \dots + y_{n-2} y_1),$$

a recurrence formula which can be simplified by adding the second term to the last, the third to the last but one, and so on, using the identity ${}^{n-2}C_r + {}^{n-2}C_{r+1} = {}^{n-1}C_{r+1}$. The last term in the simplified form will depend upon whether n is odd or even.

III. For small values of n, we can find y_n as follows:

$$y_1 = \sec^2 x = 1+y^2,$$

$$y_2 = 2yy_1 = 2y(1+y^2) = 2(y+y^3),$$

$$y_3 = 2(1+3y^2)\,y_1 = 2(1+3y^2)\,(1+y^2),$$

$$y_4 = 2(8y+12y^3)\,y_1 = 8(2y+3y^3)\,(1+y^2), \dots.$$

IV. If we put $y_{n+1} = (1+y^2)\,\phi_n(y)$, where $\phi_n(y)$ is a polynomial of degree n, then we find

$$y_{n+2} = (1+y^2)\,\phi_{n+1}(y) = (1+y^2)\,\{2y\phi_n(y) + (1+y^2)\,\phi_n'(y)\}$$

5

and hence the recurrence formula

$$\phi_{n+1} = 2y\phi_n + (1+y^2)\,\phi_n'$$

from which the polynomials ϕ_n can be successively calculated.

3.10 Change of variables

It is sometimes a convenience to transform $y', y'', y''', \ldots$ by changing the dependent or independent variable or both.

Example

Change the independent variable in the equation

$$(1-x^2)\frac{d^2y}{dx^2} - x\frac{dy}{dx} + n^2y = 0$$

from x to t by means of the substitution $x = \sin t$.

To obtain the substitution for dy/dx we use §2.22, regarding x and y as functions of t, thus;

$$\frac{dy}{dx} = \frac{dy/dt}{dx/dt} = \frac{1}{\cos t}\frac{dy}{dt}. \quad \text{Symbolically,} \quad \frac{d}{dx} = \frac{1}{\cos t}\frac{d}{dt}.$$

Applying this a second time,

$$\frac{d}{dx}\frac{dy}{dx} = \frac{1}{\cos t}\frac{d}{dt}\left(\frac{1}{\cos t}\frac{dy}{dt}\right) = \frac{1}{\cos^2 t}\frac{d^2y}{dt^2} + \frac{\sin t}{\cos^3 t}\frac{dy}{dt}.$$

On substituting in the given equation and putting $x = \sin t$, the equation reduces to $d^2y/dt^2 + n^2y = 0$.

3.11 The Wronskian

Definition. Let $y_1, y_2, \ldots, y_n$ be n functions each $(n-1)$-times differentiable with respect to x. The *Wronskian* of $y_1, y_2, \ldots, y_n$ with respect to x is denoted by $W_x(y_1, y_2, \ldots, y_n)$ or briefly by W, and is defined by

$$W = W_x(y_1, y_2, \ldots, y_n) = \begin{vmatrix} y_1 & y_2 & y_3 & \cdots & y_n \\ y_1' & y_2' & y_3' & \cdots & y_n' \\ y_1'' & y_2'' & y_3'' & \cdots & y_n'' \\ \cdots\cdots\cdots\cdots\cdots\cdots\cdots\cdots \\ y_1^{(n-1)} & y_2^{(n-1)} & y_3^{(n-1)} & \cdots & y_n^{(n-1)} \end{vmatrix}. \quad (31)$$

3.12   Linear dependence of functions of one variable

Definition. The n functions $y_1, y_2, ..., y_n$ are said to be *linearly dependent* if they satisfy an identity in x of the form

$$c_1 y_1 + c_2 y_2 + ... + c_n y_n \equiv 0, \tag{32}$$

where the coefficients $c_1, c_2, ..., c_n$ are independent of x and are not all zero.

In what follows it will be assumed that the functions $y_1, y_2, ..., y_n$ are $(n-1)$-times differentiable.

Theorem 1. A *necessary* condition that $y_1, y_2, ..., y_n$ should be linearly dependent is $W_x(y_1, y_2, ..., y_n) \equiv 0$.

Proof. Let the identity (32) be differentiated $n-1$ times. Then, since the coefficients $c_1, c_2, ..., c_n$ are independent of x, we have a set of n homogeneous linear equations satisfied by $c_1, c_2, ..., c_n$, not all zero. It follows, by a well-known theorem in algebra, that the determinant of the set must be zero; that is, $W \equiv 0$ throughout any interval in which (32) holds good.

Corollary. The functions $y_1, y_2, ..., y_n$ are linearly independent in any interval in which $W \not\equiv 0$.

Theorem 2. The condition $W \equiv 0$ is *sufficient* to ensure the linear dependence of the functions $y_1, y_2, ..., y_n$ in any interval in which the Wronskian of any set of $n-1$ of the functions does not vanish for any value of x.

Proof. The general case will be exemplified by taking $n = 3$.

Let y_1, y_2, y_3 be three functions each twice-differentiable in an interval $[a, b]$. Suppose that, in this interval, $W(y_1, y_2, y_3) = 0$ for every value of x and that $W(y_1, y_2) \neq 0$ for any value of x.

Now consider as equations in c_1, c_2 the three linear equations

$$c_1 y_1 + c_2 y_2 + y_3 = 0 \quad \text{(i)},$$

$$c_1 y_1' + c_2 y_2' + y_3' = 0 \quad \text{(ii)},$$

$$c_1 y_1'' + c_2 y_2'' + y_3'' = 0 \quad \text{(iii)}.$$

Since $W(y_1, y_2) \neq 0$, equations (i), (ii) can be solved for c_1, c_2. The values of c_1, c_2 thus found will satisfy (iii) since $W(y_1, y_2, y_3) \equiv 0$. Consequently, if these values be substituted in (i), (ii), (iii), then all three equations will become identities in x. By differentiating the first of these identities and subtracting the second, and differentiating the second and subtracting the third, remembering that c_1, c_2 are now presumably functions of x, we get

$$c_1'y_1 + c_2'y_2 \equiv 0, \quad c_1'y_1' + c_2'y_2' \equiv 0.$$

Hence follow

$$(y_1y_2' - y_1'y_2)\, c_1' \equiv 0, \quad (y_1y_2' - y_1'y_2)\, c_2' \equiv 0.$$

But $y_1y_2' - y_1'y_2 = W(y_1, y_2) \neq 0$ for any value of x in $[a, b]$. Consequently, $c_1' \equiv 0$ and $c_2' \equiv 0$, and therefore c_1 and c_2 are constants.

We have thus proved that there exists an identity of the form $c_1y_1 + c_2y_2 + c_3y_3 \equiv 0$, where c_1, c_2, c_3 are not all zero, since $c_3 = 1 \neq 0$. Accordingly, by definition, y_1, y_2, y_3 are linearly dependent.

Note. The condition $W(y_1, y_2, ..., y_n) \equiv 0$ is not alone sufficient to ensure a linear relation between $y_1, y_2, ..., y_n$.

For instance, let y_1, y_2, y_3 be defined in the interval $[-1, 1]$ by

$$y_1 = x^3, \quad y_2 = x^3 + x^4, \quad y_3 = -x^4, \quad -1 \leqslant x \leqslant 0,$$
$$y_1 = 0, \quad y_2 = x^3, \quad y_3 = x^4, \quad 0 < x \leqslant 1.$$

It may at once be verified that y_1, y_2, y_3 are twice-differentiable and that $W(y_1, y_2, y_3) \equiv 0$ throughout the interval $[-1, 1]$. But there is no relation of the form $c_1y_1 + c_2y_2 + c_3y_3 \equiv 0$ which holds good throughout the interval, with c_1, c_2, c_3 not all zero; although $y_1 - y_2 - y_3 \equiv 0$ is such a relation over the first half of the interval, and $y_1 \equiv 0$ is another over the second half.

It may be added, however, that the condition $W \equiv 0$ is *necessary and sufficient* when the functions $y_1, y_2, ..., y_n$ are *analytic* (i.e. are expansible in convergent Taylor series) at every point of the interval concerned. In the above example, none of the functions y_1, y_2, y_3 is analytic at $x = 0$ (see §4.18).

Examples 3 A

(1) Show that:

(i) $\dfrac{d^4}{dx^4}(1 - x^2)^{\frac{3}{2}} = \dfrac{9}{(1 - x^2)^{\frac{5}{2}}},$

(ii) $\dfrac{d^3}{dx^3}\log(1 + \sin x) = \dfrac{\cos x}{(1 + \sin x)^2},$

(iii) $\dfrac{d^3}{dx^3}\{x + \sqrt{(1 + x^2)}\}^2 = \dfrac{6}{(1 + x^2)^{\frac{5}{2}}}.$

(2) If $y^3 + 3y = 2x$, show that $(x^2 + 1)\, y'' + xy' = \frac{1}{9}y.$

(3) If $\cot^2 y = \tanh \frac{1}{2}x$, show that $y'' + y' + 4y'^3 = 0.$

(4) If $y = f(v)$ and $x = g(u)$ and v is a function of u, show that

$$\frac{d^2y}{dx^2} = \frac{1}{g'(u)} \frac{d}{du} \left\{ \frac{f'(v)}{g'(u)} \frac{dv}{du} \right\}.$$

(5) If $y = f(x)$, show that

$$\frac{d^2x}{dy^2} = -\frac{y''}{y'^3}, \quad \frac{d^3x}{dy^3} = \frac{3y''^2 - y'y'''}{y'^5}.$$

(6) Show that the substitution $x = \sinh t$ transforms the equation

$$(1+x^2)\frac{d^2y}{dx^2} + x\frac{dy}{dx} - n^2y = 0 \quad \text{into} \quad \frac{d^2y}{dt^2} - n^2y = 0.$$

(7) If in the differential equation

$$\frac{d^2y}{dx^2} + \frac{1}{x}\frac{dy}{dx} + \left(1 - \frac{n^2}{x^2}\right)y = 0$$

the dependent variable is changed from y to z by means of the substitution $z = y\sqrt{x}$, show that

$$\frac{d^2z}{dx^2} + \left(1 - \frac{n^2 - \frac{1}{4}}{x^2}\right)z = 0.$$

(8) If $(1-x^2)\,d^2y/dx^2 - (2n+3)\,x\,dy/dx - (n+1)^2y = 0$ and we make the substitution $z = y(1-x^2)^{n+\frac{1}{2}}$, show that

$$(1-x^2)\,d^2z/dx^2 + (2n-1)\,x\,dz/dx - n^2z = 0.$$

(9) Show that the equation

$$u^2\,d^2v/du^2 + u\,dv/du + (u^2 - n^2)\,v = 0$$

is transformed into

$$\frac{d^2y}{dx^2} - \frac{2a-1}{x}\frac{dy}{dx} + \left(b^2c^2x^{2c-2} + \frac{a^2 - n^2c^2}{x^2}\right)y = 0$$

by the substitutions $y = vx^a$, $u = bx^c$.

Examples 3B

(1) Find the nth derivatives of the following functions:

(i) $\dfrac{x^2}{1-x}$,

(ii) $\dfrac{ax+b}{cx+d}$,

(iii) $\dfrac{1}{\sqrt{(1-x)}}$,

(iv) $\cos^2 x$,

(v) $\cos x \cos 2x$,

(vi) $\cos x \cos 2x \cos 3x$,

(vii) $\sin^3 x$,

(viii) $\sin^2 x \cos x$,

(ix) $e^x \cos x$,

(x) $\cosh x$,

(xi) $\sinh ax$,

(xii) $e^{3x}(\cos x + \sin x)^2$,

(xiii) $x^3 e^{ax}$,

(xiv) $x^n e^x$,

(xv) $x^2 \sin x$,

(xvi) $\dfrac{e^x}{x}$,

(xvii) $x^3 \log\dfrac{a}{x}$,

(xviii) $\log\dfrac{(x-a)^p}{(b-x)^q}$.

(2) Show that:

(i) $D^n e^{x\cos\alpha}\cos(x\sin\alpha) = e^{x\cos\alpha}\cos(x\sin\alpha+n\alpha)$,

(ii) $D^n e^{x\cos\alpha}\sin(x\sin\alpha) = e^{x\cos\alpha}\sin(x\sin\alpha+n\alpha)$,

(iii) $D^n e^{x\cot\alpha}\cos x = e^{x\cot\alpha}\cos(x+n\alpha)/\sin^n\alpha$,

(iv) $D^n e^{x\cot\alpha}\sin x = e^{x\cot\alpha}\sin(x+n\alpha)/\sin^n\alpha$.

(3) Show that $D^n(\cos x\cosh x) =$

$$2^{\frac{1}{2}n}\{\cos\tfrac{1}{4}n\pi\cos(x+\tfrac{1}{2}n\pi)\cosh x+\sin\tfrac{1}{4}n\pi\sin(x+\tfrac{1}{2}n\pi)\sinh x\}.$$

Find similar expressions for

$$D^n(\sin x\sinh x), \quad D^n(\sin x\cosh x), \quad D^n(\cos x\sinh x).$$

(4) Show that if $y = \tanh^{-1}x$ $(-1 < x < 1)$, then

$$D^n(\tanh^{-1}x) = (n-1)!\cosh^n y\cosh ny \quad (n\,\text{odd}),$$
$$= (n-1)!\cosh^n y\sinh ny \quad (n\,\text{even}).$$

Show that if $y = \coth^{-1}x$ $(|x| > 1)$, then

$$D^n(\coth^{-1}x) = (-)^n(n-1)!\sinh^n y\sinh ny.$$

(5) If $u+iv = \frac{1}{2}\log(x^2+a^2)+i\tan^{-1}(a/x)$, show that

$$\frac{du}{dx}+i\frac{dv}{dx} = \frac{1}{x+ia}.$$

Deduce that

$$D^n\log(x^2+a^2) = 2(-)^{n-1}(n-1)!e^{-nu}\cos nv,$$
$$D^n\tan^{-1}(a/x) = (-)^n(n-1)!e^{-nu}\sin nv.$$

(6) If $u = (1+x^2)^{\frac{1}{2}m}\cos(m\tan^{-1}x)$, $v = (1+x^2)^{\frac{1}{2}m}\sin(m\tan^{-1}x)$, show that $u+iv = (1+ix)^m$. Deduce that $y = u$ and $y = v$ satisfy the equation

$$(1+x^2)y'' - 2(m-1)xy' + m(m-1)y = 0.$$

Also, find the nth derivatives of u and v.

(7) If $\tan\theta = x\sin\alpha/(a-x\cos\alpha)$, show that

$$D^n\tan^{-1}\left(\frac{a+x}{a-x}\tan\frac{\alpha}{2}\right) = \frac{(n-1)!\sin n(\theta+\alpha)}{(a^2-2ax\cos\alpha+x^2)^{\frac{1}{2}n}}.$$

(8) Prove that

(i) $(\sin x\,d/dx)^n(\tan\frac{1}{2}x)^s = s^n(\tan\tfrac{1}{2}x)^s$,

(ii) $\{\sqrt{(x^2+1)}\,d/dx\}^n\{\sqrt{(x^2+1)}-x\}^s = (-s)^n\{\sqrt{(x^2+1)}-x\}^s$.

(9) If u is a real function of x, and $D = d/dx$, prove that

$$D^n(u\cos x) = \{\cos x\phi(D)-\sin x\psi(D)\}u,$$
$$D^n(u\sin x) = \{\sin x\phi(D)+\cos x\psi(D)\}u,$$

where $\phi(D)+i\psi(D) = (D+i)^n$. Show that

$$D^n(\tan x) = \sec x\,\psi(D)\sec x,$$
$$D^n(\sec x) = \sec x\{\cos\tfrac{1}{2}n\pi+\psi(D)\tan x\}.$$

(10) If $y = \sec x + \tan x$, show that $y_{n+1} = \frac{1}{2}(y^2 + 1)\,\phi_n(y)$, where ϕ_n denotes a polynomial of degree n. Also show that

$$\phi_{n+1}(y) = y\phi_n(y) + \frac{1}{2}(y^2 + 1)\,\phi_n'(y).$$

(11) If $s = \sec x$, $t = \tan x$ and $s_n = D^n \sec x$, show that

$$s_{2n} = sf_n(s^2), \quad s_{2n+1} = stg_n(s^2),$$

where f_n, g_n denote polynomials of degree n.

Also show that $f_n(z)$, $g_n(z)$ can be successively calculated from the recurrence formulae

$$g_n(z) = f_n(z) + 2zf_n'(z),$$

$$f_{n+1}(z) = (2z - 1)\,g_n(z) + 2z(z - 1)\,g_n'(z).$$

(12) Show that if $D = d/dx$ and $u = f(x)$, then

$$D^n(u e^{ix}) = e^{ix}(D + i)^n u = e^{i(x + \frac{1}{2}n\pi)}(1 - iD)^n u.$$

Deduce that

$$\frac{d^n}{dx^n}\left(\frac{\cos x}{x}\right) = \frac{n!}{x^{n+1}}\{P\cos(x + \tfrac{1}{2}n\pi) - Q\sin(x + \tfrac{1}{2}n\pi)\},$$

$$\frac{d^n}{dx^n}\left(\frac{\sin x}{x}\right) = \frac{n!}{x^{n+1}}\{P\sin(x + \tfrac{1}{2}n\pi) + Q\cos(x + \tfrac{1}{2}n\pi)\},$$

where

$$P = \frac{x^n}{n!} - \frac{x^{n-2}}{(n-2)!} + \dots, \quad Q = \frac{x^{n-1}}{(n-1)!} - \frac{x^{n-3}}{(n-3)!} + \dots.$$

(13) Prove that if $D = d/dx$ and $y = f(x)$, then

$$x^m D^m x^s D^n y = x^s D^n x^{m+n-s} D^m x^{s-n} y.$$

(14) Prove that

(i) $D^n(x^{n-1}\log x) = (n-1)!/x$,

(ii) $D^n(x^n\log x) = n!(\log x + 1 + \frac{1}{2} + \frac{1}{3} + \dots + 1/n)$,

(iii) $D^n\left(\dfrac{\log x}{x}\right) = \dfrac{(-)^n n!}{x^{n+1}}(\log x - 1 - \tfrac{1}{2} - \dots - 1/n)$,

(iv) $D^n(\log x)^2 = \dfrac{2(-)^{n-1}(n-1)!}{x^n}\left(\log x - 1 - \tfrac{1}{2} - \dots - \dfrac{1}{n-1}\right) \quad (n > 1)$,

(v) $D^n\{x^{n-1}(\log x)^2\} = \dfrac{2(n-1)!}{x}\left(\log x + 1 + \tfrac{1}{2} + \dots + \dfrac{1}{n-1}\right) \quad (n > 1)$,

(vi) $\dfrac{1}{n!}D^n(x\log x)^n = 1 + s_1\log x + \dfrac{s_2}{2!}(\log x)^2 + \dots + \dfrac{s_n}{n!}(\log x)^n$,

where s_r is the sum of the products, r at a time, of the integers $1, 2, 3, \dots, n$. [Murphy's formula.]

(15) If $y = (A\cos x + B\sin x)e^x + (C\cos x + D\sin x)e^{-x}$, shew that

$$y_{n+4} = -4y_n.$$

(16) If $y = (px + q)/(ax^2 + 2bx + c)$, show that if $n > 1$

$$(ax^2 + 2bx + c)\,y_n + 2n(ax + b)\,y_{n-1} + n(n-1)\,a y_{n-2} = 0.$$

(17) If $y = 1/\sqrt{(ax^2 + 2bx + c)}$, show that if $n > 1$

$$(ax^2 + 2bx + c)\, y_n + (2n - 1)\, (ax + b)\, y_{n-1} + (n - 1)^2\, ay_{n-2} = 0.$$

(18) If $y = (\sin^{-1} x)^2$ show that:

   (i) $(1 - x^2)\, y_1^2 = 4y,$ (ii) $(1 - x^2)\, y_2 - xy_1 = 2,$

   (iii) $(1 - x^2)\, y_{n+2} - (2n + 1)\, xy_{n+1} - n^2 y_n = 0$ $(n \geqslant 1).$

(19) If $y = (\sinh^{-1} x)^2$ show that

$$(1 + x^2)\, y_{n+2} + (2n + 1)\, xy_{n+1} + n^2 y_n = 0 \ (n \geqslant 1).$$

(20) Show that, if n and s are positive integers and $n > s$, then

$$[D^n\{(x - a)^s f(x)\}]_{x=a} = n(n - 1)\ldots(n - s + 1)f^{(n-s)}(a).$$

(21) If $f(x) = a_0 + a_1 x + a_2 x^2 + \ldots + a_n x^n$ prove that, if s is a positive integer, and $a_0, a_1, \ldots$ are constants,

$$1^s a_1 + 2^s a_2 + 3^s a_3 + \ldots + n^s a_n = [(x\, d/dx)^s f(x)]_{x=1}.$$

(22) Prove that, if n and s are positive integers,

$$n^s - \binom{n}{1}(n - 1)^s + \binom{n}{2}(n - 2)^s - \ldots = 0 \quad (s < n),$$
$$= n! \quad (s = n).$$

(23) Prove that

$$n^2 + \binom{n}{1}(n - 2)^2 + \binom{n}{2}(n - 4)^2 + \ldots \text{ to } n + 1 \text{ terms} = 2^n n.$$

[*Hint.* Evaluate $(d/dx)\{x\, d/dx(x + x^{-1})^n\}$ when $x = 1$, in two ways.]

(24) If $y = e^{a \sin x}$ prove that

$$(y_{n+1})_0 = a\{(y_n)_0 - {}^nC_2(y_{n-2})_0 + {}^nC_4(y_{n-4})_0 - \ldots\}.$$

(25) If $y = e^{c \tan^{-1} x}$ prove that

$$(1 + x^2)\, y_{n+1} + (2nx - c)\, y_n + n(n - 1)\, y_{n-1} = 0.$$

(26) Show that, if $y = \sqrt[3]{\{x + \sqrt{(1 + x^2)}\}} + \sqrt[3]{\{x - \sqrt{(1 + x^2)}\}}$, then

   (i) $(1 + x^2)\, y'' + xy' = \tfrac{1}{9}y;$

   (ii) $(1 + x^2)\, y_{n+2} + (2n + 1)\, xy_{n+1} + (n^2 - \tfrac{1}{9})\, y_n = 0;$

   (iii) $y_0 = 0, \quad (y_1)_0 = \tfrac{2}{3}, \quad (y_2)_0 = 0, \quad (y_{n+2})_0 = -(n^2 - \tfrac{1}{9})\, (y_n)_0;$

   (iv) $(y_{2n})_0 = 0, \quad (y_{2n-1})_0 = (-)^{n-1}\{(2n - 3)^2 - \tfrac{1}{9}\}\ldots(1^2 - \tfrac{1}{9})\tfrac{2}{3}.$

(27) Show that $D^n e^{ax^2} = \phi_n(x)\, e^{ax^2}$, where

$$\phi_n(x) = a^n(2x)^n + \frac{n(n - 1)}{1!}\, a^{n-1}\, (2x)^{n-2} + \ldots$$

Deduce that $D^n e^{ix^2} = (P - iQ) \exp\{i(x^2 + \tfrac{1}{2}n\pi)\}$ where P, Q are polynomials in x of degrees n, $n - 2$, respectively.

Hence find expressions for $D^n \cos(x^2)$, $D^n \sin(x^2)$.

(28) If $y_n = D^n e^{-x^2} = \phi_n(x) e^{-x^2}$, show that

(i) $y_n + 2xy_{n-1} + 2(n-1)y_{n-2} = 0$;

(ii) $y_n'' + 2xy_n' + 2(n+1)y_n = 0$,

(iii) $\phi_n'' - 2x\phi_n' + 2n\phi_n = 0$;

(iv) $\phi_{2n}(0) = (-)^n (2n)!/n!$, $\phi_{2n}'(0) = 0$;

(v) the polynomial equation $\phi_n(x) = 0$ has n distinct real roots, separated by those of the equation $\phi_{n-1}(x) = 0$.

[*Note.* The Hermite polynomials $H_n(x)$ are defined by

$$H_n(x) = (-)^n e^{x^2} D^n e^{-x^2}.]$$

(29) Prove that

$$\frac{d^n}{dx^n} f\left(\frac{1}{x}\right) = (-)^n \left\{ \frac{1}{x^{2n}} f^{(n)}\left(\frac{1}{x}\right) + \frac{n(n-1)}{x^{2n-1}} f^{(n-1)}\left(\frac{1}{x}\right) \right.$$
$$\left. + \frac{n(n-1)^2(n-2)}{2! \, x^{2n-2}} f^{(n-2)}\left(\frac{1}{x}\right) + \ldots \right\}.$$

(30) Prove that, if $x \neq 0$, $D^n e^{-1/x}$

$$= \frac{e^{-1/x}}{x^n} \left\{ \frac{1}{x^n} - \frac{n(n-1)}{x^{n-1}} + \frac{n(n-1)^2(n-2)}{2! \, x^{n-2}} - \ldots \pm \frac{n!}{x} \right\}.$$

If $f(x) = e^{-1/x}$, $0 < x$, and $f(0) = 0$, show that $f^{(n)}(0+) = 0$, for all n.

(31) (i) If $y = \tanh x$, show that $y_n = \phi_{n-1}(y) \operatorname{sech}^2 x$, where ϕ_{n-1} denotes a polynomial of degree $n-1$.

(ii) If $f(x) = \tanh(1/x)$ $(0 < x)$, and $f(0) = 1$, show that $f^{(n)}(0+) = 0$, for all n.

(32) Show that, if $x \neq 0$,

$$\frac{d^n}{dx^n} \exp\left(\frac{i}{x}\right) = \frac{P - iQ}{x^n} \exp\left\{ i\left(\frac{1}{x} - \frac{n\pi}{2}\right) \right\},$$

where $P = \frac{1}{x^n} - \frac{n(n-1)^2(n-2)}{2! \, x^{n-2}} + \ldots, \quad Q = \frac{n(n-1)}{x^{n-1}} - \ldots.$

Deduce expressions for $D^n \cos(1/x)$, $D^n \sin(1/x)$.

(33) I. If $y = P_n(x) = \frac{1}{2^n n!} \frac{d^n}{dx^n}(x^2-1)^n$ (i)

show that $(x^2-1)\frac{d^2 y}{dx^2} + 2x\frac{dy}{dx} - n(n+1)y = 0.$ (ii)

[Put $z = (x^2-1)^n$, and hence $(x^2-1)z_1 = 2nxz$, and differentiate $n+1$ times by Leibniz's rule.

Here $P_n(x)$ is Legendre's polynomial of degree n, if n is a positive integer.]

II. Show directly from (i), and also from the differential equation (ii), that

$$P_n(x) = \frac{(2n)!}{2^n (n!)^2} \left\{ x^n - \frac{n(n-1)}{1!(2n-1)} \frac{x^{n-2}}{2} + \frac{n(n-1)(n-2)(n-3)}{2!(2n-1)(2n-3)} \frac{x^{n-4}}{2^2} - \ldots \right\}.$$

III. Show that the equation $P_n(x) = 0$ has n distinct real roots between -1 and $+1$.

(34) The polynomial $T_n(x)$ of degree n defined by

$$T_n(x) = \cos{(n \cos^{-1} x)}/2^{n-1} \quad (|x| \leqslant 1),$$

is called Tschebyscheff's polynomial.

Show that $y = T_n = T_n(x)$ satisfies the equation

$$(1-x^2)\, y'' - xy' + n^2 y = 0.$$

Show also that

(i) $T_0 = 2$, $T_1 = x$, $T_2 = x^2 - \frac{1}{2}$, $T_3 = x^3 - \frac{3}{4}x$, $T_4 = x^4 - x^2 + \frac{1}{8}, \dots$,

(ii) $T_{n+1} - xT_n + \frac{1}{4}T_{n-1} = 0.$

(35) The polynomial $L_n(x)$ of degree n defined by

$$L_n(x) = e^x D^n(x^n e^{-x}),$$

with $L_0 = 1$, is called Laguerre's polynomial.

Show that $y = L_n = L_n(x)$ satisfies the equation

$$xy'' + (1-x)\, y' + ny = 0.$$

Show also that:

(i) $L_1 = 1 - x$, $L_2 = 2 - 4x + x^2$, $L_3 = 6 - 18x + 9x^2 - x^3, \dots$,

(ii) $L_n = (D-1)^n x^n$,   (iii) $L'_n = n(L'_{n-1} - L_{n-1})$,

(iv) $L_{n+1} = xL'_n + (n+1-x)\, L_n$,   (v) $xL'_n = nL_n - n^2 L_{n-1}$,

(vi) $L_{n+1} = (2n+1-x)\, L_n - n^2 L_{n-1}.$

(36) Prove that

(i) $D^{m+1} \sin{(m \cos^{-1} x)} = -\dfrac{mM}{(1-x^2)^{m+\frac{1}{2}}}$,

(ii) $D^{2m}(1-x^2)^{m-\frac{1}{2}} = (-)^m \dfrac{M^2}{(1-x^2)^{m+\frac{1}{2}}}$,

(iii) $D^{m-1}(1-x^2)^{m-\frac{1}{2}} = (-)^{m-1} \dfrac{M}{m} \sin{(m \cos^{-1} x)}$,

where $M = 1.3.5\dots(2m-1)$.

(37) Show that

$$D^{2m}(ax^2 + 2bx + c)^{m-\frac{1}{2}} = \left\{\frac{(2m)!}{2^m m!}\right\}^2 \frac{(ac-b^2)^m}{(ax^2 + 2bx + c)^{m+\frac{1}{2}}}.$$

(38) If $y = x^{n-1} e^{1/x}$, $z = d^n y/dx^n = \phi(x)\, e^{1/x}$, prove that

$$x^2\, dy/dx = \{(n-1)\, x - 1\}\, y,$$

$$x^2\, dz/dx + \{(n+1)\, x + 1\}\, z = 0,$$

$$x\, d\phi/dx + (n+1)\, \phi = 0.$$

Hence show that $D^n(x^{n-1} e^{1/x}) = (-)^n e^{1/x}/x^{n+1}.$

Also prove this result by the method of induction and verify by expanding $e^{1/x}$ in ascending powers of $1/x$.

(39) If $y = \{x + \sqrt{(1+x^2)}\}^m$, prove that

$$(1+x^2)\,y_{n+2} + (2n+1)\,xy_{n+1} + (n^2 - m^2)\,y_n = 0.$$

By putting $n = m$, deduce that if m is a positive integer,

$$y_{m+1} = C_m/(1+x^2)^{m+\frac{1}{2}}$$

where $\qquad C_m = \{m^2 - (m-1)^2\}\,\{m^2 - (m-3)^2\}\,\{m^2 - (m-5)^2\}\ldots$

ending with $(m^2 - 2^2)\,m^2$ if m is odd, or $(m^2 - 1^2)\,m$ if m is even.

(40) If $u = u(x)$ is a function of x, and $D = d/dx$, prove that

$$D^n f(u) = f'(u)\,X_1/1! + f''(u)\,X_2/2! + \ldots + f^{(n)}(u)\,X_n/n! \qquad \text{(i)}$$

where $\qquad X_s = [(d/dh)^n\,\{u(x+h) - u(x)\}^s]_{h=0} \qquad \text{(ii)}$

$$= D^n u^s - \binom{s}{1} u D^n u^{s-1} + \binom{s}{2} u^2 D^n u^{s-2} - \ldots$$

and in particular $\qquad X_1 = D^n u, \quad X_n = n!(Du)^n. \qquad \text{(iii)}$

Proof. It is plain by induction that a formula of the type (i) will hold good, where $X_1, X_2, \ldots$ do not depend on the function f.

Consider a particular value of x, say $x = a$. When $x = a$ let $u = b, X_s = A_s$; then

$$[D^n f(u)]_{x=a} = f'(b)\,A_1/1! + f''(b)\,A_2/2! + \ldots + f^{(n)}(b)\,A_n/n!.$$

Now put $f(u) = (u-b)^s$; then every term on the right vanishes except the sth term, and we get

$$A_s = [D^n(u-b)^s]_{x=a}$$

$$= \left[D^n u^s - \binom{s}{1} b D^n u^{s-1} + \binom{s}{2} b^2 D^n u^{s-2} - \ldots \right]_{x=a}$$

by the binomial theorem. Hence follows (ii) with a change of notation.

Example. If $u = x^2$, find $D^n f(u)$, where $D = d/dx$.

We have $u(x+h) - u(x) = (x+h)^2 - x^2 = h(2x+h)$, and hence

$$X_n = [(d/dh)^n\,h^n(2x+h)^n]_{h=0} = n!(2x)^n,$$

$$X_{n-1} = [(d/dh)^n\,h^{n-1}(2x+h)^{n-1}]_{h=0} = n!(n-1)\,(2x)^{n-2},$$

and so on. Substituting in (i), we obtain the same result as in §3.8, Example 1.

(41) If $y = \sinh^{-1} x = \log\{x + \sqrt{(1+x^2)}\}$, show that

$$-y_4(1+x^2)^{\frac{7}{2}} = \begin{vmatrix} x & 1+x^2 & 0 \\ 1 & 3x & 1+x^2 \\ 0 & 4 & 5x \end{vmatrix}.$$

Obtain the generalized form of this result, giving y_n.

Also show that $\qquad (1+x^2)\,(2x^2-1)\,y_4 = 3x(3-2x^2)\,y_3.$

(42) Show that, the determinant being of order n,

$$D^n e^{-\frac{1}{2}x^2} = (-)^n e^{-\frac{1}{2}x^2} \begin{vmatrix} x & 1 & 0 & 0 & \ldots \\ 1 & x & 1 & 0 & \ldots \\ 0 & 2 & x & 1 & \ldots \\ 0 & 0 & 3 & x & \ldots \\ \multicolumn{5}{c}{\ldots\ldots\ldots\ldots\ldots} \end{vmatrix}.$$

(43) If u is a function of x, and $D = d/dx$, show that

$$-D^n f(u) = \begin{vmatrix} 0 & f'(u)/1! & f''(u)/2! & \cdots & f^{(n)}(u)/n! \\ D^n(u) & 1 & 0 & \cdots & 0 \\ D^n(u^2) & 2u & 1 & \cdots & 0 \\ \cdots\cdots\cdots\cdots\cdots\cdots\cdots \\ D^n(u^n) & {}^nC_1 u^{n-1} & {}^nC_2 u^{n-2} & \cdots & 1 \end{vmatrix}.$$

[In Example 40 the coefficients X_s do not depend upon the function f. Put $f(u) = u, u^2, \ldots, u^n$ in turn, and eliminate the coefficients X_s from the $n+1$ equations thus obtained.]

(44) If $x = f(t), y = g(t), x_s = d^s x/dt^s, y_s = d^s y/dt^s$, show that

$$\frac{d^n y}{dx^n} = \frac{\begin{vmatrix} x_1 & 0 & 0 & \cdots & y_1 \\ x_2 & A_{22} & 0 & \cdots & y_2 \\ x_3 & A_{32} & A_{33} & \cdots & y_3 \\ \cdots\cdots\cdots\cdots\cdots\cdots\cdots \\ x_n & A_{n2} & A_{n3} & \cdots & y_n \end{vmatrix}}{1!\,2!\,3!\ldots(n-1)!\,(x_1)^{\frac{1}{2}n(n+1)}}$$

where
$$A_{ns} = [(d/dh)^n \{f(t+h) - f(t)\}^s]_{h=0}$$
$$= \left(\frac{d}{dt}\right)^n x^s - \binom{s}{1} x \left(\frac{d}{dt}\right)^n x^{s-1} + \binom{s}{2} x^2 \left(\frac{d}{dt}\right)^n x^{s-2} - \cdots.$$

[Put $y = \phi(x)$; differentiate n times wo t, and eliminate
$$\phi'(x), \phi''(x), \ldots, \phi^{(n-1)}(x).]$$

(45) If $y = f(x)$ and $y_s = f^{(s)}(x)$, show that

$$(-)^{n-1}\frac{d^n x}{dy^n} = \frac{\begin{vmatrix} y_2 & A_{22} & 0 & \cdots & 0 \\ y_3 & A_{32} & A_{33} & \cdots & 0 \\ \cdots\cdots\cdots\cdots\cdots\cdots\cdots \\ y_{n-1} & A_{n-1\,2} & A_{n-1,3} & \cdots & A_{n-1,\,n-1} \\ y_n & A_{n,2} & A_{n,3} & \cdots & A_{n,\,n-1} \end{vmatrix}}{1!\,2!\,3!\ldots(n-1)!\,(y_1)^{\frac{1}{2}n(n+1)}}$$

where
$$A_{n,s} = [(d/dh)^n \{f(x+h) - f(x)\}^s]_{h=0}$$
$$= \left(\frac{d}{dx}\right)^n y^s - \binom{s}{1} y \left(\frac{d}{dx}\right)^n y^{s-1} + \binom{s}{2} y^2 \left(\frac{d}{dx}\right)^n y^{s-2} - \cdots.$$

(46) If $(1/y)\,dy/dx = u$ and $y_1 = dy/dx, u_1 = du/dx, \ldots$ show that

$$\frac{y_4}{y} = \begin{vmatrix} u & -1 & 0 & 0 \\ u_1 & u & -1 & 0 \\ u_2 & 2u_1 & u & -1 \\ u_3 & 3u_2 & 3u_1 & u \end{vmatrix}.$$

If $D^n (\log y)$ can be found, show how to express $D^n y$ as a determinant.

Examples 3C

(1) Show that $W(e^{ax}, e^{bx}, e^{cx})$ does not vanish for any value of x if a, b, c are constants and $a \neq b \neq c \neq a$. Generalize this.

(2) Show that $W(e^{ax}, x e^{ax}, x^2 e^{ax})$ does not vanish for any value of x if a is a constant.

(3) Show that $\sin(x - \alpha)$, $\sin(x - \beta)$, $\sin(x - \gamma)$ are linearly dependent functions of x. Find a linear relation connecting them.

(4) Show that the necessary and sufficient condition that the three quadratics

$$a_1 x^2 + b_1 x + c_1, \quad a_2 x^2 + b_2 x + c_2, \quad a_3 x^2 + b_3 x + c_3$$

should be linearly dependent is the vanishing of the determinant $|a_1 b_2 c_3|$.

In Examples 5–8, the existence of any derivatives involved may be assumed.

(5) If $\lambda, y_1, y_2, \ldots, y_n$ are functions of x, prove by Leibniz's rule for differentiating a product, and by the properties of determinants, that

$$W(\lambda y_1, \lambda y_2, \ldots, \lambda y_n) = \lambda^n W(y_1, y_2, \ldots, y_n).$$

Show that

(i) $W\{f(x), x f(x), x^2 f(x)\} = 2\{f(x)\}^3$,

(ii) $W(\sin x, \sin x \cos x, \sin x \cos^2 x) = -2 \sin^6 x$,

(iii) $W\left(\dfrac{1}{y_2 y_3}, \dfrac{1}{y_3 y_1}, \dfrac{1}{y_1 y_2}\right) = \dfrac{1}{(y_1 y_2 y_3)^3} W(y_1, y_2, y_3)$.

(6) If $y_1, y_2, \ldots, y_n$ are functions of u, and u is a function of x, show that

$$W_x(y_1, y_2, \ldots, y_n) = W_u(y_1, y_2, \ldots, y_n)(du/dx)^{\frac{1}{2}n(n-1)}.$$

In particular, show that

$$W_x(u, u^2, \ldots, u^n) = 1! \, 2! \, 3! \ldots (n-1)! \, u^n (du/dx)^{\frac{1}{2}n(n-1)}.$$

(7) If $W = W_x(y_1, y_2, \ldots, y_n)$, show that

$$\frac{dW}{dx} = \begin{vmatrix} y_1 & y_2 & \cdots & y_n \\ y_1' & y_2' & \cdots & y_n' \\ \cdots\cdots\cdots\cdots\cdots\cdots\cdots\cdots \\ y_1^{(n-2)} & y_2^{(n-2)} & \cdots & y_n^{(n-2)} \\ y_1^{(n} & y_2^{(n)} & \cdots & y_n^{(n)} \end{vmatrix}$$

so that dW/dx may be written down by replacing the elements of the last row of W by their derivatives.

(8) If y_1, y_2, y_3 are three solutions of the linear equation

$$y''' + P(x) \cdot y'' + Q(x) \cdot y' + R(x) \cdot y = 0$$

and if $W(x) = W(y_1, y_2, y_3)$, prove that

$$W(x) = W(c) \exp\left\{-\int_c^x P(t)\, dt\right\}$$

where c is arbitrary. Deduce that $W(x)$ either vanishes for no value of x or for all values of x. Generalize this.

4

TAYLOR'S THEOREM

4.1 Properties of the nth derivative

I. The existence of the nth derivative $f^{(n)}(a)$ at $x = a$ implies (§2.11, II) that there is an interval $(a - \delta \leqslant x \leqslant a + \delta)$ in which $f^{(n-1)}(x)$ exists, and that $f^{(n-1)}(x)$ is continuous at a. Further, since $f^{(n-1)}(x)$ exists in this interval, it follows that $f^{(n-2)}(x), \ldots, f'(x), f(x)$ all exist and are continuous in it.

Note. The interval must be replaced by $(a - \delta \leqslant x \leqslant a)$ or by $(a \leqslant x \leqslant a + \delta)$ if $f^{(n)}(a)$ exists only as a left-hand or right-hand derivative, respectively.

II. Let $f^{(n-1)}(x)$ be continuous in $(a \leqslant x \leqslant b)$ and differentiable in $(a < x < b)$.

Theorem. If $f(x)$ and its first $n - 1$ derived functions all vanish at $x = a$, none of them can vanish in $(a < x \leqslant b)$ if $f^{(n)}(x) \neq 0$ in

$$(a < x < b).$$

Proof. Since $f^{(n-1)}(a) = 0$ and $f^{(n)}(x) \neq 0$ in $(a < x < b)$ it follows from §2.13, Corollary that $f^{(n-1)}(x) \neq 0$ in $(a < x \leqslant b)$; hence also $f^{(n-2)}(x) \neq 0$ in $(a < x \leqslant b)$; and so on.

Corollary. If $f(x)$ and its first $n - 1$ derived functions all vanish at $x = a$, none of them can vanish at any other point in a small enough neighbourhood of a if $f^{(n)}(a) \neq 0$ exists.

Proof. By §2.11, III, Corollary 3, since $f^{(n)}(a) \neq 0$, there is a neighbourhood of a in which $f^{(n-1)}(x) \neq 0$, $x \neq a$. It follows from the present theorem that also neither $f(x)$ nor any of its first $n - 2$ derived functions can vanish in this neighbourhood, except at $x = a$.

III. Let $f^{(n-1)}(x)$ be continuous in $(a \leqslant x \leqslant b)$ and differentiable in $(a < x < b)$.

Theorem. If $f(x)$ and its first $n - 1$ derived functions all vanish at $x = a$, and if $f^{(n)}(x) > 0$ in $(a < x < b)$, then $f(x)$ and its first $n - 1$ derived functions are all positive in $(a < x \leqslant b)$ and strictly increasing in $(a \leqslant x \leqslant b)$.

Proof. By §2.15, Corollary 2, since $f^{(n)}(x) > 0$ in $(a < x < b)$, therefore $f^{(n-1)}(x)$ is strictly increasing in $(a \leqslant x \leqslant b)$. Hence, and since

$$f^{(n-1)}(a) = 0, \quad f^{(n-1)}(x) \text{ is positive in } (a < x \leqslant b).$$

Similarly, $f^{(n-2)}(x)$ is strictly increasing in $(a \leqslant x \leqslant b)$ and positive in $(a < x \leqslant b)$; and so on.

Example. If $0 < x$ prove that

$$f(x) \equiv e^{-x} - 1 + \frac{x}{1!} - \frac{x^2}{2!} + \frac{x^3}{3!} - \dots (-)^n \frac{x^{n-1}}{(n-1)!} > 0 \quad \text{or} \quad < 0$$

according as n is even or odd.

Proof. We find $f(0) = f'(0) = \dots = f^{(n-1)}(0) = 0$, and

$$f^{(n)}(x) = (-)^n e^{-x} > 0 \quad \text{or} \quad < 0$$

according as n is even or odd. Hence, by the theorem, if n is even $f(x)$ and its first $n-1$ derived functions are all positive if $0 < x$. If n is odd, then $-f(x)$ and its first $n-1$ derivatives are all positive if $0 < x$, and therefore $f(x)$ and its first $n-1$ derivatives are all negative if $0 < x$.

4.2   Extension of L'Hospital's first rule (§2.19)

Theorem. Let $F(x)$, $G(x)$ and their first $n-1$ derived functions all vanish at $x = a$. Let $F^{(n)}(a)$, $G^{(n)}(a)$ exist, $G^{(n)}(a) \neq 0$. Then

$$\lim_{x \to a} \frac{F(x)}{G(x)} = \frac{F^{(n)}(a)}{G^{(n)}(a)}. \tag{1}$$

Proof. By repeated applications of l'Hospital's second rule

$$\lim_{x \to a} \frac{F(x)}{G(x)} = \lim_{x \to a} \frac{F'(x)}{G'(x)} = \dots = \lim_{x \to a} \frac{F^{(n-1)}(x)}{G^{(n-1)}(x)}$$

provided that the last limit exists. But, by l'Hospital's first rule, the last limit exists and is equal to $F^{(n)}(a)/G^{(n)}(a)$. This proves the theorem.

4.3   Taylor's formula for $f(x)$ to base a

Let $f^{(n)}(a)$ exist, and let $P_n(x)$ denote the special polynomial of degree n defined by

$$P_n(x) = f(a) + (x-a)f'(a) + (x-a)^2 f''(a)/2! + \dots + (x-a)^n f^{(n)}(a)/n!. \tag{2}$$

This polynomial is constructed so that it and its first n derivatives have the values $f(a), f'(a), \dots, f^{(n)}(a)$ at $x = a$.

Theorem. The function $f(x)$ can be expressed in the form

$$f(x) = f(a) + (x-a)f'(a) + (x-a)^2 f''(a)/2! + \dots$$
$$+ (x-a)^n f^{(n)}(a)/n! + \eta(x-a)^n, \tag{3}$$

that is, $\qquad\qquad\qquad f(x) = P_n(x) + \eta(x-a)^n, \tag{4}$

where η is a function of x which, with n fixed, tends to zero when $x \to a$.

Proof. Put
$$f(x) = P_n(x) + R(x) \tag{5}$$
so that

$$R(x) = f(x) - f(a) - (x-a)f'(a) - \ldots - (x-a)^n f^{(n)}(a)/n! \tag{6}$$

Also put
$$G(x) = (x-a)^n.$$

Then $R(x)$, $G(x)$ and their first $n-1$ derived functions all vanish at $x = a$. Also $R^{(n)}(a) = 0$, $G^{(n)}(a) = n! \neq 0$. Hence, by (1),

$$\lim_{x \to a} \frac{R(x)}{G(x)} = \frac{R^{(n)}(a)}{G^{(n)}(a)} = \frac{0}{n!} = 0,$$

and therefore we can put

$$\frac{R(x)}{G(x)} = \frac{R(x)}{(x-a)^n} = \eta, \quad R(x) = \eta(x-a)^n, \tag{7}$$

where $\eta \to 0$ when $x \to a$. The theorem follows.

Corollary 1. The ratio of $R(x)$ to $(x-a)^n$ is as small as we please when $x - a$ is small enough.

Corollary 2. The ratio of $R(x)$ to the last term in the polynomial $P_n(x)$ is as small as we please when $x-a$ is small enough, provided that $f^{(n)}(a) \neq 0$.

Note. Formula (3) is called *Taylor's formula* to base a. It expresses $f(x)$ as the sum of a polynomial $P_n(x)$ of special form and a *complementary term* or '*remainder*' $R(x)$. The remainder can be put in other forms (§4.10 et seq.); the form $\eta(x-a)^n$ is called *Young's* form.

4.4 Maclaurin's formula

Taylor's formula when $a = 0$ is called '*Maclaurin's formula*'. Thus, Maclaurin's formula reads

$$f(x) = f(0) + xf'(0) + x^2 f''(0)/2! + \ldots + x^n f^{(n)}(0)/n! + \eta x^n, \tag{8}$$

where η is a function of x which, with n fixed, tends to zero when $x \to 0$. This assumes only that $f^{(n)}(0)$ exists.

Taylor's formula can also be regarded as a Maclaurin formula. For if we put $x = a+h$, $x-a = h$, Taylor's formula becomes

$$f(a+h) = f(a) + hf'(a) + \ldots + h^n f^{(n)}(a)/n! + \eta h^n, \tag{9}$$

where η is now a function of h which, with n fixed, tends to zero when $h \to 0$. This is the Maclaurin formula for the function $f(a+h)$ regarded as a function of h.

It is sometimes convenient to write

$$f(x+h) = f(x) + hf'(x) + \ldots + h^n f^{(n)}(x)/n! + \eta h^n, \tag{10}$$

where, temporarily, x is fixed. Here, η is a function of x and h which, with x and n fixed, tends to zero as $h \to 0$. The only assumption is that the nth derivative exists at the point x.

Notation. We shall generally take it for granted that, in such terms as $\eta x^n, \alpha h^n, \beta k^n$, the letters $\eta, \alpha, \beta, \ldots$ denote functions of $x, h, k, \ldots$ that tend to zero when $x, h, k, \ldots$ tend to zero, respectively. In such a term as $\eta(x-a)^n$, the letter η will denote a function of x that tends to zero when $x - a \to 0$, or $x \to a$; while in η/x^n, $\eta \to 0$ when $1/x \to 0$, or $x \to \infty$.

In the following examples, use is made of some of the nth derivatives listed in §3.2.

Examples

(1) $$f(x) = e^x,$$

$$f^{(n)}(x) = e^x, \quad f^{(n)}(0) = e^0 = 1,$$

$$e^x = 1 + \frac{x}{1!} + \frac{x^2}{2!} + \frac{x^3}{3!} + \ldots + \frac{x^n}{n!} + \eta x^n. \tag{11}$$

(2) $$f(x) = \sin x,$$

$$f^{(n)}(x) = \sin\left(x + \tfrac{1}{2}n\pi\right), \quad f^{(n)}(a) = \sin\left(a + \tfrac{1}{2}n\pi\right),$$

$$\sin(a+h) = \sin a + h \cos a + \ldots + \frac{h^n}{n!}\sin\left(a + \tfrac{1}{2}n\pi\right) + \eta h^n. \tag{12}$$

Putting $a = 0$, $h = x$, we have in particular

$$\sin x = \frac{x}{1!} - \frac{x^3}{3!} + \frac{x^5}{5!} + \ldots + \sin\frac{n\pi}{2}\frac{x^n}{n!} + \eta x^n. \tag{13}$$

Again, by putting $a = \tfrac{1}{2}\pi$, $h = x$,

$$\cos x = 1 - \frac{x^2}{2!} + \frac{x^4}{4!} - \ldots + \cos\frac{n\pi}{2}\frac{x^n}{n!} + \eta x^n. \tag{14}$$

(3) $$f(x) = \log(1+x) \quad (-1 < x),$$

$$f^{(n)}(x) = \frac{(-)^{n-1}(n-1)!}{(1+x)^n}, \quad f^{(n)}(0) = (-)^{n-1}(n-1)!,$$

$$\log(1+x) = \frac{x}{1} - \frac{x^2}{2} + \frac{x^3}{3} - \ldots(-)^{n-1}\frac{x^n}{n} + \eta x^n. \tag{15}$$

(4) $$f(x) = (1+x)^m,$$

$$f^{(n)}(x) = m_n(1+x)^{m-n}, \quad f^{(n)}(0) = m_n$$

where $m_n = m(m-1)...(m-n+1)$. Hence,

$$(1+x)^m = 1 + \binom{m}{1}x + \binom{m}{2}x^2 + ... + \binom{m}{n}x^n + \eta x^n. \qquad (16)$$

Here we suppose $x > -1$ if m is not an integer.

(5) $f(x) = e^{-1/x^2}$ $(x \neq 0)$, $f(0) = 0$.

If $x \neq 0$, we find $f'(x) = (2/x^3)e^{-1/x^2}$ and hence, with $1/x = v$,

$$\lim_{x \to 0} f'(x) = \lim_{v \to \infty} \frac{2v^3}{e^{v^2}} = 0.$$

Since $f(x)$ is continuous at $x = 0$, it follows from §2.17 that $f'(0)$ exists and that $f'(0) = 0$.

Similarly, we find that $f''(0), f'''(0), ...$ all exist and that

$$f''(0) = 0, \quad f'''(0) = 0, ..., f^{(n)}(0) = 0,$$

where n is any positive integer. Consequently, the Maclaurin formula for the function e^{-1/x^2} is

$$e^{-1/x^2} = 0 + 0.x + 0.x^2 + ... + 0.x^n + \eta x^n. \qquad (17)$$

(6) From (11) and (17) follows

$$e^x + e^{-1/x^2} = 1 + \frac{x}{1!} + \frac{x^2}{2!} + \frac{x^3}{3!} + ... + \frac{x^n}{n!} + \eta x^n. \qquad (18)$$

Note. The last two examples warn us that, even when the Maclaurin *series* derived from $f(x)$ converges (§4.16), the sum of the infinite series is not necessarily $f(x)$.

(7) $$y = f(x) = (\sin^{-1} x)^2.$$

Having regard to Examples 3B, 18, we have

$$y_0 = 0, \quad (y_1)_0 = 0, \quad (y_2)_0 = 2, \quad (y_{n+2})_0 = n^2(y_n)_0 \quad (n \geqslant 1).$$

Putting $n = 1, 3, 5, ...$ we find $(y_3)_0 = (y_5)_0 = ... = 0$.

Put $n = 2, 4, 6, ...$ we find

$$(y_4)_0 = 2^2(y_2)_0 = 2^2.2, \quad (y_6)_0 = 4^2(y_4)_0 = 4^2.2^2.2, ...$$

Hence, by Maclaurin's formula,

$$\frac{(\sin^{-1}x)^2}{2} = \frac{x^2}{2!} + \frac{2^2 x^4}{4!} + \frac{2^2.4^2 x^6}{6!} + \frac{2^2.4^2.6^2 x^8}{8!} + ... + \eta x^n. \qquad (19)$$

The law of formation of the successive terms is sufficiently indicated by the terms given. The dots at the end indicate that the polynomial part continues as far as the term in x^n (the coefficient of which will be zero if n is odd).

(8) If $f''(a)$ exists uniquely, prove that

$$\lim_{h\to 0} \frac{f(a+h)+f(a-h)-2f(a)}{h^2} = f''(a). \qquad (20)$$

Since $f''(a)$ exists as a unique derivative, we have both

$$f(a+h) = f(a) + hf'(a) + \tfrac{1}{2}h^2 f''(a) + \alpha h^2$$

and $\qquad\qquad f(a-h) = f(a) - hf'(a) + \tfrac{1}{2}h^2 f''(a) + \beta h^2,$

where $\alpha \to 0$, $\beta \to 0$, when $h \to 0$. Hence

$$f(a+h) + f(a-h) - 2f(a) = h^2 f''(a) + (\alpha + \beta) h^2$$

from which the required result follows after division by h^2.

4.5   Differentiation and integration of Taylor or Maclaurin formulae

Let $f(x)$ be continuous in an interval $[-h, h]$.

We shall assume that the definite integral $\int_0^x f(t)\, dt$ exists (§9.9) and has $f(x)$ for its first derived function (§9.16). It will follow from §4.4 that, if $f^{(n)}(0)$ exists, the three Maclaurin formulae

$$f(x) = f(0) + xf'(0) + \frac{x^2}{2!}f''(0) + \dots + \frac{x^n}{n!}f^{(n)}(0) + \eta x^n, \qquad (21)$$

$$f'(x) = f'(0) + xf''(0) + \dots + \frac{x^{n-1}}{(n-1)!}f^{(n)}(0) + \alpha x^{n-1}, \qquad (22)$$

$$\int_0^x f(t)\, dt = xf(0) + \frac{x^2}{2!}f'(0) + \dots + \frac{x^{n+1}}{(n+1)!}f^{(n)}(0) + \beta x^{n+1} \qquad (23)$$

all hold good, the second because $f^{(n)}(0)$ is the $(n-1)$th derivative of $f'(x)$ at $x = 0$, the third because $f^{(n)}(0)$ is the $(n+1)$th derivative of the integral at $x = 0$. But the polynomial part of the second formula is the derivative of that of the first, and the polynomial part of the third formula is the integral of that of the first. We say therefore that *a Maclaurin formula can be differentiated or integrated.*

Consequently, the Maclaurin formula for $f(x)$ can be found by integrating that for $f'(x)$ or by differentiating that for $\int_0^x f(t)\, dt$. This is useful when the formula for $f'(x)$ or for the integral is known or can be found more easily than that for $f(x)$.

If $f^{(n)}(0)$ exists for *all* values of n, we can evidently differentiate or integrate any number of times, continuing the polynomial part of any of the resulting formulae as far as we please.

Similar considerations apply to Taylor formulae. Thus, if $f(x)$ is continuous in an interval $[a-h, a+h]$ and if $f^{(n)}(a)$ exists as a unique derivative, then the formulae corresponding to (21), (22), (23) will be

$$f(x) = f(a) + (x-a)f'(a) + \dots + (x-a)^n f^{(n)}(a)/n! + \eta(x-a)^n, \quad (21')$$

$$f'(x) = f'(a) + \dots + (x-a)^{n-1} f^{(n)}(a)/(n-1)! + \alpha(x-a)^{n-1}, \quad (22')$$

$$\int_a^x f(t)\, dt = (x-a)f(a) + \dots + (x-a)^{n+1} f^{(n)}(a)/(n+1)! + \beta(x-a)^{n+1}. \quad (23')$$

Note. If $f(x)$ is continuous in $[a, a+h]$, $0 < h$, and if $f^{(n)}(a)$ exists as a right-hand derivative at $x = a$, then the Taylor formulae will hold good if $a < x$ but not necessarily if $x < a$.

By putting $a = 0$, we see that if $f(x)$ is continuous in $[0, h]$, $0 < h$, and if $f^{(n)}(0)$ exists as a right-hand derivative at $x = 0$, then the Maclaurin formulae will hold good if $0 < x$, but not necessarily if $x < 0$.

Examples

(1) If $f(x) = 1/(1-x)$, then $f^{(n)}(x) = n!/(1-x)^{n+1}$ and hence

$$f^{(n)}(0) = n!$$

Consequently, by (21), the formula

$$1/(1-x) = 1 + x + x^2 + x^3 + \dots + x^n + \eta x^n,$$

where $x \neq 1$ and $\eta \to 0$ when $x \to 0$, holds good for all values of n.

By applying (22) and (23), we find, respectively,

$$1/(1-x)^2 = 1 + 2x + 3x^2 + \dots + nx^{n-1} + \alpha x^{n-1},$$

$$\log \frac{1}{1-x} = \frac{x}{1} + \frac{x^2}{2} + \frac{x^3}{3} + \dots + \frac{x^{n+1}}{n+1} + \beta x^{n+1} \quad (x < 1).$$

Further differentiation would give

$$2/(1-x)^3 = 1 \cdot 2 + 2 \cdot 3x + \dots + n(n-1)x^{n-2} + \gamma x^{n-2},$$

and so on. Since $f^{(n)}(0)$ exists for all values of n, we could rewrite the last three formulae so that they ended with

$$(n+1)x^n + \alpha x^n, \quad x^n/n + \beta x^n, \quad (n+2)(n+1)x^n + \gamma x^n,$$

respectively. As indicated previously (§4.4), the letters α, β, γ denote functions of x that tend to zero as $x \to 0$. In this example it would be easy to find them explicitly, since we know, by algebra, that

$$\eta = x/(1-x).$$

(2) To find the Maclaurin formula for $y = \sin^{-1}x$.

We have, using (16), if $x^2 < 1$,

$$\frac{dy}{dx} = \frac{1}{\sqrt{(1-x^2)}} = 1 + \frac{x^2}{2} + \frac{1.3}{2.4}x^4 + \frac{1.3.5}{2.4.6}x^6 + \ldots + \eta x^n.$$

By integration, and since $y = 0$ when $x = 0$,

$$\sin^{-1}x = x + \frac{1}{2}\frac{x^3}{3} + \frac{1.3}{2.4}\frac{x^5}{5} + \frac{1.3.5}{2.4.6}\frac{x^7}{7} + \ldots + \alpha x^n.$$

(3) By differentiating (19), we find the formula

$$\frac{\sin^{-1}x}{\sqrt{(1-x^2)}} = x + \tfrac{2}{3}x^3 + \frac{2.4}{3.5}x^5 + \frac{2.4.6}{3.5.7}x^7 + \ldots + \alpha x^n.$$

4.6 Properties of Maclaurin's (or Taylor's) formula

I. *Theorem.* There cannot exist more than one formula of the type

$$f(x) = c_0 + c_1 x + c_2 x^2 + \ldots + c_n x^n + \eta x^n, \tag{24}$$

holding good in any interval including $x = 0$, where $c_0, c_1, \ldots, c_n$ are independent of x, and $\eta \to 0$ as $x \to 0$.

Proof. If possible, let

$$f(x) = b_0 + b_1 x + b_2 x^2 + \ldots + b_n x^n + \alpha x^n \tag{25}$$

be a different formula of the same type. Then there must be an integer $s \leqslant n$, such that $b_r = c_r$ $(r < s)$, $b_s \neq c_s$, and hence, by subtraction, such that

$$0 = (b_s - c_s) x^s + \ldots + (\alpha - \eta) x^n.$$

Dividing by x^s and letting $x \to 0$, we get $0 = b_s - c_s$, which is impossible since $b_s \neq c_s$. It follows that (25) cannot differ from (24).

Corollary 1. Maclaurin's formula is unique: that is, if $f^{(n)}(0)$ exists, Maclaurin's formula is the only one of the type (24).

Corollary 2. If a formula of the type (24) is obtainable by any method, and if $f^{(n)}(0)$ is known to exist, the formula must be Maclaurin's formula and from it will follow $c_s = f^{(s)}(0)/s!$ or $f^{(s)}(0) = s!\,c_s\,(s \leqslant n)$.

In the same kind of way, it can be shown that a Taylor formula is unique: that is to say, if $f^{(n)}(a)$ exists, then Taylor's formula (3) exists, and no other formula exists of the type

$$f(x) = c_0 + c_1(x-a) + \ldots + c_n(x-a)^n + \eta(x-a)^n,$$

where $c_0, c_1, \ldots, c_n$ are independent of x, and $\eta \to 0$ when $x \to a$.

It follows that, if a formula of this type is obtained by any method and if $f^{(n)}(a)$ is known to exist, the formula can only be Taylor's formula and from it we can deduce that $f^{(s)}(a) = s!\,c_s\,(s \leqslant n)$.

II. *Theorem.* If a formula of the type (24) holds good in an interval $[-h, h]$ and if $f(x)$ is an even (odd) function, then the polynomial part of the formula must consist entirely of even (odd) powers of x.

Proof. Let $f(x)$ be an even function, and assume that

$$f(x) = c_0 + c_2 x^2 + c_4 x^4 + \ldots + c_s x^s + \ldots + \eta x^n,$$

where $c_s x^s (c_s \neq 0)$ is the first term in which s is odd. Changing the sign of x, we have also

$$f(-x) = c_0 + c_2 x^2 + c_4 x^4 + \ldots - c_s x^s + \ldots + \eta_1 x^n.$$

Now $f(-x) = f(x)$ since $f(x)$ is even; hence, by subtraction,

$$0 = 2c_s x^s + \ldots + (\eta - \eta_1) x^n.$$

Dividing by x^s and letting $x \to 0$ we get $2c_s = 0$, contrary to the assumption that $c_s \neq 0$. Consequently, there cannot be any odd powers of x in the polynomial.

A similar proof applies when $f(x)$ is an odd function.

Corollary. If $f(x)$ is an even (odd) function, and if $f^{(n)}(0)$ exists as a unique nth derivative, the polynomial part of Maclaurin's formula consists entirely of even (odd) powers of x.

4.7　Algebraic operations on Taylor or Maclaurin formulae

Let $u = u(x)$, $v = v(x)$ be functions of x whose nth derivatives exist at x. Let u_s, v_s denote the sth derivatives of u, v respectively ($s \leqslant n$). Then $u(x+h)$, $v(x+h)$ have the following Taylor formulae to base x:

$$u(x+h) = u + u_1 h + u_2 h^2/2! + \ldots + u_n h^n/n! + \alpha h^n, \qquad (26)$$

$$v(x+h) = v + v_1 h + v_2 h^2/2! + \ldots + v_n h^n/n! + \beta h^n, \qquad (27)$$

where, with x and n fixed, α and β tend to zero when $h \to 0$.

I. *Sum and product.* The Taylor formula for the sum of $u(x+h)$ and $v(x+h)$ can be found by adding their Taylor formulae. The Taylor formula for their product can be found by multiplying their Taylor formulae.

Proof. The case of the sum is left to the reader.

Consider the product $u(x+h) \cdot v(x+h)$. From (26), (27), by multiplication we get a product which can evidently be arranged in the form

$$u(x+h) \cdot v(x+h) = c_0 + c_1 h + c_2 h^2 + \ldots + c_n h^n + \eta h^n, \qquad (28)$$

where c_s is independent of h. By §4.6, I, this can only be the Taylor formula for the product, since the nth derivative of uv is known to exist.

Verification. The coefficient of h^s in the algebraic product is found to be

$$c_s = \frac{u_s v}{s!} + \frac{u_{s-1} v_1}{(s-1)!\,1!} + \dots + \frac{u v_s}{s!}$$

which, by Leibniz's theorem, is $(uv)_s/s!$.

Examples

(1) To find the Maclaurin formula for $\cos^4 x$.

First, express $\cos^4 x$ as a sum. By trigonometry

$$\cos^4 x = \tfrac{3}{8} + \tfrac{1}{2}\cos 2x + \tfrac{1}{8}\cos 4x.$$

Then, by (14), we have the Maclaurin formulae

$$\frac{3}{8} + \frac{1}{2}\cos 2x = \frac{3}{8} + \frac{1}{2}\left\{1 - \frac{(2x)^2}{2!} + \frac{(2x)^4}{4!} - \frac{(2x)^6}{6!} + \dots + \alpha x^n\right\},$$

$$\frac{1}{8}\cos 4x = \frac{1}{8}\left\{1 - \frac{(4x)^2}{2!} + \frac{(4x)^4}{4!} - \frac{(4x)^6}{6!} + \dots + \beta x^n\right\}.$$

By addition

$$\cos^4 x = 1 - 2x^2 + \frac{2^2+1}{2}\frac{(2x)^4}{4!} - \frac{2^4+1}{2}\frac{(2x)^6}{6!} + \dots + \eta x^n.$$

(2) $$f(x) = \tfrac{1}{2}\{\log(1+x)\}^2, \quad f'(x) = \frac{\log(1+x)}{1+x},$$

$$\log(1+x) = x/1 - x^2/2 + x^3/3 - \dots (-)^{n-1}x^n/n + \alpha x^n,$$

$$1/(1+x) = 1 - x + x^2 - x^3 + \dots (-)^n x^n + \beta x^n.$$

By multiplication

$$\frac{\log(1+x)}{1+x} = x - (1 + \tfrac{1}{2})x^2 + (1 + \tfrac{1}{2} + \tfrac{1}{3})x^3 - \dots + \gamma x^n.$$

By integration (§4.5) and since $f(0) = 0$,

$$\tfrac{1}{2}\{\log(1+x)\}^2 = \frac{x^2}{2} - (1 + \tfrac{1}{2})\frac{x^3}{3} + (1 + \tfrac{1}{2} + \tfrac{1}{3})\frac{x^4}{4} - \dots + \eta x^n.$$

(3) Prove that

$$\tan x = x + \frac{x^3}{3} + \frac{2x^5}{15} + \frac{17x^7}{315} + \frac{62x^9}{2835} + \eta x^{10}. \tag{29}$$

Deduce similar expressions for $\sec^2 x$ and $\log(\sec x)$.

Put $y = \tan x$. Then $dy/dx = 1 + y^2$. Since $(y_n)_0$ is known to exist, and since y is an odd function and $(y_1)_0 = 1$, we may put (§4.6, II)

$$y = x + ax^3 + bx^5 + cx^7 + dx^9 + 0 \cdot x^{10} + \eta x^{10}.$$

Then by §§4.5 and 4.7

$$dy/dx = 1 + 3ax^2 + 5bx^4 + 7cx^6 + 9dx^8 + \alpha x^9,$$

$$1 + y^2 = 1 + x^2 + 2ax^4 + (a^2 + 2b)x^6 + 2(c + ab)x^8 + \beta x^9.$$

By equating coefficients and solving for a, b, c, d,

$$a = \tfrac{1}{3}, \quad b = \tfrac{2}{15}, \quad c = \tfrac{17}{315}, \quad d = \tfrac{62}{2835}.$$

By differentiating (29), we find

$$\sec^2 x = 1 + x^2 + \frac{2x^4}{3} + \frac{17x^6}{45} + \frac{62x^8}{315} + \gamma x^9. \tag{30}$$

By integrating (29), and since log $(\sec x) = 0$ at $x = 0$,

$$\log(\sec x) = \frac{x^2}{2} + \frac{x^4}{12} + \frac{x^6}{45} + \frac{17x^8}{2520} + \frac{31x^{10}}{14175} + \delta x^{11}. \tag{31}$$

II. *Reciprocal and quotient.* The Taylor formula for the reciprocal $1/v(x+h)$ can be found algebraically from that of $v(x+h)$, provided that $v(x) \neq 0$. Thus, by (27), we can put

$$\frac{1}{v(x+h)} = \frac{1}{v+k},$$

where $k = v_1 h + v_2 h^2/2! + \ldots + v_n h^n/n! + \beta h^n$ and, with x fixed, $\beta \to 0$ when $h \to 0$.

But, by Taylor's formula for $1/(v+k)$, with x fixed, and provided that $v = v(x) \neq 0$, we have also

$$\frac{1}{v+k} = \frac{1}{v} - \frac{k}{v^2} + \frac{k^2}{v^3} - \ldots(-)^n \frac{k^n}{v^{n+1}} + \gamma k^n,$$

where $\gamma \to 0$ when $k \to 0$. But, with x fixed, $k \to 0$ when $h \to 0$, and therefore also $\gamma \to 0$ when $h \to 0$. Consequently, when we substitute the value of k in the last formula, and rearrange in ascending powers of h, we get a result of the form (24), and this result must be the Taylor formula for $1/v(x+h)$, since the nth derivative of $1/v(x)$ with respect to x is known to exist.

Corollary. The Taylor formula for the quotient $u(x+h)/v(x+h)$ can be found algebraically from the Taylor formulae for $u(x+h)$ and $v(x+h)$, provided that $v(x) \neq 0$. For the quotient can be expressed as the product of $u(x+h)$ and $1/v(x+h)$.

Note. In practice we can use the method indicated in the proof, or the method of ordinary 'long division', or the method of 'undetermined coefficients'.

Examples

(1) Obtain the Maclaurin formula

$$\sec x = 1 + \frac{x^2}{2} + \frac{5x^4}{24} + \frac{61x^6}{720} + \eta x^7, \tag{32}$$

and deduce corresponding formulae for

$$\sec x \tan x \quad \text{and} \quad \log(\sec x + \tan x).$$

By actual division we find

$$\sec x = \frac{1}{\cos x} = \left(1 - \frac{x^2}{2!} + \frac{x^4}{4!} - \frac{x^6}{6!} + \alpha x^7\right)^{-1}$$

$$= 1 + \frac{x^2}{2} + \frac{5x^4}{24} + \frac{61x^6}{720} + \eta x^7.$$

By differentiation,

$$\sec x \tan x = x + \frac{5x^3}{6} + \frac{61x^5}{120} + \eta_1 x^6.$$

By integration of (32) and finding the constant of integration by putting $x = 0$,

$$\log(\sec x + \tan x) = x + \frac{x^3}{6} + \frac{x^5}{24} + \frac{61x^7}{5040} + \eta_2 x^8.$$

(2) Prove that

$$\sec x = 1 + \frac{x^2}{2} + \frac{5x^4}{24} + \frac{61x^6}{720}$$

$$+ \begin{vmatrix} 1 & 1 & 0 & 0 \\ 1 & 6 & 1 & 0 \\ 1 & 15 & 15 & 1 \\ 1 & 28 & 70 & 28 \end{vmatrix} \frac{x^8}{8!} + \begin{vmatrix} 1 & 1 & 0 & 0 & 0 \\ 1 & 6 & 1 & 0 & 0 \\ 1 & 15 & 15 & 1 & 0 \\ 1 & 28 & 70 & 28 & 1 \\ 1 & 45 & 210 & 210 & 45 \end{vmatrix} \frac{x^{10}}{10!} + \ldots + \eta x^n.$$

Because $\sec x$ is an even function, by §4.6, II, we may put

$$\sec x = \frac{1}{\cos x} = 1 + \frac{c_1 x^2}{2!} + \frac{c_2 x^4}{4!} + \frac{c_3 x^6}{6!} + \ldots + \eta x^n.$$

Then, by the Maclaurin formula for $\cos x$,

$$1 = \left(1 - \frac{x^2}{2!} + \frac{x^4}{4!} - \ldots + \alpha x^n\right)\left(1 + \frac{c_1 x^2}{2!} + \frac{c_2 x^4}{4!} + \ldots + \eta x^n\right).$$

The coefficients of x^2, x^4, x^6, ... in the product on the R.H.S. are easily written down by keeping Leibniz's theorem in mind. All of these coefficients must be zero, by comparison with the L.H.S.; hence

$$c_1 - 1 = 0,$$

$$-c_2 + 6c_1 - 1 = 0,$$

$$c_3 - 15c_2 + 15c_1 - 1 = 0,$$

$$-c_4 + 28c_3 - 70c_2 + 28c_1 - 1 = 0, \dots .$$

These equations determine the coefficients $c_1, c_2, c_3, \dots$ which can therefore be expressed as determinants whose elements are binomial coefficients. In particular, c_4 and c_5 are found to have the values indicated in the formula to be proved.

III. *Function of a function.* The Taylor formula for $f\{u(x+h)\}$ can be found algebraically from that of $u(x+h)$ and that of $f(u+k)$, provided that $u_n(x)$ and $f_n(u)$ exist.

The method has been given in §3.8. A particular case is that of the reciprocal of a function, considered in II above.

Example

Show that

$$\log \tan^{-1}(1+h) = \log\frac{\pi}{4} + 2\frac{h}{\pi} - (\pi+2)\frac{h^2}{\pi^2} + \eta h^2,$$

where $\eta \to 0$ when $h \to 0$.

Put $u(x) = \tan^{-1} x$. Then we find $u(1) = \frac{1}{4}\pi$, $u'(1) = \frac{1}{2}$ and

$$u''(1) = -\tfrac{1}{2}.$$

We can therefore put

$$\tan^{-1}(1+h) = \tfrac{1}{4}\pi + \tfrac{1}{2}h - \tfrac{1}{4}h^2 + \alpha h^2,$$

where $\alpha \to 0$ when $h \to 0$.

Also, if we put $f(u) = \log u$, we find $f(\frac{1}{4}\pi) = \log(\frac{1}{4}\pi)$, $f'(\frac{1}{4}\pi) = 4/\pi$, $f''(\frac{1}{4}\pi) = -16/\pi^2$ and therefore

$$\log\left(\frac{\pi}{4} + k\right) = \log\frac{\pi}{4} + \frac{4}{\pi}k - \frac{8}{\pi^2}k^2 + \beta k^2,$$

where $\beta \to 0$ when $k \to 0$.

Putting $k = \frac{1}{2}h - \frac{1}{4}h^2 + \alpha h^2$ and rearranging, we obtain the formula required.

IV. *Inverse function. Inversion of a Taylor formula.* Let $y = f(x)$, $b = f(a)$. Let $f^{(n)}(a)$ exist $(n > 1)$, $f'(a) \neq 0$.

Put $h = x - a$, $k = y - b$; then, since $f^{(n)}(a)$ exists,

$$k = c_1 h + c_2 h^2 + \ldots + c_n h^n + \alpha h^n, \tag{33}$$

where $c_s = f^{(s)}(a)/s!$.

Again, by §4.1, I, $f'(x)$ is continuous at $x = a$, since $f''(a)$ exists; also $f'(a) \neq 0$; hence the equation $y = f(x)$ defines the inverse function $x = \phi(y)$ in the neighbourhood of $y = b$, by §2.21, with a derivative given by the formula $\phi'(y) = 1/f'(x)$.

Moreover, since $f^{(n)}(a)$ exists, it follows that $f^{(s)}(x)$ $(s < n)$ exists near $x = a$, and hence, by successively differentiating $\phi'(y) = 1/f'(x)$ with respect to y, that $\phi^{(n)}(b)$ exists. Hence

$$h = C_1 k + C_2 k^2 + \ldots + C_n k^n + \beta k^n, \tag{34}$$

where $C_s = \phi^{(s)}(b)/s!$

Since (33) and (34) both exist and are unique, if either has been determined the other can be found by the method of undetermined coefficients. Thus, if (33) has been determined, the coefficients C_s in (34) can be found by substituting for k from (33) in (34) and equating coefficients of like powers of h. We thus obtain a set of linear equations giving $C_1, C_2, \ldots, C_n$.

Example

Let $y = 4x - x^2$. Then, when $x = 1$, $y = 3$ and $dy/dx = 2 \neq 0$. Consequently, the equation $y = 4x - x^2$ defines x as a function of y in the neighbourhood of $y = 3$.

Put $h = x - 1$, $k = y - 3$. Then we find

$$k = 2h - h^2$$

and for the inverse function we may put

$$h = C_1 k + C_2 k^2 + \ldots + C_n k^n + \beta k^n,$$

where $\beta \to 0$ when $k \to 0$ and therefore when $h \to 0$.

By substituting $k = 2h - h^2$ in the last formula, and rearranging in ascending powers of h,

$$h = 2C_1 h + (-C_1 + 4C_2) h^2 + (-4C_2 + 8C_3) h^3 + \ldots + \eta h^n,$$

and hence $2C_1 = 1$, $-C_1 + 4C_2 = 0$, $-4C_2 + 8C_3 = 0, \ldots$ and therefore $C_1 = \frac{1}{2}, C_2 = \frac{1}{8}, C_3 = \frac{1}{16}, \ldots$.

Consequently, taking $n = 3$, for instance, we get

$$h = \tfrac{1}{2}k + \tfrac{1}{8}k^2 + \tfrac{1}{16}k^3 + \eta k^3$$

or, $x = 1 + \tfrac{1}{2}(y-3) + \tfrac{1}{8}(y-3)^2 + \tfrac{1}{16}(y-3)^3 + \eta(y-3)^3,$

where $\eta \to 0$ when $y - 3 \to 0$.

The result can easily be verified in this simple case, because x is that root of the quadratic equation $y = 4x - x^2$ which equals 1 when $y = 3$. Thus,
$$x = 2 - \sqrt{(4-y)}, \quad \text{or} \quad h = 1 - \sqrt{(1-k)},$$
where $h = x - 1$, $k = y - 3$; and the expansion of h in powers of k can at once be found from the binomial expansion of $(1-k)^{\frac{1}{2}}$.

4.8   Use of Taylor's formula to find limits
Examples

(1)
$$\lim_{x \to 0} \frac{\{\log(1+x)\}^3}{x - \sin x} = 6.$$

Proof. If $x \neq 0$, by (13) and (15),
$$\frac{\{\log(1+x)\}^3}{x - \sin x} = \frac{(x + \alpha x)^3}{x - (x - \frac{1}{6}x^3 + \beta x^3)} = \frac{(1+\alpha)^3}{\frac{1}{6} - \beta},$$
where $\alpha \to 0$ and $\beta \to 0$ when $x \to 0$. The required result follows.

(2)
$$\lim_{x \to 0} (\cos 2x)^{\operatorname{cosec}^2 x} = \frac{1}{e^2}.$$

Proof. Put $y = (\cos 2x)^{\operatorname{cosec}^2 x}$. Then
$$\log y = \operatorname{cosec}^2 x \log(\cos 2x) = \log(\cos 2x)/\sin^2 x.$$
Hence, by (14) with $n = 2$, (13) with $n = 1$, (15) with $n = 1$,
$$\log y = \log(1 - 2x^2 + \alpha x^2)/(x + \beta x)^2$$
$$= (-2x^2 + \eta x^2)/x^2(1+\beta)^2 = (-2+\eta)/(1+\beta)^2,$$
where $\alpha, \beta, \eta \to 0$ when $x \to 0$. It follows that
$$\log y \to -2 \quad \text{when} \quad x \to 0$$
and hence that $y \to e^{-2}$.

(3) If a is independent of x,
$$\lim_{x \to \pm\infty} \left(1 + \frac{a}{x}\right)^x = e^a.$$

Proof. If $a/x > -1$, we can put, using (15) with $n = 1$,
$$\log\left(1 + \frac{a}{x}\right)^x = x \log\left(1 + \frac{a}{x}\right) = x\left(\frac{a}{x} + \frac{\eta}{x}\right) = a + \eta,$$
where $\eta \to 0$ when $x \to \pm\infty$. The required result follows.

4.9   The remainder in Taylor's formula

From §4.3, assuming that $f^{(n-1)}(a)$ exists, we can put

$$f(x) = f(a) + \frac{x-a}{1!}f'(a) + \ldots + \frac{(x-a)^{n-1}}{(n-1)!}f^{(n-1)}(a) + R_n(x)$$

where

$$R_n(x) = \eta(x-a)^{n-1}$$

and $\eta \to 0$ when $x \to a$. This form (Young's form) of the remainder $R_n(x)$ gives no indication of its magnitude for any given values of x and n. Accordingly, other forms of $R_n(x)$ must be sought. These can be found if we make further assumptions.

We shall continue the discussion with special reference to $R_n(x)$ in the Maclaurin formula

$$f(x) = f(0) + \frac{x}{1!}f'(0) + \frac{x^2}{2!}f''(0) + \ldots + \frac{x^{n-1}}{(n-1)!}f^{(n-1)}(0) + R_n(x). \quad (35)$$

This will involve no loss of generality because Taylor's formula can be regarded as a Maclaurin formula (§4.4). It will be possible, in fact, to transform the results obtained for the Maclaurin formula into the corresponding results for the Taylor formula by considering the Maclaurin formula for $f(a+h)$ as a function of h.

4.10

Theorem. (i) Let $f^{(n-1)}(x)$ be continuous in $[-h, h]$;
(ii) let $f^{(n)}(x)$ exist in $(-h, h)$.
Then $R_n(x)$ in (35) can be expressed in the form

$$R_n(x) = \frac{x^n}{p(n-1)!}(1-\theta)^{n-p}f^{(n)}(\theta x), \quad (36)$$

where $-h \leqslant x \leqslant h$, $0 < p$, $0 < \theta < 1$.

This is called *Schlömilch's* form of the remainder. In it, p denotes any positive number, and '$0 < \theta < 1$' is intended briefly to indicate that, when x, n and p satisfy the given conditions, then at least one number θ exists between 0 and 1, excluding 0 and 1 themselves, for which (36) holds good.

Proof. Suppose x fixed, $0 < x \leqslant h$, and consider the following two functions of t in the interval $0 \leqslant t \leqslant x$:

$$F(t) = f(x) - f(t) - (x-t)f'(t) - \ldots - \frac{(x-t)^{n-1}}{(n-1)!}f^{(n-1)}(t),$$

$$G(t) = (x-t)^p; \quad G'(t) \neq 0 \text{ in } 0 < t < x.$$

The function $F(t)$ is constructed so that $F(0) = R_n(x)$.

Both $F(t)$ and $G(t)$ are continuous in the interval $[0, x]$ and differentiable in $(0, x)$, and both vanish at $t = x$, that is, at one end of the interval $[0, x]$. Hence, by §2.18, (26), the ratio of their values at the other end $t = 0$ is equal to the ratio of their gradients (derivatives) for some number ξ between 0 and x. The terms arising from the differentiation of $F(t)$ cancel in pairs, leaving only one term, and hence we find, for at least one such number ξ,

$$\frac{F(0)}{G(0)} = \frac{F'(\xi)}{G'(\xi)} = \frac{-\dfrac{(x-\xi)^{n-1}}{(n-1)!}f^{(n)}(\xi)}{-p(x-\xi)^{p-1}} = \frac{(x-\xi)^{n-p}}{p(n-1)!}f^{(n)}(\xi).$$

It follows, after putting $\xi = \theta x$, where θ is some number between 0 and 1, and since $F(0) = R_n(x)$, $G(0) = x^p$, that

$$\frac{R_n(x)}{x^p} = \frac{x^n(1-\theta)^{n-p}}{p(n-1)!\,x^p}f^{(n)}(\theta x)$$

which at once reduces to (36) for $0 < x \leqslant h$, by cancelling x^p on both sides.

It remains to consider $-h \leqslant x < 0$. In this case, we consider the interval $x \leqslant t \leqslant 0$ and, remembering that p may be fractional, now put $G(t) = (t-x)^p$. We then, by the same kind of reasoning, arrive at the result

$$\frac{R_n(x)}{(-x)^p} = \frac{x^n(1-\theta_1)^{n-p}}{p(n-1)!\,(-x)^p}f^{(n)}(\theta_1 x)$$

where $0 < \theta_1 < 1$. This again gives (36) after cancelling $(-x)^p$. Thus $R_n(x)$ can be put in the form given by (36) for all values of x in the interval $[-h, h]$.

Corollary 1. Putting $p = n$, we get

$$R_n(x) = \frac{x^n}{n!}f^{(n)}(\theta x) \tag{37}$$

which is called *Lagrange's* form of the remainder.

Corollary 2. Putting $p = 1$

$$R_n(x) = \frac{x^n(1-\theta)^{n-1}}{(n-1)!}f^{(n)}(\theta x) \tag{38}$$

which is called *Cauchy's* form.

Note. The conditions imposed upon $f(x)$ are *sufficient* but not *necessary* for (36).

A brief sufficient condition is that $f^{(n+1)}(0)$ should exist. For if so, by §2.11, II, there is an interval, including $x = 0$, in which $f^{(n)}(x)$ exists and therefore in which $f^{(n-1)}(x)$ exists and is continuous.

4.11

To find another form of the remainder, the *integral form*, let there be an interval $[-h, h]$ in which $f^{(n)}(x)$ exists and is continuous, and hence in which the integral

$$\int_0^x (x-t)^{n-1} f^{(n)}(t)\, dt$$

exists. Then, if x belongs to this interval, $R_n(x)$ can be put in the form

$$R_n(x) = \frac{x^n}{(n-1)!} \int_0^1 (1-u)^{n-1} f^{(n)}(ux)\, du. \tag{39}$$

Proof. Let x and t belong to the interval $[-h, h]$, with x fixed and t variable. Consider $F(t)$, the function of t defined by

$$F(t) = f(x) - f(t) - (x-t)f'(t) - \ldots - \frac{(x-t)^{n-1}}{(n-1)!} f^{(n-1)}(t).$$

As in §4.10, by differentiation wo t, we find

$$F'(t) = -\frac{(x-t)^{n-1}}{(n-1)!} f^{(n)}(t),$$

and hence, by integrating both sides from $t = 0$ to $t = x$, since

$$F(x) = 0 \quad \text{and} \quad F(0) = R_n(x),$$

$$-R_n(x) = -\int_0^x \frac{(x-t)^{n-1}}{(n-1)!} f^{(n)}(t)\, dt.$$

After changing the variable of integration from t to u by the substitution $t = ux$, $dt = x\, du$, we obtain (39).

Note. A brief sufficient condition is that $f^{(n+2)}(0)$ should exist. For if so, there is an interval, including $x = 0$, in which $f^{(n+1)}(x)$ exists, and therefore in which $f^{(n)}(x)$ exists and is continuous.

4.12

The corresponding expressions for $R_n(x)$ in Taylor's formula to base a may be found as indicated in §4.9. They are:

$$R_n(x) = \eta(x-a)^{n-1}, \quad \eta \to 0 \quad \text{when} \quad x \to a \quad \text{(Young)}, \tag{40}$$

$$R_n(x) = \frac{(x-a)^n}{n!} f^{(n)}\{a + \theta(x-a)\} \quad \text{(Lagrange)}, \tag{41}$$

$$R_n(x) = \frac{(x-a)^n}{(n-1)!} (1-\theta)^{n-1} f^{(n)}\{a + \theta(x-a)\} \quad \text{(Cauchy)}, \tag{42}$$

$$R_n(x) = \frac{(x-a)^n}{p(n-1)!}(1-\theta)^{n-p}f^{(n)}\{a+\theta(x-a)\} \quad (0 < p) \quad \text{(Schlömilch)},$$

(43)

$$R_n(x) = \frac{(x-a)^n}{(n-1)!}\int_0^1 (1-u)^{n-1}f^{(n)}\{a+u(x-a)\}\,du.$$

(44)

In the first of these expressions, η is a function of x, a and n, which, with a and n fixed, tends to zero when $x \to a$.

In the next three expressions, θ denotes a function of x, a and n, and in the third case, of p; it will not be the same in all cases; its essential property is that it lies between 0 and 1, so that $a+\theta(x-a)$ will lie between a and x.

4.13

With $x-a = h$, $x = a+h$, Taylor's formula will read

$$f(x) = f(a+h) = f(a)+hf'(a)+\ldots+\frac{h^{n-1}}{(n-1)!}f^{(n-1)}(a)+R_n$$

and the remainder R_n will take one of the forms:

$$R_n = \eta h^{n-1}, \quad \eta \to 0 \quad \text{when} \quad h \to 0 \quad \text{(Young)},$$

(45)

$$R_n = \frac{h^n}{n!}f^{(n)}(a+\theta h) \quad \text{(Lagrange)},$$

(46)

$$R_n = \frac{h^n}{(n-1)!}(1-\theta)^{n-1}f^{(n)}(a+\theta h) \quad \text{(Cauchy)},$$

(47)

$$R_n = \frac{h^n}{p(n-1)!}(1-\theta)^{n-p}f^{(n)}(a+\theta h) \quad (0 < p) \quad \text{(Schlömilch)},$$

(48)

$$R_n = \frac{h^n}{(n-1)!}\int_0^1 (1-u)^{n-1}f^{(n)}(a+uh)\,du.$$

(49)

The proper fraction θ will now be a function of h, a, n and in (48) of p.

Example

In Lagrange's form of $R_n(x)$, prove that $\theta = 1/(n+1)+\eta$, where $\eta \to 0$ when $x \to a$; provided that $f^{(n+1)}(a)$ exists and is not zero.

Proof. Take $n = 1$ and $a = 0$ as sufficiently illustrative. Then we are given that $f''(0)$ exists and is not zero. Since $f''(0)$ exists, we can put

$$f(x) = f(0)+xf'(0)+\tfrac{1}{2}x^2f''(0)+\alpha x^2,$$

(i)

where $\alpha \to 0$ when $x \to 0$. Also we can put (§4.10, Note)

$$f(x) = f(0)+xf'(\theta x) \quad (0 < \theta < 1)$$

and hence, by applying §2.11, (18), to $f'(\theta x)$,

$$f(x) = f(0) + x[f'(0) + \theta x\{f''(0) + \beta\}], \tag{ii}$$

where $\beta \to 0$ when $x \to 0$. After subtracting (ii) from (i), dividing by $x^2 f''(0) \neq 0$, and solving for θ, we find

$$\theta = \frac{\tfrac{1}{2} + \alpha/f''(0)}{1 + \beta/f''(0)} \to \tfrac{1}{2} \quad \text{when} \quad x \to 0.$$

4.14　The magnitude of the remainder

Formula (37) was proved on the assumption that $f^{(n)}(x)$ exists in an open interval $(-h, h)$. This implies that, if $|x| < h$, then $f^{(n)}(t)$ is bounded in the interval $0 \leqslant t \leqslant x$ or $x \leqslant t \leqslant 0$, according as x is positive or negative.

If also $f^{(n)}(t)$ is monotonic, then $f^{(n)}(\theta x)$ lies between $f^{(n)}(0)$ and $f^{(n)}(x)$, and we see from (37) that $R_n(x)$ lies between $x^n f^{(n)}(0)/n!$ and $x^n f^{(n)}(x)/n!$

Example

Taking $f(x) = (1+x)^{\frac{1}{2}}$ and $n = 2$, formula (37) gives

$$(1+x)^{\frac{1}{2}} = 1 + \tfrac{1}{2}x + R_2(x), \quad \text{where} \quad R_2(x) = \frac{x^2}{2}\frac{-1}{4(1+\theta x)^{\frac{3}{2}}}$$

and θ is some number between 0 and 1.

If $x > 0$, $1/(1+t)^{\frac{3}{2}}$ decreases from 1 to $1/(1+x)^{\frac{3}{2}}$ as t increases from 0 to x. If $-1 < x < 0$, $1/(1+t)^{\frac{3}{2}}$ decreases from $1/(1+x)^{\frac{3}{2}}$ to 1 as t increases from x to 0.

Consequently, $R_2(x)$ lies between $-\tfrac{1}{8}x^2$ and $(-\tfrac{1}{8}x^2)/(1+x)^{\frac{3}{2}}$ for all values of x greater than -1.

Corollary. (i) If $0 < x$, then $|R_2(x)| < \tfrac{1}{8}x^2$.

(ii) If $-1 < x < 0$, then

$$|R_2(x)| < \frac{x^2}{8}\frac{1}{(1-|x|)^{\frac{3}{2}}}.$$

For instance, if we put $x = \tfrac{1}{10}$ and $x = -\tfrac{1}{10}$ in turn, we find

$$(1+\tfrac{1}{10})^{\frac{1}{2}} = 1 + \tfrac{1}{20} - R_2, \quad \text{where} \quad 0 < R_2 < \tfrac{1}{800};$$

$$(1-\tfrac{1}{10})^{\frac{1}{2}} = 1 - \tfrac{1}{20} - R_2',$$

where $\quad 0 < R_2' < \dfrac{1}{800}\dfrac{1}{(1-\tfrac{1}{10})^{\frac{3}{2}}} = \dfrac{1}{800}\left(\dfrac{10}{9}\right)^{\frac{3}{2}} < \dfrac{1}{800}\cdot\dfrac{4}{3} = \dfrac{1}{600}.$

4.15　Infinite series

An expression of the form

$$u_0 + u_1 + u_2 + u_3 + \ldots \tag{50}$$

is called an *infinite series* if u_r is defined for all integral values of r.

If the series has the property that the sum U_n of the first n terms tends to a limit U when $n \to \infty$, the series is said to *converge*, U is called its *sum to infinity*, or simply its *sum*, and we write

$$U = \lim_{n \to \infty} U_n = u_0 + u_1 + u_2 + u_3 + \ldots. \tag{51}$$

An infinite series of the type

$$c_0 + c_1 x + c_2 x^2 + c_3 x^3 + \ldots, \tag{52}$$

where the c's are independent of x, is called an infinite *power series in* x. If it converges and if $S(x)$ denotes its sum, and $S_n(x)$ the sum of the first n terms, then

$$S(x) = \lim_{n \to \infty} S_n(x) = c_0 + c_1 x + c_2 x^2 + c_3 x^3 + \ldots. \tag{53}$$

If $f(x)$ is a function of x and if the power-series (52) has the property that, for every value of n,

$$f(x) = c_0 + c_1 x + c_2 x^2 + \ldots + c_n x^n + \eta x^n, \tag{54}$$

where $\eta \to 0$ when $x \to 0$, the series is said to be *asymptotic* to $f(x)$ at $x = 0$. This may be expressed by

$$f(x) \sim c_0 + c_1 x + c_2 x^2 + c_3 x^3 + \ldots. \tag{55}$$

Every convergent power-series (53) is asymptotic to its sum $S(x)$ when $x \to 0$ (G. H. Hardy, loc. cit. §201). In other words, the class of asymptotic power-series includes convergent power-series.

Similar remarks apply to an expansion of the form

$$f(x) = c_0 + c_1(x-a) + \ldots + c_n(x-a)^n + \eta(x-a)^n, \tag{56}$$

the series being asymptotic to $f(x)$ at $x = a$ if $\eta \to 0$ when $x \to a$.

The term 'asymptotic series' is most frequently used in connection with expansions of the form

$$f(x) = c_0 + c_1/x + \ldots + c_n/x^n + \eta/x^n, \tag{57}$$

where $\eta \to 0$ when $1/x \to 0$, i.e. when $x \to \infty$.

A power-series may be asymptotic to a function only for positive (negative) values of x; thus (cf. §4.4, Example 5) it may be shown that, if $0 < x$,

$$e^{-1/x} \sim 0 + 0 \cdot x + 0 \cdot x^2 + 0 \cdot x^3 + \ldots. \tag{58}$$

4.16   Maclaurin's series

Definition. Let $f(x)$ be a function of x such that $f^{(n)}(0)$ exists for all values of n. Then the infinite series

$$f(0) + xf'(0) + \frac{x^2}{2!}f''(0) + \frac{x^3}{3!}f'''(0) + \cdots \tag{59}$$

is called *Maclaurin's series derived from $f(x)$.*

Since $f^{(n)}(0)$ may exist as a unique derivative, as a right-hand derivative only, or as a left-hand derivative only (§2.1), Maclaurin's series may have a meaning for both positive and negative values of x, for positive values only, or for negative values only.

Properties of Maclaurin's series

I.   *Theorem.* Maclaurin's series is asymptotic to $f(x)$ at $x = 0$.

Proof. Let $S_n(x)$ denote the sum of the first n terms of Maclaurin's series derived from $f(x)$. Then, by Maclaurin's formula (§4.4), we can put, for all n,
$$f(x) = S_n(x) + R_n(x), \tag{60}$$

where $R_n(x) = \eta x^{n-1}$, since $f^{(n)}(0)$ exists for all n. The theorem follows at once from the definition of asymptotic series (§4.15).

II.   Since $f^{(n)}(0)$ exists for all n, any of the forms of $R_n(x)$ in §4.10, §4.11 will apply. In particular it follows from §4.14 that if $0 < x$ and if $|f^{(n)}(t)|$ is a decreasing function of t in the interval $0 < t < x$, then
$$|R_n(x)| < x^n |f^{(n)}(0)|/n!$$

that is, $|R_n(x)|$ is numerically less than the $(n+1)$th term of the series (whether the series converges or not). For instance, see §4.14, Example, and §4.27.

III.   By supposing n to tend to infinity in (60), we deduce that

$$\lim S_n(x) = f(x) - \lim R_n(x)$$

provided that $\lim R_n(x)$ exists. It follows that, in order that Maclaurin's series should converge it is necessary and sufficient that $R_n(x)$ should tend to a limit when $n \to \infty$, and that *a necessary and sufficient condition that the Maclaurin series derived from $f(x)$ should converge to the sum $f(x)$ is that* $\lim R_n(x) = 0$ when $n \to \infty$.

IV.   We may note that a necessary condition that $R_n(x)$ should tend to a limit when $n \to \infty$ is that $x^n f^{(n)}(0)/n!$ should tend to zero.

For, if $\lim R_n(x)$ exists, then $R_{n+1}(x)$ has the same limit. But, by (60),

$$R_n(x) - R_{n+1}(x) = S_{n+1}(x) - S_n(x) = x^n f^{(n)}(0)/n!$$

and hence, when $n \to \infty$,

$$\lim x^n f^{(n)}(0)/n! = \lim R_n(x) - \lim R_{n+1}(x) = 0.$$

4.17 Taylor's series

If $f^{(n)}(a)$ exists for all n, the infinite series

$$f(a) + (x-a)f'(a) + \frac{(x-a)^2}{2!}f''(a) + \frac{(x-a)^3}{3!}f'''(a) + \dots \qquad (61)$$

or, if $h = x - a$,

$$f(a) + hf'(a) + \frac{h^2}{2!}f''(a) + \frac{h^3}{3!}f'''(a) + \dots \qquad (62)$$

is called *Taylor's series* to base a derived from $f(x)$. It may be regarded as the Maclaurin series for $f(a+h)$ considered as a function of h.

4.18 Analytic functions

Definition: A function $f(x)$ is said to be *analytic* at $x = a$ if $f^{(n)}(a)$ exists for all values of n, and if there is an open interval $(a-h, a+h)$ in which the Taylor series for $f(x)$ to base a converges to the sum $f(x)$ for all x in the interval; in other words, in which, for every fixed x in the interval, $R_n(x) \to 0$ when $n \to \infty$.

In §4.25 it will be shown that $R_n(x) \to 0$ for some of the standard 'elementary functions'. Further discussion of analytic functions belongs properly to the study of functions of a complex variable.

4.19 Infinite sequences

A set of numbers $u_1, u_2, u_3, \dots, u_n, \dots$ is said to be an *infinite sequence* if u_n is defined for every positive integer n. Such a sequence may be denoted briefly by (u_n), so that we may put

$$(u_n) = u_1, u_1, u_3, \dots, u_n, \dots. \qquad (63)$$

It is said:

(i) to *converge* if u_n tends to a limit as $n \to \infty$,

(ii) to *diverge* if $u_n \to +\infty$ or if $u_n \to -\infty$ as $n \to \infty$,

(iii) to *oscillate* if it neither converges nor diverges.

In what follows it will be tacitly assumed, where necessary, that $n \to \infty$.

Theorem. If $a_r > 0$ $(r = 1, 2, \dots, n)$ and if $n > 1$, then

$$(1+a_1)(1+a_2)\dots(1+a_n) > 1 + a_1 + a_2 + \dots + a_n. \qquad (64)$$

Proof.
$$(1+a_1)(1+a_2) = 1+a_1+a_2+a_1a_2 > 1+a_1+a_2 \qquad \text{(i)}$$

since $a_1 a_2 > 0$. By applying this twice

$$(1+a_1)(1+a_2)(1+a_3) > (1+a_1+a_2)(1+a_3) > 1+a_1+a_2+a_3 \qquad \text{(ii)}$$

since $(a_1+a_2)a_3 > 0$. By applying (ii) and then (i),

$$(1+a_1)(1+a_2)(1+a_3)(1+a_4) > (1+a_1+a_2+a_3)(1+a_4)$$
$$> 1+a_1+a_2+a_3+a_4$$

since $(a_1+a_2+a_3)a_4 > 0$. Evidently (64) follows by induction.

Corollary 1. If $a_1+a_2+\ldots+a_n \to \infty$ and if $a_r > 0$ for all r, then $(1+a_1)(1+a_2)\ldots(1+a_n) \to \infty$.

Corollary 2. If $a > 0$, then $(1+a)^n \to +\infty$.

This follows by putting $a_1 = a_2 = \ldots = a_n = a$.

4.20

Theorem. If $a_1+a_2+\ldots+a_n \to \infty$ and if there exists an integer p, independent of n, such that $0 < a_r < 1$ for all $r > p$, then

$$(1-a_1)(1-a_2)\ldots(1-a_n) \to 0.$$

Proof. If $r > p$, then $1-a_r^2 < 1$ and so $1-a_r < 1/(1+a_r)$, since $1+a_r > 0$. Hence, if $n > p$,

$$|(1-a_1)(1-a_2)\ldots(1-a_n)| < \frac{|(1-a_1)(1-a_2)\ldots(1-a_p)|}{(1+a_{p+1})\ldots(1+a_n)}.$$

But, by (64), the denominator

$$(1+a_{p+1})\ldots(1+a_n) > 1+a_{p+1}+\ldots+a_n$$
$$\equiv (1+a_1+a_2+\ldots+a_n)-(a_1+a_2+\ldots+a_p)$$

of which the first term tends to ∞ (given) and the second term is independent of n. The theorem follows.

4.21   Behaviour of the sequence

$$(u_n) = x, x^2, x^3, \ldots, x^n, \ldots.$$

where x is independent of n.

(i) $x > 1$; $x^n \to +\infty$.

Proof. Put $x = 1+a$. Then $x^n = (1+a)^n$ where $a > 0$ and a is independent of n. By §4.19, Corollary 2, it follows that $x^n \to +\infty$.

(ii) $x = 1$; $x^n = 1$, $x^n \to 1$.

(iii) $0 < x < 1$; $x^n \to 0$.

Proof of (iii). If $0 < x < 1$, then $1/x > 1$ and $(1/x)^n \to +\infty$, by (i); that is, $1/x^n \to +\infty$ and hence $x^n \to 0$.

(iv) $x = 0$; $x^n = 0$, $x^n \to 0$.

(v) $-1 < x < 0$; $x^n \to 0$.

Proof of (v). If $-1 < x < 0$, then $0 < -x < 1$ and $(-x)^n \to 0$, by (iii); that is, $(-)^n x^n \to 0$ and hence $x^n \to 0$.

(vi) $x = -1$; $x^n = (-1)^n = -1$ if n is odd, $+1$ if n is even; x^n oscillates 'finitely'.

(vii) $x < -1$; x^n oscillates 'infinitely'.

Proof of (vii). If $x < -1$, then $-x > 1$ and $(-x)^n \to +\infty$, by (i); that is, $(-)^n x^n \to +\infty$ and hence $x^n \to -\infty$ if n is odd, $+\infty$ if n is even.

Corollary. If c is independent of n, then

(i) $cx^n \to 0$ if $-1 < x < 1$,

(ii) $|cx^n| = |c|$ if $|x| = 1$,

(iii) $|cx^n| \to +\infty$ if $|x| > 1$ $(c \neq 0)$.

4.22

Theorem. Let a sequence (u_n) be given and let $\lim (u_{n+1}/u_n)$ exist. Let $|\lim (u_{n+1}/u_n)| = l$.

(i)   If $l < 1$, then $u_n \to 0$.

(ii)   If $l > 1$, then $|u_n| \to \infty$.

Proof. Since $\lim (u_{n+1}/u_n)$ exists, there must be a positive integer N such that $u_n \neq 0$ for all $n \geqslant N$. Then, if $n > N$, we can put identically

$$u_n = \frac{u_n}{u_{n-1}} \frac{u_{n-1}}{u_{n-2}} \dots \frac{u_{N+1}}{u_N} u_N. \tag{65}$$

(i)   If $l < 1$, put $r = \frac{1}{2}(1+l)$. Then $0 \leqslant l < r < 1$. By definition of a limit (cf. §1.6), since $l < r$, the number N can be chosen so that also $|u_{n+1}/u_n| < r$ for all $n \geqslant N$. Hence, by (65), if $n > N$,

$$|u_n| < r^{n-N} |u_N| = cr^n,$$

where $c = |u_N|/r^N$, independent of n, and $0 < r < 1$. It follows by §4.21, Corollary, that $u_n \to 0$.

(ii)   If $l > 1$, again put $r = \frac{1}{2}(1+l)$. Then $1 < r < l$. By definition of a limit, since $r < l$ the number N can be chosen so that

$$|u_{n+1}/u_n| > r \quad \text{for all} \quad n > N.$$

Then by (65), if $n > N$,

$$|u_n| > r^{n-N} |u_N| = cr^n,$$

where $c = |u_N|/r^N$, independent of n, and $r > 1$. It follows by §4.21, Corollary, that $|u_n| \to \infty$.

Examples

(1) Prove that the sequence

$$(u_n) = x/1!, \; x^2/2!, \; x^3/3!, \ldots, x^n/n!, \ldots$$

tends to zero if x has any value independent of n.

Proof.
$$\frac{u_{n+1}}{u_n} = \frac{x^{n+1}}{(n+1)!} \Big/ \frac{x^n}{n!} = \frac{x}{n+1}$$

from which it follows that $u_{n+1}/u_n \to 0$ if x is independent of n, and hence that $u_n \to 0$, by part (i) of the theorem.

(2) In the sequence

$$(u_n) = x/1, \; x^2/2, \; x^3/3, \ldots, x^n/n, \ldots$$

prove that $u_n \to 0$ if $|x| \leqslant 1$, and that $|u_n| \to \infty$ if $|x| > 1$.

Proof. If $|x| \leqslant 1$, $|u_n| = |x^n|/n \leqslant 1/n \to 0$.

If $|x| > 1$, since
$$\frac{u_{n+1}}{u_n} = \frac{x^{n+1}}{n+1} \Big/ \frac{x^n}{n} = \frac{nx}{n+1} \to x,$$

it follows that $|u_n| \to \infty$, by part (ii) of the theorem, provided that x is independent of n.

(3) Prove that the sequence

$$(u_n) = 1^r x, \; 2^r x^2, \; 3^r x^3, \ldots, n^r x^n, \ldots$$

tends to zero if $|x| < 1$, provided that x and r are independent of n.

Proof.
$$\frac{u_{n+1}}{u_n} = \frac{(n+1)^r x^{n+1}}{n^r x^n} = \left(1 + \frac{1}{n}\right)^r x$$

from which it follows, if x and r are independent of n, that $u_{n+1}/u_n \to x$, and hence that $u_n \to 0$ if $|x| < 1$, by part (i) of the theorem.

4.23

Theorem. If p is positive and independent of n,

$$\frac{1}{p} + \frac{1}{p+1} + \frac{1}{p+2} + \ldots + \frac{1}{p+n} \to +\infty. \tag{i}$$

Proof. By the mean value theorem, there is a number θ between 0 and 1 such that, if a is positive,

$$\log(a+1) - \log a = 1/(a+\theta) < 1/a.$$

By putting $a = p, \; p+1, \; p+2, \ldots, p+n$ in turn, and adding,

$$\log(p+n+1) - \log p < \frac{1}{p} + \frac{1}{p+1} + \ldots + \frac{1}{p+n}$$

from which the theorem follows, since $\log (p+n+1) \to \infty$ when $n \to \infty$.

Corollary. The theorem remains true when p is any negative number except a negative integer.

For if $p < 0$ and $n > -p$, there will be a finite number of negative terms at the beginning of series (i), followed by positive terms. The sum of the negative terms will be a finite negative number; the sum of the positive terms will tend to ∞ when $n \to \infty$.

4.24   Behaviour of the infinite sequence

$$(u_n) = \binom{m}{1} x, \ \binom{m}{2} x^2, \ \binom{m}{3} x^3, \ ..., \ \binom{m}{n} x^n, \ ...$$

where x and m are independent of n.

If m is a positive integer, the sequence ends at $n = m$. In the following theorem it is assumed that m is not a positive integer or zero.

Theorem. If m is not a positive integer or zero:

(i) $u_n \to 0$ if $|x| < 1$, $-\infty < m < \infty$; and if $|x| = 1$, $-1 < m < \infty$.

(ii) $|u_n| = 1$ if $|x| = 1$, $m = -1$.

(iii) $|u_n| \to \infty$ if $|x| > 1$, $-\infty < m < \infty$; and if $|x| = 1$, $m < -1$.

Proof.

$$u_n = \binom{m}{n} x^n = \frac{m(m-1)(m-2) \ldots (m-n+1)}{1 \cdot 2 \cdot 3 \ldots n} x^n,$$

$$\frac{u_n}{u_{n-1}} = \frac{m-n+1}{n} x = \left(-1 + \frac{m+1}{n}\right) x \to -x.$$

It follows from §4.22 that

$$u_n \to 0 \quad \text{if} \quad |x| < 1 \quad (-\infty < m < \infty),$$

$$|u_n| \to \infty \quad \text{if} \quad |x| > 1 \quad (-\infty < m < \infty).$$

It remains to consider $|x| = 1$. In this case,

$$|u_n| = \left| \left(1 - \frac{m+1}{1}\right) \left(1 - \frac{m+1}{2}\right) \ldots \left(1 - \frac{m+1}{n}\right) \right|.$$

If $m = -1$, evidently $|u_n| = 1$.

If $m < -1$, then $-(m+1) > 0$ and $|u_n|$ is of the form

$$(1 + a_1)(1 + a_2) \ldots (1 + a_n),$$

where $a_r > 0$ and $a_1 + a_2 + \ldots + a_n \to \infty$, by §4.23 with $p = 1$. It follows from §4.19, Corollary 1, that $|u_n| \to \infty$.

If $m > -1$, then $m + 1 > 0$ and $|u_n|$ is of the form

$$|(1 - a_1)(1 - a_2)\dots(1 - a_n)|,$$

where $a_r > 0$ and $a_1 + a_2 + \dots + a_n \to \infty$. Moreover, if p is a positive integer greater than $m + 1$, then $0 < a_r < 1$ for all $r > p$. It follows from §4.20 that $u_n \to 0$.

4.25 Standard Maclaurin power-series

I. *The exponential series.* For all values of x

$$e^x = 1 + \frac{x}{1!} + \frac{x^2}{2!} + \frac{x^3}{3!} + \dots + \frac{x^n}{n!} + \dots. \tag{66}$$

Proof. For all n, $f(x) = e^x$, $f^{(n)}(x) = e^x$, $f^{(n)}(0) = 1$. We can therefore put, using Lagrange's form of the remainder,

$$e^x = 1 + \frac{x}{1!} + \frac{x^2}{2!} + \dots + \frac{x^{n-1}}{(n-1)!} + R_n(x),$$

where $R_n(x) = x^n e^{\theta x}/n!$ $(0 < \theta < 1)$. Consequently,

$$0 < R_n(x) < x^n e^x/n! \quad \text{if} \quad x > 0; \quad |R_n(x)| < |x^n|/n! \quad \text{if} \quad x < 0.$$

It follows (§4.22, Example 1) that $R_n(x) \to 0$ when $n \to \infty$, whatever the value of x, and hence that the series converges to e^x, by §4.16, III.

Further, we note that, when $x < 0$, the sum of the first n terms differs from e^x by less than the $(n+1)$th term numerically.

II. *The sine and cosine series.* For all values of a and h

$$\sin(a+h) = \sin a + h \cos a + \dots + \frac{h^n}{n!} \sin(a + \tfrac{1}{2}n\pi) + \dots. \tag{67}$$

Proof. Put $f(x) = \sin x$, $f^{(n)}(a) = \sin(a + \tfrac{1}{2}n\pi)$, by §3.2 (10); then

$$\sin(a+h) = \sin a + h \cos a - \dots + \frac{h^{n-1}}{(n-1)!} \sin\{a + \tfrac{1}{2}(n-1)\pi\} + R_n$$

where, with Lagrange's form of remainder,

$$R_n = (h^n/n!) \sin(a + \theta h + \tfrac{1}{2}n\pi) \quad (0 < \theta < 1).$$

Since $|\sin x| \leqslant 1$, it follows that $|R_n| \leqslant |h^n|/n!$ and hence that $R_n \to 0$ for all values of h when $n \to \infty$, by §4.22, Example 1. The series therefore converges to $\sin(a+h)$, by §4.16, III.

Corollary. For all values of x, by putting $a = 0$, $h = x$,

$$\sin x = \frac{x}{1!} - \frac{x^3}{3!} + \frac{x^5}{5!} - \frac{x^7}{7!} + \dots \tag{68}$$

and by putting $a = \frac{1}{2}\pi$, $h = x$,

$$\cos x = 1 - \frac{x^2}{2!} + \frac{x^4}{4!} - \frac{x^6}{6!} + \dots \qquad (69)$$

In these two series the remainder after any number of terms is numerically less than the next term.

III. *The logarithmic series.* If $-1 < x \leqslant 1$,

$$\log(1+x) = \frac{x}{1} - \frac{x^2}{2} + \frac{x^3}{3} - \dots (-)^{n+1}\frac{x^n}{n} + \dots \qquad (70)$$

Proof. Put

$$f(x) = \log(1+x), \quad f^{(n)}(x) = (-)^{n-1}(n-1)!/(1+x)^n,$$

by §3.2 (3); then $f^{(n)}(0) = (-)^{n-1}(n-1)!$ and

$$\log(1+x) = \frac{x}{1} - \frac{x^2}{2} + \frac{x^3}{3} - \dots (-)^n \frac{x^{n-1}}{n-1} + R_n(x).$$

If $0 \leqslant x \leqslant 1$, by Lagrange's form (37) of $R_n(x)$

$$|R_n(x)| = \frac{x^n}{n} \frac{1}{(1+\theta x)^n} \leqslant \frac{x^n}{n} \leqslant \frac{1}{n} \to 0 \quad \text{when} \quad n \to \infty.$$

If $-1 < x < 0$ Lagrange's form is not helpful. Putting $x = -x_1$ $(0 < x_1 < 1)$ we then have, by Cauchy's form (38),

$$|R_n(x)| = \frac{x_1^n}{1 - \theta x_1}\left(\frac{1-\theta}{1-\theta x_1}\right)^{n-1} \quad (0 < \theta < 1)$$

and hence, since $1 - \theta < 1 - \theta x_1$,

$$|R_n(x)| < x_1^n/(1 - \theta x_1) < x_1^n/(1 - x_1) \to 0 \quad \text{when} \quad n \to \infty,$$

by §4.21, (iii). Hence follows (70), by §4.16, III.

Note 1. By the more precise integral form (39), we have

$$R_n(x) = (-)^{n-1}x^n \int_0^1 \frac{(1-u)^{n-1}}{(1+ux)^n}\,du,$$

and hence, putting

$$v = (1-u)/(1+ux), \quad dv = -(1+x)\,du/(1+ux)^2,$$

$$R_n(x) = (-)^{n-1}x^n \int_0^1 \frac{v^{n-1}}{1+vx}\,dv.$$

It follows that, if $x > 0$,

$$|R_n(x)| < x^n \int_0^1 v^{n-1} \, dv = \frac{x^n}{n}$$

as above; while if $x = -x_1 \, (0 < x_1 < 1)$,

$$|R_n(x)| < x_1^n \int_0^1 \frac{v^{n-1}}{1-x_1} \, dv = \frac{x_1^n}{n(1-x_1)}$$

an improvement upon the value found from Cauchy's form.

Note 2. When $x = -1$, the series becomes $-1 - \frac{1}{2} - \frac{1}{3} - \dots$, which diverges to $-\infty$, by §4.23; while $\log(1+x)$ is not defined at $x = -1$, though $\log(1+x) \to -\infty$ when $x \to -1+$.

When $|x| > 1$, the nth term of the series tends to infinity in numerical value, by §4.22, Example 2.

IV. *The binomial series.* The expansion

$$(1+x)^m = 1 + \binom{m}{1} x + \binom{m}{2} x^2 + \binom{m}{3} x^3 + \dots + \binom{m}{n} x^n + \dots \quad (71)$$

consists of $m+1$ terms if m is a positive integer or zero. For other values of m the series on the R.H.S. is an infinite series, and the expansion is valid, i.e. the sum of the series is $(1+x)^m$, when

$$-1 < x < 1, \quad \text{all } m; \quad x = 1, \ -1 < m < \infty; \quad x = -1, \ 0 < m < \infty.$$

Proof. All values of x and m will be considered, except $m = $ a positive integer or zero. With these exceptions it is convenient to separate the following cases:

(i) $|x| > 1$, all values of m;   (ii) $x = 1, m < -1$;

(iii) $x = 1, m = -1$;   (iv) $x = -1, m < 0$;

(v) $0 \leqslant x < 1$, all values of m;   (vi) $x = 1, -1 < m$;

(vii) $-1 < x < 0$, all values of m; (viii) $x = -1, 0 < m$.

Let u_n be the $(n+1)$th term of the series.

In cases (i), (ii), (iii), $R_n(x)$ cannot tend to zero because u_n does not tend to zero (§4.16, IV). For, by §4.24, in cases (i) and (ii), $|u_n| \to \infty$; in case (iii), $|u_n| = 1$.

In case (iv), the L.H.S. of (71) has no meaning, so the question of the validity of the expansion does not arise.

For the remaining cases we note that if

$$f(x) = (1+x)^m, \quad \text{then} \quad f^{(n)}(x) = m_n (1+x)^{m-n},$$

where $m_n = m(m-1)(m-2)\dots(m-n+1).$

Cases (v), (vi). By Lagrange's form (37) of the remainder, there is a number θ between 0 and 1 such that

$$R_n(x) = \frac{x^n}{n!} \frac{m_n}{(1+\theta x)^{n-m}} = \frac{u_n}{(1+\theta x)^{n-m}}.$$

Now, by §4.24, (i), in these cases $|u_n| \to 0$; and since

$$1/(1+\theta x)^{n-m} \leqslant 1 \quad \text{if} \quad n > m,$$

it follows that $R_n(x) \to 0$.

Case (vii). Since $-1 < x < 0$, therefore $1/(1+\theta x) > 1$ and so Lagrange's form of $R_n(x)$ is not helpful. Cauchy's form (38) gives

$$R_n(x) = \frac{x^n (1-\theta)^{n-1}}{(n-1)!} \frac{m_n}{(1+\theta x)^{n-m}} = mxMN,$$

where $\quad M = \binom{m-1}{n-1} x^{n-1}, \quad N = (1+\theta x)^{m-1} \left(\frac{1-\theta}{1+\theta x}\right)^{n-1}.$

Now, since $-1 < x < 0$, therefore $0 < 1-\theta < 1+\theta x$ and so

$$0 < (1-\theta)/(1+\theta x) < 1.$$

Also, if $m > 1$, then $(1+\theta x)^{m-1} < 1$; if $m < 1$, then

$$(1+\theta x)^{m-1} < (1+x)^{m-1}.$$

It follows that N remains bounded when $n \to \infty$, whatever the value of m.

Further, by §4.24 (i), $M \to 0$ when $n \to \infty$, whatever the value of m. Since $M \to 0$ and N is bounded, therefore $R_n(x) \to 0$.

Case (viii). Since $0 < m$, we can put $p = m$ in Schlömilch's form (36), obtaining

$$R_n(x) = \frac{x^n (1-\theta)^{n-m}}{m(n-1)!} m_n (1+\theta x)^{m-n}$$

and with $x = -1$ this gives

$$R_n(-1) = (-)^n \binom{m-1}{n-1}$$

which, by §4.24, (i), tends to zero when $n \to \infty$, provided that

$$-1 < m-1 < \infty,$$

which is true since $0 < m < \infty$. This completes the proof.

In Fig. 6 the coordinates are (x, m). In order that the binomial expansion may be valid, the point (x, m) must belong to the unshaded region, including the faint, but not the heavy, parts of the boundary lines.

4.26

If we differentiate both sides of (68), differentiating the series term by term, we get

$$\cos x = 1 - \frac{x^2}{2!} + \frac{x^4}{4!} - \frac{x^6}{6!} + \dots.$$

This, by (69), is correct for all values of x.

Again, if we integrate both sides of (68) from $x = 0$ to $x = \theta$, we obtain

$$1 - \cos\theta = \frac{\theta^2}{2!} - \frac{\theta^4}{4!} + \frac{\theta^6}{6!} - \dots$$

which is correct for all values of θ.

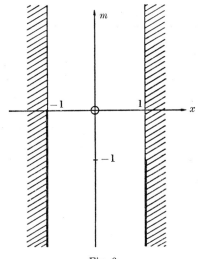

Fig. 6.

We say therefore that the Maclaurin series for $\sin x$ can be differentiated or integrated *term by term*, the sum of the new series thus obtained being equal to the derivative or integral of $\sin x$, as the case may be, allowing for the correct value of the constant of integration.

It may be verified that, in the same sense, the other series in §4.25 can be differentiated or integrated, for all values of x in the case of e^x or $\cos x$, for all values of x within the range of convergence $(-1 < x < 1)$ in the case of $\log(1 + x)$ or $(1 + x)^m$.

These examples illustrate a property of all analytic functions (§4.18), that their Taylor series may be differentiated or integrated term by term within their ranges of convergence.

4.27   Asymptotic but not convergent series

An example of a series which is asymptotic to the function from which it is derived, but which is not convergent, will be given. It will be assumed in this example that $f^{(n)}(x)$ can be found by differentiating n times under the integral sign (§10.16).

Example

If $x > 0$,
$$f(x) = \int_0^\infty \frac{e^{-v}dv}{1+xv} \sim 1 - x + 2! \, x^2 - 3! \, x^3 + \dots. \tag{72}$$

Proof. By differentiating n times wo x under the integral sign and then putting $x = 0$,
$$f^{(n)}(x) = \int_0^\infty \frac{(-)^n \, n! \, v^n \, e^{-v} \, dv}{(1+xv)^{n+1}}, \quad f^{(n)}(0) = (-)^n \, n! \int_0^\infty v^n \, e^{-v} \, dv.$$

Hence $f^{(n)}(0) = (-)^n \, (n!)^2$ from which follows (72) by §4.16, I.

Since $|f^{(n)}(x)|$ is a decreasing function of x, we see from §4.16, II that $|R_n(x)|$ is numerically less than the $(n+1)$th term of the series, i.e. $|R_n(x)| < n! \, x^n$.

If we put $|u_{n+1}| = n! \, x^n$, then $|u_{n+1}/u_n| = nx \to \infty$ if $x > 0$, which shows that $|u_n| \to \infty$ whatever the value of x. Consequently, the series does not converge for any $x > 0$.

Nevertheless, the series can be used to approximate to $f(x)$ when x is small enough, to any degree of approximation. For if n is fixed $|R_n(x)|$ is as small as we please when x is small enough. Thus, if $x = 1/4!$ and $n = 4$, then approximately $f(x) = 1 - x + 2! \, x^2 - 3! \, x^3$, with an error numerically less than $4! \, (1/4!)^4 = (1/4!)^3 = 0 \cdot 00007 \dots$.

4.28   Extension of Rolle's theorem

Let $[a, b, c]$ denote the closed interval that extends from the least to the greatest of three distinct numbers a, b, c.

Theorem. Let $\phi(x)$ vanish and be continuous at a, b, c and let $\phi''(x)$ exist in the open interval (a, b, c). Then there exists at least one number ξ, belonging to this interval, such that $\phi''(\xi) = 0$.

Proof. Without loss of generality, we may suppose $a < b < c$.

Since $\phi''(x)$ exists in (a, b) it necessarily follows that $\phi'(x)$ exists in (a, b). Since also $\phi(x)$ is continuous at a and b and $\phi(a) = \phi(b) = 0$, it follows by Rolle's theorem that at least one number ξ_1 exists $(a < \xi_1 < b)$ such that $\phi'(\xi_1) = 0$. Similarly, at least one number ξ_2 exists $(b < \xi_2 < c)$ such that $\phi'(\xi_2) = 0$.

Again, since $\phi''(x)$ exists in $[\xi_1, \xi_2]$ it follows that $\phi'(x)$ is continuous in $[\xi_1, \xi_2]$. And since $\phi'(\xi_1) = \phi'(\xi_2) = 0$, a second application of Rolle's theorem shows that at least one number ξ exists ($\xi_1 < \xi < \xi_2$) such that $\phi''(\xi) = 0$.

Generalization. Let $[a_1, a_2, ..., a_n]$ denote the interval extending from the least to the greatest of n distinct numbers $a_1, a_2, ..., a_n$.

Let $\phi(x)$ vanish and be continuous at $a_1, a_2, ..., a_n$. Let $\phi^{(n-1)}(x)$ exist in the open interval $(a_1, a_2, ..., a_n)$. Then there exists at least one number ξ, belonging to this interval, such that $\phi^{(n-1)}(\xi) = 0$.

The proof is left to the reader.

4.29   Extension of Cauchy's mean value theorem

Theorem. Let $F(x)$, $G(x)$ vanish at a and at b and be continuous at a, b and c. Also let $F''(x)$, $G''(x)$ exist in the open interval (a, b, c).

Then, if (i) $G(c) \neq 0$, (ii) $F''(x)$, $G''(x)$ do not vanish together, there exists at least one number ξ, belonging to the open interval, such that

$$F(c)/G(c) = F''(\xi)/G''(\xi). \tag{73}$$

Proof. Put $\phi(x) = F(c)\,G(x) - F(x)\,G(c)$. Then $\phi(x)$ satisfies all the conditions of §4.28. Hence a number ξ exists, belonging to the open interval (a, b, c), such that

$$\text{(iii)} \quad \phi''(\xi) = F(c)\,G''(\xi) - F''(\xi)\,G(c) = 0.$$

Now from (i), (ii), (iii) it follows that $G''(\xi) \neq 0$; hence follows (73) on dividing (iii) by $G''(\xi)\,G(c)$.

Note. A sufficient condition is that $G''(x)$ should not vanish in the open interval (a, b, c).

Generalization. Let $F(x)$, $G(x)$ vanish at $a_1, a_2, ..., a_n$ and be continuous at $a_1, a_2, ..., a_n, b$. Also let $F^{(n)}(x)$, $G^{(n)}(x)$ exist in the open interval $(a_1, a_2, ..., a_n, b)$.

Then, if (i) $G(b) \neq 0$, (ii) $F^{(n)}(x)$, $G^{(n)}(x)$ do not vanish together, there exists at least one number ξ, belonging to the open interval, such that

$$F(b)/G(b) = F^{(n)}(\xi)/G^{(n)}(\xi). \tag{74}$$

The proof is left to the reader.

Note. A sufficient condition is that $G^{(n)}(x)$ should not vanish in the open interval $(a_1, a_2, ..., a_n, b)$.

4.30   Rule of proportional parts, or linear interpolation formula

Theorem. Let $f(t)$ be continuous at a, b, x and let $f''(t)$ exist in the open interval (a, b, x). Then there exists a number ξ, belonging to this interval, such that

$$f(x) = \frac{(b-x)f(a) + (x-a)f(b)}{b-a} + \tfrac{1}{2}(x-a)(x-b)f''(\xi). \qquad (75)$$

Proof. Note that the first term on the R.H.S. is the linear function of x which has the value $f(a)$ when $x = a$ and the value $f(b)$ when $x = b$.
Put

$$F(t) = f(t) - \frac{(b-t)f(a) + (t-a)f(b)}{b-a},$$

$$G(t) = (t-a)(t-b).$$

Then $F(t)$, $G(t)$ vanish at $t = a$, $t = b$, and are continuous in the closed interval $[a, b, x]$; also $F''(t)$, $G''(t)$ exist in the open interval. Further, $G(x) \neq 0$ if x is distinct from a and b, and $F''(t)$, $G''(t)$ cannot vanish together, since $G''(t) = 2 \neq 0$. Hence, by (73), there exists at least one number ξ, belonging to the open interval, such that

$$\frac{F(x)}{G(x)} = \frac{f(x) - \dfrac{(b-x)f(a) + (x-a)f(b)}{b-a}}{(x-a)(x-b)} = \frac{f''(\xi)}{2}.$$

Hence the theorem.

Corollary. If $f(x)$ is continuous at a and $a+h$ and if $f''(x)$ exists in $(a < x < a+h)$, then

$$f(a+\lambda h) = f(a) + \lambda \Delta f(a) - \tfrac{1}{2}\lambda(1-\lambda)h^2 f''(\xi), \qquad (76)$$

where $0 \leqslant \lambda \leqslant 1$, $\Delta f(a) = f(a+h) - f(a)$, $a < \xi < a+h$.

Note that $f(a) + \lambda \Delta f(a)$ is the linear function of λ which has the value $f(a)$ when $\lambda = 0$ and the value $f(a+h)$ when $\lambda = 1$.

Formula (76) follows from (75) by putting $b-a = h$, $x-a = \lambda h$, $b-x = (1-\lambda)h$.

Alternatively, we could put

$$F(t) = f(a+th) - f(a) - t\Delta f(a), \quad G(t) = t(1-t),$$

$$\frac{F(\lambda)}{G(\lambda)} = \frac{f(a+\lambda h) - f(a) - \lambda \Delta f(a)}{\lambda(1-\lambda)} = \frac{h^2 f''(\xi)}{-2},$$

by (73), and hence obtain (76) directly for $0 < \lambda < 1$; for $\lambda = 0$ or $\lambda = 1$, (76) is evidently an identity.

Note. From (76) we can put

$$f(a+\lambda h) = f(a)+\lambda\Delta f(a) - R, \tag{77}$$

where R lies between $\frac{1}{2}\lambda(1-\lambda)h^2f''(a)$ and $\frac{1}{2}\lambda(1-\lambda)h^2f''(a+h)$, provided that $f''(x)$ is monotonic.

4.31

Example

If $f''(x)$ exists in $(a-h \leqslant x \leqslant a+h)$, prove that a number ξ exists $(a-h < \xi < a+h)$ such that

$$f(a+h)+f(a-h)-2f(a) = h^2f''(\xi). \tag{78}$$

Proof. Since $f''(x)$ exists throughout the given closed interval, $f'(x)$ and $f(x)$ are both continuous in the closed interval. Two methods of proof will be given:

First method. Put

$$F(x) = f(a+x)+f(a-x)-2f(a), \quad G(x) = x^2.$$

Then $F(x), G(x)$ both vanish at $x = 0$ and so, by §2.18 (26), a number ζ exists $(0 < \zeta < h)$ such that

$$\frac{F(h)}{G(h)} = \frac{F'(\zeta)}{G'(\zeta)} = \frac{f'(a+\zeta)-f'(a-\zeta)}{2\zeta}.$$

But, by §2.14 (21), we can put $f'(a+\zeta)-f'(a-\zeta) = 2\zeta f''(\xi)$, where $a-\zeta < \xi < a+\zeta$; hence

$$F(h)/h^2 = 2\zeta f''(\xi)/2\zeta = f''(\xi)$$

from which (78) follows.

Second method. Let Ax^2+Bx+C be a quadratic function of x which is equal to $f(x)$ at $x = a-h$, $x = a$, $x = a+h$; and put

$$\phi(x) = f(x)-(Ax^2+Bx+C).$$

Then $\phi(x)$ vanishes at $x = a-h$, $x = a$, $x = a+h$. Hence by §4.28, a number ξ exists $(a-h < \xi < a+h)$ such that

$$\phi''(\xi) = f''(\xi)-2A = 0. \tag{i}$$

Now A, B, C are determined by the equations

$$\begin{aligned}
f(a-h) &= A(a-h)^2+B(a-h)+C, \\
f(a) &= Aa^2 \qquad +Ba \qquad +C, \\
f(a+h) &= A(a+h)^2+B(a+h)+C,
\end{aligned}$$

from which follows, by (i),

$$f(a+h)+f(a-h)-2f(a) = 2Ah^2 = h^2f''(\xi).$$

4.32 The notation of finite differences

This will be mentioned here for the sake of a few of Examples 4F that follow.

If $f(x)$ is a function of x, it is usual to put

$$\Delta f(a) = f(a+h) - f(a),$$

$$\Delta^2 f(a) = \Delta\Delta f(a) = \Delta\{f(a+h) - f(a)\}$$

$$= \Delta f(a+h) - \Delta f(a)$$

$$= f(a+2h) - 2f(a+h) + f(a).$$

Similarly,

$$\Delta^3 f(a) = \Delta^2 f(a+h) - \Delta^2 f(a)$$

$$= f(a+3h) - 3f(a+2h) + 3f(a+h) - f(a),$$

....

$$\Delta^n f(a) = \Delta^{n-1} f(a+h) - \Delta^{n-1} f(a)$$

$$= f(a+nh) - \binom{n}{1} f\{a+(n-1)h\} + \ldots (-)^n f(a). \tag{79}$$

When $f(a), f(a+h), \ldots, f(a+nh)$ are given, e.g. in a table of values of the function $f(x)$, the differences $\Delta f(a), \Delta^2 f(a), \ldots, \Delta^n f(a)$ can be readily expressed by the kind of scheme indicated below for $n = 4$:

$f(a)$

         $\Delta f(a)$

$f(a+h)$ $\Delta^2 f(a)$

         $\Delta f(a+h)$ $\Delta^3 f(a)$

$f(a+2h)$ $\Delta^2 f(a+h)$ $\Delta^4 f(a)$

         $\Delta f(a+2h)$ $\Delta^3 f(a+h)$

$f(a+3h)$ $\Delta^2 f(a+2h)$

         $\Delta f(a+3h)$

$f(a+4h)$

The entries in each column after the first are calculated by simple subtractions from entries in the preceding column. The differences $\Delta f(a), \Delta^2 f(a), \Delta^3 f(a), \ldots$ are called first, second, third, ... *forward differences* of $f(x)$.

4.33 Newton's method for approximating to a root of an equation

Let $x = r$ be a root of the equation $f(x) = 0$, so that $f(r) = 0$. Let a be an approximation to r. Then, provided that $f'(x)$ is continuous in the interval $[r, a]$ and that $f''(x)$ exists in (r, a), we can put, by Taylor's formula,

$$0 = f(r) = f(a) + (r-a)f'(a) + \tfrac{1}{2}(r-a)^2 f''(\xi),$$

where ξ is some number between r and a. Now, if $f'(a) \neq 0$, the last term is as small as we please compared with the next term before it when $r - a$ is small enough. Neglecting the last term, therefore, we infer that r will then be given approximately by $f(a) + (r-a)f'(a) = 0$, or $r = a - f(a)/f'(a)$.

Consequently, if a is near enough to r, and if we put $b = a - f(a)/f'(a)$, then $r \doteqdot b$ will be a better approximation than $r \doteqdot a$. Sufficient conditions for this to be so are given in the following theorem:

Theorem. Let there be an interval $r \leqslant x \leqslant r+h$ in which $f(r) = 0$, $f'(r) > 0$, and $f''(x) > 0$ for all x. Let $x = a$ belong to the interval $r < x < r+h$ and put $b = a - f(a)/f'(a)$.

Then b will be greater than r and less than a, that is $r < b < a$, so that b will be a closer approximation to r than a is.

Proof. As above, we can put

$$0 = f(r) = f(a) + (r-a)f'(a) + \tfrac{1}{2}(r-a)^2 f''(\xi), \tag{80}$$

where ξ lies between r and a. Also we have, by definition of b,

$$0 = f(a) + (b-a)f'(a). \tag{81}$$

By subtracting (81) from (80)

$$0 = (r-b)f'(a) + \tfrac{1}{2}(r-a)^2 f''(\xi). \tag{82}$$

Now, from the data of the theorem we deduce at once that $f'(a) > 0$ and $f(a) > 0$. It follows from (81) that $a > b$ and from (82) that $b > r$, and therefore that $r < b < a$.

Note 1. In practice, having found a first approximation a, we calculate b from the formula $b = a - f(a)/f'(a)$. We can then improve upon b by putting $c = b - f(b)/f'(b)$, and so on.

Note 2. In the proof, we assumed that $r < a < r+h$, so that $a > r$ and $f(a) > 0$. If the first approximation a were less than r, so that $a < r$ and $f(a) < 0$, then from (82) would follow $b > r$ and b would not necessarily be closer to r than a was. But since $b > r$ the next approximation c will be closer to r than b is, the next closer than c is, and so on.

Note 3. We also assumed $f'(r)$ and $f''(x)$ to be positive. If they are both negative or of opposite signs, the appropriate modifications can easily be made. The truth of the theorem and its modifications can be inferred from graphical considerations, provided that $f'(x)$ and $f''(x)$ are one-signed.

[See Examples 4 G.]

Examples 4 A

(1) $f(x) = \tan^{-1}x$. Show that $f''(x) < 0$ for all positive x.

(2) $f(x) = x\tan^{-1}x$. Show that $f''(x) > 0$ for all x.

(3) $f(x) = x\tanh^{-1}x$. Show that $f''(x) > 0$, $-1 < x < 1$.

(4) $f(x) = x + 2 + (x-2)\,e^x$.
Prove that $f(x) > 0$ when $x > 0$.

(5) $f(x) = 1 - \tfrac{1}{2}x^2 + \tfrac{1}{3}x^3 - (1+x)\,e^{-x}$.
Prove that $f(x), f'(x)$ are both positive when $x > 0$.

(6) $f(x) = (x+2)\log(1+x) - 2x$.
Prove that $f(x), f'(x), f''(x)$ are all positive when $x > 0$.

(7) If $0 < a < x$, prove that
$$x\log x - a\log a - (x-a)\left(1 + \log\frac{a+x}{2}\right) < 0.$$

Examples 4 B

(1) Obtain the following formulae in which η, α, β have the meanings assigned to them in § 4.4:

    (i) $\cosh x = 1 + x^2/2! + x^4/4! + \ldots + x^{2n}/(2n)! + \alpha x^{2n+1}$;

    (ii) $\sinh x = x/1! + x^3/3! + \ldots + x^{2n-1}/(2n-1)! + \beta x^{2n}$;

    (iii) $e^{ax}\cos bx = 1 + ax + (a^2 - b^2)\,x^2/2! + \ldots + c^n\cos n\gamma\,.\,x^n/n! + \alpha x^n$,

          $e^{ax}\sin bx = bx + abx^2 + \ldots + c^n\sin n\gamma\,.\,x^n/n! + \beta x^n$, where

$$c = |a + ib|, \quad \gamma = \arg(a + ib);$$

    (iv) $e^x\cos x = 1 + x - 2x^3/3! - \ldots + 2^{\frac{1}{2}n}\cos\tfrac{1}{4}n\pi\,.\,x^n/n! + \alpha x^n$;

    (v) $e^x\sin x = x + x^2 + 2x^3/3! - \ldots + 2^{\frac{1}{2}n}\sin\tfrac{1}{4}n\pi\,.\,x^n/n! + \beta x^n$;

    (vi) $\cos x\cosh x = 1 - x^4/6 + x^8/2520 - \ldots$
$$+ 2^{\frac{1}{2}n}\cos\tfrac{1}{4}n\pi\cos\tfrac{1}{2}n\pi\,.\,x^n/n! + \alpha x^n;$$

    (vii) $\sin x\sinh x = x^2 - x^6/90 + \ldots - 2^{\frac{1}{2}n}\sin\tfrac{1}{4}n\pi\cos\tfrac{1}{4}n\pi\,.\,x^n/n! + \beta x^n$;

    (viii) $\sec x = 1 + \tfrac{1}{2}x^2 + \eta x^3$;   (ix) $\tan x = x + \tfrac{1}{3}x^3 + \eta x^4$;

    (x) $(x+3)\,e^{-2x} + 4x\,e^{-x} + x - 3 = x^5/30 - x^6/30 + \eta x^6$;

    (xi) $\tan(a+b) = \tan a + b\sec^2 a + b^2\sec^2 a\tan a + \eta b^2$;

    (xii) $\tan^{-1}(x+h) = \tan^{-1}x + \dfrac{h}{1+x^2} - \dfrac{xh^2}{(1+x^2)^2} + \eta h^2$;

    (xiii) $\tfrac{1}{2}\log\dfrac{1+x}{1-x} = \dfrac{x}{1} + \dfrac{x^3}{3} + \dfrac{x^5}{5} + \ldots + \dfrac{x^{2n-1}}{2n-1} + \eta x^{2n}$;

    (xiv) $\tan^{-1}x = \dfrac{x}{1} - \dfrac{x^3}{3} + \dfrac{x^5}{5} - \ldots(-)^{n-1}\dfrac{x^{2n-1}}{2n-1} + \eta x^{2n}$;

    (xv) $\tan^{-1}\dfrac{a+x}{1-ax} = \tan^{-1}a + \dfrac{x}{1} - \dfrac{x^3}{3} + \dfrac{x^5}{5} - \ldots(-)^{n-1}\dfrac{x^{2n-1}}{2n-1} + \eta x^{2n}$;

(xvi) $\tan^{-1}\dfrac{2x}{1-x^2} = 2\left(\dfrac{x}{1} - \dfrac{x^3}{3} + \dfrac{x^5}{5} - \ldots(-)^{n-1}\dfrac{x^{2n-1}}{2n-1} + \eta x^{2n}\right);$

(xvii) $x\tan^{-1}x - \tfrac{1}{2}\log(1+x^2) = \displaystyle\sum_{s=1}^{n}(-)^{s-1}\dfrac{x^{2s}}{(2s-1)2s} + \eta x^{2n+1};$

(xviii) $\sinh^{-1}x = \dfrac{x}{1} - \dfrac{1}{2}\dfrac{x^3}{3} + \dfrac{1.3}{2.4}\dfrac{x^5}{5} - \dfrac{1.3.5}{2.4.6}\dfrac{x^7}{7} + \ldots + \eta x^n.$

(2) If $y = (\sinh^{-1}x)^2 = [\log\{x+\sqrt{(1+x^2)}\}]^2$, prove that $(y_{n+2})_0 = -n^2(y_n)_0$, where $n \geqslant 1$. Deduce that

$$(\sinh^{-1}x)^2 = x^2 - \dfrac{2}{3}\dfrac{x^4}{2} + \dfrac{2.4}{3.5}\dfrac{x^6}{3} - \dfrac{2.4.6}{3.5.7}\dfrac{x^8}{4} + \ldots + \eta x^n,$$

$$\dfrac{\log\{x+\sqrt{(1+x^2)}\}}{\sqrt{(1+x^2)}} = x - \tfrac{2}{3}x^3 + \dfrac{2.4}{3.5}x^5 - \dfrac{2.4.6}{3.5.7}x^7 + \ldots + \eta x^n.$$

(3) If $y = \sqrt[3]{\{x+\sqrt{(1+x^2)}\}} + \sqrt[3]{\{x-\sqrt{(1+x^2)}\}}$, prove that [see Examples 3B, 26]

$$\tfrac{3}{2}y = x - \dfrac{3^2-1}{3^2}\dfrac{x^3}{3!} + \dfrac{(9^2-1)(3^2-1)}{3^4}\dfrac{x^5}{5!} - \ldots + \eta x^n.$$

(4) If $y = \exp(c\sin^{-1}x)$, prove that:

    (i) $(1-x^2)y_2 - xy_1 - c^2y = 0;$

    (ii) $(y_{n+2})_0 = (c^2+n^2)(y_n)_0;$

    (iii) $\exp(c\sin^{-1}x) = 1 + cx + c^2x^2/2! + c(c^2+1)x^3/3!$
$$\qquad\qquad + c^2(c^2+2^2)x^4/4! + c(c^2+1)(c^2+3^2)x^5/5! + \ldots + \eta x^n.$$

Deduce that:

    (a) $\cos(m\sin^{-1}x) = 1 - m^2x^2/2! + m^2(m^2-2^2)x^4/4! - \ldots + \eta x^n,$

    (b) $\sin(m\sin^{-1}x) = mx - m(m^2-1^2)x^3/3! + \ldots + \eta x^n,$

    (c) $\cos m\theta = 1 - m^2\sin^2\theta/2! + m^2(m^2-2^2)\sin^4\theta/4! - \ldots + \eta\sin^n\theta,$

    (d) $\sin m\theta = m\sin\theta - m(m^2-1^2)\sin^3\theta/3! + \ldots + \eta\sin^n\theta,$

    (e) $\dfrac{\cos m\theta}{\cos\theta} = 1 - \dfrac{(m^2-1^2)}{2!}\sin^2\theta + \dfrac{(m^2-1^2)(m^2-3^2)}{4!}\sin^4\theta - \ldots + \eta\sin^n\theta,$

    (f) $\dfrac{\sin m\theta}{\cos\theta} = \dfrac{m}{1!}\sin\theta - \dfrac{m(m^2-2^2)}{3!}\sin^3\theta + \ldots + \eta\sin^n\theta.$

Examples 4C

(1) Obtain the following formulae by algebraic or partly algebraic methods:

    (i) $e^{\cos x} = e(1 - x^2/2 + x^4/6 - 31x^6/720 + \eta x^7);$

    (ii) $e^{-\cos x} = e^{-1}(1 + x^2/2 + x^4/12 + x^6/720 + \eta x^7);$

    (iii) $e^{\sin x} = 1 + x + x^2/2 - x^4/8 - x^5/15 - x^6/240 + \eta x^6;$

    (iv) $\cosh(\sin x) = 1 + x^2/2 - x^4/8 - x^6/240 + \eta x^7;$

    (v) $\sin(\sin x) = x - x^3/3 + x^5/10 - 8x^7/315 + \eta x^8;$

    (vi) $\cos^n x = 1 - nx^2/2 + n(3n-2)x^4/24 + \eta x^5;$

    (vii) $\log(1+x+x^2) = x/1 + x^2/2 - 2x^3/3 + x^4/4 + x^5/5 - 2x^6/6 + \eta x^6;$

(viii) $\log(1+\sin x) = x - x^2/2 + x^3/6 - x^4/12 + x^5/24 + \eta x^5$;

(ix) $\sin^3 x - x^3 \cos x = x^7/15 + \eta x^8$;

(x) $\sin x - x \sqrt[3]{(\cos x)} = x^5/45 + \eta x^6$;

(xi) $1/x - \cot x = \tfrac{1}{3}x + \eta x^2$;

(xii) $(1+x)^{1/x} = e(1 - x/2 + 11x^2/24 + \eta x^2)$,

$$\left(1 + \frac{1}{n}\right)^n = e(1 - 1/2n + 11/24n^2 + \eta/n^2).$$

(2) Show that $\log(1+e^{2x}) - x$ is an even function and that
$$\log(1+e^{2x}) = \log 2 + x + x^2/2 - x^4/12 + \eta x^5.$$

(3) If $y = ax + bx^2 + cx^3$ $(a \neq 0)$ and $x = 0$ when $y = 0$, show that
$$x = y/a - by^2/a^3 + (2b^2 - ac) y^3/a^5 + \eta y^3.$$

(4) If $y = e^x$ show that
$$x = (y-1)/1 - (y-1)^2/2 + (y-1)^3/3 - \ldots + \eta(y-1)^n.$$

(5) If $y \log y + y - 1 = x$, and $y = 1$ when $x = 0$, show that
$$y = 1 + \tfrac{1}{2}x - \tfrac{1}{16}x^2 + \eta x^2.$$

(6) If $x^{2+h} = e^2$ and $x = e$ when $h = 0$, show that
$$x = e(1 - \tfrac{1}{2}h + \tfrac{3}{8}h^2 + \eta h^2).$$

(7) Given that y and p are connected by the equation
$$\cos H \cos(a-y) \sin p = \sin a - \sin(a-y) \cos p$$
and that $y \to 0$ when $p \to 0$, show that
$$y = p \cos H - \tfrac{1}{2}p^2 \sin^2 H \tan a + \eta p^2.$$

(8) If $y = \cot x$, show that $dy/dx + 1 + y^2 = 0$.
Assuming that $y = x^{-1} + ax + bx^3 + cx^5 + \eta x^6$ satisfies this equation, show that
$$\cot x = \frac{1}{x} - \frac{x}{3} - \frac{x^3}{45} - \frac{2x^5}{945} + \eta x^6.$$
Deduce that
$$\log \frac{\sin x}{x} = -\frac{x^2}{6} - \frac{x^4}{180} - \frac{x^6}{2835} + \eta x^7.$$

(9) Show that, for any integer n,

(i) $e^{-x} = \eta/x^n$, where $\eta \to 0$ when $x \to +\infty$,
$e^x = \eta/x^n$, where $\eta \to 0$ when $x \to -\infty$;

(ii) $\tanh x = x - x^3/3 + 2x^5/15 + \eta x^6$,
$\tanh x = 1 - \eta/x^n$, where $\eta \to 0$ when $x \to +\infty$;

(iii) $\log \cosh x = x - \log 2 + \eta/x^n$, where $\eta \to 0$, $x \to +\infty$;

(iv) $\cosh x = \tfrac{1}{2}e^x(1 + \eta/x^n)$, where $\eta \to 0$ when $x \to +\infty$;

(v) $\sinh^{-1} x = \log(2x) + \dfrac{1}{2}\dfrac{1}{2x^2} + \dfrac{\eta}{x^3}$, where $x > 0$ and $\eta \to 0$ when $x \to \infty$;

(vi) $\sqrt{(x^2+x+1)} - \sqrt{(x^2-x+1)} = x(1 - 3x^2/8 + \eta x^3)$,

$$= 1 - 3/(8x^2) + \alpha/x^3 \ (\alpha \to 0, x \to +\infty),$$

$$= -1 + 3/(8x^2) + \beta/x^3 \ (\beta \to 0, x \to -\infty);$$

(vii) $\dfrac{ae^x+b}{ce^x+d} = a/c + \alpha/x^n \quad (c \neq 0, \alpha \to 0 \text{ when } x \to +\infty),$

$$= b/d + \beta/x^n \quad (d \neq 0, \beta \to 0 \text{ when } x \to -\infty).$$

(10) Show that, if $b \neq 0$,

$$\frac{px+q}{ax+b} = \frac{q}{b} - (aq - pb)\left(\frac{x}{b^2} - \frac{ax^2}{b^3} + \frac{a^2x^3}{b^4} - \ldots + \alpha x^n\right)$$

and that, if $a \neq 0$,

$$\frac{px+q}{ax+b} = \frac{p}{a} + (aq - pb)\left(\frac{1}{a^2 x} - \frac{b}{a^3 x^2} + \frac{b^2}{a^4 x^3} - \ldots + \frac{\beta}{x^n}\right).$$

(11) Show that, if $c \neq 0$,

$$\frac{1}{ax^2+bx+c} = \frac{1}{c} - \frac{bx}{c^2} - \frac{ac - b^2}{c^3} x^2 + \alpha x^2,$$

and that, if $a \neq 0$,

$$\frac{1}{ax^2+bx+c} = \frac{1}{ax^2} - \frac{b}{a^2 x^3} - \frac{ac - b^2}{a^3 x^4} + \frac{\beta}{x^4}.$$

(12) Show that, if $c > 0$,

$$\log(ax^2+bx+c) = \log c - \left(\frac{1}{\alpha} + \frac{1}{\beta}\right)x - \left(\frac{1}{\alpha^2} + \frac{1}{\beta^2}\right)\frac{x^2}{2} - \ldots + \eta x^n$$

where α, β are the roots of the equation $ax^2+bx+c = 0.$

Show also that, if $a > 0$,

$$\log(ax^2+bx+c) = \log a + \log(x^2) - \frac{\alpha + \beta}{x} - \frac{\alpha^2 + \beta^2}{2x^2} - \ldots + \frac{\eta_1}{x^n}.$$

(13) Show that, if $a_2 \neq 0$,

$$\frac{bx^2+b_1x+b_2}{ax^2+a_1x+a_2} = \frac{b_2}{a_2} - \begin{vmatrix} b_2 & a_2 \\ b_1 & a_1 \end{vmatrix}\frac{x}{a_2^2} - \begin{vmatrix} b_2 & 0 & a_2 \\ b_1 & a_2 & a_1 \\ b & a_1 & a \end{vmatrix}\frac{x^2}{a_2^3} + \alpha x^2,$$

and that, if $a \neq 0$,

$$\frac{bx^2+b_1x+b_2}{ax^2+a_1x+a_2} = \frac{b}{a} - \begin{vmatrix} b & a \\ b_1 & a_1 \end{vmatrix}\frac{1}{a^2 x} - \begin{vmatrix} b & 0 & a \\ b_1 & a & a_1 \\ b_2 & a_1 & a_2 \end{vmatrix}\frac{1}{a^3 x^2} + \frac{\beta}{x^2}.$$

(14) If $\qquad f(x) = a_0 + a_1 x + a_2 x^2 + \ldots + a_n x^n + \eta x^n,$

and $\qquad \dfrac{f(x)}{1-x} = s_0 + s_1 x + s_2 x^2 + \ldots + s_n x^n + \alpha x^n,$

show that $\qquad s_n = a_0 + a_1 + a_2 + \ldots + a_n.$

(15) If $\qquad (1-x)^{-m} = 1 + c_1 x + c_2 x^2 + \ldots + c_n x^n + \eta x^n$

show that $\qquad 1 + c_1 + c_2 + \ldots + c_n = (m+1)(m+2)\ldots(m+n)/n!.$

(16) Show that

$$\cosh^3 x = \frac{3}{4}\left\{\frac{4}{3} + (3+1)\frac{x^2}{2!} + (3^3+1)\frac{x^4}{4!} + (3^5+1)\frac{x^6}{6!} + \ldots + \alpha x^n\right\},$$

$$\sinh^3 x = \frac{3}{4}\left\{(3^2-1)\frac{x^3}{3!} + (3^4-1)\frac{x^5}{5!} + (3^6-1)\frac{x^7}{7!} + \ldots + \beta x^n\right\}.$$

(17) If $a_0 \neq 0$ and

$$\frac{b_0 + b_1 x/1! + b_2 x^2/2! + \ldots + \beta x^n}{a_0 + a_1 x/1! + a_2 x^2/2! + \ldots + \alpha x^n} = c_0 + c_1 x/1! + c_2 x^2/2! + \ldots + \eta x^n,$$

show that
$$a_0^{s+1} c_s = \begin{vmatrix} a_0 & 0 & 0 & \ldots & b_0 \\ a_1 & a_0 & 0 & \ldots & b_1 \\ a_2 & 2a_1 & a_0 & \ldots & b_2 \\ \hdotsfor{5} \\ a_s & {}^sC_1 a_{s-1} & {}^sC_2 a_{s-2} & \ldots & b_s \end{vmatrix}.$$

(18) Show that

$$\tan x = x + x^3/3 + 2x^5/15$$

$$+ \begin{vmatrix} 1 & 1 & 0 & 0 \\ 1 & 3 & 1 & 0 \\ 1 & 5 & 10 & 1 \\ 1 & 7 & 35 & 21 \end{vmatrix} \frac{x^7}{7!} + \begin{vmatrix} 1 & 1 & 0 & 0 & 0 \\ 1 & 3 & 1 & 0 & 0 \\ 1 & 5 & 10 & 1 & 0 \\ 1 & 7 & 35 & 21 & 1 \\ 1 & 9 & 84 & 126 & 36 \end{vmatrix} \frac{x^9}{9!} + \eta x^{10}.$$

(19) Show that

$$\frac{\sin nx}{n \sin x} = 1 - \begin{vmatrix} 1 & 1 \\ 1 & n^2 \end{vmatrix} \frac{x^2}{3!} + \begin{vmatrix} 1 & 0 & 1 \\ 1 & 3 & n^2 \\ 1 & 10 & n^4 \end{vmatrix} \frac{x^4}{3 \cdot 5!} + \eta x^5.$$

(20) Show that, if n is a positive integer,

$$\frac{e^{nx} - 1}{e^x - 1} = n + s_1 x + \frac{s_2 x^2}{2!} + \frac{s_3 x^3}{3!} + \ldots + \frac{s_k x^k}{k!} + \eta x^k$$

where
$$s_k = 1^k + 2^k + 3^k + \ldots + (n-1)^k.$$

Deduce that
$$s_{k-1} = \frac{1}{k!} \begin{vmatrix} 1 & 0 & 0 & \ldots & 0 & n \\ 1 & {}^2C_1 & 0 & \ldots & 0 & n^2 \\ 1 & {}^3C_1 & {}^3C_2 & \ldots & 0 & n^3 \\ \hdotsfor{6} \\ 1 & {}^kC_1 & {}^kC_2 & \ldots & {}^kC_{k-2} & n^k \end{vmatrix}.$$

(21) *Euler's numbers* $E_1, E_2, E_3, \ldots$ are defined by

$$\sec x = 1 + E_1 \frac{x^2}{2!} + E_2 \frac{x^4}{4!} + E_3 \frac{x^6}{6!} + \ldots + E_n \frac{x^{2n}}{(2n)!} + \eta x^{2n}.$$

Show that $E_1 = 1, E_2 = 5, E_3 = 61, E_4 = 1385, \ldots$

$$E_n = \begin{vmatrix} 1 & 1 & 0 & \ldots & 0 \\ 1 & {}^4C_2 & 1 & \ldots & 0 \\ 1 & {}^6C_2 & {}^6C_4 & \ldots & 0 \\ \hdotsfor{5} \\ 1 & {}^{2n}C_2 & {}^{2n}C_4 & \ldots & {}^{2n}C_{2n-2} \end{vmatrix}.$$

(22) *Bernoulli's numbers* B_1, B_2, B_3, ... may be defined by

$$\frac{x}{2}\cot\frac{x}{2} = 1 - B_1\frac{x^2}{2!} - B_2\frac{x^4}{4!} - \ldots - B_n\frac{x^{2n}}{(2n)!} + \eta x^{2n}.$$

Show that $B_1 = \frac{1}{6}$, $B_2 = \frac{1}{30}$, $B_3 = \frac{1}{42}$, $B_4 = \frac{1}{30}$, $B_5 = \frac{5}{66}$, ...

$$\frac{2^{n-1}(2n+1)!}{n!}B_n = \begin{vmatrix} 1 & 3 & 0 & \ldots & 0 \\ 2 & 10 & 5 & \ldots & 0 \\ 3 & 21 & 35 & \ldots & 0 \\ \hdotsfor{5} \\ n & {}^{2n+1}C_2 & {}^{2n+1}C_4 & \ldots & {}^{2n+1}C_{2n-2} \end{vmatrix}.$$

(23) Show that

$$\frac{\theta}{\sin\theta} = \theta\cot\tfrac{1}{2}\theta - \theta\cot\theta$$

$$= 1 + (2^2 - 2)B_1\frac{\theta^2}{2!} + (2^4 - 2)B_2\frac{\theta^4}{4!} + (2^6 - 2)B_3\frac{\theta^6}{6!} + \eta\theta^7$$

(24) Show that

$$\tan\theta = \cot\theta - 2\cot 2\theta$$

$$= 2^2(2^2 - 1)B_1\frac{\theta}{2!} + 2^4(2^4 - 1)B_2\frac{\theta^3}{4!} + 2^6(2^6 - 1)B_3\frac{\theta^5}{6!} + \eta\theta^6.$$

Examples 4D

(1) A function $f(x)$ is defined in the interval $(-\frac{1}{2} \leqslant x \leqslant \frac{1}{2})$ by

$$f(x) = \frac{1}{x(1+x)} - \frac{\log(1+x)}{x^2} \quad (x \neq 0), \quad f(0) = -\tfrac{1}{2}.$$

Show that $f(x)$ is continuous at $x = 0$.

(2) Find
$$\lim_{x \to 1}\frac{1}{(x-1)^3}\int_1^x \log(2t - t^2)\,dt.$$

(3) Show that
$$\lim_{x \to 0}\frac{\sinh(\sin x) - \sin(\sinh x)}{x^7}$$

exists and find it. Sketch the graphs of $\sinh(\sin x)$ and $\sin(\sinh x)$.

(4) If $f''(a)$ exists show that

$$\int_a^{a+x} f(t)\,dt = \frac{x}{1!}f(a) + \frac{x^2}{2!}f'(a) + \frac{x^3}{3!}f''(a) + \eta x^3$$

where $\eta \to 0$ when $x \to 0$. Show that

$$\int_3^{3+x}(t^2 + 16)^{\frac{3}{2}}\,dt = 125x + 45x^2/2 + 17x^3/5 + \eta x^3.$$

(5) If x and y are small, and $f(t)$ is continuous, show that approximately

$$\int_x^{a+y}\{f(t) - f(a - t)\}\,dt = \{f(a) - f(0)\}(x + y).$$

Show that approximately

$$\int_x^{1+y} [\exp(3t^2) - \exp\{3(1-t)^2\}]\,dt = 19(x+y).$$

(6) If $f(x)$ is an odd continuous function and if a, b are small, show that

$$\int_{-h+a}^{h+b} f(x)\,dx \doteqdot (a+b)f(h).$$

Show that, if θ, ϕ are small,

$$\int_{-\alpha+\theta}^{\alpha+\phi} \sin mx \cos nx\,dx \doteqdot (\theta+\phi)\sin m\alpha \cos n\alpha.$$

Examples 4E

(1) In the formula $f(x) = f(0) + xf'(\theta x)$ $(0 < \theta < 1)$, show that if

(i) $f(x) = x^2$, then $\theta = \frac{1}{2}$;   (ii) $f(x) = x^3$, then $\theta = 1/\sqrt{3}$;

(iii) $f(x) = x^{\frac{1}{2}}$, then $\theta = \frac{1}{4}$.

Show that these values of θ are not inconsistent with §4.13, Example.

(2) In the formula $f(x) = f(0) + xf'(\theta x)$: if $f''(0) = 0$ and $f'''(0) \neq 0$ exists, prove that $\theta = 1/\sqrt{3} + \eta$ $(\eta \to 0, x \to 0)$.

(3) In the formula $f(x) = f(0) + xf'(\theta x)$: if $f''(0) \neq 0$ and $f'''(0)$ exists, prove that

$$\theta = \tfrac{1}{2} + \tfrac{1}{24}xf'''(0)/f''(0) + \eta x \quad (\eta \to 0, x \to 0).$$

(4) If the formula $f(x) = f(0) + xf'(0) + R_2$ is applied to the function

$$f(x) = x^{3/2} \quad (0 < x),$$

show that when R_2 is expressed

(i) in Lagrange's form, $\theta = \frac{9}{64} \doteqdot 0.14$,

(ii) in Cauchy's form, $\theta = \{17 - \sqrt{(208)}\}/9 \doteqdot 0.3$,

(iii) in the integral form, then the integral works out to $x^{\frac{3}{2}}$.

(5) Let $f(x)$ be defined by

$$f(x) = x\sin(1/x) \quad (x \neq 0), \quad f(0) = 0.$$

Show that $f'(0)$ does not exist.

Show that the formula $f(x) = f(0) + xf'(\theta x)$, when $x = 1$, becomes

$$\sin 1 = \sin(1/\theta) - (1/\theta)\cos(1/\theta),$$

and that there is an infinity of values of θ between 0 and 1 that satisfy this equation.

(6) If $f(x) = e^x$, show that the value of θ in the formula

$$f(x+h) - f(x) = hf'(x+\theta h)$$

is independent of x and that

$$\theta = \tfrac{1}{2} + h/24 - h^3/2880 + \eta h^3 \quad (\eta \to 0, h \to 0).$$

(7) In the formula $\log(n+1) = \log n + 1/(n+\theta)$, prove that

$$\theta = \frac{1}{2} - \frac{1}{12n} + \frac{1}{24n^2} + \frac{\eta}{n^2} \quad (\eta \to 0, n \to +\infty).$$

(8) If

$$e^x = 1 + x + x^2/2! + \ldots + x^{n-1}/(n-1)! + x^n e^{\theta x}/n!$$

show that

$$\theta = \frac{1}{n+1} + \frac{nx}{2(n+1)^2(n+2)} + \eta x \quad (\eta \to 0, x \to 0).$$

(9) If $f'(t)$ is continuous in $(0 \leqslant t \leqslant x)$ and $f''(t)$ exists in $(0 < t < x)$, show that

$$f'(x) = f'(0) + xf''(\theta x),$$

$$\int_0^x f(t)\, dt = xf(0) + \tfrac{1}{2}x^2 f'(0) + \tfrac{1}{6}x^3 f''(\theta_1 x),$$

where θ and θ_1 lie between 0 and 1.

(10) If $u_{n+1} = \sqrt{(2u_n + 8)}$ for $n \geqslant 1$, and if $u_1 > -4$, show that the sequence (u_n) converges monotonically to the limit 4.

(11) Putting $^mC_0 = 1$, $^mC_n = \binom{m}{n}$ for $n \geqslant 1$, show that, if $m > -1$ and m is not a positive integer,

   (i) ^mC_n is positive as long as $n < m+1$ and is then alternately negative and positive,

   (ii) $^mC_n > {}^mC_{n-1}$ as long as $n < \tfrac{1}{2}(m+1)$,

   (iii) $|{}^mC_n| < |{}^mC_{n-1}|$ when $n > \tfrac{1}{2}(m+1)$,

   (iv) ^mC_n is greatest when n is the largest integer in $\tfrac{1}{2}(m+1)$.

(12) In the binomial series when $x = -1$, $m > 0$, prove that

$$R_n = (-)^n \binom{m-1}{n-1}, \quad S_n = (-)^{n+1}\binom{m-1}{n-1}.$$

(13) In the binomial series, prove that, if $x \neq -1$, $R_n(x)$ can be expressed in the form

$$R_n(x) = n\binom{m}{n} x^n (1+x)^m \int_0^1 \frac{v^{n-1}\, dv}{(1+vx)^{m+1}}.$$

Deduce that $R_n(1) \to 0$ if $m > -1$.

(14) If $N > 0$ and $h = N - a^2$, show that $\sqrt{N} = \tfrac{1}{2}(a + N/a) - R$, where R lies between $h^2/8a^3$ and $h^2/8N^{3/2}$.

If $a = 1.732$ show that $\tfrac{1}{2}(a + 3/a)$ exceeds $\sqrt{3}$ by about 7.5×10^{-10}.

(15) If a is an approximation to $\sqrt[3]{N}$ prove that approximately

$$\tfrac{1}{3}(2a + N/a^2) - \sqrt[3]{N} = a(N - a^3)^2/9N^2.$$

If $a = 1.71$ show that $\tfrac{1}{3}(2a + 5/a^2)$ exceeds $\sqrt[3]{5}$ by about 3×10^{-10}.

(16) If $0 < x < 1$, prove that

$$\log\frac{1+x}{1-x} = 2\left(\frac{x}{1} + \frac{x^3}{3} + \frac{x^5}{5} + \ldots + \frac{x^{2n-1}}{2n-1}\right) + R,$$

where $\quad 0 < R < \dfrac{x^{2n+1}}{2n+1}\left(1 + \dfrac{1}{1-x}\right).$ [See §4.25, III, Note 1.]

Deduce that if $p > q > 0$, then

$$\log \frac{p}{q} = 2\left\{\frac{p-q}{p+q} + \frac{1}{3}\left(\frac{p-q}{p+q}\right)^3 + \dots + \frac{1}{2n-1}\left(\frac{p-q}{p+q}\right)^{2n-1}\right\} + R,$$

where
$$0 < R < \frac{1}{2n+1}\frac{p+3q}{2q}\left(\frac{p-q}{p+q}\right)^{2n+1}.$$

(17) If $x > 0$ show that

  (i) $e^x + e^{-1/x} \sim 1 + x + x^2/2! + x^3/3! + \dots$,

  (ii) $e^{x-1/x} \sim 0 + 0.x + 0.x^2 + 0.x^3 + \dots$.

(18) In the Maclaurin formula for $e^{x \cot \alpha} \sin x$, where α is not a multiple of π, show that one form of R_n is

$$R_n = (1/n!)(x \operatorname{cosec} \alpha)^n e^{\theta x \cot \alpha} \sin(\theta x + n\alpha),$$

where $0 < \theta < 1$. Deduce that for all x,

$$e^{x \cot \alpha} \sin x = \frac{\sin \alpha}{1!}\frac{x}{\sin \alpha} + \frac{\sin 2\alpha}{2!}\left(\frac{x}{\sin \alpha}\right)^2 + \frac{\sin 3\alpha}{3!}\left(\frac{x}{\sin \alpha}\right)^3 + \dots.$$

(19) If $y = \tan^{-1} x$, show that the remainder in the Maclaurin formula for y can be written in the form

$$R_n = (x^n/n)\cos^n \theta y \sin n(\tfrac{1}{2}\pi + \theta y)$$

where $0 < \theta < 1$. Deduce that

$$\tan^{-1} x = x/1 - x^3/3 + x^5/5 - x^7/7 + \dots \quad (-1 \leqslant x \leqslant 1).$$

(20) Show that
$$\left(\frac{d}{dx}\right)^n \log(1 + x e^{i\alpha}) = \frac{(-)^{n-1}(n-1)!}{(x + e^{-i\alpha})^n}.$$

Hence prove that, when $-1 < x < 1$,

$$\tfrac{1}{2}\log(1 + 2x\cos\alpha + x^2) = \frac{x}{1}\cos\alpha - \frac{x^2}{2}\cos 2\alpha + \frac{x^3}{3}\cos 3\alpha - \dots,$$

$$\tan^{-1}\frac{x\sin\alpha}{1 + x\cos\alpha} = \frac{x}{1}\sin\alpha - \frac{x^2}{2}\sin 2\alpha + \frac{x^3}{3}\sin 3\alpha - \dots.$$

(21) Prove that

$$\cos x \cosh x = 1 - 2^2 x^4/4! + 2^4 x^8/8! - \dots(-)^n 2^{2n} x^{4n}/(4n)! + R$$

where, if $x > 0$, $|R| < 2^{2n} x^{4n+1} e^x/(4n+1)!$.

(22) If $f(x)$ has a second derivative at every point of a closed interval, and if x, x_1 are points of the interval, show that a number ξ exists between x and x_1 such that
$$f(x_1) = f(x) + (x_1 - x)f'(x) + \tfrac{1}{2}(x_1 - x)^2 f''(\xi).$$

Deduce that, if $f''(x) \geqslant 0$ at every point of the interval and if $x_1, x_2, \dots, x_n$ are points of the interval, then

$$\frac{f(x_1) + f(x_2) + \dots + f(x_n)}{n} \geqslant f\left(\frac{x_1 + x_2 + \dots + x_n}{n}\right). \tag{83}$$

Further, if $m_1, m_2, ..., m_n$ are positive, show that

$$\frac{m_1 f(x_1) + m_2 f(x_2) + ... + m_n f(x_n)}{m_1 + m_2 + ... + m_n} \geqslant f\left(\frac{m_1 x_1 + m_2 x_2 + ... + m_n x_n}{m_1 + m_2 + ... + m_n}\right). \quad (84)$$

If $f''(x) \leqslant 0$, the signs of inequality must be reversed.

A function $f(x)$ is said to be *convex* in an interval throughout which $f''(x) \geqslant 0$, *concave* if $f''(x) < 0$.

(23) By putting $f(x) = x^p$ in (83), prove that if $p > 1$ and $x_1, x_2, ..., x_n$ are positive, then

$$\frac{x_1^p + x_2^p + ... + x_n^p}{n} \geqslant \left(\frac{x_1 + x_2 + ... + x_n}{n}\right)^p. \quad (85)$$

In particular show that, except when $x_1 = x_2 = ... = x_n$, the average of their squares exceeds the square of their average.

(24) If $q > p > 1$ and $x_1, x_2, ..., x_n$ are positive, show that

$$\left(\frac{x_1^q + x_2^q + ... + x_n^q}{n}\right)^{1/q} \geqslant \left(\frac{x_1^p + x_2^p + ... + x_n^p}{n}\right)^{1/p}. \quad (86)$$

(25) Noting that $\log x$ is a concave function, by putting $f(x) = \log x$ in (83), prove that

$$\frac{x_1 + x_2 + ... + x_n}{n} \geqslant (x_1 x_2 ... x_n)^{1/n}, \quad (87)$$

that is $A \geqslant G$, where A denotes the arithmetic mean and G the geometric mean of the n positive numbers $x_1, x_2, ..., x_n$.

More generally, by putting $q_r = m_r / \Sigma m_r$ $(r = 1, 2, ..., n)$ in (84), so that $0 < q_r < 1$ and $\Sigma q_r = 1$, prove, after putting $f(x) = \log x$, that

$$q_1 x_1 + q_2 x_2 + ... + q_n x_n \geqslant x_1^{q_1} x_2^{q_2} ... x_n^{q_n}, \quad (88)$$

which is known as 'the general theorem of the means' (Hardy, *Course of Pure Mathematics*, Appendix I).

(26) Let x, y, p, q be all positive and $p + q = 1$. Then, from (88), with $n = 2$,

$$x^p y^q \leqslant px + qy. \quad (89)$$

Also, let $a_1, a_2, ..., a_n$ and $b_1, b_2, ..., b_n$ be two sets of n positive numbers, and put $A = \Sigma a_r$, $B = \Sigma b_r$.

Now, putting $x = a_r/A$, $y = b_r/B$ in (89), we have, for each pair a_r, b_r,

$$\left(\frac{a_r}{A}\right)^p \left(\frac{b_r}{B}\right)^q \leqslant p\frac{a_r}{A} + q\frac{b_r}{B}.$$

Summing wo r, we get

$$\frac{\Sigma(a_r^p b_r^q)}{A^p B^q} \leqslant p\frac{\Sigma a_r}{A} + q\frac{\Sigma b_r}{B} = p + q = 1$$

and hence

$$\Sigma(a_r^p b_r^q) \leqslant A^p B^q = (\Sigma a_r)^p (\Sigma b_r)^q. \quad (90)$$

(27) From (90), by putting $p = \frac{1}{2}, q = \frac{1}{2}$ and replacing a_r by a_r^2 and b_r by b_r^2, deduce Cauchy's inequality

$$(\Sigma a_r b_r)^2 \leqslant (\Sigma a_r^2)(\Sigma b_r^2). \quad (91)$$

(28) From (90) by replacing a_r by $a_r^{1/p}$ and b_r by $b_r^{1/q}$, deduce that, if $0 < p < 1$, $0 < q < 1$, and $p + q = 1$, then

$$\Sigma a_r b_r \leqslant (\Sigma a_r^{1/p})^p (\Sigma b_r^{1/q})^q. \tag{92}$$

If we put $p = 1/k$, $q = 1/l$, so that $k > 1$, $l > 1$, and $1/k + 1/l = 1$, we obtain Hölder's inequality

$$\Sigma a_r b_r \leqslant (\Sigma a_r^k)^{1/k}(\Sigma b_r^l)^{1/l}. \tag{93}$$

Examples 4 F

(1) If $f(x), g(x), h(x)$ are continuous at a and b and differentiable in $a < x < b$, prove that a number ξ exists ($a < \xi < b$) such that

$$\begin{vmatrix} f(a) & f(b) & f'(\xi) \\ g(a) & g(b) & g'(\xi) \\ h(a) & h(b) & h'(\xi) \end{vmatrix} = 0.$$

Deduce Cauchy's mean value theorem.

(2) If $f(t), g(t), h(t)$ are continuous at a, b, x and twice-differentiable in the open interval (a, b, x), prove that a number ξ exists, belonging to this interval, such that

$$\begin{vmatrix} f(a) & f(b) & f(x) \\ g(a) & g(b) & g(x) \\ h(a) & h(b) & h(x) \end{vmatrix} = \tfrac{1}{2}(x-a)(x-b)\begin{vmatrix} f(a) & f(b) & f''(\xi) \\ g(a) & g(b) & g''(\xi) \\ h(a) & h(b) & h''(\xi) \end{vmatrix}.$$

[Use § 4.29]. Deduce the rule of proportional parts.

(3) If $0 < n, 0 < \lambda < 1$, show that

(i) $\log(n+\lambda) - \log n = \lambda \{\log(n+1) - \log n\} + R$,

where $\qquad \tfrac{1}{2}\lambda(1-\lambda)/(n+1)^2 < R < \tfrac{1}{2}\lambda(1-\lambda)/n^2;$

(ii) $\sqrt{(n+\lambda)} - \sqrt{n} = \lambda\{\sqrt{(n+1)} - \sqrt{n}\} + R$,

where $\qquad \tfrac{1}{8}\lambda(1-\lambda)/(n+1)^{\frac{3}{2}} < R < \tfrac{1}{8}\lambda(1-\lambda)/n^{\frac{3}{2}}.$

(4) Let a, b, c be three distinct numbers, and let $Q(x)$ be the quadratic function which has the same values as $f(x)$ at $x = a, b, c$. Show that $Q(x)$ can be expressed in the form

$$Q(x) \equiv \Sigma \frac{(x-b)(x-c)}{(a-b)(a-c)} f(a). \tag{94}$$

If $f(x)$ is continuous at a, b, c and $f''(x)$ exists in the open interval (a, b, c), prove that a number ξ exists, belonging to this interval, such that

$$\frac{f(a)}{(a-b)(a-c)} + \frac{f(b)}{(b-c)(b-a)} + \frac{f(c)}{(c-a)(c-b)} = \tfrac{1}{2}f''(\xi). \tag{95}$$

Deduce the rule of proportional parts.

[*Hint.* Consider the function $\phi(x) \equiv f(x) - Q(x)$.]

Also prove that, if $f(t)$ is continuous at a, b, c, x and $f'''(t)$ exists in the open interval (a, b, c, x), then a number ξ exists, belonging to this interval, such that

$$f(x) = Q(x) + \tfrac{1}{6}(x-a)(x-b)(x-c)f'''(\xi). \tag{96}$$

[Put $F(t) = f(t) - Q(t)$, $G(t) = (t-a)(t-b)(t-c)$, and use (74) with $n = 3$.]

(5) *Quadratic interpolation formula.* If $f'''(x)$ exists in $a \leqslant x \leqslant a+2h$, prove that

$$f(a+\lambda h) = f(a) + \lambda\Delta f(a) + \tfrac{1}{2}\lambda(\lambda-1)\,\Delta^2 f(a) + R, \qquad (97)$$

where

$$R = \tfrac{1}{6}\lambda(\lambda-1)\,(\lambda-2)\,h^3 f'''(\xi),$$

$0 \leqslant \lambda \leqslant 2, a < \xi < a+2h$.

[Put

$$F(t) = f(a+th) - f(a) - t\Delta f(a) - \tfrac{1}{2}t(t-1)\,\Delta^2 f(a),$$

$$G(t) = t(t-1)\,(t-2).$$

Then $F(t)$, $G(t)$ vanish at $t = 0, 1, 2$. Use (74) with $n = 3$.]

Note that, if $f'''(x)$ is monotonic, then R lies between

$$\tfrac{1}{6}\lambda(\lambda-1)\,(\lambda-2)\,h^3 f'''(a) \quad \text{and} \quad \tfrac{1}{6}\lambda(\lambda-1)\,(\lambda-2)\,h^3 f'''(a+2h).$$

Given that $\cosh 2\cdot0 = 3\cdot7622$, $\cosh 2\cdot1 = 4\cdot1443$, $\cosh 2\cdot2 = 4\cdot5679$, show that approximately

$$\cosh 2\cdot05 = 3\cdot9483, \quad \cosh 2\cdot15 = 4\cdot3507.$$

(6) Let $a_1, a_2, \ldots, a_n$ be n distinct numbers, and let $Q_{n-1}(x)$ be the polynomial of degree $n-1$ which has the same values as $f(x)$ when $x = a_1, a_2, \ldots, a_n$.

If $f(t)$ is continuous at $a_1, a_2, \ldots, a_n, x$, and if $f^{(n)}(t)$ exists in the open interval $(a_1, a_2, \ldots, a_n, x)$, prove that a number ξ exists, belonging to this interval, such that

$$f(x) = Q_{n-1}(x) + R, \qquad (98)$$

where

$$R = (x-a_1)\,(x-a_2)\ldots(x-a_n)f^{(n)}(\xi)/n!$$

(7) *Lagrange's interpolation formula.* Put

$$g_n(x) = (x-a_1)\,(x-a_2)\ldots(x-a_n).$$

Then the polynomial $Q_{n-1}(x)$ in (98) can be expressed in the form

$$Q_{n-1}(x) \equiv \sum_{s=1}^{s=n} \frac{f(a_s)}{g_n'(a_s)}\,\frac{g_n(x)}{x-a_s} \qquad (99)$$

and when $Q_{n-1}(x)$ is thus expressed, (98) is called *Lagrange's interpolation formula*, which therefore reads

$$f(x) = \sum_{s=1}^{s=n} \frac{g_n(x)}{x-a_s}\,\frac{f(a_s)}{g_n'(a_s)} + \frac{g_n(x)}{n!}\,f^{(n)}(\xi). \qquad (100)$$

(8) *Newton's interpolation formula.* The polynomial $Q_{n-1}(x)$ in (98) can also be expressed in the form

$$Q_{n-1}(x) \equiv c_0 + c_1(x-a_1) + c_2(x-a_1)\,(x-a_2) + \ldots + c_{n-1}(x-a_1)\,(x-a_2)\ldots(x-a_{n-1})$$

$$= c_0 + c_1 g_1(x) + c_2 g_2(x) + \ldots + c_{n-1}g_{n-1}(x),$$

where $g_r(x) = (x-a_1)\ldots(x-a_r)$. The coefficients $c_0, c_1, c_2, \ldots$ will then be given by the equations

$$f(a_1) = c_0,$$

$$f(a_2) = c_0 + c_1(a_2-a_1),$$

$$f(a_3) = c_0 + c_1(a_3-a_1) + c_2(a_3-a_1)\,(a_3-a_2),$$

....

It follows that, for $1 \leqslant r \leqslant n$,

$$Q_{r-1} = c_0 + c_1 g_1(x) + c_2 g_2(x) + \ldots + c_{r-1} g_{r-1}(x).$$

Comparing this with $Q_{r-1}(x)$ expressed in the form (99), and equating co-efficients of x^{r-1}, we see that

$$c_{r-1} = \sum_{s=1}^{s=r} \frac{f(a_s)}{g_r'(a_s)}. \tag{101}$$

With the form of $Q_{n-1}(x)$ thus obtained, (98) is called *Newton's interpolation formula*, which therefore reads

$$f(x) = \sum_{r=0}^{r=n-1} g_r(x) \sum_{s=1}^{s=r+1} \frac{f(a_s)}{g_{r+1}'(a_s)} + \frac{g_n(x)}{n!} f^{(n)}(\xi). \tag{102}$$

(9) If $a_1 = a$, $a_2 = a+h$, $a_3 = a+2h$, ..., $a_n = a+(n-1)h$, show that

$$g_n'(a_s) = (-)^{n-s}(s-1)!(n-s)!h^{n-1} \tag{103}$$

and that Lagrange's formula can then be written (with $0! = 1$)

$$f(x) = \frac{1}{h^{n-1}} \sum_{s=1}^{s=n} \frac{g_n(x)}{x-a_s} \frac{(-)^{n-s} f(a_s)}{(s-1)!(n-s)!} + \frac{g_n(x)}{n!} f^{(n)}(\xi). \tag{104}$$

(10) Using (79), show that

$$c_r = \sum_{k=0}^{k=r} \frac{(-)^{r-k} f(a+kh)}{k!(r-k)!h^r} = \frac{\Delta^r f(a)}{r!h^r} \tag{105}$$

and hence, putting $x = a + \lambda h$, that Newton's formula can be written

$$f(a+\lambda h) = f(a) + \binom{\lambda}{1} \Delta f(a) + \ldots + \binom{\lambda}{n-1} \Delta^{n-1} f(a) + \binom{\lambda}{n} h^n f^{(n)}(\xi) \tag{106}$$

where $a < \xi < a+nh$. Given that $\sqrt[3]{8} = 2$, $\sqrt[3]{9} = 2 \cdot 0801$, $\sqrt[3]{10} = 2 \cdot 1544$, and taking $n = 3$, estimate the values of $\sqrt[3]{(8 \cdot 5)}$ and $\sqrt[3]{(9 \cdot 5)}$.

(11) If $f^{(n)}(a)$ exists, prove that when $h \to 0$, $\Delta^n f(a)/h^n \to f^{(n)}(a)$.

(12) Let $\phi(x)$ and its first $n-1$ derived functions all vanish and be continuous at $x = a$; let $\phi(x)$ vanish and be continuous at $x = b$; and let $\phi^{(n)}(x)$ exist in $a < x < b$.

Prove that a number ξ exists $(a < \xi < b)$ such that $\phi^{(n)}(\xi) = 0$.

(13) Let $F(x)$, $G(x)$ and their first $n-1$ derived functions all vanish and be continuous at $x = a$. Let $F(x)$, $G(x)$ be continuous at $x = b$, and $G(b) \neq 0$. Also let $F^{(n)}(x)$, $G^{(n)}(x)$ exist and not vanish together in $a < x < b$.

Prove that a number ξ exists $(a < \xi < b)$ such that

$$F(b)/G(b) = F^{(n)}(\xi)/G^{(n)}(\xi).$$

(14) Let

$$\phi(a) = 0, \quad \phi^{(s)}(a) = 0 \quad (s = 1, 2, \ldots, p-1),$$

$$\phi(b) = 0, \quad \phi^{(s)}(b) = 0 \quad (s = 1, 2, \ldots q-1),$$

$$\phi(c) = 0, \quad \phi^{(s)}(c) = 0 \quad (s = 1, 2, \ldots r-1),$$

where $p+q+r = n$. Also, let $\phi^{(n-1)}(x)$ exist in the closed interval $[a, b, c]$. Show that a number ξ exists, belonging to the open interval, such that $\phi^{(n-1)}(\xi) = 0$.

(15) Let $Q(x) = Ax^{n-1} + \ldots$ be a polynomial of degree $n-1$, and $f(x)$ a function of x such that

$$f(a) = Q(a), \quad f^{(s)}(a) = Q^{(s)}(a) \quad (s = 1, 2, \ldots, p-1),$$

$$f(b) = Q(b), \quad f^{(s)}(b) = Q^{(s)}(b) \quad (s = 1, 2, \ldots, q-1),$$

$$f(c) = Q(c), \quad f^{(s)}(c) = Q^{(s)}(c) \quad (s = 1, 2, \ldots, r-1),$$

where $p + q + r = n$. Also, let $f^{(n-1)}(x)$ exist in the closed interval $[a, b, c]$. Prove that a number ξ exists, belonging to the open interval, such that

$$A = f^{(n-1)}(\xi)/(n-1)!$$

Also, if $f^{(n)}(t)$ exists in the closed interval $[a, b, c, x]$, prove that a number ξ exists, belonging to the open interval, such that

$$f(x) = Q(x) + (x-a)^p (x-b)^q (x-c)^r f^{(n)}(\xi)/n!.$$

(16) If $f'''(t)$ exists in $-x \leqslant t \leqslant x$, prove that

$$\frac{f(x) - f(-x)}{2} = xf'(0) + \frac{x^3}{3!} f'''(\xi) \quad (-x < \xi < x).$$

Proof. Let $Q(x) = f(0) + xf'(0) + Bx^2 + Ax^3$ be the cubic such that

$$Q(0) = f(0), \quad Q'(0) = f'(0), \quad Q(h) = f(h), \quad Q(-h) = f(-h).$$

Then A, B are given by the two equations

$$f(h) = f(0) + hf'(0) + Bh^2 + Ah^3, \quad f(-h) = f(0) - hf'(0) + Bh^2 - Ah^3.$$

Only A is needed. By subtraction we find

$$f(h) - f(-h) = 2hf'(0) + 2Ah^3.$$

But $A = f'''(\xi)/3!$, where ξ is some number belonging to the interval $(-h, h)$, by Example 15. The result follows on dividing by 2 and substituting x for h.

(17) If $f^{(2n+1)}(t)$ exists in $-x \leqslant t \leqslant x$, prove that

$$\frac{f(x) - f(-x)}{2} = xf'(0) + \frac{x^3}{3!} f'''(0) + \ldots + \frac{x^{2n-1}}{(2n-1)!} f^{(2n-1)}(0) + \frac{x^{2n+1}}{(2n+1)!} f^{(2n+1)}(\theta x),$$

where $-1 < \theta < 1$. Apply this to (i) $f(x) = e^x$, (ii) $f(x) = \log(1+x)$.

(18) If $f^{(2n)}(t)$ exists in $-x \leqslant t \leqslant x$, prove that

$$\frac{f(x) + f(-x)}{2} = f(0) + \frac{x^2}{2!} f''(0) + \frac{x^4}{4!} f^{iv}(0) + \ldots + \frac{x^{2n-2}}{(2n-2)!} f^{(2n-2)}(0) + \frac{x^{2n}}{(2n)!} f^{(2n)}(\theta x),$$

where $-1 < \theta < 1$. Apply this to (i) $f(x) = e^x$, (ii) $f(x) = \cosh x$, (iii) $f(x) = x^3$.

(19) If $f^{(5)}(t)$ exists in $-x \leqslant t \leqslant x$, prove that

$$\frac{f(x) - f(-x)}{2} = \frac{x}{6}\{f'(x) + 4f'(0) + f'(-x)\} - \frac{x^5}{180} f^{(5)}(\theta x),$$

where $-1 < \theta < 1$. Apply this to

(i) $f(x) = e^x$, (ii) $f(x) = \sin x$, (iii) $f(x) = \log(1+x)$, (iv) $f(x) = x^5$.

9

(20) If $f'''(t)$ exists in $-x \leqslant t \leqslant x$, prove that

$$\frac{f(x)-f(-x)}{2} = \frac{x}{2}\{f'(x)+f'(-x)\} - \frac{x^3}{3}f'''(\theta x),$$

where $-1 < \theta < 1$. Apply this to (i) $f(x) = x^5$, (ii) $f(x) = \log(1+x)$.

(21) The coordinates of three points P, Q, R on a curve $y = f(x)$ are respectively

$$(3\cdot250, 8\cdot526), \quad (3\cdot500, 8\cdot910), \quad (3\cdot750, 9\cdot239).$$

Find, approximately, the values of dy/dx and d^2y/dx^2 at Q, and deduce an approximate value of the radius of the circle PQR.

[Use the approximate formula

$$f(a+h) = f(a) + hf'(a) + \tfrac{1}{2}h^2 f''(a).]$$

(22) On the curve $y = \tfrac{1}{2}x^2 + \tfrac{1}{3}ax^3 + \tfrac{1}{4}bx^4$, let the radius and the centre of curvature be ρ, C at the point (x, y) and ρ_0, C_0 at the point $(0, 0)$. Prove that

$$\{(\rho-\rho_0)^2 - CC_0^2\}y_2^2 = \tfrac{1}{3}a^2x^4 + \eta x^4 \quad (\eta \to 0, x \to 0),$$

where $y_2 = d^2y/dx^2$ at (x, y). Deduce that, if $a \neq 0$, 'consecutive' circles of curvature do not intersect in real points.

(23) Let $s = f(\psi)$ be the intrinsic equation of a curve touching the axis of x at the origin. Assuming the existence of the derivatives concerned, show that

$$x = s_1\psi + \tfrac{1}{2}s_2\psi^2 + \tfrac{1}{6}(s_3 - s_1)\psi^3 + \tfrac{1}{24}(s_4 - 3s_2)\psi^4 + \alpha\psi^4,$$

$$y = \tfrac{1}{2}s_1\psi^2 + \tfrac{1}{3}s_2\psi^3 + \tfrac{1}{24}(3s_3 - s_1)\psi^4 + \beta\psi^4,$$

where s_1, s_2, s_3, s_4 denote derivatives of s wo ψ at $\psi = 0$.

(24) Let $P(x, y)$ be a point on a curve touching the axis of x at the origin O. Let s be the length of the arc OP. Let κ be the curvature at P, expressed in terms of s. Assuming the existence of the derivatives concerned, show that

$$x = s - \tfrac{1}{6}\kappa_0^2 s^3 - \tfrac{1}{8}\kappa_0\kappa_1 s^4 + \alpha s^4,$$

$$y = \tfrac{1}{2}\kappa_0 s^2 + \tfrac{1}{6}\kappa_1 s^3 + \tfrac{1}{24}(\kappa_2 - \kappa_0^3)s^4 + \beta s^4,$$

where $\kappa_0, \kappa_1, \kappa_2$ denote the values of $\kappa, d\kappa/ds, d^2\kappa/ds^2$ respectively at $s = 0$.

(25) Use the expansions in Example 23 or 24 to prove that in general 'consecutive' circles of curvature do not intersect in real points.

(26) The point $P(a, b)$ lies on the curve $y = f(x)$. A chord parallel to and close to the tangent at P meets the curve at Q and R near P. Prove that the gradient of the line joining P to the mid-point of QR is $f'(a) - 3\{f''(a)\}^2/f'''(a)$, approximately, assuming that $f''(a) \neq 0$ and $f'''(a) \neq 0$.

(27) In the expansion of $(ax^2 + 2bx + c)^{m-\frac{1}{2}}$ in ascending powers of x, show that the coefficient of x^{2m} is

$$\frac{(2m)!}{(2^m m!)^2} \frac{(ac-b^2)^m}{c^{m+\frac{1}{2}}}$$

where m is a positive integer, and $c > 0$. [See Examples 3B, 37.]

Examples 4G

(1) The point P lies on the curve $y = \cos x$ between $x = 0$ and $x = \tfrac{1}{2}\pi$. Perpendiculars PM, PN are drawn to the axes Ox, Oy respectively. Find approximately the coordinates of P when the area of the rectangle $OMPN$ is a maximum, and find the maximum area.

(2) On the graph of $y = x \tanh x$, show that points of inflexion occur where x satisfies the equation $x = \coth x$. Find x approximately.

(3) Find approximately the distance from the point $(0, 5)$ to the nearest point on the curve $y = \log x$.

(4) Find approximate values of x_1 and x_2 such that
  (i) $\log(1+x) \geqslant x - x^2$ for all $x > x_1$,
  (ii) $x - \frac{1}{3}x^2 \geqslant \log(1+x)$ for all x in the interval $-1 < x < x_2$.

(5) Find approximately the value of x between 0 and π at which the curve $y = (\sin x)/x$ has a point of inflexion.

(6) Find approximately the gradient of the tangent drawn from the origin to touch the curve $y = \sin^2 x$ between $x = 0$ and $x = \frac{1}{2}\pi$.

(7) Show that $x = 2 \cdot 5$ is an approximate solution of the equation $x^x = 10$, and that a closer approximation is $x = 2 \cdot 506$.

(8) Find an approximation to the positive root of the equation $x \sinh x = 1$.

(9) Find an approximation to the smallest positive root of the equation $x = \tan x$.
  [Consider $f(x) = \sin x - x \cos x = 0$.]

(10) Find an approximation to the least positive root of the equation

$$1 + x \cos x = 0.$$

Show that an approximation to the root near $(2n+1)\frac{1}{2}\pi$,

$$\text{for} \quad n = 1, \pm 2, \pm 3, \dots, \quad \text{is} \quad \alpha + (-)^n/\alpha$$

where $\alpha = (2n+1)\frac{1}{2}\pi$.

(11) Show that the positive roots of the equation $\tan x = \tanh x$ are close to the values $\pi + \frac{1}{4}\pi$, $2\pi + \frac{1}{4}\pi$, $3\pi + \frac{1}{4}\pi$, ..., and that a closer approximation to the nth positive root is $\alpha - e^{-2\alpha}$ where $\alpha = (n + \frac{1}{4})\pi$.

(12) If $0 = f(a) + hf'(a) + \frac{1}{2}h^2f''(a) + \frac{1}{6}h^3f'''(a) + \eta h^3$, and if $f'(a) \neq 0$, show that

$$h = -\frac{f}{f'} - \frac{f''f^2}{2f'^3} - \frac{(3f''^2 - f'f''')f^3}{6f'^5} + \eta_1 f^3$$

where $f, f', \dots,$ denote $f(a), f'(a), \dots.$

5

PARTIAL DERIVATIVES

5.1 Region. Neighbourhood

Let a pair of rectangular axes Ox, Oy be drawn in a plane. A portion of the plane bounded by a finite number of finite straight lines or arcs of simple curves, and such that the boundary does not intersect itself anywhere, is called a *region* or *domain* of the pair of variables x, y. (By an arc of a simple curve is meant here an arc along which y is a continuous monotonic function of x.) An *interior* point of a region has the property that a circle can be drawn with this point as centre so that every point of the circle belongs to the region. A point is called a *boundary point* of the region if every circle drawn with the point as centre contains points belonging to the region and points not belonging to the region. A region is said to be *closed* if it includes all its boundary points; it is said to be *open* if it includes no boundary points.

A circular region with a point (a, b) as centre is called a *neighbourhood* of (a, b).

A region is said to be *connected* if any two points of it can be connected by a finite number of segments of straight lines without passing out of the region. A connected region is said to be *simply-connected* if every closed curve drawn through points of the region can be shrunk to a point without passing out of the region; otherwise, the region is said to be *multiply-connected*.

5.2 Functions of two or more variables

For the sake of brevity, functions of two variables will usually be considered. The necessary extensions to more than two variables can often be made without difficulty.

A function of two independent variables x, y is denoted by

$$f(x, y), \ \phi(x, y), \ u(x, y), \dots$$

or simply by f, ϕ, u, ... as convenient.

A single-valued function $f(x, y)$ is said to be *defined at a point* (a, b) if there is sufficient information to determine the value of the function uniquely when $x = a$, $y = b$. The function is said to be *defined in a region* when it is defined at every point of the region.

A function defined in a closed region is either bounded or unbounded. A bounded function has a least upper bound M and a greatest lower bound m, which may or may not be values of the function (§1.15); they are also called the 'closest bounds' of the function. Their difference $M - m$ will be called the *spread* of the function in the region (§1.17).

5.3 Limit of a function of two variables at a point

Let $f(x, y)$ be defined at all points in the neighbourhood of a point (a, b) with the possible exception of the point (a, b) itself. Put

$$\rho = \sqrt{\{(x - a)^2 + (y - b)^2\}}.$$

Definition. We say that $f(x, y)$ approaches a limit l at (a, b) if, for every positive ϵ, a number δ exists such that, at every point (x, y) belonging to the circular region $0 < \rho < \delta$,

$$|f(x, y) - l| < \epsilon$$

and then we may write

$$\lim f(x, y) = l \quad \text{when} \quad (x, y) \to (a, b),$$

or $f(x, y) \to l$ when $(x, y) \to (a, b)$, or $f(x, y) \to l$ when

$$\sqrt{\{(x - a)^2 + (y - b)^2\}} \to 0, \quad \text{or when} \quad |x - a| + |y - b| \to 0;$$

or we may say that $f(x, y) \to l$ when $(x, y) \to (a, b)$ along an *arbitrary* path, or *in any manner*, in the region in which $f(x, y)$ is defined. The point (a, b) here may be an interior point or a boundary point.

Example

$$f(x, y) = y \sin (1/x) \quad (x \neq 0), \quad f(0, y) = y.$$

Consider this function near $(0, 0)$. Given any positive ϵ, choose δ so that $0 < \delta < \epsilon$. Then

(i) if $0 < |x| < \delta$ and $0 < |y| < \delta$, since $|\sin 1/x| \leqslant 1$,

$$|f(x, y)| = |y \sin 1/x| \leqslant |y| < \delta < \epsilon:$$

(ii) if $x = 0$ and $0 < |y| < \delta$, $|f(x, y)| = |y| < \delta < \epsilon$.

It follows that $\lim (y \sin 1/x) = 0$ when $(x, y) \to (0, 0)$. Here $f(x, y)$ is defined and is continuous at $(0, 0)$. (See §5.4.)

5.4 Continuity at a point

Let $f(x, y)$ be defined at all points near (a, b), including the point (a, b) itself. The following three equivalent definitions of the *continuity* of $f(x, y)$ *at the point* (a, b) correspond to the definitions in §1.18:

I. The function $f(x, y)$ is continuous at (a, b) if

$$\lim f(x, y) = f(a, b), \quad (x, y) \to (a, b),$$

a necessary condition being that the limit should exist (§5.3).

II. The function $f(x, y)$ is continuous at (a, b) if the spread of $f(x, y)$ in a circular region of radius δ, centre (a, b), tends to zero when $\delta \to 0$.

III. The function $f(x, y)$ is continuous at (a, b) if, for every positive ϵ, a number δ exists such that

$$|f(x, y) - f(a, b)| < \epsilon$$

for all (x, y) such that $\sqrt{\{(x - a)^2 + (y - b)^2\}} < \delta$.

Note. Above and occasionally elsewhere the words 'in the neighbourhood of' are abbreviated to the word 'near'.

5.5 Continuity in a region

The following two definitions of *continuity in a closed region* can be proved to be equivalent (cf. §1.20):

I. A function is continuous in a closed region if it is continuous at every point of the region, including boundary points.

II. A function is continuous in a closed region if, for every positive ϵ, a number δ exists such that the spread of the function is less than ϵ in every subregion bounded by a circle of radius δ.

5.6 Vanishing functions

If $f(a, b) = 0$ and $f(x, y)$ is continuous at (a, b), it will be convenient to say that $f(x, y)$ *vanishes* at (a, b), or that $f(x, y)$ *vanishes with* $(x - a, y - b)$.

Thus, the statement '$f(x, y)$ vanishes at $(0, 0)$' or '$f(x, y)$ vanishes with (x, y)' will mean that $f(0, 0) = 0$ and that $f(x, y) \to 0$ when $(x, y) \to (0, 0)$ in any manner, whereas the statement '$f(x, y) \to 0$ when $(x, y) \to (0, 0)$' does not imply that $f(0, 0)$ is even defined.

5.7 Properties of a continuous function of two variables (cf. §1.22)

I. If $f(x, y)$ is continuous at (a, b) it follows from §5.4, I, that we can put

$$f(x, y) = f(a, b) + \eta,$$

where η vanishes at (a, b); or we can put

$$f(a + h, b + k) = f(a, b) + \eta,$$

where η vanishes with (h, k).

II. If $f(x,y)$ is continuous at (a,b) and $f(a,b) \neq 0$, there is a neighbourhood of (a,b) in which $f(x,y)$ has the same sign as $f(a,b)$.

This follows from I, as in §1.22, II.

III. The sum and product of two, and hence of a finite number of continuous functions define continuous functions. The quotient of two continuous functions defines a continuous function, except at points where the denominator vanishes.

IV. A function which is continuous in a closed region is bounded; the l.u.b. and the g.l.b. are values of the function.

V. A continuous function $f(x,y)$ may be called a *continuous function of position* in a plane in which x,y are Cartesian coordinates. For instance, to say that $r^2 + \cos\theta$ is a continuous function of r and θ is the same as to say that it is a continuous function of position in a plane in which r, θ are Cartesian coordinates; it is not a continuous function of position at the pole in a plane in which r, θ are polar coordinates, for, in any neighbourhood of the pole, however small, $\cos\theta$ varies between 1 and -1.

While the idea of a continuous function of position may be a useful geometrical aid in thinking of a function of two, or three, variables, it will not be helpful when there are more than three. In any case, the analytical definitions of §§5.4 and 5.5 are fundamental.

VI. A continuous function cannot change sign without passing through zero (cf. §1.22, VI).

More precisely, if a continuous function $f(x,y)$ has opposite signs at two points $P_1(x_1,y_1)$ and $P_2(x_2,y_2)$, then on any continuous path joining P_1 and P_2 there must be at least one point $P_0(x_0,y_0)$ such that $f(x_0,y_0) = 0$.

For example, the function $ax+by+c$, where a, b, c are constants, is a continuous function of (x,y). If ax_1+by_1+c and ax_2+by_2+c have opposite signs, then on any continuous path joining (x_1,y_1) and (x_2,y_2) there must be at least one point (x_0,y_0) such that

$$ax_0+by_0+c = 0,$$

i.e. the path must cross the straight line $ax+by+c = 0$ at least once.

VII. It must be emphasized that continuity in both variables x and y means much more than continuity in each variable while the other remains fixed.

For example, we could define the value of $f(x,y)$ to be unity if $xy = 0$ is zero, but to be zero if $xy \neq 0$. At the origin, the function

$f(x, y)$ would then be continuous in x along the line $y = 0$, continuous in y along the line $x = 0$, but not continuous in both variables, i.e. not a continuous function of position at $(0, 0)$.

Examples

(1) $$f(x, y) = \frac{xy^2}{x^2 + y^2}, \quad x^2 + y^2 \neq 0; \quad f(0, 0) = 0.$$

Since $y^2 \leqslant x^2 + y^2$ and $f(0, 0) = 0$, therefore $|f(x, y)| \leqslant |x|$ for all (x, y). It follows, by any one of the three definitions of §5.4, that $f(x, y)$ is continuous at $(0, 0)$.

It is also continuous at every other point, being the quotient of two continuous functions, of which the denominator vanishes only at $(0, 0)$.

(2) $$f(x, y) = \frac{x^2 y}{x^4 + 2y^2}, \quad x^2 + y^2 \neq 0; \quad f(0, 0) = 0.$$

Only the point $(0, 0)$ needs consideration, as the function is evidently continuous at every other point.

Near the point $(0, 0)$ put $x = \lambda t$, $y = \mu t$; then if $t \neq 0$,

$$f(x, y) = \lambda^2 \mu t / (\lambda^4 t^2 + 2\mu^2),$$

the limit of which is zero when $t \to 0$. Thus $f(x, y) \to f(0, 0)$ when $(x, y) \to (0, 0)$ from any fixed direction. But if we put $x = t$, $y = t^2$, we get $f(x, y) = \frac{1}{3}$ $(t \neq 0)$, which shows that within any neighbourhood of $(0, 0)$, however small, there are points (x, y) such that

$$f(x, y) - f(0, 0) = \tfrac{1}{3} - 0 = \tfrac{1}{3}.$$

It follows that $f(x, y)$ is not continuous at $(0, 0)$.

[If we put $z = x^2 y / (x^4 + 2y^2)$ and suppose this equation represented by a surface, the axis of z being vertical, the contour lines will be parabolas whose projections on the xy plane are of the form $ay = x^2$.]

5.8 Partial derivatives

Let $f(x, y)$ be a function of two independent variables x and y, defined in the neighbourhood of a point $P(x, y)$. Suppose P fixed and let $(x + h, y)$ be a variable point on the line through P parallel to the axis of x. Then the *partial derivative* of $f(x, y)$ at P with respect to x is defined by

$$\frac{\partial f}{\partial x} = f_x(x, y) = \lim_{h \to 0} \frac{f(x + h, y) - f(x, y)}{h} \tag{1}$$

provided that the limit exists. Further, if $(x, y + k)$ is a variable point on the line through P parallel to the axis of y, the partial derivative at (x, y) with respect to y is defined by

$$\frac{\partial f}{\partial y} = f_y(x, y) = \lim_{k \to 0} \frac{f(x, y+k) - f(x, y)}{k} \tag{2}$$

provided that the limit exists.

A necessary condition for the existence of the first of these limits is that $f(x+h, y)$ should be a continuous function of h at $h = 0$ (§2.3) and a necessary condition for the existence of the second limit is that $f(x, y+k)$ should be a continuous function of k at $k = 0$; but

$$f(x+h, y+k)$$

need not be a continuous function of (h, k) at $h = 0, k = 0$ (§5.7, V, VII): in other words, $f(x, y)$ need not be a continuous function of position at $(0, 0)$.

The rules of ordinary differentiation apply to partial differentiation.

Examples

(1) $z = x^2 + 2y^2 - 3xy, \quad \dfrac{\partial z}{\partial x} = 2x - 3y, \quad \dfrac{\partial z}{\partial y} = 4y - 3x.$

(2) $x = r\cos\theta, \quad \dfrac{\partial x}{\partial r} = \cos\theta, \quad \dfrac{\partial x}{\partial \theta} = -r\sin\theta.$

5.9   Notation

Other ways of denoting the partial derivatives of the function $z = f(x, y)$ are $z_x', z_y'; z_x, z_y; z_1, z_2; f_x', f_y'; f_1, f_2; D_x z, D_y z; D_1 z, D_2 z$; etc.

It may be necessary to be more precise and instead of $\partial z / \partial x$, for instance, to write $(\partial z / \partial x)_y$ or $(dz/dx)_y$ when y is the only other independent variable. For example, if x, y are the usual Cartesian coordinates and r, θ the usual polar coordinates of a point P in the first quadrant, then (i) $x = r\cos\theta$, (ii) $x = \sqrt{(r^2 - y^2)}$. From (i) follows $(\partial x / \partial r)_\theta = \cos\theta$, but from (ii) follows

$$(\partial x / \partial r)_y = r/\sqrt{(r^2 - y^2)} = r/x = \sec\theta.$$

Thus $(\partial x / \partial r)_y \neq (\partial x / \partial r)_\theta$ showing that $\partial x / \partial r$ is meaningless unless the second independent variable is expressed or understood. Usually it is understood from the context.

5.10   Partial derivatives of a function of a function of two independent variables

Let $u = \phi(z)$, $z = f(x, y)$ and let $\partial z / \partial x$, $\partial z / \partial y$ exist at $P(x, y)$. Also let $\phi(z)$ be differentiable (§2.2) at the value of z corresponding to P.

Let x receive an increment δx while y remains constant. Let

$$\delta z = f(x + \delta x, y) - f(x, y)$$

be the resulting increment in z and δu the resulting increment in u. Since $\phi(z)$ is differentiable we can put, by §2.7 (13),

$$\delta u = \{\phi'(z) + \alpha\}\delta z$$

where $\alpha \to 0$ when $\delta z \to 0$. Division by δx gives

$$\delta u / \delta x = \{\phi'(z) + \alpha\}\delta z / \delta x.$$

Now let $\delta x \to 0$: then $\delta z / \delta x \to \partial z / \partial x$ since $\partial z / \partial x$ exists, and $\alpha \to 0$ since $\delta z \to 0$. Hence and similarly

$$\frac{\partial u}{\partial x} = \phi'(z)\frac{\partial z}{\partial x} = \frac{du}{dz}\frac{\partial z}{\partial x}, \quad \frac{\partial u}{\partial y} = \phi'(z)\frac{\partial z}{\partial y} = \frac{du}{dz}\frac{\partial z}{\partial y}. \tag{3}$$

Example

If $u = \phi(ax^2 + by^2)$ and $\phi(z)$ is a differentiable function of z, find $\partial u / \partial x$ and $\partial u / \partial y$.

Put $z = ax^2 + by^2$; then $u = \phi(z)$ and

$$\frac{\partial u}{\partial x} = \frac{du}{dz}\frac{\partial z}{\partial x} = \phi'(z) \cdot 2ax = 2ax\phi'(z), \quad \frac{\partial u}{\partial y} = 2by\,\phi'(z),$$

often written (§3.1 *Note*), to avoid introducing the variable z,

$$\partial u / \partial x = 2ax\phi'(ax^2 + by^2), \quad \partial u / \partial y = 2by\phi'(ax^2 + by^2).$$

5.11 Differentiability at a point

Let $f(x, y)$ be a function of two independent variables x and y, defined in the neighbourhood of a point $P(x, y)$. Suppose P fixed and let $(x + h, y + k)$ be a variable point. The function is said to be *differentiable at P* if the difference $f(x + h, y + k) - f(x, y)$ can be expressed in the form
$$f(x + h, y + k) - f(x, y) = ph + qk + \alpha h + \beta k, \tag{4}$$

where p, q are independent of h, k and α, β both vanish with (h, k).

Note. It is merely a convenience to suppose that $\alpha = 0$ and $\beta = 0$ at $(h, k) = (0, 0)$; it would be sufficient for the definition of differentiability that α and β should each tend to the limit zero as $(h, k) \to (0, 0)$, for the same kind of reason as in §2.4.

Examples

(1) If $f(x,y) = y/x$, we find

$$\frac{y+k}{x+h} - \frac{y}{x} = ph + qk + \alpha h + \beta k,$$

where

$$p = -y/x^2, \quad q = 1/x, \quad \alpha = hy/(x^3 + x^2 h), \quad \beta = -h/(x^2 + xh).$$

Here p, q are independent of h, k and α, β both vanish with (h, k) provided that $x \neq 0$. Thus the function is differentiable at (x, y) if $x \neq 0$.

(2) $f(x,y) = (x^2 + y^2) \sin (y/x) \quad (x \neq 0), \quad f(0, y) = 0$.

This function is differentiable at $(0, 0)$. For we can write

$$f(0+h, 0+k) - f(0, 0) = 0 \cdot h + 0 \cdot k + \alpha h + \beta k,$$

where α, β are defined by

$$\alpha = h \sin (k/h), \quad \beta = k \sin (k/h), \quad \text{if} \quad h \neq 0;$$
$$\alpha = 0, \quad \beta = 0, \quad \text{if} \quad h = 0;$$

thus α, β both vanish with (h, k).

5.12

Theorem. Necessary conditions that a function should be differentiable at P are that the function should be continuous and that the partial derivatives should exist at P.

Proof. From (4) it follows that $f(x+h, y+k) \to f(x, y)$ when $(h, k) \to (0, 0)$; therefore $f(x, y)$ is continuous at P.

Again from (4), putting $k = 0$ and dividing by $h \, (\neq 0)$,

$$\{f(x+h, y) - f(x, y)\}/h = p + \alpha,$$

where $\alpha \to 0$ when $h \to 0$; it follows that the limit of the L.H.S. when $h \to 0$ exists and is equal to p: that is, $\partial f/\partial x$ exists and is equal to p. Similarly, $\partial f/\partial y$ exists and is equal to q.

Corollary. If $f(x, y)$ is differentiable at P we can put

$$f(x+h, y+k) - f(x, y) = h \, \partial f/\partial x + k \, \partial f/\partial y + \alpha h + \beta k, \tag{5}$$

where α, β both vanish with (h, k).

Note. The existence of the partial derivatives is not a *sufficient* reason why the function should be differentiable: it may not even be continuous (§5.7, VII). See §5.15.

5.13　A mean value theorem

This theorem assumes that one partial derivative exists throughout a neighbourhood (§ 5.1) of (a, b) and that the other exists at (a, b).

Theorem. Let $f_x(x, y)$ exist throughout a neighbourhood of the point (a, b), and let $f_y(a, b)$ exist. Let $(a + h, b + k)$ be any point of the neighbourhood. Then we can put

$$f(a + h, b + k) - f(a, b) = hf_x(a + \theta h, b + k) + k\{f_y(a, b) + \eta\},$$

where θ, a function of h and k, lies between 0 and 1, and η is a function of k which vanishes with k.

Proof. First, since $f_x(x, b + k)$ exists in the closed interval $[a, a + h]$, the function $f(x, b + k)$ is a function of x which is continuous and differentiable in this interval. We can therefore put, by § 2.14 (22),

$$f(a + h, b + k) - f(a, b + k) = hf_x(a + \theta h, b + k) \quad (0 < \theta < 1).$$

Secondly, since $f_y(a, b)$ exists, we can put, by § 2.2 (2),

$$f(a, b + k) - f(a, b) = k\{f_y(a, b) + \eta\},$$

where $\eta \to 0$ when $k \to 0$. The theorem follows by addition.

Corollary. If $f_x(x, y)$ is continuous at (a, b) and $f_y(a, b)$ exists, we can put

$$f(a + h, b + k) - f(a, b) = h\{f_x(a, b) + \alpha\} + k\{f_y(a, b) + \eta\},$$

where α vanishes with (h, k) and η with k.

5.14　Sufficient conditions for continuity

Theorem. Sufficient conditions that a function $f(x, y)$ should be continuous at a point (a, b) are that one partial derivative should exist and be bounded in the neighbourhood of (a, b) and that the other should exist at (a, b).

Proof. Let $f_x(x, y)$ exist and be bounded in the neighbourhood of (a, b) and let $f_y(a, b)$ exist. Then it follows from the theorem of § 5.13 that $f(a + h, b + k) \to f(a, b)$ when $(h, k) \to (0, 0)$, and hence that $f(x, y)$ is continuous at (a, b).

Corollary. Sufficient conditions that a function should be continuous in a closed region are that both partial derivatives should exist and be bounded throughout the region.

Note that these conditions are sufficient but not necessary. For example, consider the hemisphere $z = \sqrt{(c^2 - x^2 - y^2)}$. The function z is continuous at every point on the circumference of the circle $x^2 + y^2 = c^2$, although at least one of the partial derivatives tends to infinity as the point (x, y) approaches any point on the circumference.

5.15 Sufficient conditions for differentiability

Theorem. Sufficient conditions that a function $f(x, y)$ should be differentiable at a point (a, b) are that one partial derivative should be continuous at (a, b) and that the other should exist at (a, b).

This follows at once from §5.11 and §5.13, Corollary.

Corollary. Sufficient conditions that a function $f(x, y)$ should be differentiable at every point of a closed region are that both partial derivatives should be continuous throughout the region.

Example

$$f(x, y) = \frac{x^3}{x^2 + y^2}, \quad x^2 + y^2 \neq 0; \quad f(0, 0) = 0.$$

Prove that:

(i) $f(x, y)$ is continuous at $(0, 0)$,

(ii) $f_x(0, 0)$ and $f_y(0, 0)$ both exist,

(iii) $f(x, y)$ is not differentiable at $(0, 0)$.

Proof.

(i) Continuity at $(0, 0)$ follows as in §5.7, Example 1.

(ii) Since $f(0 + h, 0) = h$, and $f(0, 0) = 0$, therefore

$$f_x(0, 0) = \lim (h - 0)/h = 1.$$

Since $f(0, 0 + k) = 0$, and $f(0, 0) = 0$, therefore

$$f_y(0, 0) = \lim (0 - 0)/k = 0.$$

Thus both partial derivatives exist at $(0, 0)$.

(iii) It follows from §5.12, Corollary that if $f(x, y)$ is differentiable at $(0, 0)$, then there must be an identity of the form

$$f(0 + h, 0 + k) - f(0, 0) = h^3/(h^2 + k^2) \equiv h + \alpha h + \beta k,$$

where α, β both vanish with (h, k). But this is not so, as we see, e.g. by putting $k = h \neq 0$, and dividing by h.

Note. Since $f_x(0, 0)$ and $f_y(0, 0)$ both exist, it follows further, from the present theorem, that neither $f_x(x, y)$ nor $f_y(x, y)$ can be continuous at $(0, 0)$. We leave this to be verified.

5.16 The total differential

Let $z = f(x, y)$ be differentiable at (x, y). Let $\delta x, \delta y$ be arbitrary increments in x, y within the region in which z is defined, and let δz be the resulting increment in z, so that $\delta z = f(x + \delta x, y + \delta y) - f(x, y)$. Then by (5), we can put

$$\delta z = \frac{\partial z}{\partial x} \delta x + \frac{\partial z}{\partial y} \delta y + \alpha \, \delta x + \beta \, \delta y, \tag{6}$$

where α, β both vanish with $(\delta x, \delta y)$.

Definition. The first part of the expression on the R.H.S. of (6), consisting of the first two terms, is called the *total differential* of z and is denoted by dz, so that

$$dz = \frac{\partial z}{\partial x}\,\delta x + \frac{\partial z}{\partial y}\,\delta y. \tag{7}$$

When, in particular, we put $z = x$, $z = y$ in turn, we find $dx = \delta x$, $dy = \delta y$, and so it is usual to write

$$dz = \frac{\partial z}{\partial x}\,dx + \frac{\partial z}{\partial y}\,dy. \tag{8}$$

5.17

Theorem. The total differential dz is an approximation to the increment δz in the sense that the difference $\delta z - dz$ is small compared with $\sqrt{\{(\delta x)^2 + (\delta y)^2\}}$ when $\delta x, \delta y$ are small enough (cf. §2.7).

Proof. Put $\Delta s = \sqrt{\{(\delta x)^2 + (\delta y)^2\}}$, $\delta x = \Delta s \cos \psi$, $\delta y = \Delta s \sin \psi$. Then, by subtracting (7) from (6) and dividing by Δs,

$$(\delta z - dz)/\Delta s = \alpha \cos \psi + \beta \sin \psi \tag{9}$$

which vanishes with $(\delta x, \delta y)$. This proves the theorem.

5.18   Small errors

Let $z = f(x, y)$ and let $\partial z/\partial x$, $\partial z/\partial y$ be continuous, so that z is differentiable (§5.15).

Suppose the value of z is to be calculated from measured values of x, y. Let the measured values be $x + \delta x, y + \delta y$, exceeding (algebraically) the true values x, y by $\delta x, \delta y$. Let the calculated value of z be $z + \delta z$, exceeding the true value by δz. Then by §5.17

$$\delta z \doteq dz = (\partial z/\partial x)\,\delta x + (\partial z/\partial y)\,\delta y, \tag{10}$$

which gives the approximate error dz in the calculated value of z due to the errors $\delta x, \delta y$ in the measured values of x, y.

The *relative error* or *proportional error* is defined to be $\delta z/z$; consequently, the relative error is given approximately by

$$\frac{\delta z}{z} \doteq \delta(\log z) \doteq \frac{\partial(\log z)}{\partial x}\,\delta x + \frac{\partial(\log z)}{\partial y}\,\delta y. \tag{11}$$

In practice, the maximum errors to be allowed for in the measurements of x, y will often be known, but not their signs. If the maximum errors to be expected are $\pm e_1, \pm e_2$, where $e_1 > 0$, $e_2 > 0$, the magnitude of the greatest possible error in the calculated value of z will be approximately

$$e_1|\partial z/\partial x| + e_2|\partial z/\partial y| \tag{12}$$

and the maximum relative error will be

$$e_1 |\partial(\log z)/\partial x| + e_2 |\partial(\log z)/\partial y|. \tag{13}$$

The *percentage error* or *error per cent.* is defined to be one hundred times the relative error.

5.19 Directed derivatives

Let $z = f(x, y)$ be defined near the point $P(x, y)$. Let

$$(x + \delta s \cos \psi, \; y + \delta s \sin \psi) \quad (\delta s > 0),$$

be a variable point on the line through P in the direction making an angle ψ with the axis of x. Then *the derivative at P in the direction ψ* is defined by

$$\left(\frac{dz}{ds}\right)_\psi = \lim_{\delta s \to 0} \frac{f(x + \delta s \cos \psi, y + \delta s \sin \psi) - f(x, y)}{\delta s} \tag{14}$$

provided that the limit exists.

5.20 A differentiable function has a derivative in every direction

Proof. Let $z = f(x, y)$ be differentiable at (x, y) and let δz be the increment in z when x, y receive increments

$$\delta x = \delta s \cos \psi, \quad \delta y = \delta s \sin \psi.$$

Then $\qquad \delta z = f(x + \delta s \cos \psi, \; y + \delta s \sin \psi) - f(x, y)$

and hence, by (5), since z is differentiable,

$$\delta z = (\partial z/\partial x + \alpha) \, \delta s \cos \psi + (\partial z/\partial y + \beta) \, \delta s \sin \psi, \tag{15}$$

where α, β vanish with δs. Dividing by δs and letting $\delta s \to 0$, we see that $\lim (\delta z/\delta s)$ exists, i.e. $(dz/ds)_\psi$ exists, and that

$$\left(\frac{dz}{ds}\right)_\psi = \frac{\partial z}{\partial x} \cos \psi + \frac{\partial z}{\partial y} \sin \psi. \tag{16}$$

In particular, the partial derivatives $\partial z/\partial x$, $\partial z/\partial y$ are the derivatives in the directions $\psi = 0$, $\psi = \frac{1}{2}\pi$, respectively.

5.21

Geometrical meaning. Let the equation $z = f(x, y)$ be represented by a surface of which z is the vertical ordinate, the xy plane being horizontal. Then the partial derivatives $\partial z/\partial x$, $\partial z/\partial y$ are represented at any point P on the surface by the gradients of the cross-sections of the surface made by vertical planes through P parallel to the xz and yz planes respectively; the directed derivative $(dz/ds)_\psi$ is represented by the gradient of the cross-section made by a vertical plane through P at an angle ψ with the xz plane.

The direction in which the gradient is zero, i.e. the direction of the *contour line* through P, is determined by putting $(dz/ds)_\psi = 0$ in (16): this gives

$$(\partial z/\partial x)\cos\psi + (\partial z/\partial y)\sin\psi = 0. \tag{17}$$

The direction in which the gradient at P is a maximum, i.e. the direction of *the line of greatest slope* at P, is found by differentiating (16) with respect to ψ and equating the result to zero: this gives

$$-(\partial z/\partial x)\sin\psi + (\partial z/\partial y)\cos\psi = 0. \tag{18}$$

The directions determined by (17) and (18) are evidently perpendicular : the line of greatest slope and the contour line through any point P cut at right angles.

The differentiability (§5.11) of the function $f(x,y)$ at (x,y) is the analytical equivalent of the existence of a tangent plane to the surface $z = f(x,y)$ at the point $\{x,y,f(x,y)\}$.

5.22 Total derivative of a function of two functions of one variable

Theorem. Let $z = f(x,y)$ be a differentiable function of x, y and let x, y be differentiable functions of a single independent variable t. Then z, through x and y, is a differentiable function of the single variable t and the derivative dz/dt is given by

$$\frac{dz}{dt} = \frac{\partial z}{\partial x}\frac{dx}{dt} + \frac{\partial z}{\partial y}\frac{dy}{dt}. \tag{19}$$

Proof. Let t receive an arbitrary increment δt, and let δx, δy, δz be the consequent increments in x, y, z. Then, by §2.7(13), since x,y are differentiable functions of t,

$$\delta x = (dx/dt)\,\delta t + \eta_1\delta t, \quad \delta y = (dy/dt)\,\delta t + \eta_2\delta t,$$

where η_1, η_2 vanish with δt. On substituting these values of δx, δy in (6), which holds good for all increments δx, δy since z is differentiable, we obtain a result expressible in the form

$$\delta z = \left(\frac{\partial z}{\partial x}\frac{dx}{dt} + \frac{\partial z}{\partial y}\frac{dy}{dt}\right)\delta t + \gamma\,\delta t, \tag{20}$$

where γ vanishes with δt. It follows (end of §2.2) that dz/dt exists and is given by (19).

Corollary 1. The differential dz is given by

$$dz = \left(\frac{\partial z}{\partial x}\frac{dx}{dt} + \frac{\partial z}{\partial y}\frac{dy}{dt}\right)dt \tag{21}$$

or $$dz = (\partial z/\partial x)\,dx + (\partial z/\partial y)\,dy, \tag{22}$$

where dx, dy, dz denote differentials with respect to t.

Note that (22) has exactly the same form as when x, y are independent variables.

Corollary 2. If $z = f(x, y)$ is a differentiable function of x, y and if y is a differentiable function of the single variable x, then z is a differentiable function of x and the derivative dz/dx is given by

$$\frac{dz}{dx} = \frac{\partial z}{\partial x} + \frac{\partial z}{\partial y}\frac{dy}{dx}. \tag{23}$$

Proof. Put $t = x$ in the theorem; then (23) follows.

Corollary 3. If $z = f(u, v)$ is a differentiable function of u, v and u, v are differentiable functions of the single variable x, then z, through u and v, is a differentiable function of x and the derivative dz/dx is given by

$$\frac{dz}{dx} = \frac{\partial z}{\partial u}\frac{du}{dx} + \frac{\partial z}{\partial v}\frac{dv}{dx}. \tag{24}$$

This is merely a re-statement of the theorem in another notation. Particular cases are the *product rule* when $f(u, v) = uv$, and the *quotient rule* when $f(u, v) = u/v$.

Corollary 4. If x, y are differentiable functions of t, but equation (19) does not hold good for some value of t, it follows that the function $z = f(x, y)$ cannot be differentiable at the corresponding point (x, y). For example, consider the function

$$z = xy/(x^2 + y^2), \quad xy \neq 0; \quad z = 0, \quad xy = 0.$$

If $x = t$, $y = t^2$, we find $z = t/(1 + t^2)$ for all values of t, including $t = 0$.

If $t \neq 0$, it may be verified that (19) holds good.

But if $t = 0$, the corresponding point (x, y) is $(0, 0)$; and here we find $f_x(0, 0) = 0, f_y(0, 0) = 0, dx/dt = 1, dy/dt = 0$, so that the value of the R.H.S. of (19) is 0, whereas that of the L.H.S. is 1.

It follows that the function $xy/(x^2 + y^2)$ is not differentiable at $(0, 0)$. Actually it is not even continuous, for its value at $x = r\cos\theta$, $y = r\sin\theta, r \neq 0$, is $\frac{1}{2}\sin 2\theta$, so that its closest bounds are $M = \frac{1}{2}$, $m = -\frac{1}{2}$ in any neighbourhood of $(0, 0)$, however small.

5.23　Partial derivatives of a function of two functions of two variables

Theorem. Let $z = f(u, v)$ be a differentiable function of u, v and let u, v be differentiable functions of x, y. Then z, through u and v, is a differentiable function of x, y and the partial derivatives $\partial z/\partial x$, $\partial z/\partial y$ are given by

$$\frac{\partial z}{\partial x} = \frac{\partial z}{\partial u}\frac{\partial u}{\partial x} + \frac{\partial z}{\partial v}\frac{\partial v}{\partial x}, \quad \frac{\partial z}{\partial y} = \frac{\partial z}{\partial u}\frac{\partial u}{\partial y} + \frac{\partial z}{\partial v}\frac{\partial v}{\partial y}. \tag{25}$$

Proof. Let x, y receive arbitrary increments δx, δy, and let δu, δv, δz be the consequent increments in u, v, z. Then, since u and v are differentiable, we can put

$$\delta u = (\partial u/\partial x)\,\delta x + (\partial u/\partial y)\,\delta y + \alpha_1\,\delta x + \beta_1\,\delta y,$$

$$\delta v = (\partial v/\partial x)\,\delta x + (\partial v/\partial y)\,\delta y + \alpha_2\,\delta x + \beta_2\,\delta y,$$

where $\alpha_1, \beta_1, \alpha_2, \beta_2$ all vanish with $(\delta x, \delta y)$. Also, since z is a differentiable function of u, v, we can put

$$\delta z = (\partial z/\partial u)\,\delta u + (\partial z/\partial v)\,\delta v + \alpha_3\,\delta u + \beta_3\,\delta v,$$

where α_3, β_3 both vanish with $(\delta u, \delta v)$. Substituting the above expressions for δu, δv, we get a result expressible in the form

$$\delta z = \left(\frac{\partial z}{\partial u}\frac{\partial u}{\partial x} + \frac{\partial z}{\partial v}\frac{\partial v}{\partial x}\right)\delta x + \left(\frac{\partial z}{\partial u}\frac{\partial u}{\partial y} + \frac{\partial z}{\partial v}\frac{\partial v}{\partial y}\right)\delta y + \alpha_4\,\delta x + \beta_4\,\delta y,$$

where α_4, β_4 both vanish with $(\delta x, \delta y)$. It follows, by §5.12, that z is a differentiable function of x, y and that $\partial z/\partial x$, $\partial z/\partial y$ are given by (25).

Corollary. By §5.16, the total differential dz is given by

$$dz = \left(\frac{\partial z}{\partial u}\frac{\partial u}{\partial x} + \frac{\partial z}{\partial v}\frac{\partial v}{\partial x}\right)dx + \left(\frac{\partial z}{\partial u}\frac{\partial u}{\partial y} + \frac{\partial z}{\partial v}\frac{\partial v}{\partial y}\right)dy, \tag{26}$$

which may be written

$$dz = (\partial z/\partial u)\,du + (\partial z/\partial v)\,dv, \tag{27}$$

where du, dv denote total differentials with respect to x, y. Note that (27) has exactly the same form as when u, v are independent variables.

5.24.　Extension to more than two variables

Where proofs are needed for the following statements, they can usually be constructed without difficulty on the lines already indicated.

Let $\phi = f(x, y, z, \ldots)$ be a function of the independent variables $x, y, z, \ldots$.

We assume that definitions of continuity of $\phi(x, y, z, \ldots)$ at a point and in a closed region can be formulated on the lines of §§5.4 and 5.5.

The partial derivative of ϕ with respect to x at a point $P(x, y, z, \ldots)$ is defined by

$$\frac{\partial \phi}{\partial x} = \frac{\partial f}{\partial x} = f_x = \lim_{\delta x \to 0} \frac{f(x + \delta x, y, z, \ldots) - f(x, y, z, \ldots)}{\delta x}$$

provided that the limit exists.

The existence of this limit, and consequently of $\partial \phi / \partial x$, at any point P requires as a necessary condition that at P the function ϕ should vary continuously as x varies while the other variables $y, z, \ldots$ remain fixed. It does not require that ϕ should be a continuous function of position at P, i.e. it does not require that the difference between the values of ϕ at P and P' should be infinitesimal for *every* point P' in a sufficiently small neighbourhood of the point P.

Partial derivatives with respect to $y, z, \ldots$ are defined in a similar way.

5.25

Let $\delta x, \delta y, \delta z, \ldots$ denote arbitrary increments in the independent variables and let

$$\delta \phi = f(x + \delta x, y + \delta y, \ldots) - f(x, y, \ldots)$$

be the consequent increment in ϕ. The function ϕ is said to be *differentiable* at $(x, y, z, \ldots)$ if $\delta \phi$ can be expressed in the form

$$\delta \phi = p_1 \delta x + p_2 \delta y + p_3 \delta z + \ldots + \alpha_1 \delta x + \alpha_2 \delta y + \alpha_3 \delta z + \ldots,$$

where $p_1, p_2, p_3, \ldots$ are independent of $\delta x, \delta y, \delta z, \ldots$ and $\alpha_1, \alpha_2, \alpha_3, \ldots$ all vanish with $(\delta x, \delta y, \delta z, \ldots)$.

Necessary conditions that ϕ should be differentiable at a point $P(x, y, z \ldots)$ are that ϕ should be continuous at P and that all the partial derivatives should exist at P (see §5.12). We can then put

$$p_1 = \partial \phi / \partial x, \quad p_2 = \partial \phi / \partial y, \quad p_3 = \partial \phi / \partial z, \ldots$$

and if ϕ is differentiable,

$$\delta \phi = \frac{\partial \phi}{\partial x} \delta x + \frac{\partial \phi}{\partial y} \delta y + \frac{\partial \phi}{\partial z} \delta z + \ldots + \alpha_1 \delta x + \alpha_2 \delta y + \alpha_3 \delta z + \ldots. \quad (28)$$

Sufficient conditions that ϕ should be differentiable at P are that all the partial derivatives but one should be continuous at P and that the remaining one should exist at P (see §5.15).

Sufficient conditions that ϕ should be differentiable at every point of a closed region are that all the partial derivatives should be continuous throughout the region.

5.26

If ϕ is differentiable, the *total differential $d\phi$* is defined by

$$d\phi = \frac{\partial \phi}{\partial x} \delta x + \frac{\partial \phi}{\partial y} \delta y + \frac{\partial \phi}{\partial z} \delta z + \dots . \tag{29}$$

Putting $\phi = x$, $\phi = y$, $\phi = z, \dots$ in turn, we see that $dx = \delta x$, $dy = \delta y$, $dz = \delta z, \dots$ and so we usually write

$$d\phi = \frac{\partial \phi}{\partial x} dx + \frac{\partial \phi}{\partial y} dy + \frac{\partial \phi}{\partial z} dz + \dots . \tag{30}$$

5.27

Let ϕ be a differentiable function of $x, y, z, \dots$ and let $x, y, z, \dots$ be differentiable functions of a single variable t. Then ϕ, through $x, y, z, \dots$, is a differentiable function of t and the derivative $d\phi/dt$ is given by

$$\frac{d\phi}{dt} = \frac{\partial \phi}{\partial x} \frac{dx}{dt} + \frac{\partial \phi}{\partial y} \frac{dy}{dt} + \frac{\partial \phi}{\partial z} \frac{dz}{dt} + \dots \tag{31}$$

(see §5.22). Hence also the differential $d\phi$ with respect to t is given by

$$d\phi = \frac{\partial \phi}{\partial x} dx + \frac{\partial \phi}{\partial y} dy + \frac{\partial \phi}{\partial z} dz + \dots , \tag{32}$$

where $dx, dy, dz, \dots$ denote differentials with respect to t. Note that (32) has the same form as (30) where $dx, dy, dz, \dots$ denoted arbitrary increments in $x, y, z, \dots$ respectively.

5.28

Let ϕ be a differentiable function of m variables $u, v, w, \dots$ and let $u, v, w, \dots$ be differentiable functions of n variables $x, y, z, \dots$. Then ϕ, through the m functions $u, v, w, \dots$ is a differentiable function of the n variables $x, y, z, \dots$ and the n partial derivatives $\partial\phi/\partial x, \partial\phi/\partial y, \dots$ are given by

$$\frac{\partial \phi}{\partial x} = \frac{\partial \phi}{\partial u} \frac{\partial u}{\partial x} + \frac{\partial \phi}{\partial v} \frac{\partial v}{\partial x} + \frac{\partial \phi}{\partial w} \frac{\partial w}{\partial x} + \dots , \tag{33}$$

$$\frac{\partial \phi}{\partial y} = \frac{\partial \phi}{\partial u} \frac{\partial u}{\partial y} + \frac{\partial \phi}{\partial v} \frac{\partial v}{\partial y} + \frac{\partial \phi}{\partial w} \frac{\partial w}{\partial y} + \dots \tag{34}$$

(see §5.23). The total differential of ϕ with respect to $x, y, z, \ldots$ is given by

$$d\phi = \frac{\partial \phi}{\partial x}dx + \frac{\partial \phi}{\partial y}dy + \frac{\partial \phi}{\partial z}dz + \ldots \tag{35}$$

On multiplying (33), (34), ... by dx, dy, ... respectively and adding by columns, we see that

$$d\phi = \frac{\partial \phi}{\partial u}du + \frac{\partial \phi}{\partial v}dv + \frac{\partial \phi}{\partial w}dw + \ldots, \tag{36}$$

where $d\phi$, du, dv, $dw, \ldots$ denote total differentials with respect to $x, y, z, \ldots$. Note that (36) has the same form as though $u, v, w, \ldots$ were independent variables.

Particular cases of the above results previously discussed were concerned with $m = 1$, $n = 2$ in §5.10; $m = 2$, $n = 1$ in §5.22; $m = 2$, $n = 2$ in §5.23.

5.29 Euler's theorem on homogeneous functions

Definition. A function $f(x, y, z, \ldots)$ is said to be *homogeneous of degree n* if it satisfies the identity

$$f(tx, ty, tz, \ldots) \equiv t^n f(x, y, z \ldots). \tag{37}$$

Euler's theorem. If ϕ is a differentiable homogeneous function of $x, y, z, \ldots$ of degree n, then

$$x\frac{\partial \phi}{\partial x} + y\frac{\partial \phi}{\partial y} + z\frac{\partial \phi}{\partial z} + \ldots = n\phi. \tag{38}$$

Conversely, any differentiable solution of this equation is homogeneous of degree n in x, y, $z, \ldots$.

Proof. Let $\phi = f(x, y, z, \ldots)$ be a given homogeneous function of $x, y, z, \ldots$ of degree n and differentiable. Then the function satisfies the identity (37), and by differentiation of both sides with respect to t, we get

$$\Sigma x f_x(tx, ty, tz, \ldots) \equiv nt^{n-1} f(x, y, z, \ldots)$$

from which follows (38) by putting $t = 1$.

Secondly, let the equation (38) be given and let $\phi = f(x, y, z, \ldots)$ be a solution. Then

$$\Sigma x f_x(x, y, z, \ldots) \equiv nf(x, y, z, \ldots)$$

will be an identity. We can therefore replace x by tx, y by ty, z by $tz, \ldots$. The result can be written $t\,\partial T/\partial t = nT$, where $T = f(tx, ty, tz, \ldots)$.

Hence follows $T/t^n = C$ where C is independent of t. Putting $t = 1$ gives $C = f(x, y, z, ...)$ and therefore $T = t^n f(x, y, z, ...)$, that is,

$$f(tx, ty, tz, ...) \equiv t^n f(x, y, z, ...),$$

which was to be proved.

5.30   The Jacobian

Let ϕ, ψ be differentiable functions of x, y.

Definition. The Jacobian J of ϕ, ψ with respect to x, y is defined by

$$J = J(\phi, \psi) = \frac{\partial(\phi, \psi)}{\partial(x, y)} = \begin{vmatrix} \partial\phi/\partial x & \partial\phi/\partial y \\ \partial\psi/\partial x & \partial\psi/\partial y \end{vmatrix}. \tag{39}$$

The single letter J is used when the variables are known from the context; $J(\phi, \psi)$ may be used when the independent variables are known.

Theorem. If ϕ, ψ are differentiable functions of u, v and u, v are differentiable functions of x, y, so that ϕ, ψ are differentiable functions of x, y through u, v, then

$$\frac{\partial(\phi, \psi)}{\partial(x, y)} = \frac{\partial(\phi, \psi)}{\partial(u, v)} \frac{\partial(u, v)}{\partial(x, y)}. \tag{40}$$

Proof. By (25) we have

$$\begin{vmatrix} \dfrac{\partial\phi}{\partial x} & \dfrac{\partial\phi}{\partial y} \\[2mm] \dfrac{\partial\psi}{\partial x} & \dfrac{\partial\psi}{\partial y} \end{vmatrix} = \begin{vmatrix} \dfrac{\partial\phi}{\partial u}\dfrac{\partial u}{\partial x} + \dfrac{\partial\phi}{\partial v}\dfrac{\partial v}{\partial x} & \dfrac{\partial\phi}{\partial u}\dfrac{\partial u}{\partial y} + \dfrac{\partial\phi}{\partial v}\dfrac{\partial v}{\partial y} \\[2mm] \dfrac{\partial\psi}{\partial u}\dfrac{\partial u}{\partial x} + \dfrac{\partial\psi}{\partial v}\dfrac{\partial v}{\partial x} & \dfrac{\partial\psi}{\partial u}\dfrac{\partial u}{\partial y} + \dfrac{\partial\psi}{\partial v}\dfrac{\partial v}{\partial y} \end{vmatrix}$$

from which (40) follows, by the rule for multiplication of determinants.

5.31

Extension. The Jacobian of a set of n functions $y_1, y_2, ..., y_n$ of n variables $x_1, x_2, ..., x_n$ is defined by

$$J = \frac{\partial(y_1, y_2, ..., y_n)}{\partial(x_1, x_2, ..., x_n)} = \begin{vmatrix} \partial y_1/\partial x_1 & \cdots & \partial y_1/\partial x_n \\ \cdots\cdots\cdots\cdots\cdots\cdots \\ \partial y_n/\partial x_1 & \cdots & \partial y_n/\partial x_n \end{vmatrix}, \tag{41}$$

the element in the rth row and sth column of the determinant being $\partial y_r/\partial x_s$.

If $y_1, y_2, ..., y_n$ are differentiable functions of $u_1, u_2, ..., u_n$ and $u_1, u_2, ..., u_n$ are differentiable functions of $x_1, x_2, ..., x_n$, it follows as above, by the rule for multiplying determinants, that

$$\frac{\partial(y_1, y_2, ..., y_n)}{\partial(x_1, x_2, ..., x_n)} = \frac{\partial(y_1, y_2, ..., y_n)}{\partial(u_1, u_2, ..., u_n)} \frac{\partial(u_1, u_2, ..., u_n)}{\partial(x_1, x_2, ..., x_n)} \tag{42}$$

which, when convenient, may be abbreviated to

$$\frac{\partial(y)}{\partial(x)} = \frac{\partial(y)}{\partial(u)}\frac{\partial(u)}{\partial(x)}. \tag{43}$$

If there are two sets of intermediate variables, by applying (43) twice, we find

$$\frac{\partial(y)}{\partial(x)} = \frac{\partial(y)}{\partial(v)}\frac{\partial(v)}{\partial(u)}\frac{\partial(u)}{\partial(x)}. \tag{44}$$

Examples 5 A

(1) Indicate in a sketch the region to which the point (x, y) belongs when x, y satisfy the inequalities:

(i) $x^2 + y^2 \leqslant a^2$;

(ii) $y^2 \leqslant 4x$;

(iii) $\sqrt{\{(x-a)^2 + (y-b)^2\}} < \delta$;

(iv) $3x + 4y > 5$;

(v) $|x-a| + |y-b| < \delta$;

(vi) $-3x - 4y > 5$;

(vii) $x^2 + y^2 > 4, \quad x > y$;

(viii) $y^2 > 8x, \quad y - x < 2$;

(ix) $x^2 + y^2 < 25, \quad xy < 12$;

(x) $|x-a| < \alpha, \quad |y-b| < \beta$;

(xi) $x^2 > y^2, \quad 0 < x < 1$;

(xii) $xy < 0, \quad x - y - 1 < 0$.

(2) (i) $f(x, y) = \dfrac{x^2 + y^2}{x^2 - y^2}$,   (ii) $f(x, y) = \dfrac{x + y + x^2 + y^2}{x - y}$.

In each case, find the limit of $f(x, y)$ when $(x, y) \to (0, 0)$ along the line $y = mx$; deduce that $\lim f(x, y)$ when $(x, y) \to (0, 0)$ in any manner does not exist.

(3) $u = \dfrac{\sin x + \sin 2y}{\tan 2x + \tan y}$.

Find the limit of u when

(i) first $x \to 0 \,(y \neq 0)$, then $y \to 0$;   (ii) first $y \to 0 \,(x \neq 0)$, then $x \to 0$;

           (iii) $(x, y) \to (0, 0)$ along the line $y = mx$.

Also, find the limit of u when (i) first $x \to \pi \,(y \neq \pi)$, then $y \to \pi$; (ii) first $y \to \pi \,(x \neq \pi)$, then $x \to \pi$; (iii) $(x, y) \to (\pi, \pi)$ along the line $y = x$.

(4) $f(x, y) = y^x \quad (x \geqslant 0, y \geqslant 0, x + y \neq 0)$.

Find the limit of $f(x, y)$ when (i) first $y \to 0 \,(x \neq 0)$, then $x \to 0$; (ii) $(x, y) \to (0, 0)$ along the line $y = mx \,(m \neq 0)$.

(5) $f(x, y) = \dfrac{(x-y)^2}{1 - 2xy + y^2}$.

Find the limit of $f(x, y)$ when (i) first $x \to 1 \,(y \neq 1)$, then $y \to 1$; (ii) first $y \to 1 \,(x \neq 1)$, then $x \to 1$. Deduce that $f(x, y)$ cannot be continuous at $(1, 1)$ however $f(1, 1)$ is defined.

(6) Show that each of the functions

(i) $(x+y)^3/(x^2 + y^2)$,   (ii) $xy/\sqrt{(x^2 + y^2)}$,   (iii) $(ax + by) \sin(x/y)$

is continuous at $(0, 0)$, given that its value at $(0, 0)$ is zero in cases (i) and (ii), and that in case (iii) its value at $(x, 0)$ is zero for all values of x.

(7) Show that

$$\lim_{y \to 0} \left(\lim_{x \to 0} \frac{y}{x^2 + y^2} \right) \text{ does not exist,} \quad \lim_{x \to 0} \left(\lim_{y \to 0} \frac{y}{x^2 + y^2} \right) = 0.$$

(8) $$f(x, \alpha) = \frac{\sin^n \alpha}{1 - 2x \cos \alpha + x^2}, \quad f(1, 0) = 0.$$

Show that $f(x, \alpha)$ is continuous at $(1, 0)$ if $n > 2$, but discontinuous if $n \leqslant 2$.

(9) If the functions $f(x, y)$ and $g(x, y)$ are continuous at (a, b), prove that the same is true of the functions $f \pm g, fg$, and f/g if $g(a, b) \neq 0$. [Use §5.7, I.]

(10) Show that, if

$$r = \sqrt{(x^2 + y^2)}, \quad \cos \theta = x/r, \quad \sin \theta = y/r,$$

(i) r is continuous over the whole xy plane,

(ii) $\cos \theta, \sin \theta$ are not continuous at the origin.

Examples 5 B

In the following examples assume, where necessary, that the given functions are differentiable.

(1) Find $\partial z/\partial x$ and $\partial z/\partial y$ and verify that $(\partial/\partial x)(\partial z/\partial y) = (\partial/\partial y)(\partial z/\partial x)$ when

(i) $z = \dfrac{\sin(xy)}{x}$,   (ii) $z = \dfrac{x+y}{1-xy}$,   (iii) $z = x \tan^{-1} \dfrac{y}{x}$,

(iv) $z = (x^2 + y^2)/\sin(2x - 3y)$,   (v) $z = e^x(x \cos y - y \sin y)$.

(2) Verify the given results:

(i)	$z = ax^2 + 2hxy + by^2$,	$x\,\partial z/\partial x + y\,\partial z/\partial y = 2z$;
(ii)	$u = ax^2 + by^2 + cz^2 + 2fyz + 2gzx + 2hxy$,	$\Sigma x\,\partial u/\partial x = 2u$;
(iii)	$z = \sin(ax + by + c)$,	$b\,\partial z/\partial x = a\,\partial z/\partial y$;
(iv)	$u = (y-z)(z-x)(x-y)$,	$\Sigma \partial u/\partial x = 0, \Sigma x\,\partial u/\partial x = 3u$;
(v)	$\phi = (x^2 + y^2 + z^2)^{-n}$,	$\Sigma x\,\partial \phi/\partial x + 2n\phi = 0$;
(vi)	$u = \sqrt{(x^2 - y^2)}\sin^{-1}(y/x)$,	$x\,\partial u/\partial x + y\,\partial u/\partial y = u$;
(vii)	$u = \{x + \sqrt{(x^2 + y^2)}\}^{\frac{1}{2}}$,	$x\,\partial u/\partial x + y\,\partial u/\partial y = \tfrac{1}{2}u$;
(viii)	$V = z^n \exp(-y/x)$,	$\Sigma x\,\partial V/\partial x = nV$;
(ix)	$z = f(x + ay)$,	$\partial z/\partial y = a\,\partial z/\partial x$;
(x)	$V = \phi(y/x)/(ax + by + cz)$,	$\Sigma x\,\partial V/\partial x + V = 0$;
(xi)	$u = f(x^2 + y^2)$,	$y\,\partial u/\partial x = x\,\partial u/\partial y$;
(xii)	$u = f(x^2 + y^2) + \phi(y/x)$,	$y(\partial/\partial x)(x\,\partial u/\partial x + y\,\partial u/\partial y)$ $= x(\partial/\partial y)(x\,\partial u/\partial x$ $+ y\,\partial u/\partial y).$

(3) If $\theta = t^n \exp(-\frac{1}{4}r^2/t)$, show that

$$\frac{\partial \theta}{\partial t} = \frac{1}{r^2} \frac{\partial}{\partial r}\left(r^2 \frac{\partial \theta}{\partial r} \right)$$

provided that the constant n has a certain value: find this value.

(4) If $V = (1 - 2xr + r^2)^{-\frac{1}{2}}$, verify that

$$\frac{\partial}{\partial r}\left(r^2 \frac{\partial V}{\partial r}\right) + \frac{\partial}{\partial x}\left\{(1 - x^2)\frac{\partial V}{\partial x}\right\} = 0.$$

(5) If P, Q, R are functions of x, y, z that satisfy the equation

$$P(\partial Q/\partial z - \partial R/\partial y) + Q(\partial R/\partial x - \partial P/\partial z) + R(\partial P/\partial y - \partial Q/\partial x) = 0,$$

verify that $\lambda P, \lambda Q, \lambda R$ satisfy the same equation, where λ is also a function of x, y, z.

(6) $$f(x, y) = \frac{xy}{x^2 + y^2} \quad (x^2 + y^2 \neq 0), \quad f(0, 0) = 0.$$

Show that $f_x(0, 0)$ and $f_y(0, 0)$ both exist, but that $f(x, y)$ is not continuous at $(0, 0)$.

(7) $$f(x, y) = \frac{xy(x^2 - y^2)}{x^2 + y^2} \quad (x^2 + y^2 \neq 0), \quad f(0, 0) = 0.$$

Show that $f(x, y)$ is differentiable at $(0, 0)$.

(8) Verify the statement made in the Note at the end of §5.15.

Examples 5C

(1) Obtain the given differentials:

(i)	$z = xy$,	$dz = x\,dy + y\,dx$;
(ii)	$z = y/x$,	$dz = (x\,dy - y\,dx)/x^2$;
(iii)	$z = \tan\dfrac{y}{x}$,	$dz = \dfrac{x\,dy - y\,dx}{x^2}\sec^2\dfrac{y}{x}$;
(iv)	$z = \tan^{-1}(y/x)$,	$dz = (x\,dy - y\,dx)/(x^2 + y^2)$;
(v)	$z = \sin^m\theta \cos^n\phi$,	$dz = z(m\cot\theta\,d\theta - n\tan\phi\,d\phi)$;
(vi)	$z = \tan^2\theta/\tan^2\phi$,	$dz = 4z(\operatorname{cosec} 2\theta\,d\theta - \operatorname{cosec} 2\phi\,d\phi)$;
(vii)	$z = \log\{xy/(x+y)\}$,	$dz = (x^2\,dy + y^2\,dx)/(x^2 y + xy^2)$;
(viii)	$T = 2\pi\sqrt{(l/g)}$,	$dT/T = \frac{1}{2}(dl/l - dg/g)$;
(ix)	$V = \frac{1}{3}\pi h(a^2 + ab + b^2)$,	$\dfrac{dV}{V} = \dfrac{dh}{h} + \dfrac{(2a+b)\,da + (a+2b)\,db}{a^2 + ab + b^2}$.

(2) The values of a, b, c are measured and the value of V is calculated from the formula $V = \frac{1}{3}\pi c(a^2 + ab + b^2)$. If there are percentage errors e_1, e_2, e_3 in the measured values of a, b, c, respectively, show that the percentage error in the value of V will be

$$e_1 + e_2 + e_3 + (a^2 - b^2)(e_1 - e_2)/(a^2 + ab + b^2).$$

(3) The weight of a piece of metal was measured and found to be W in air and W' in water. From these measurements the specific gravity s was calculated. If $\delta W, \delta W'$ were the errors made in measuring W and W', show that the percentage error in the calculated value of s was approximately

$$100(s - 1)(\delta W'/W' - \delta W/W).$$

(4) An angle ABC was once drawn on a sheet of parchment so that BA, BC made angles θ, $\theta+\phi$ degrees with the length of the sheet. The parchment has since shrunk α per cent. lengthways and β per cent. sideways. Show that the angle ABC now measures approximately in degrees

$$\phi+(9/5\pi)\,(\alpha-\beta)\cos\,(2\theta+\phi)\sin\phi.$$

(5) Verify formula (19) when

(i) $z=x\tan\,(y/x), x=at^2, y=2at$; provided that $t\neq 4/(2n+1)\,\pi$ where n is an integer;

(ii) $z=\tan^{-1}(y/x), x=\cos t, y=\sin t$; provided that z is defined so as to vary continuously with t.

(6) (i) Verify (24) when $z=u^3+v^3+3uv, u=\sin x, v=e^x$.

(ii) Verify (25) when $z=u^2-v^2, u=x\cos y, v=x\sin y$.

(7) If V is a differentiable function of x,y and $x=r\cos\theta, y=r\sin\theta$, show that

$$\frac{\partial V}{\partial r}=\cos\theta\,\frac{\partial V}{\partial x}+\sin\theta\,\frac{\partial V}{\partial y},\quad \frac{\partial V}{r\,\partial\theta}=-\sin\theta\,\frac{\partial V}{\partial x}+\cos\theta\,\frac{\partial V}{\partial y}.$$

If $V=x^3-3xy^2$, $x=r\cos\theta, y=r\sin\theta$, show that

$$\frac{\partial V}{\partial r}=3r^2\cos 3\theta,\quad \frac{\partial V}{\partial\theta}=-3r^3\sin 3\theta.$$

(8) Find the maximum gradient at the point where $x=2, y=1$ on the surface $2z=2x^2+3y^2$, the xy plane being horizontal.

(9) By applying the ordinary mean-value theorem, show that

$$F(1)=F(0)+F'(\theta)\quad (0<\theta<1),$$

where $F(t)$ is a continuous function at $t=0$ and $t=1$, and $F'(t)$ exists in the interval $0<t<1$.

By putting $F(t)=f(x,y)$, $x=a+ht$, $y=b+kt$, deduce the following mean-value theorem for a differentiable function of two variables:

$$f(a+h,b+k)=f(a,b)+hf_x(\xi,\eta)+kf_y(\xi,\eta)$$

where $\qquad\qquad \xi=a+\theta h,\quad \eta=b+\theta k\quad (0<\theta<1).$

(10) If $z=\sqrt{|xy|}, x=t, y=t+t^3$, verify that (19) does not hold good at $t=0$.

Examples 5 D

(1) Verify Euler's theorem for the homogeneous functions:

(i) $ax+by+cz$ $(n=1)$,   (ii) $ax^2+2hxy+by^2$ $(n=2)$,

(iii) $\tan^{-1}(y/x)$ $(n=0)$,   (iv) $(x+y)/(\sqrt{x}+\sqrt{y})$ $(n=\frac{1}{2})$,

(v) $(x^2+y^2+z^2)^{-\frac{1}{2}}$ $(n=-1)$.

(2) Show that a homogeneous function of x,y,z of degree n can be expressed in the form $x^n\phi(y/x,z/x)$. Prove Euler's theorem, using this form.

(3) If $x=r\cos\theta, y=r\sin\theta$, show that

$$\frac{\partial(x,y)}{\partial(r,\theta)}=r.$$

(4) If $xu = yv = x^2 + y^2$, show that

$$\frac{\partial(u, v)}{\partial(x, y)} = -\frac{uv}{xy}.$$

(5) If $x = f(u) \operatorname{sech} v, y = f(u) \tanh v$, show that

$$\frac{\partial(x, y)}{\partial(u, v)} = f(u)f'(u) \operatorname{sech} v.$$

(6) If $x = aX + bY, y = cX + dY$ and f, g are functions of x, y, show that

$$\frac{\partial(f, g)}{\partial(X, Y)} = M \frac{\partial(f, g)}{\partial(x, y)}, \quad M = ad - bc.$$

(7) If $x = r \sin \theta \cos \phi, y = r \sin \theta \sin \phi, z = r \cos \theta$, show that

$$\frac{\partial(x, y, z)}{\partial(r, \theta, \phi)} = r^2 \sin \theta.$$

(8) If $xu = yv = zw = x^2 + y^2 + z^2$, show that

$$\frac{\partial(u, v, w)}{\partial(x, y, z)} = \frac{uvw}{xyz}.$$

(9) If $u = yz/x, v = zx/y, w = xy/z$, show that

$$\frac{\partial(u, v, w)}{\partial(x, y, z)} = 4.$$

(10) If α, β, γ are the roots of $x^3 + px^2 + qx + r = 0$, show that

$$\frac{\partial(p, q, r)}{\partial(\alpha, \beta, \gamma)} = (\alpha - \beta)(\alpha - \gamma)(\beta - \gamma).$$

(11) If $u/x = v/y = w/z = 1/(x^2 + y^2 + z^2)$, show that

$$\frac{\partial(u, v, w)}{\partial(x, y, z)} = -\frac{1}{(x^2 + y^2 + z^2)^3}.$$

(12) If ϕ, ψ are differentiable functions of x, y, z, show that

$$\frac{\partial \phi}{\partial x} \frac{\partial(\phi, \psi)}{\partial(y, z)} + \frac{\partial \phi}{\partial y} \frac{\partial(\phi, \psi)}{\partial(z, x)} + \frac{\partial \phi}{\partial z} \frac{\partial(\phi, \psi)}{\partial(x, y)} = 0.$$

(13) If ϕ, ψ are differentiable functions of u, v, w and u, v, w are differentiable functions of x, y, prove that

$$\frac{\partial(\phi, \psi)}{\partial(x, y)} = \frac{\partial(\phi, \psi)}{\partial(v, w)} \frac{\partial(v, w)}{\partial(x, y)} + \frac{\partial(\phi, \psi)}{\partial(w, u)} \frac{\partial(w, u)}{\partial(x, y)} + \frac{\partial(\phi, \psi)}{\partial(u, v)} \frac{\partial(u, v)}{\partial(x, y)}.$$

(14) If x, y, z are differentiable functions of u, v, show that

$$\frac{\partial(y, z)}{\partial(u, v)} dx + \frac{\partial(z, x)}{\partial(u, v)} dy + \frac{\partial(x, y)}{\partial(u, v)} dz = 0.$$

(15) Prove that a common factor of two homogeneous functions of the same degree is a squared factor of their Jacobian. In fact, if $f = \lambda u, g = \lambda v$, where f, g are homogeneous of degree n and λ is of degree m, prove that

$$\frac{\partial(f, g)}{\partial(x, y)} = \frac{n\lambda^2}{n - m} \frac{\partial(u, v)}{\partial(x, y)}.$$

(16) Prove that, if λ, u, v are differentiable functions of x, y, then

$$\frac{\partial(\lambda u, \lambda v)}{\partial(x, y)} = 2\lambda^2 \frac{\partial(u, v)}{\partial(x, y)} - \lambda \begin{vmatrix} \lambda & \partial\lambda/\partial x & \partial\lambda/\partial y \\ u & \partial u/\partial x & \partial u/\partial y \\ v & \partial v/\partial x & \partial v/\partial y \end{vmatrix}.$$

(17) Prove that, if λ, u, v, w are differentiable functions of x, y, z, then

$$J(\lambda u, \lambda v, \lambda w) = 2\lambda^3 J(u, v, w) - \lambda^2 K(\lambda, u, v, w),$$

where x, y, z are independent variables and

$$K(\lambda, u, v, w) = \begin{vmatrix} \lambda & \partial\lambda/\partial x & \partial\lambda/\partial y & \partial\lambda/\partial z \\ u & \partial u/\partial x & \partial u/\partial y & \partial u/\partial z \\ v & \partial v/\partial x & \partial v/\partial y & \partial v/\partial z \\ w & \partial w/\partial x & \partial w/\partial y & \partial w/\partial z \end{vmatrix}.$$

(18) Prove that, if λ is a differentiable function of x, y, z,

$$\frac{\partial(\lambda x, \lambda y, \lambda z)}{\partial(x, y, z)} = \lambda^3 + \lambda^2 \left(x\frac{\partial\lambda}{\partial x} + y\frac{\partial\lambda}{\partial y} + z\frac{\partial\lambda}{\partial z} \right).$$

6

IMPLICIT FUNCTIONS

6.1

The equation $y^2 = x$, with the condition that y is continuous and $y = 1$ when $x = 1$, defines y implicitly as a single-valued function $(y = \sqrt{x})$ in the neighbourhood of $x = 1$. (Cf. §1.4, Example 7.)

The equation $(y-1)^2 = x - 1$, with the same condition does not define a single-valued function, since both $y = 1 + \sqrt{(x-1)}$ and $y = 1 - \sqrt{(x-1)}$ would satisfy the conditions, provided that $x \geqslant 1$.

The equation $x^2 + y^3 = 0$, with $y = 0$ when $x = 0$ and x and y real, defines y as a single-valued function of x near $x = 0$.

The equation $x^2 + y^2 = 0$, with $y = 0$ when $x = 0$ and x and y real, does not define y as a function of x, since $(0,0)$ is the only pair of real values that satisfy this equation.

The equation $(x-1)^2 + (y-2)^2 + 4 = 0$ is not satisfied by any pair of real values of x and y.

In the first theorem below (§6.2) sufficient conditions are stated in order that the equation $F(x,y) = 0$ may define y as a single-valued continuous real function, say $y = \phi(x)$, in the neighbourhood of some particular value of x. The second theorem (§6.3) states sufficient conditions for this function to have a derived function given by the usual rule

$$\frac{dy}{dx} = \phi'(x) = -\frac{\partial F/\partial x}{\partial F/\partial y}. \tag{1}$$

The third theorem (§6.4) discusses $F(x,y,z) = 0$ as an equation defining z implicitly as a function of the independent variables x and y; the fourth theorem (§6.5) states the generalization of these theorems. It is assumed that only real values of x and y are under consideration.

6.2

Theorem. (i) Let $x = a$, $y = b$ satisfy the equation $F(x,y) = 0$, so that $F(a,b) = 0$.

In the neighbourhood of (a,b) let the function $F(x,y)$ be

(ii) a continuous function of position (§5.7, V),

(iii) strictly monotonic wo y when x is fixed.

Then the equation $F(x, y) = 0$, together with the condition $y = b$ when $x = a$, defines y as a single-valued continuous function of x, say $y = \phi(x)$, in the neighbourhood of $x = a$.

As a consequence, when we put $y = \phi(x)$ in the equation $F(x, y) = 0$, it will become an identity $F\{x, \phi(x)\} \equiv 0$ in the neighbourhood of a.

[Geometrically, the equation $F(x, y) = 0$ then represents a curve which passes through the point (a, b) and is continuous near this point.]

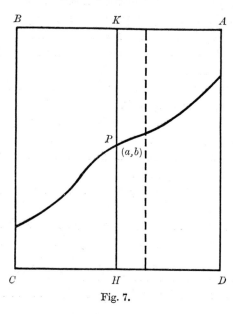

Fig. 7.

Proof. Let P be the point (a, b) in figure 7. Then, by (ii) and (iii), about P as centre we can describe a circle enclosing a region R within which $F(x, y)$ is a continuous function of position and, let us say, strictly increasing wo y at every point.

[If $F(x, y)$ were decreasing, we could consider $-F(x, y)$, which would be increasing.]

Let $ABCD$ be the rectangle $(a - h \leqslant x \leqslant a + h, b - k \leqslant y \leqslant b + k)$, entirely included in R, with its centre at P, and let K, H be the mid-points of AB, CD, respectively.

The function $y = \phi(x)$ is defined and single-valued

Since $F(a, y)$ strictly increases from $F(a, b - k)$ at H to $F(a, b + k)$ at K, passing through zero at P, therefore $F(a, b - k) < 0 < F(a, b + k)$, by §2.10. Hence, and by (ii), if h is small enough, $F(x, y)$ is positive along AB, negative along CD.

Now let x be fixed. Then $F(x, y)$ strictly increases from a negative value at $(x, b - k)$ to a positive value at $(x, b + k)$ and therefore vanishes for one and only one value of y between $b - k$ and $b + k$, by § 1.22, VI, Corollary.

The equation $F(x, y) = 0$ is thus satisfied by one and only one value of y in the interval $b - k < y < b + k$ for every value of x in the interval $a - h \leqslant x \leqslant a + h$, and so defines a single-valued function, $y = \phi(x)$, near $x = a$.

The function $y = \phi(x)$ is continuous at $x = a$

For, given any positive ϵ, we could have chosen k so that $k < \epsilon$, and then have determined h. Thus, given ϵ, we could have found h so that $|y - b| < \epsilon$ when $|x - a| \leqslant h$; consequently, since $y = b$ when $x = a$, the function $y = \phi(x)$ is continuous at $x = a$ (§ 1.18, II).

The function $y = \phi(x)$ is continuous near $x = a$

Let $x = a'$, where $a - h \leqslant a' \leqslant a + h$, and let $b' = \phi(a')$. Then $F(a', b') = 0$, by the first part of the proof. Consequently, conditions (i), (ii), (iii) are all satisfied for the point (a', b'), and so, by the second part of the proof, the function $y = \phi(x)$ is continuous at $x = a'$.

Example

The equation $y^3 - x = 0$ defines the single-valued continuous function $y = x^{\frac{1}{3}}$ near $x = 0$ (and, in fact, near any other value of x). The equation $y^2 - x = 0$ does not define such a function near $x = 0$; in this case, condition (iii) is not satisfied, since $y^2 - x$ is not monotonic wo y in the neighbourhood of $(0, 0)$.

6.3

Theorem. Let $x = a, y = b$ satisfy the equation $F(x, y) = 0$. Also, let the partial derivatives $F_x(x, y)$ and $F_y(x, y)$ be continuous near (a, b), with $F_y(a, b) \neq 0$.

Then the equation $F(x, y) = 0$, together with the condition $y = b$ when $x = a$, defines y as a differentiable function of x, say $y = \phi(x)$, in the neighbourhood of $x = a$, with a continuous derivative given by the usual rule

$$\frac{dy}{dx} = \phi'(x) = -\frac{\partial F / \partial x}{\partial F / \partial y} = -\frac{F_x}{F_y}. \tag{2}$$

[Geometrically, the equation $F(x, y) = 0$ then represents a curve which passes through the point (a, b) and has a continuously varying tangent near this point.]

Proof. Since F_x and F_y are continuous, $F(x, y)$ is differentiable in the neighbourhood of (a, b), by §5.15, Corollary.

Since $F_y(a, b) \neq 0$ and F_y is continuous, there is a neighbourhood of (a, b) in which F_y is one-signed: suppose $F_y > 0$.

Then we can draw a circle, centre (a, b), enclosing a region R in which $F(x, y)$ is (i) differentiable and therefore (ii) continuous, and in which (iii) $F_y > 0$ so that $F(x, y)$ strictly increases wo y when x is fixed.

By (ii) and (iii) and §6.2, the equation $F(x, y) = 0$, together with the condition $y = b$ when $x = a$, defines y as a single-valued continuous function, $y = \phi(x)$, near $x = a$.

Now let (x, y) and $(x + \delta x, y + \delta y)$ be points in the region R; then by (i) and §5.12 (5),

$$F(x + \delta x, y + \delta y) - F(x, y) = (F_x + \alpha)\, \delta x + (F_y + \beta)\, \delta y,$$

where α and β vanish with $(\delta x, \delta y)$. In particular, if x and $x + \delta x$ belong to the interval in which $y = \phi(x)$ is defined and continuous, and if

$$y = \phi(x), \quad y + \delta y = \phi(x + \delta x),$$

then $F(x, y) = 0$, $F(x + \delta x, y + \delta y) = 0$, and $\delta y \to 0$ when $\delta x \to 0$, and we have

$$0 = (F_x + \alpha)\, \delta x + (F_y + \beta)\, \delta y,$$

where α and β now vanish with δx. Hence and since $F_y \neq 0$, if δx is so small that $F_y + \beta \neq 0$,

$$\frac{\delta y}{\delta x} = \frac{\phi(x + \delta x) - \phi(x)}{\delta x} = -\frac{F_x + \alpha}{F_y + \beta}$$

from which (2) follows when $\delta x \to 0$.

The derivative $\phi'(x)$ is continuous, since F_x and F_y are continuous.

6.4

Theorem. Let $x = a$, $y = b$, $z = c$ satisfy the equation $F(x, y, z) = 0$ so that $F(a, b, c) = 0$. Also, let the partial derivatives F_x, F_y, F_z be continuous near (a, b, c), with $F_z(a, b, c) \neq 0$.

Then the equation $F(x, y, z) = 0$, together with the condition $z = c$ when $x = a$, $y = b$, defines z as a differentiable function of the independent variables x and y, say $z = \phi(x, y)$, in the neighbourhood of (a, b), with continuous partial derivatives given by

$$\frac{\partial z}{\partial x} = -\frac{\partial F/\partial x}{\partial F/\partial z} = -\frac{F_x}{F_z}, \quad \frac{\partial z}{\partial y} = -\frac{\partial F/\partial y}{\partial F/\partial z} = -\frac{F_y}{F_z}. \tag{3}$$

[Geometrically, the equation $F(x, y, z) = 0$ then represents a surface which passes through the point (a, b, c) and has a continuously varying tangent plane near this point.]

Proof. Since F_x, F_y, F_z are continuous, $F(x, y, z)$ is differentiable, and therefore continuous, in the neighbourhood of (a, b, c), by §5.25.

Since $F_z(a, b, c) \neq 0$ and F_z is continuous, F_z is one-signed near (a, b, c): suppose $F_z > 0$.

It follows, by reasoning as in §6.2, that there exists a cylinder $\sqrt{\{(x-a)^2 + (y-b)^2\}} \leqslant \delta \quad (c-l \leqslant z \leqslant c+l)$, such that the function $F(x, y, z)$ is positive over the end $z = c+l$, negative over the end $z = c-l$, and strictly increasing wo z at every point in the cylinder. Consequently, along every generator of every coaxial cylinder of smaller radius, the function $F(x, y, z)$ strictly increases from a negative value where this generator meets the end $z = c-l$ to a positive value where it meets the end $z = c+l$, and therefore vanishes at one and only one point between the ends.

The equation $F(x, y, z) = 0$ is thus satisfied by one and only one value of z between $c-l$ and $c+l$ for every pair of values of x and y in the neighbourhood of $x = a$, $y = b$, and so defines a single-valued function $z = \phi(x, y)$, near (a, b). That this function is continuous near (a, b) follows by reasoning as in §6.2.

Now, since $F(x, y, z)$ is differentiable, we can put (§5.25)

$$F(x + \delta x, y + \delta y, z + \delta z) - F(x, y, z)$$
$$= (F_x + \alpha)\,\delta x + (F_y + \beta)\,\delta y + (F_z + \gamma)\,\delta z,$$

where α, β, γ all vanish with $(\delta x, \delta y, \delta z)$. In particular, if (x, y) and $(x + \delta x, y + \delta y)$ belong to the neighbourhood in which $z = \phi(x, y)$ is continuous, and if

$$z = \phi(x, y), \quad z + \delta z = \phi(x + \delta x, y + \delta y)$$

then $F(x, y, z) = 0$, $F(x + \delta x, y + \delta y, z + \delta z) = 0$, and δz vanishes with $(\delta x, \delta y)$ and we can put

$$0 = (F_x + \alpha)\,\delta x + (F_y + \beta)\,\delta y + (F_z + \gamma)\,\delta z,$$

where α, β, γ now all vanish with $(\delta x, \delta y)$. Putting $\delta y = 0$, we have

$$0 = (F_x + \alpha)\,\delta x + (F_z + \gamma)\,\delta z,$$

where α, γ both vanish with δx. Hence, and since $F_z \neq 0$, if δx is so small that $F_z + \gamma \neq 0$,

$$\left(\frac{\delta z}{\delta x}\right)_y = \frac{\phi(x + \delta x, y) - \phi(x, y)}{\delta x} = -\frac{F_x + \alpha}{F_z + \gamma}$$

from which follows the first of the two equations (3) when $\delta x \to 0$. The second is obtained in a similar way.

The partial derivatives $\partial z/\partial x$, $\partial z/\partial y$ are continuous because F_x, F_y, F_z are continuous.

6.5

Theorem. Let $x_1 = a_1, ..., x_m = a_m, y = b$ satisfy the equation $F(x_1, ..., x_m, y) = 0$. Also let the partial derivatives of the function F wo $x_1, ..., x_m, y$ be continuous near $(a_1, ..., a_m, b)$, with

$$F_y(a_1, ..., a_m, b) \neq 0.$$

Then the equation $F(x_1, ..., x_m, y) = 0$, together with the condition $y = b$ when $x_1 = a_1, ..., x_m = a_m$, defines y as a differentiable function of the m independent variables $x_1, ..., x_m$, say $y = \phi(x_1, ..., x_m)$, in the neighbourhood of $(a_1, ..., a_m)$, with continuous partial derivatives given by

$$\frac{\partial y}{\partial x_1} = -\frac{\partial F/\partial x_1}{\partial F/\partial y}, ..., \quad \frac{\partial y}{\partial x_m} = -\frac{\partial F/\partial x_m}{\partial F/\partial y}. \tag{4}$$

The proof follows the same lines as that of §6.4, though without any geometrical picture being available for the sake of illustration.

6.6 Rules for the differentiation of an implicit function

The object of the preceding theorems has been to prove, under certain conditions, the existence, continuity and differentiability of a function defined implicitly by one equation.

Once having shown that the function is defined and differentiable, we can obtain its derivative or derivatives by simple practical rules. For example, if the equation $F(x, y) = 0$ defines a differentiable function $y = \phi(x)$, we can obtain the derivative $dy/dx = \phi'(x)$ by one of the following rules:

I. Differentiate the equation $F(x, y) = 0$ wo x, regarding y as a function of x; thus, by §5.22 (23), and since $F_y \neq 0$,

$$\frac{\partial F}{\partial x} + \frac{\partial F}{\partial y} \frac{dy}{dx} = 0, \quad \frac{dy}{dx} = -\frac{\partial F/\partial x}{\partial F/\partial y} = -\frac{F_x}{F_y}. \tag{5}$$

II. Differentiate the equation totally; thus, by §5.22 (22),

$$F_x dx + F_y dy = 0$$

and hence $$dy = -\frac{F_x}{F_y} dx, \quad \frac{dy}{dx} = -\frac{F_x}{F_y}. \tag{6}$$

Again, if it is known that the equation $F(x, y, z) = 0$ defines a differentiable function $z = \phi(x, y)$, we can use either of the rules:

I. Differentiate the equation $F(x, y, z) = 0$ partially wo x, regarding

z as a function of the independent variables x and y; thus, remembering that y is independent of x, and that $\partial F/\partial z \neq 0$, we get

$$\frac{\partial F}{\partial x} + \frac{\partial F}{\partial z}\frac{\partial z}{\partial x} = 0, \quad \frac{\partial z}{\partial x} = -\frac{\partial F/\partial x}{\partial F/\partial z} \tag{7}$$

and similarly, by partial differentiation wo y,

$$\frac{\partial F}{\partial y} + \frac{\partial F}{\partial z}\frac{\partial z}{\partial y} = 0, \quad \frac{\partial z}{\partial y} = -\frac{\partial F/\partial y}{\partial F/\partial z}. \tag{8}$$

II. Differentiate the equation $F(x,y,z) = 0$ totally; then, by §5.27 (32),

$$F_x dx + F_y dy + F_z dz = 0$$

and hence

$$dz = -\frac{F_x}{F_z}dx - \frac{F_y}{F_z}dy, \tag{9}$$

and by §5.16(8),

$$\frac{\partial z}{\partial x} = -\frac{F_x}{F_z}, \quad \frac{\partial z}{\partial y} = -\frac{F_y}{F_z}. \tag{10}$$

6.7   Inverse function of one variable

As a deduction from §6.3, we have the following theorem:

Let $y = f(x)$ and let $y = b$ when $x = a$, so that $b = f(a)$. Also, let $f'(x)$ be continuous near $x = a$, and $f'(a) \neq 0$.

Then the equation $y = f(x)$ defines x as a differentiable function of y, say $x = \phi(y)$, in the neighbourhood of $y = b$, with a continuous derivative given by

$$\frac{dx}{dy} = \phi'(y) = \frac{1}{f'(x)} = \frac{1}{dy/dx}. \tag{11}$$

The function ϕ is called the *inverse* of the function f: its derivative was considered in §2.21.

6.8   Implicit functions: simultaneous equations

Theorem. Let $x = a$, $y = b$, $z = c$ satisfy the two equations

$$F(x,y,z) = 0, \quad G(x,y,z) = 0. \tag{12}$$

Also, let the partial derivatives of the two functions F, G wo x,y,z be continuous in the neighbourhood of the point (a,b,c) and at this point suppose

$$\partial(F,G)/\partial(y,z) \neq 0. \tag{13}$$

Then equations (12), together with the conditions $y = b$, $z = c$ when $x = a$, define y and z as single-valued functions of x, say $y = \phi(x)$,

$z = \psi(x)$, with continuous derivatives $\phi'(x)$, $\psi'(x)$ in the neighbourhood of $x = a$.

[In three-dimensional coordinate geometry, equations (12) then represent a curve which passes through the point (a, b, c) and has a continuously varying tangent near this point.]

Proof. By (13), $\partial F/\partial y$ and $\partial F/\partial z$ cannot both vanish at (a, b, c); suppose

$$F_z(a, b, c) \neq 0. \tag{14}$$

Then, by §6.4, the equation $F(x, y, z) = 0$ defines z as a function of x and y near (a, b), say

$$z = f(x, y) \tag{15}$$

with continuous partial derivatives. Let this function be substituted for z in equations (12). The first equation becomes an identity

$$F(x, y, f) \equiv 0 \tag{16}$$

in x and y, the second becomes an equation of the form

$$G(x, y, f) \equiv \Phi(x, y) = 0. \tag{17}$$

This equation is satisfied by $x = a$, $y = b$; also, the function Φ has continuous partial derivatives, since G and f have continuous partial derivatives. Consequently, by §6.3, the equation $\Phi(x, y) = 0$ will define y as a function of x near $x = a$, with a continuous derivative, provided $\partial\Phi/\partial y \neq 0$ at (a, b). Now from (16) and (17) we have, differentiating partially wo y,

$$\frac{\partial F}{\partial y} + \frac{\partial F}{\partial z}\frac{\partial f}{\partial y} \equiv 0, \quad \frac{\partial G}{\partial y} + \frac{\partial G}{\partial z}\frac{\partial f}{\partial y} \equiv \frac{\partial \Phi}{\partial y},$$

and on multiplying the first of these by $\partial G/\partial z$ and the second by $-\partial F/\partial z$ and adding, we get

$$\frac{\partial(F, G)}{\partial(y, z)} \equiv -\frac{\partial \Phi}{\partial y}\frac{\partial F}{\partial z}.$$

It follows, by (13) and (14), that $\partial\Phi/\partial y$ has a finite non-zero value at (a, b). We can therefore put $y = \phi(x)$, where $\phi(x)$ has a continuous derivative near $x = a$; and by substituting $y = \phi(x)$ in (15) we have further $z = f\{x, \phi(x)\} \equiv \psi(x)$, where $\psi(x)$ also has a continuous derivative, since f and ϕ have continuous derivatives; hence the theorem.

6.9

Knowing that the derivatives of $\phi(x)$ and $\psi(x)$ exist, we can obtain them by the rule: Differentiate the equations $F = 0$, $G = 0$ wo x, regarding y and z as functions of x, and solve the resulting equations for dy/dx and dz/dx, thus

$$\frac{\partial F}{\partial x} + \frac{\partial F}{\partial y}\frac{dy}{dx} + \frac{\partial F}{\partial z}\frac{dz}{dx} = 0, \quad \frac{\partial G}{\partial x} + \frac{\partial G}{\partial y}\frac{dy}{dx} + \frac{\partial G}{\partial z}\frac{dz}{dx} = 0,$$

from which follow

$$\frac{dy}{dx} = \frac{\partial(F, G)}{\partial(z, x)} \Big/ \frac{\partial(F, G)}{\partial(y, z)}, \quad \frac{dz}{dx} = \frac{\partial(F, G)}{\partial(x, y)} \Big/ \frac{\partial(F, G)}{\partial(y, z)}. \tag{18}$$

We could alternatively use the notation of differentials, thus

$$F_x dx + F_y dy + F_z dz = 0,$$

$$G_x dx + G_y dy + G_z dz = 0,$$

from which
$$\frac{dx}{\dfrac{\partial(F, G)}{\partial(y, z)}} = \frac{dy}{\dfrac{\partial(F, G)}{\partial(z, x)}} = \frac{dz}{\dfrac{\partial(F, G)}{\partial(x, y)}}. \tag{19}$$

6.10   General case of implicit functions

Theorem. Let $x_1 = a_1, \ldots, x_m = a_m, y_1 = b_1, \ldots, y_n = b_n$ satisfy the n equations
$$F_r(x_1, \ldots, x_m, y_1, \ldots, y_n) = 0 \quad (r = 1, 2, \ldots, n). \tag{20}$$

Also, let the partial derivatives of the n functions

$$F_1, F_2, \ldots, F_n \text{ with respect to } x_1, \ldots, x_m, y_1, \ldots, y_n$$

be continuous near the point $(a_1, \ldots, a_m, b_1, \ldots, b_n)$ and at this point suppose
$$J \equiv \frac{\partial(F_1, \ldots, F_n)}{\partial(y_1, \ldots, y_n)} \neq 0. \tag{21}$$

Then the n equations (20), together with the conditions

$$y_1 = b_1, \ldots, y_n = b_n \quad \text{when} \quad x_1 = a_1, \ldots, x_m = a_m,$$

define n functions

$$y_1 = \phi_1(x_1, \ldots, x_m), \ldots, y_n = \phi_n(x_1, \ldots, x_m) \tag{22}$$

and these functions have continuous derivatives near $(a_1, \ldots, a_m)$.

Proof. The case $n = 1$ has been noticed in §6.5. The proof for $n = 2$ follows the same lines as that of §6.8. We shall therefore assume the theorem to be true for $n = 1$ and $n = 2$, and outline the proof for $n = 3$. The general case can then be inferred by induction.

When $n = 3$ the equations are

$$F_r(x_1, \ldots, x_m, y_1, y_2, y_3) = 0 \quad (r = 1, 2, 3) \tag{23}$$

and the functions F_r satisfy the condition

$$J \equiv \frac{\partial(F_1, F_2, F_3)}{\partial(y_1, y_2, y_3)} \neq 0 \tag{24}$$

at the point $(a_1, \ldots, a_m, b_1, b_2, b_3)$.

Since the determinant J does not vanish, not all its first minors can vanish: suppose

$$\frac{\partial(F_1, F_2)}{\partial(y_1, y_2)} \neq 0. \tag{25}$$

Then, since the theorem is true for $n = 2$, the equations $F_1 = 0$, $F_2 = 0$ can be solved for y_1 and y_2, giving, say,

$$y_1 = f_1(x_1, \ldots, x_m, y_3), \quad y_2 = f_2(x_1, \ldots, x_m, y_3), \tag{26}$$

where f_1 and f_2 have continuous derivatives. When these values of y_1 and y_2 are substituted in (23), we obtain results of the form

$$F_1(x_1, \ldots, x_m, f_1, f_2, y_3) \equiv 0, \tag{27}$$

$$F_2(x_1, \ldots, x_m, f_1, f_2, y_3) \equiv 0, \tag{28}$$

$$F_3(x_1, \ldots, x_m, f_1, f_2, y_3) \equiv \Phi(x_1, \ldots, x_m, y_3) = 0. \tag{29}$$

By §6.5 the last equation will determine y_3 as a function of $x_1, \ldots, x_m$ with continuous derivatives, provided that $\partial\Phi/\partial y_3 \neq 0$. Now from (27), (28), (29), by differentiating partially wo y_3, we have

$$\frac{\partial F_1}{\partial y_1}\frac{\partial f_1}{\partial y_3} + \frac{\partial F_1}{\partial y_2}\frac{\partial f_2}{\partial y_3} + \frac{\partial F_1}{\partial y_3} \equiv 0,$$

$$\frac{\partial F_2}{\partial y_1}\frac{\partial f_1}{\partial y_3} + \frac{\partial F_2}{\partial y_2}\frac{\partial f_2}{\partial y_3} + \frac{\partial F_2}{\partial y_3} \equiv 0,$$

$$\frac{\partial F_3}{\partial y_1}\frac{\partial f_1}{\partial y_3} + \frac{\partial F_3}{\partial y_2}\frac{\partial f_2}{\partial y_3} + \frac{\partial F_3}{\partial y_3} \equiv \frac{\partial \Phi}{\partial y_3}$$

and on multiplying these equations by the cofactors of the respective elements of the third column of J, and adding, we find

$$\frac{\partial(F_1, F_2, F_3)}{\partial(y_1, y_2, y_3)} \equiv \frac{\partial \Phi}{\partial y_3}\frac{\partial(F_1, F_2)}{\partial(y_1, y_2)}$$

from which $\partial\Phi/\partial y_3$ has a finite non-zero value, by (24) and (25).

We can therefore put $y_3 = \phi_3(x_1, \ldots, x_m)$ where ϕ_3 has continuous partial derivatives, and by substitution in (26) we obtain, say,

$$y_1 = \phi_1(x_1, \ldots, x_m), \quad y_2 = \phi_2(x_1, \ldots, x_m)$$

where ϕ_1 and ϕ_2 have continuous derivatives. This completes the proof.

6.11

When we know that equations (23) define y_1, y_2, y_3 as differentiable functions of $x_1, \ldots, x_m$, we can obtain their partial derivatives by differentiating equations (23), regarding y_1, y_2, y_3 as functions of the independent variables $x_1, \ldots, x_m$. Thus, the three equations

$$\frac{\partial F_r}{\partial x_s} + \frac{\partial F_r}{\partial y_1}\frac{\partial y_1}{\partial x_s} + \frac{\partial F_r}{\partial y_2}\frac{\partial y_2}{\partial x_s} + \frac{\partial F_r}{\partial y_3}\frac{\partial y_3}{\partial x_s} = 0 \quad (r = 1, 2, 3)$$

determine the derivatives $\partial y_1/\partial x_s$, $\partial y_2/\partial x_s$, $\partial y_3/\partial x_s$ $(s = 1, 2, \ldots, m)$.

6.12   Inverse pair of functions

Theorem. Let the pair of functions

$$u = f(x, y), \quad v = g(x, y) \tag{30}$$

be satisfied by $x = x_0$, $y = y_0$, $u = u_0$, $v = v_0$. Also let the partial derivatives of the functions f and g be continuous near the point (x_0, y_0), and at this point suppose

$$J \equiv \partial(u, v)/\partial(x, y) \neq 0. \tag{31}$$

Then equations (30) define x and y inversely as functions of u and v, as independent variables, near (u_0, v_0), say

$$x = \phi(u, v), \quad y = \psi(u, v) \tag{32}$$

and these functions have continuous partial derivatives.

The pair of functions ϕ, ψ may be called the *inverse* of the pair f, g.

Proof. Put

$$F(x, y, u) = u - f(x, y), \quad G(x, y, v) = v - g(x, y).$$

Then the equations $F = 0$, $G = 0$ are satisfied by

$$x = x_0, \quad y = y_0, \quad u = u_0, \quad v = v_0;$$

the functions F, G have continuous partial derivatives near

$$(x_0, y_0, u_0, v_0)$$

and at this point $\dfrac{\partial(F, G)}{\partial(x, y)} \equiv \dfrac{\partial(f, g)}{\partial(x, y)} \neq 0.$

The theorem follows as a particular case of §6.10.

6.13   Derivatives of the inverse pair of functions

In order to find the derivatives of the inverse pair of functions $x = \phi(u, v)$, $y = \psi(u, v)$ from the original pair of equations (30), we can differentiate equations (30) partially wo u and v in turn, regarding x and y as functions of u and v, and solve the resulting four equations which will be linear in $\partial x/\partial u$, $\partial x/\partial v$, $\partial y/\partial u$, $\partial y/\partial v$. Thus, by differentiating both equations partially wo u, we get

$$1 = \frac{\partial u}{\partial x}\frac{\partial x}{\partial u} + \frac{\partial u}{\partial y}\frac{\partial y}{\partial u}, \quad 0 = \frac{\partial v}{\partial x}\frac{\partial x}{\partial u} + \frac{\partial v}{\partial y}\frac{\partial y}{\partial u}, \tag{33}$$

and by differentiating both partially wo v,

$$0 = \frac{\partial u}{\partial x}\frac{\partial x}{\partial v} + \frac{\partial u}{\partial y}\frac{\partial y}{\partial v}, \quad 1 = \frac{\partial v}{\partial x}\frac{\partial x}{\partial v} + \frac{\partial v}{\partial y}\frac{\partial y}{\partial v}. \tag{34}$$

In these equations it must be borne in mind that $\partial u/\partial x$ means $(\partial u/\partial x)_y$, while $\partial x/\partial u$ means $(\partial x/\partial u)_v$, and so on.

From (33) and (34), we then find

$$\frac{\partial x}{\partial u} = \frac{1}{J}\frac{\partial v}{\partial y}, \quad \frac{\partial y}{\partial u} = -\frac{1}{J}\frac{\partial v}{\partial x}, \tag{35}$$

$$\frac{\partial x}{\partial v} = -\frac{1}{J}\frac{\partial u}{\partial y}, \quad \frac{\partial y}{\partial v} = \frac{1}{J}\frac{\partial u}{\partial x}. \tag{36}$$

If we now put
$$j \equiv \partial(x,y)/\partial(u,v), \tag{37}$$

we find from (35), (36), $\quad j = 1/J \quad$ or $\quad jJ = 1 \tag{38}$

that is, the Jacobian of the inverse pair of functions is the reciprocal of that of the original pair. This is also an immediate consequence of a property of Jacobians proved in §5.30. For if equations (30) above are known to define x,y inversely as differentiable functions of u,v, we can put $\phi = x, \psi = y$ in §5.30 (40) and obtain (38) again, viz.

$$1 = \frac{\partial(x,y)}{\partial(u,v)}\frac{\partial(u,v)}{\partial(x,y)} \quad \text{or} \quad 1 = jJ. \tag{39}$$

6.14   General case of inverse functions

Theorem. Let the set of n equations

$$y_r = f_r(x_1, ..., x_n) \quad (r = 1, 2, ..., n) \tag{40}$$

be satisfied by $x_1 = a_1, ..., x_n = a_n, y_1 = b_1, ..., y_n = b_n$. Also, let the partial derivatives of the functions f_r be continuous near the point $(a_1, ..., a_n)$, and at this point suppose

$$J \equiv \partial(y_1, ..., y_n)/\partial(x_1, ..., x_n) \neq 0. \tag{41}$$

Then the equations (40) define an inverse set of functions

$$x_s = \phi_s(y_1, ..., y_n) \quad (s = 1, 2, ..., n) \tag{42}$$

with continuous partial derivatives near the point $(b_1, ..., b_n)$.

This theorem is a generalization of that of §6.12, and also a particular case of that of §6.10.

The relation between the Jacobian j of the inverse functions and the Jacobian J of the original set of functions is $jJ = 1$, as in §6.13, that is,

$$\frac{\partial(x_1, \ldots, x_n)}{\partial(y_1, \ldots, y_n)} \frac{\partial(y_1, \ldots, y_n)}{\partial(x_1, \ldots, x_n)} = 1, \quad \text{or} \quad \frac{\partial(x)}{\partial(y)} = \frac{1}{\partial(y)/\partial(x)}. \tag{43}$$

6.15   Partial derivatives of a function of inverse functions

Let T be a function of u, v and let u, v be functions of x, y defined inversely by the equations

$$x = f(u, v), \; y = g(u, v); \quad J = \partial(x, y)/\partial(u, v) \neq 0. \tag{44}$$

Then T may be regarded as a function of x, y through u, v. It is required to find the partial derivatives $\partial T/\partial x$, $\partial T/\partial y$.

Note that it is now convenient to interchange the roles of x, y and u, v as compared with §§6.12, 6.13, so that here $J = \partial(x, y)/\partial(u, v)$ and

$$\frac{\partial u}{\partial x} = \frac{1}{J}\frac{\partial y}{\partial v}, \quad \frac{\partial u}{\partial y} = -\frac{1}{J}\frac{\partial x}{\partial v}, \quad \frac{\partial v}{\partial x} = -\frac{1}{J}\frac{\partial y}{\partial u}, \quad \frac{\partial v}{\partial y} = \frac{1}{J}\frac{\partial x}{\partial u}. \tag{45}$$

First method. By §5.23, we have

$$\frac{\partial T}{\partial x} = \frac{\partial T}{\partial u}\frac{\partial u}{\partial x} + \frac{\partial T}{\partial v}\frac{\partial v}{\partial x} = \frac{y_2}{J}\frac{\partial T}{\partial u} - \frac{y_1}{J}\frac{\partial T}{\partial v}, \tag{46}$$

$$\frac{\partial T}{\partial y} = \frac{\partial T}{\partial u}\frac{\partial u}{\partial y} + \frac{\partial T}{\partial v}\frac{\partial v}{\partial y} = \frac{x_1}{J}\frac{\partial T}{\partial v} - \frac{x_2}{J}\frac{\partial T}{\partial u}, \tag{47}$$

where $x_1 = \partial x/\partial u$, $x_2 = \partial x/\partial v$, ...

Second method. Using §5.30 (40), we have

$$\frac{\partial T}{\partial x} \equiv \frac{\partial(T, y)}{\partial(x, y)} = \frac{\partial(T, y)}{\partial(u, v)}\frac{\partial(u, v)}{\partial(x, y)} = \frac{1}{J}\frac{\partial(T, y)}{\partial(u, v)} = \frac{1}{J}\begin{vmatrix} T_1 & T_2 \\ y_1 & y_2 \end{vmatrix},$$

$$\frac{\partial T}{\partial y} \equiv \frac{\partial(x, T)}{\partial(x, y)} = \frac{\partial(x, T)}{\partial(u, v)}\frac{\partial(u, v)}{\partial(x, y)} = \frac{1}{J}\frac{\partial(x, T)}{\partial(u, v)} = \frac{1}{J}\begin{vmatrix} x_1 & x_2 \\ T_1 & T_2 \end{vmatrix},$$

where $T_1 = \partial T/\partial u$, $T_2 = \partial T/\partial v$; agreeing with (46), (47).

Example

If T is a function of r, θ and if

$$x = r\cos\theta, \quad y = r\sin\theta, \quad J = \partial(x, y)/\partial(r, \theta) = r \neq 0,$$

then we have

$$\frac{\partial T}{\partial x} = \frac{1}{r}\begin{vmatrix} T_1 & T_2 \\ \sin\theta & r\cos\theta \end{vmatrix} = \cos\theta\frac{\partial T}{\partial r} - \frac{\sin\theta}{r}\frac{\partial T}{\partial\theta},$$

$$\frac{\partial T}{\partial y} = \frac{1}{r}\begin{vmatrix} \cos\theta & -r\sin\theta \\ T_1 & T_2 \end{vmatrix} = \sin\theta\frac{\partial T}{\partial r} + \frac{\cos\theta}{r}\frac{\partial T}{\partial\theta}.$$

6.16

Let T be a function of u, v, w, where u, v, w are functions of x, y, z defined inversely by the equations

$$x = f(u, v, w), \quad y = g(u, v, w), \quad z = h(u, v, w), \Big\}$$
$$J = \partial(x, y, z)/\partial(u, v, w) \neq 0. \qquad \qquad \qquad (48)$$

Then T can be regarded as a function of x, y, z through u, v, w. Let it be required to find the partial derivatives $\partial T/\partial x$, $\partial T/\partial y$, $\partial T/\partial z$.

Using the second method, we have

$$\frac{\partial T}{\partial x} \equiv \frac{\partial(T, y, z)}{\partial(x, y, z)} = \frac{1}{J} \frac{\partial(T, y, z)}{\partial(u, v, w)} = \frac{1}{J} \begin{vmatrix} T_1 & T_2 & T_3 \\ y_1 & y_2 & y_3 \\ z_1 & z_2 & z_3 \end{vmatrix}$$

or

$$\frac{\partial T}{\partial x} = \frac{X_1}{J} \frac{\partial T}{\partial u} + \frac{X_2}{J} \frac{\partial T}{\partial v} + \frac{X_3}{J} \frac{\partial T}{\partial w} \qquad (49)$$

where $x_1 = \partial x/\partial u, ..., X_1 = y_2 z_3 - y_3 z_2, ...$

In particular, putting $T = u, v, w$ in turn, we see that

$$\frac{\partial u}{\partial x} = \frac{X_1}{J}, \quad \frac{\partial v}{\partial x} = \frac{X_2}{J}, \quad \frac{\partial w}{\partial x} = \frac{X_3}{J}. \qquad (50)$$

Similarly, we could find $\partial T/\partial y$, $\partial T/\partial z$, $\partial u/\partial y$, $\partial u/\partial z, ...$

6.17 Jacobian of implicit functions

Theorem. Let n functions $y_1, ..., y_n$ of n independent variables $x_1, ..., x_n$ be defined implicitly (see §6.10) by the n equations

$$F_r(x_1, ..., x_n, y_1, ..., y_n) = 0 \quad (r = 1, 2, ..., n). \qquad (51)$$

Then the Jacobian of $y_1, ..., y_n$ wo $x_1, ..., x_n$ is given by

$$J = \frac{\partial(y_1, ..., y_n)}{\partial(x_1, ..., x_n)} = (-)^n \frac{\partial(F_1, ..., F_n)}{\partial(x_1, ..., x_n)} \Big/ \frac{\partial(F_1, ..., F_n)}{\partial(y_1, ..., y_n)}. \qquad (52)$$

Proof. Differentiating equations (51) partially wo x_s we obtain n^2 equations $(r = 1, 2, ..., n; s = 1, 2, ..., n)$ expressible in the form

$$-\frac{\partial F_r}{\partial x_s} = \frac{\partial F_r}{\partial y_1} \frac{\partial y_1}{\partial x_s} + \frac{\partial F_r}{\partial y_2} \frac{\partial y_2}{\partial x_s} + ... + \frac{\partial F_r}{\partial y_n} \frac{\partial y_n}{\partial x_s}.$$

By the rule for the multiplication of determinants, it follows that

$$(-)^n \frac{\partial(F_1, ..., F_n)}{\partial(x_1, ..., x_n)} = \frac{\partial(F_1, ..., F_n)}{\partial(y_1, ..., y_n)} \frac{\partial(y_1, ..., y_n)}{\partial(x_1, ..., x_n)},$$

which is equivalent to (52).

Examples

(1) If λ, μ are the roots of the equation

$$\frac{x^2}{a^2+\theta}+\frac{y^2}{b^2+\theta}=1$$

considered as an equation in θ, find $\partial(x,y)/\partial(\lambda,\mu)$.

The equation, quadratic in θ, may be written

$$x^2(b^2+\theta)+y^2(a^2+\theta)-(a^2+\theta)(b^2+\theta)=0.$$

Considering the sum and product of the roots, we may put

$$F(\lambda,\mu,x,y)\equiv\lambda+\mu+a^2+b^2-x^2-y^2=0,$$

$$G(\lambda,\mu,x,y)\equiv\lambda\mu+b^2x^2+a^2y^2-a^2b^2=0,$$

and by (52) follows

$$\frac{\partial(x,y)}{\partial(\lambda,\mu)}=\frac{\partial(F,G)}{\partial(\lambda,\mu)}\Big/\frac{\partial(F,G)}{\partial(x,y)}=-\frac{\lambda-\mu}{4(a^2-b^2)xy}.$$

(2) If

$$y_1=r\sin\theta_1\sin\theta_2...\sin\theta_{n-1}$$

$$y_2=r\sin\theta_1\sin\theta_2...\cos\theta_{n-1}$$

$$y_3=r\sin\theta_1\sin\theta_2...\cos\theta_{n-2}$$

$$\cdots\cdots\cdots\cdots\cdots\cdots\cdots\cdots\cdots$$

$$y_n=r\cos\theta_1$$

prove that

$$\frac{\partial(y_1,y_2,...,y_n)}{\partial(r,\theta_1,...,\theta_{n-1})}=(-)^{\frac{1}{2}n(n-1)}r^{n-1}\sin^{n-2}\theta_1\sin^{n-3}\theta_2...\sin\theta_{n-2}.$$

Squaring and adding the given n equations, we get

$$F_1\equiv y_1^2+y_2^2+...+y_n^2-r^2=0.$$

Also, write the last $n-1$ of the given equations in the form

$$F_2\equiv y_2-rs_1s_2...c_{n-1}=0,$$

$$F_3\equiv y_3-rs_1s_2...c_{n-2}=0,$$

$$\cdots\cdots\cdots\cdots\cdots\cdots\cdots\cdots\cdots$$

$$F_n\equiv y_n-rc_1\qquad\qquad=0,$$

where $s_1=\sin\theta_1,\ldots$. If we now apply (52) to this new set of n equations, the required result follows without difficulty.

6.18   Dependence of functions

Let $u, v, \ldots$ be functions of $x, y, \ldots$. The functions $u, v, \ldots$ are said to be *not independent* if they satisfy an equation of the form

$$F(u, v, \ldots) = 0, \tag{53}$$

which does not involve $x, y, \ldots$ and is not an identity in u, v. They are said to be *independent* if they do not satisfy such an equation.

For example, the functions $u = xyz, v = x^2y^2z^2$ are not independent, for they satisfy the equation $v = u^2$; the functions $u = x+y, v = xy$ are independent.

Note that the equation $F(u, v, \ldots) = 0$, though not an identity in $u, v, \ldots$ becomes an identity in $x, y, \ldots$ when $u, v, \ldots$ are replaced by their expressions in terms of $x, y, \ldots$.

6.19

Theorem 1. Let u, v be functions of x, y defined by the equations

$$u = f(x, y), \quad v = g(x, y), \tag{54}$$

where the functions f, g have continuous partial derivatives.

A necessary condition that u, v should be connected by an equation

$$F(u, v) = 0, \tag{55}$$

the function $F(u, v)$ being differentiable and not involving x or y explicitly, is that the Jacobian of u, v should vanish identically, that is,

$$J = \partial(u, v)/\partial(x, y) \equiv 0. \tag{56}$$

Proof. Given that an equation $F(u, v) = 0$ exists, we have to prove that $J \equiv 0$.

Now when we put $u = f(x, y), v = g(x, y)$, the equation $F(u, v) = 0$ becomes an identity $F(f, g) \equiv 0$ in x and y, and hence also

$$\frac{\partial F}{\partial u}\frac{\partial f}{\partial x} + \frac{\partial F}{\partial v}\frac{\partial g}{\partial x} \equiv 0, \quad \frac{\partial F}{\partial u}\frac{\partial f}{\partial y} + \frac{\partial F}{\partial v}\frac{\partial g}{\partial y} \equiv 0$$

are identities when expressed in terms of x and y.

Eliminating $\partial F/\partial u$, $\partial F/\partial v$, which are not identically zero since $F(u, v) = 0$ is not an identity in u and v, we obtain (56).

Corollary. If $J \not\equiv 0$, the functions u, v cannot be connected by such an equation.

Theorem 2. Sufficient conditions that u and v should be connected by such an equation in the neighbourhood N of a point (a, b) are

(i) $J \equiv 0$ in N, (ii) one of the partial derivatives of u or v not zero at (a, b).

Proof. Let $u_0 = f(a, b)$, $v_0 = g(a, b)$. Suppose $f_x(a, b) \neq 0$; then, since f_x is continuous, there will be a neighbourhood of (a, b) in which $f_x \neq 0$ at any point.

Let N be a neighbourhood of (a, b) in which both $J \equiv 0$ and $f_x \neq 0$. Put

$$F_1 = u - f(x, y), \quad F_2 = v - g(x, y);$$

then

$$\frac{\partial(F_1, F_2)}{\partial(x, v)} = \begin{vmatrix} -f_x & 0 \\ -g_x & 1 \end{vmatrix} = -f_x;$$

consequently, at (a, b, u_0, v_0)

$$\partial(F_1, F_2)/\partial(x, v) = -f_x(a, b) \neq 0.$$

It follows, from the theorem of implicit functions (§ 6.10), that the equations $F_1 = 0$, $F_2 = 0$, that is, $u = f(x, y)$, $v = g(x, y)$, define x and v as differentiable functions of y and u near $y = b$, $u = u_0$. Let these functions be

$$x = \phi(y, u), \quad v = \psi(y, u);$$

then, identically in y and u,

$$u \equiv f(\phi, y), \quad \psi \equiv g(\phi, y). \tag{57}$$

We now show that $\partial \psi / \partial y \equiv 0$, and therefore that the equation $v = \psi(y, u)$ is in fact of the form $v = \psi(u)$.

By differentiating (57) partially wo y, we obtain, identically in u and y,

$$0 \equiv f_x \frac{\partial \phi}{\partial y} + f_y, \quad \frac{\partial \psi}{\partial y} \equiv g_x \frac{\partial \phi}{\partial y} + g_y$$

and hence the partial derivatives $\partial \phi / \partial y$, $\partial \psi / \partial y$, with u constant. Solving for $\partial \psi / \partial y$, we find, by condition (i),

$$f_x \frac{\partial \psi}{\partial y} \equiv \frac{\partial(f, g)}{\partial(x, y)} \equiv 0$$

and hence $\partial \psi / \partial y \equiv 0$, since $f_x \neq 0$ in N. It follows that v is a function of u only, in N.

Note. The identical vanishing of the Jacobian throughout a region R is not sufficient in itself to ensure the existence of an equation $F(u, v) = 0$ holding good throughout R.

Thus, let A_1 be the first quadrant and A_2 the second quadrant of the circle $x^2 + y^2 = 1$. In A_1 let $u = x^2 y$, $v = 0$. In A_2 let $u = 0$, $v = x^2 y^2$. Then u, v have continuous derivatives and $\partial(u, v)/\partial(x, y) \equiv 0$

throughout the semicircle $A_1 + A_2$, but there is no single equation $F(u, v) = 0$ connecting u and v throughout the semicircle (cf. §3.12, *Note*), though $v = 0$ holds good in A_1 and $u = 0$ in A_2.

6.20

Next, consider the case of three functions of three variables.

Theorem 1. Let u, v, w be functions of x, y, z defined by three equations
$$u = f(x, y, z), \quad v = g(x, y, z), \quad w = h(x, y, z), \tag{58}$$
where f, g, h have continuous derivatives.

A necessary condition that u, v, w should be connected by an equation $F(u, v, w) = 0$, the function $F(u, v, w)$ being differentiable and not involving x, y, z, is that the Jacobian of u, v, w should vanish identically, or
$$J = \partial(u, v, w)/\partial(x, y, z) \equiv 0. \tag{59}$$

Proof. The proof depends upon the same kind of reasoning as in §6.19, Theorem 1.

Corollary. If $J \not\equiv 0$, the functions u, v, w cannot be connected by such an equation.

Theorem 2. Sufficient conditions that u, v, w should be connected by such an equation in the neighbourhood N of a point (a, b, c) are (i) $J \equiv 0$ in N, and (ii) the Jacobian of two of the functions u, v, w with respect to two of the variables x, y, z not zero at (a, b, c).

Proof. Put $J_3 = \partial(u, v, w)/\partial(x, y, z)$ and $J_2 = \partial(u, v)/\partial(x, y)$. Let
$$u_0 = f(a, b, c), \quad v_0 = g(a, b, c), \quad w_0 = h(a, b, c).$$

Suppose $J_2 \neq 0$ at (a, b, c); then, since the partial derivatives are continuous, there will be a neighbourhood of (a, b, c) in which $J_2 \neq 0$ at any point.

Let N be a neighbourhood of (a, b, c) in which both $J_3 \equiv 0$ and $J_2 \neq 0$. Put
$$F_1 = u - f(x, y, z), \quad F_2 = v - g(x, y, z), \quad F_3 = w - h(x, y, z);$$

then
$$\frac{\partial(F_1, F_2, F_3)}{\partial(x, y, w)} = \begin{vmatrix} -f_x & -f_y & 0 \\ -g_x & -g_y & 0 \\ -h_x & -h_y & 1 \end{vmatrix} = J_2.$$

Since $J_2 \neq 0$ at (a, b, c) it follows from the theorem of implicit functions (§6.10) that the equations $F_1 = 0, F_2 = 0, F_3 = 0$, that is, equations (58), define x, y, w as differentiable functions of z, u, v near (c, u_0, v_0). Let these functions be denoted by
$$x = \phi(z, u, v), \quad y = \psi(z, u, v), \quad w = \chi(z, u, v);$$

then, identically in z, u, v,

$$u \equiv f(\phi, \psi, z), \quad v \equiv g(\phi, \psi, z), \quad \chi \equiv h(\phi, \psi, z). \tag{60}$$

By differentiating these identities partially wo z, we find

$$0 \equiv f_x \frac{\partial \phi}{\partial z} + f_y \frac{\partial \psi}{\partial z} + f_z,$$

$$0 \equiv g_x \frac{\partial \phi}{\partial z} + g_y \frac{\partial \psi}{\partial z} + g_z,$$

$$\frac{\partial \chi}{\partial z} \equiv h_x \frac{\partial \phi}{\partial z} + h_y \frac{\partial \psi}{\partial z} + h_z,$$

and hence the partial derivatives $\partial \phi / \partial z$, $\partial \psi / \partial z$, $\partial \chi / \partial z$ with u, v constant. Solving for $\partial \chi / \partial z$, we find $J_2 \partial \chi / \partial z \equiv J_3$

and hence $\partial \chi / \partial z \equiv 0$, since $J_3 \equiv 0$ and $J_2 \neq 0$ in N. It follows that w is a function of u and v only, in N, as was to be proved.

6.21

The following more general theorems can be proved in the same kind of way as in §6.20.

Theorem 1. Let $y_1, y_2, \ldots, y_n$ be n functions of n variables $x_1, x_2, \ldots, x_n$, defined by the n equations

$$y_r = f_r(x_1, x_2, \ldots, x_n) \quad (r = 1, 2, \ldots, n),$$

the functions f_r having continuous partial derivatives.

A necessary condition that $y_1, y_2, \ldots, y_n$ should be connected by a relation $F(y_1, y_2, \ldots, y_n) = 0$, the function F being differentiable and not involving $x_1, x_2, \ldots, x_n$, is that the Jacobian of $y_1, y_2, \ldots, y_n$ with respect to $x_1, x_2, \ldots, x_n$ should vanish identically.

Corollary. If the Jacobian does not vanish identically, no such relation can exist.

Theorem 2. Sufficient conditions that $y_1, y_2, \ldots, y_n$ should be connected by such a relation in the neighbourhood N of a particular point are that the Jacobian of $y_1, y_2, \ldots, y_n$ with respect to $x_1, x_2, \ldots, x_n$ should vanish identically in N and that the Jacobian of $n-1$ of the functions with respect to $n-1$ of the variables should not vanish at the point.

6.22　Analytic functions

Any function of the two real variables x, y may be regarded as a function of the complex variable z, $= x + iy$, in the sense that the value of the function is known when z is known, since x and y are then known. The most important class of functions from this point of view is the class of *analytic* functions which are defined as follows:

The function $f(z)$ is said to be *analytic at z* if the difference

$$f(z + \delta z) - f(z)$$

can be expressed in the form indicated by

$$f(z + \delta z) - f(z) = \zeta \delta z + \gamma \delta z, \tag{61}$$

where ζ is independent of δz, and $\gamma \to 0$ when $\delta z \to 0$ in any manner, that is, when $|\delta z| \to 0$, or $\delta x, \delta y \to 0, 0$, in any manner.

For example, z^3 is analytic at any point z, since

$$(z + \delta z)^3 - z^3 = \zeta \delta z + \gamma \delta z,$$

where $\zeta = 3z^2$, independent of δz, and $\gamma = 3z \delta z + \delta z^2$ which $\to 0$ when $\delta z \to 0$ in any manner.

A function $f(z)$ is said to be analytic in a closed region if it is analytic at every point of the region.

6.23　The derived function

If $f(z)$ is analytic in a closed region, it follows from (61) that the function $f'(z)$ defined by

$$f'(z) = \lim_{\delta z \to 0} \frac{f(z + \delta z) - f(z)}{\delta z} = \zeta \tag{62}$$

exists throughout the region. It is called the *derived function* or the *derivative* of $f(z)$.

Conversely, if the limit exists, the function is analytic. For, if the limit is equal to ζ, we can put

$$\frac{f(z + \delta z) - f(z)}{\delta z} = \zeta + \gamma, \tag{63}$$

where $\gamma \to 0$ when $\delta z \to 0$, and hence follows (61), showing that $f(z)$ is analytic, by definition.

Example

As an example of a function that is not analytic, consider

$$f(z) = (\operatorname{Re} z)^2 = x^2.$$

In this case
$$\frac{f(z+\delta z)-f(z)}{\delta z} = \frac{(x+\delta x)^2 - x^2}{\delta z} = (2x+\delta x)\frac{\delta x}{\delta z}$$

from which, if we put $\delta z = |\delta z|\, e^{i\psi}$, $\delta x = |\delta z|\cos\psi$, follows
$$\{f(z+\delta z)-f(z)\}/\delta z = (2x+\delta x)\, e^{-i\psi}\cos\psi$$

the limit of which, as $\delta z \to 0$, if ψ is constant, is $2x e^{-i\psi}\cos\psi$, which depends on ψ; thus, if $\psi = 0$ the limit is $2x$, if $\psi = \frac{1}{2}\pi$ the limit is zero.

6.24 The Riemann–Cauchy conditions

Let u, v be real continuous functions of x, y, with continuous partial derivatives.

Theorem. The necessary and sufficient conditions that $w, = u + iv$, should be an analytic function of $z, = x + iy$, in the neighbourhood N of a point z, are that u, v should satisfy the equations

$$\frac{\partial u}{\partial x} = \frac{\partial v}{\partial y}, \quad \frac{\partial u}{\partial y} = -\frac{\partial v}{\partial x}. \tag{64}$$

throughout N. These are called the *Riemann–Cauchy conditions.*

Proof. The conditions are necessary, that is, if the function is analytic, the conditions must be satisfied.

For, by (61), if the function $w, = u + iv$, is analytic, δw can be written in the form

$$\delta w = \delta u + i\delta v = (p+iq)(\delta x + i\delta y) + (\alpha + i\beta)(\delta x + i\delta y), \tag{65}$$

where p, q are real and independent of $\delta x, \delta y$ and α, β are real and vanish with $\delta x, \delta y$.

Hence, by equating real and imaginary parts,

$$\delta u = p\,\delta x - q\,\delta y + \alpha\,\delta x - \beta\,\delta y,$$
$$\delta v = q\,\delta x + p\,\delta y + \beta\,\delta x + \alpha\,\delta y,$$

from which follow (see §5.12)

$$\partial u/\partial x = p = \partial v/\partial y, \quad \partial u/\partial y = -q = -\partial v/\partial x.$$

The conditions are sufficient. The conditions being satisfied, put

$$\partial u/\partial x = \partial v/\partial y = p, \quad \partial u/\partial y = -\partial v/\partial x = -q.$$

Then, since the derivatives are continuous, u and v are differentiable (§5.15) and we can put

$$\delta u = p\,\delta x - q\,\delta y + \alpha_1\,\delta x + \beta_1\,\delta y,$$
$$\delta v = q\,\delta x + p\,\delta y + \alpha_2\,\delta x + \beta_2\,\delta y,$$

where $\alpha_1, \beta_1, \alpha_2, \beta_2$ all vanish with $\delta x, \delta y$. It follows that

$$\delta w = \delta u + i \delta v = (p + iq)(\delta x + i \delta y) + \alpha_3 \delta x + \beta_3 \delta y,$$

where α_3, β_3 vanish with $\delta x, \delta y$. This is equivalent to (65) and hence w is analytic.

Corollary. If $w, = u + iv$, is an analytic function of $z, = x + iy$, the derivative dw/dz can be expressed in the form

$$\frac{dw}{dz} = p + iq = \frac{\partial u}{\partial x} + i \frac{\partial v}{\partial x} = \frac{\partial w}{\partial x} \tag{66}$$

or in the form $$\frac{dw}{dz} = p + iq = \frac{\partial v}{\partial y} - i \frac{\partial u}{\partial y} = -i \frac{\partial w}{\partial y}. \tag{67}$$

Note. The Riemann–Cauchy conditions can be combined in the one equation

$$\frac{\partial u}{\partial x} + i \frac{\partial v}{\partial x} = -i \left(\frac{\partial u}{\partial y} + i \frac{\partial v}{\partial y} \right), \quad \text{or} \quad \frac{\partial w}{\partial x} = -i \frac{\partial w}{\partial y}. \tag{68}$$

6.25   Conjugate functions

Two real functions u, v of two independent variables x, y are said to be *conjugate* functions if $u + iv$ is an analytic function of $x + iy$.

Note that if $u + iv$ is analytic, then $v - iu$ is analytic, since

$$v - iu = -i(u + iv).$$

If u or v is given, its conjugate can be found, except for an additive constant. Thus, suppose u is given and we wish to find v.

Let $u + iv = f(z)$; then, by (64) and (66),

$$f'(z) = \frac{\partial u}{\partial x} + i \frac{\partial v}{\partial x} = \frac{\partial u}{\partial x} - i \frac{\partial u}{\partial y}. \tag{69}$$

The R.H.S. must therefore be a function of $z, = x + iy$. Putting $y = 0$, we find $f'(x)$, and by integration, $f(x)$, and hence $f(z)$, and thence v by separating the imaginary part.

Example

Given that $u + iv$ is an analytic function of $x + iy$ and that

$$u = 2 + \sin x \sinh y,$$

find v.

By finding $\partial u/\partial x$ and $\partial u/\partial y$, substituting in (69), and putting $y = 0$, we find $f'(x) = -i \sin x$ and hence $f(x) = i \cos x + A + iB$ where A, B are real constants.

Therefore $u + iv = f(z) = i \cos z + A + iB$, from which follow

$$u = \sin x \sinh y + A, \quad v = \cos x \cosh y + B.$$

From the given value of u we see that $A = 2$, while B is arbitrary.

6.26 The inverse analytic function

Let $u + iv$ be an analytic function of $x + iy$, with

$$u = \phi(x, y), \quad v = \psi(x, y). \tag{70}$$

Then we may put, by (64),

$$\frac{\partial u}{\partial x} = \frac{\partial v}{\partial y} = p, \quad \frac{\partial u}{\partial y} = -\frac{\partial v}{\partial x} = -q \tag{71}$$

and hence
$$J = \partial(u, v)/\partial(x, y) = p^2 + q^2, \tag{72}$$

which can vanish only at isolated points (x, y) which satisfy the equations $p = 0, q = 0$. Consequently, by §6.12, throughout any region from which such points are excluded, the equations (70) define x, y inversely as functions of u, v, say $x = f(u, v)$, $y = g(u, v)$. Moreover, by (35) and (36),

$$\frac{\partial x}{\partial u} = \frac{\partial y}{\partial v} = \frac{p}{J}, \quad \frac{\partial x}{\partial v} = -\frac{\partial y}{\partial u} = \frac{q}{J}, \tag{73}$$

and hence, by the Riemann–Cauchy conditions, $z, = x + iy$, is an analytic function of $w, = u + iv$, whose derivative, by (66), is given by

$$\frac{dz}{dw} = \frac{\partial z}{\partial u} = \frac{\partial x}{\partial u} + i\frac{\partial y}{\partial u} = \frac{p - iq}{J} = \frac{p - iq}{p^2 + q^2} = \frac{1}{p + iq}$$

and hence (cf. §2.21)
$$\frac{dz}{dw} = \frac{1}{dw/dz}. \tag{74}$$

Examples 6A

(1) Show that the equation $(y - x^2)^2 + (x - 2)^2 = 0$ defines a point, not a real function of x.

(2) Show that the equation $x^2 + y^2 = 1$ defines two differentiable functions of x near $x = 0$, and that the derivative of each at $x = 0$ is zero.

(3) Show that the equation $x = y(1 + y)(2 - y)$ defines three differentiable functions of x near $x = 0$. Find the derivative of each at $x = 0$.

(4) Show that the equation $x = ay + by^2 + cy^3 (a \neq 0)$, with $y = 0$ when $x = 0$, defines y as a differentiable function of x near $x = 0$, and find its derivative at $x = 0$.

(5) Show that the equation

$$ax + by + Ax^2 + Bxy + Cy^2 = 0 \quad (b \neq 0),$$

with $y = 0$ when $x = 0$, defines y as a differentiable function of x near $x = 0$. Find its derivative at $x = 0$.

(6) Show that the equation $x^2 + y^3 = 1$ defines a single-valued function of x which is differentiable except at $x = 1$ and $x = -1$. Sketch the graph of the function.

(7) If m, n are positive integers and n is odd, show that the equation

$$x^m + y^n = 1$$

defines y as a single-valued function of x. Sketch the graph when m is (i) even (ii) odd.

Show also that, if $n > 1$, y is not differentiable (i) at $x = \pm 1$ if m is even, (ii) at $x = 1$ if m is odd.

(8) If $\phi(y)$ and $\phi'(y)$ are continuous near $y = a$, show that the equation $y = a + x\phi(y)$ defines y as a differentiable function of x near $x = 0$, the value of the derivative of this function at $x = 0$ being given by $dy/dx = \phi(a)$.

Further (see §7.13), if $\phi'''(y)$ exists near $y = a$, show that we can write

$$y = a + \phi x + \phi\phi' x^2 + \tfrac{1}{2}(\phi^2\phi'' + 2\phi\phi'^2)x^3 + \eta x^3,$$

where ϕ, ϕ', ϕ'' denote $\phi(a), \phi'(a), \phi''(a)$, and $\eta \to 0$ when $x \to 0$.

(9) If y is defined as a function of x by the equation $y = a + xy^2$, with $y = a$ at $x = 0$, show that y can be expressed in the form

$$y = a + a^2 x + 2a^3 x^2 + 5a^4 x^3 + 14a^5 x^4 + \eta x^4,$$

where $\eta \to 0$ when $x \to 0$.

(10) Show that the equation $y = (\log x)/x$ defines x inversely as a function of y near $y = 0$.

Examples 6 B

(1) If x, y are rectangular Cartesian coordinates, and r, θ the usual polar coordinates, so that

$$x = r\cos\theta, \quad y = r\sin\theta, \quad r^2 = x^2 + y^2, \quad \tan\theta = y/x,$$

prove, and verify geometrically, that, if $r \neq 0$,

$$\left(\frac{\partial x}{\partial r}\right)_\theta = \cos\theta = \left(\frac{\partial r}{\partial x}\right)_y, \quad \frac{1}{r}\left(\frac{\partial x}{\partial \theta}\right)_r = -\sin\theta = r\left(\frac{\partial \theta}{\partial x}\right)_y,$$

$$\left(\frac{\partial y}{\partial r}\right)_\theta = \sin\theta = \left(\frac{\partial r}{\partial y}\right)_x, \quad \frac{1}{r}\left(\frac{\partial y}{\partial \theta}\right)_r = \cos\theta = r\left(\frac{\partial \theta}{\partial y}\right)_x.$$

(2) In the usual notation of the triangle, show that

$$\left(\frac{\partial \Delta}{\partial a}\right)_{B,c} = \frac{\Delta}{a}, \quad \left(\frac{\partial \Delta}{\partial a}\right)_{B,C} = \frac{2\Delta}{a}, \quad \left(\frac{\partial \Delta}{\partial a}\right)_{b,c} = R\cos A,$$

$$\left(\frac{\partial \Delta}{\partial A}\right)_{b,c} = \tfrac{1}{2}bc\cos A, \quad \left(\frac{\partial \Delta}{\partial A}\right)_{B,c} = \tfrac{1}{2}b^2, \quad \left(\frac{\partial \Delta}{\partial A}\right)_{b,C} = \tfrac{1}{2}c^2.$$

Show also that

    (i) if A, b, c are independent variables,

$$\frac{d\Delta}{\Delta} = \frac{db}{b} + \frac{dc}{c} + \cot A \, dA;$$

    (ii) if a, B, C are independent variables,

$$d\Delta = \frac{2\Delta}{a} da + \tfrac{1}{2}c^2 \, dB + \tfrac{1}{2}b^2 \, dC;$$

    (iii) if a, b, c are independent variables,

        (α) $d\Delta = R(da \cos A + db \cos B + dc \cos C)$,

        (β) $2\Delta \, dA = a(da - db \cos C - dc \cos B)$,

        (γ) $dA = \left(\dfrac{da}{a} - \dfrac{dc}{c}\right) \cot B + \left(\dfrac{da}{a} - \dfrac{db}{b}\right) \cot C$

and verify that $dA + dB + dC = 0$.

(3) The altitude h of a triangle ABC is calculated from measured values of the base a and the base angles B, C. If there are errors da, dB, dC in a, B, C respectively, show that approximately

$$\frac{dh}{h} = \frac{da}{a} + \frac{c \, dB}{b \sin A} + \frac{b \, dC}{c \sin A}.$$

If it is known that $|da| < \lambda$, and that $|dB|, |dC|$ are each less than ϵ, deduce that

$$\left|\frac{dh}{h}\right| < \frac{\lambda}{a} + \frac{(\sin^2 B + \sin^2 C)\epsilon}{\sin A \sin B \sin C}.$$

(4) If the radius R of the circumcircle of a triangle ABC is calculated from measurements of a, b, C, show that

$$\frac{c \, dR}{R} = da \cos B + db \cos A + 2R \cos A \cos B \, dC.$$

Examples 6 C

(1) Prove the following results. It may be assumed that the given equations define u and v as differentiable functions of x and y.

    (i) $x = u + v, \quad y = u^2 + v^2$,

$$\frac{1}{v}\frac{\partial u}{\partial x} = -2\frac{\partial u}{\partial y} = -\frac{1}{u}\frac{\partial v}{\partial x} = 2\frac{\partial v}{\partial y} = \frac{1}{v - u};$$

    (ii) $x = u + v \sin u, \quad y = v$,

$$\frac{\partial u}{\partial x} = \frac{1}{1 + v \cos u}, \quad \frac{\partial u}{\partial y} = -\frac{\sin u}{1 + v \cos u};$$

    (iii) $x = u + e^{-v} \sin u, \quad y = v + e^{-v} \cos u$,

$$\frac{\partial(u, v)}{\partial(x, y)} = \frac{e^v}{2 \sinh v}, \quad \frac{\partial u}{\partial y} = \frac{\partial v}{\partial x} = \frac{\sin u}{2 \sinh v};$$

(iv) $$ux + vy = 1, \quad \frac{x}{u} + \frac{y}{v} = 1,$$

$$\frac{\partial(u, v)}{\partial(x, y)} = -\frac{uv}{xy}, \quad \frac{\partial u}{\partial x} = \frac{u^2 + v^2}{v^2 - 1}, \quad \frac{\partial u}{\partial y} = \frac{2uv}{v^2 - 1};$$

(v) $$x = f(u) + \phi(y), \quad v = g(u) + yf'(u),$$

$$\frac{\partial(u, v)}{\partial(x, y)} = 1;$$

(vi) $$ux + vy = u^2, \quad vx - uy = v^2,$$

$$\frac{\partial(u, v)}{\partial(x, y)} = \frac{(u^2 + v^2)^2}{u^4 + v^4 - 2uv(u^2 + v^2)};$$

(vii) $$\tan u = \cos x / \sinh y, \quad \tanh v = \sin x / \cosh y,$$

$$\partial u / \partial x = \partial v / \partial y, \quad \partial u / \partial y = -\partial v / \partial x;$$

(viii) $$x = e^u \cos v, \quad v = e^y \sin x,$$

$$(\partial u / \partial y)_x = v \tan v, \quad (\partial y / \partial u)_v = -x \cot x.$$

(2) If $x = r \sin \theta \cos \phi$, $y = r \sin \theta \sin \phi$, $z = r \cos \theta$, and if $u = r \operatorname{cosec} \theta \sec \phi$, $v = r \operatorname{cosec} \theta \operatorname{cosec} \phi$, $w = r \sec \theta$, show that

$$\partial(x, y, z) / \partial(u, v, w) = \sin^4 \theta \cos^2 \theta \cos^2 \phi \sin^2 \phi.$$

(3) If T is a function of u, v, w and if

$$x = u + v + w, \quad y = u^2 + v^2 + w^2, \quad z = u^3 + v^3 + w^3,$$

and $u \neq v \neq w \neq u$, show that

$$J = \partial(x, y, z) / \partial(u, v, w) = 6(v - w)(w - u)(u - v) \neq 0,$$

$$\frac{\partial T}{\partial x} = \Sigma \frac{vw}{(u - v)(u - w)} \frac{\partial T}{\partial u},$$

$$\frac{\partial u}{\partial x} = \frac{vw}{(u - v)(u - w)}, \quad \frac{\partial v}{\partial x} = \frac{wu}{(v - u)(v - w)}, \quad \frac{\partial w}{\partial x} = \frac{uv}{(w - u)(w - v)},$$

and find similar expressions for $\partial T / \partial y$, $\partial T / \partial z$, $\partial u / \partial y$, ...

(4) If $x = r \sin \theta \cos \phi$, $y = r \sin \theta \sin \phi$, $z = r \cos \theta$, prove that

$$\frac{\partial r}{\partial x} = \sin \theta \cos \phi, \quad \frac{\partial \theta}{\partial x} = \frac{\cos \theta \cos \phi}{r}, \quad \frac{\partial \phi}{\partial x} = -\frac{\sin \phi}{r \sin \theta},$$

and find the partial derivatives of r, θ, ϕ wo y and z in terms of r, θ, ϕ.

(5) If $T = f(x, y)$, $x = cuv$, $y = c(1 + u^2)^{\frac{1}{2}}(1 - v^2)^{\frac{1}{2}}$, with c constant, show that

$$\frac{x}{y}\left(y \frac{\partial T}{\partial x} - x \frac{\partial T}{\partial y}\right) = \frac{uv}{u^2 + v^2}\left(v \frac{\partial T}{\partial u} + u \frac{\partial T}{\partial v}\right).$$

(6) If $u = x^2 - y^2 - 2xy$, $v = y$, and if z satisfies the equation

$$(x + y) \partial z / \partial x + (x - y) \partial z / \partial y = 0,$$

show that $(x - y) \partial z / \partial v = 0$. Deduce that any solution of the equation has the form $z = f(x^2 - y^2 - 2xy)$.

(7) If $x = u^3 + 3v, y = 3u + v^3, \log z = u^2 + v^2$, show that

$$\frac{\partial z}{\partial x} = \frac{2vz}{3(uv+1)}, \quad \frac{\partial z}{\partial y} = \frac{2uz}{3(uv+1)}.$$

[In Examples 8, ..., 17, it may be assumed that there exist values of the variables that satisfy the given equations, and that the given functions have continuous partial derivatives.]

(8) Show that the equations $z = f(x,y)$, $F(x,y) = 0$ define y and z as differentiable functions of x provided that $\partial F/\partial y \neq 0$. Show also that

$$\frac{dz}{dx} = \frac{\partial(f,F)}{\partial(x,y)} \bigg/ \frac{\partial F}{\partial y}.$$

(9) Show that the equations $z = f(x,y)$, $z = g(x,y)$ define y and z as differentiable functions of x provided that $\partial f/\partial y \neq \partial g/\partial y$. Show also that

$$\frac{dy}{dx} = \frac{f_x - g_x}{g_y - f_y}, \quad \frac{dz}{dx} = \frac{f_x g_y - f_y g_x}{g_y - f_y}.$$

(10) Show that the equations

$$x = f(u,v), \quad y = g(u,v), \quad z = h(u,v)$$

define z as a differentiable function of x, y provided that

$$\partial(f,g)/\partial(u,v) \neq 0.$$

Show also that

$$\frac{\partial z}{\partial x} = -\frac{\partial(y,z)}{\partial(u,v)} \bigg/ \frac{\partial(x,y)}{\partial(u,v)}, \quad \frac{\partial z}{\partial y} = -\frac{\partial(z,x)}{\partial(u,v)} \bigg/ \frac{\partial(x,y)}{\partial(u,v)}.$$

(11) If the equation $F(x,y,z) = 0$ defines each of the variables as a differentiable function of the other two, show that

$$\frac{\partial y}{\partial z} \frac{\partial z}{\partial x} \frac{\partial x}{\partial y} = -1.$$

(12) Show that the equations $u = f(x,y)$, $v = g(x,y)$ define y and v as differentiable functions of x and u if $\partial f/\partial y \neq 0$. Show also that

$$\frac{\partial y}{\partial x} = -\frac{\partial f/\partial x}{\partial f/\partial y}, \quad \frac{\partial y}{\partial u} = \frac{1}{\partial f/\partial y}, \quad \frac{\partial v}{\partial x} = -\frac{\partial(f,g)/\partial(x,y)}{\partial f/\partial y}, \quad \frac{\partial v}{\partial u} = \frac{\partial g/\partial y}{\partial f/\partial y}.$$

(13) Assuming that the equations

$$F(x,u) = 0, \quad G(u,v) = 0, \quad H(v,y) = 0$$

define u, v, y as functions of x, show that

$$\frac{dy}{dx} = -\frac{\partial F}{\partial x} \frac{\partial G}{\partial u} \frac{\partial H}{\partial v} \bigg/ \left(\frac{\partial F}{\partial u} \frac{\partial G}{\partial v} \frac{\partial H}{\partial y} \right).$$

(14) Show that the equations

$$y_1 = f_1(x_1, x_2, x_3), \quad y_2 = f_2(x_1, x_2, x_3), \quad y_3 = f_3(x_1, x_2, x_3)$$

(i) define x_1, y_2, y_3 as differentiable functions of

$$y_1, x_2, x_3 \quad \text{if} \quad \partial f_1/\partial x_1 \neq 0;$$

(ii) define x_1, x_2, y_3 as differentiable functions of

$$y_1, y_2, x_3 \quad \text{if} \quad \partial(f_1, f_2)/\partial(x_1, x_2) \neq 0.$$

(15) If y is a function of x defined by the equations

$$f(x, t) = 0, \quad g(y, t) = 0,$$

show that dy/dx is given by $\dfrac{\partial f}{\partial t}\dfrac{\partial g}{\partial y}\dfrac{dy}{dx} = \dfrac{\partial f}{\partial x}\dfrac{\partial g}{\partial t}.$

(16) If y is a function of x defined by the equations

$$f(x, u, v) = 0, \quad g(x, u, v) = 0, \quad h(y, u, v) = 0,$$

show that dy/dx is given by

$$-\frac{\partial(f, g)}{\partial(u, v)}\frac{\partial h}{\partial y}\frac{dy}{dx} = \frac{\partial f}{\partial x}\frac{\partial(g, h)}{\partial(u, v)} + \frac{\partial g}{\partial x}\frac{\partial(h, f)}{\partial(u, v)}.$$

(17) If T is defined as a differentiable function of x, y by the equations

$$T = f(u, v, x, y), \quad g(u, v, x, y) = 0, \quad h(u, v, x, y) = 0,$$

show that

$$\frac{\partial T}{\partial x}\begin{vmatrix} g_u & g_v \\ h_u & h_v \end{vmatrix} = \begin{vmatrix} f_x & f_u & f_v \\ g_x & g_u & g_v \\ h_x & h_u & h_v \end{vmatrix}, \quad \frac{\partial T}{\partial y}\begin{vmatrix} g_u & g_v \\ h_u & h_v \end{vmatrix} = \begin{vmatrix} f_y & f_u & f_v \\ g_y & g_u & g_v \\ h_y & h_u & h_v \end{vmatrix}.$$

(18) In the usual notation of a triangle ABC:

(i) If the side c and the angle C remain constant while the other sides and angles vary, prove that
$$db/da = -\cos B/\cos A.$$

(ii) If the other side and angles of the triangle are regarded as functions of A, b, c, prove that

$$\frac{\partial a}{\partial b} = -\frac{a}{b}\frac{\partial B}{\partial A} = \cos C.$$

(iii) If $\sin^2 A + \sin^2 B + \sin^2 C = $ constant, prove that

$$\frac{dB}{dA} = \frac{\tan A - \tan C}{\tan C - \tan B}.$$

(19) Assuming that the following equations define z as a function of x and y, and that sufficient conditions for differentiability are satisfied, verify the given results:

(i) If $y - bz = f(x - az)$, then $a\,\partial z/\partial x + b\,\partial z/\partial y = 1$.

(ii) If $f\left(\dfrac{x-a}{z-c}, \dfrac{y-b}{z-c}\right) = 0$, then $(x - a)\dfrac{\partial z}{\partial x} + (y - b)\dfrac{\partial z}{\partial y} = z - c$.

(iii) If $x^2 + y^2 + z^2 = f(ax + by + cz)$, then

$$(bz - cy)\frac{\partial z}{\partial x} + (cx - az)\frac{\partial z}{\partial y} = ay - bx.$$

(20) If $\qquad x + y + z = f(\xi), \quad y + z = g(\xi\eta), \quad z = h(\xi\eta\zeta),$

show that $\qquad \partial(x, y, z)/\partial(\xi, \eta, \zeta) = \xi^2\eta f'(\xi)\, g'(\xi\eta)\, h'(\xi\eta\zeta).$

(21) If λ, μ, ν are the roots of the cubic equation

$$(x - a)^3 + (x - b)^3 + (x - c)^3 = 0,$$

show that $\qquad \dfrac{\partial(\lambda, \mu, \nu)}{\partial(a, b, c)} = -2\,\dfrac{(b - c)(c - a)(a - b)}{(\mu - \nu)(\nu - \lambda)(\lambda - \mu)}.$

(22) If

$$y_0 = 1 - x_1, \quad y_1 = x_1(1 - x_2), \quad y_2 = x_1 x_2(1 - x_3), \ldots,$$

$$y_{n-1} = x_1 x_2 \ldots x_{n-1}(1 - x_n), \quad y_n = x_1 x_2 \ldots x_n,$$

prove that $\quad \dfrac{\partial(y_1, \ldots, y_n)}{\partial(x_1, \ldots, x_n)} = x_1^{n-1} x_2^{n-2} \ldots x_{n-1} = (-)^n \dfrac{\partial(y_0, \ldots, y_{n-1})}{\partial(x_1, \ldots, x_n)}.$

Examples 6 D

(1) In each of the following cases, verify that $\partial(u, v)/\partial(x, y) \equiv 0$, and find an equation connecting u and v:

(i) $\quad u = \sin^{-1}x + \sin^{-1}y, \quad v = x\sqrt{(1 - y^2)} + y\sqrt{(1 - x^2)};$

(ii) $\quad u = \tan^{-1}x + \tan^{-1}y, \quad v = (x + y)/(1 - xy);$

(iii) $\quad u = xy + \sqrt{(1 + x^2)}\sqrt{(1 + y^2)}, \quad v = x\sqrt{(1 + y^2)} + y\sqrt{(1 + x^2)};$

(iv) $\quad u = \dfrac{x + y}{1 + xy}, \quad v = \dfrac{(x + y)(1 + xy)}{(1 - x^2)(1 - y^2)}.$

(2) If $u = xy - \xi\eta, \; v = x\eta + y\xi, \; x^2 + \xi^2 = 1, \; y^2 + \eta^2 = 1$, show that

$$\partial(u, v)/\partial(x, y) \equiv 0.$$

Find a relation connecting u and v.

(3) If $u = f(x) + f(y), \; v = xy$, where $f(x) = \displaystyle\int_1^x dt/t$, verify that

$$\partial(u, v)/\partial(x, y) \equiv 0.$$

Assuming that $u = F(v)$, show that $F(v) = f(v)$.

(4) If $u = f(x) + f(y), \; v = (x\sqrt{Y} + y\sqrt{X})/(1 - k^2 x^2 y^2)$, where

$$f(x) = \int_0^x dt/\sqrt{T}, \quad T = (1 - t^2)(1 - k^2 t^2),$$

and X, Y are the same functions of x, y respectively, as T is of t, verify that $\partial(u, v)/\partial(x, y) \equiv 0$.

Assuming that $u = F(v)$, show that $F(v) = f(v)$.

(5) (i) If $u = f(x)\,\phi(y), \; v = x + y$, and if $\partial(u, v)/\partial(x, y) \equiv 0$, show that

$$f(x) = a e^{nx}, \quad \phi(x) = b e^{nx},$$

where a, b, n are constants.

(ii) If $u = f(x)\,\phi(y), v = xy$, and if $\partial(u, v)/\partial(x, y) \equiv 0$, show that $f(x) = ax^n$, $\phi(x) = bx^n$, where a, b, n are constants.

(6) Verify that $\partial(u, v, w)/\partial(x, y, z) \equiv 0$ and obtain a relation between u, v, w in each of the following cases:

(i) $\quad u = x + y + z, \quad v = x^2 + y^2 + z^2, \quad w = yz + zx + xy;$

(ii) $\quad (y - z)\,u = yz + a(y + z) + b,$

$\quad (z - x)\,v = zx + a(z + x) + b,$

$\quad (x - y)\,w = xy + a(x + y) + b, \quad$ where a, b are constants.

(7) In the usual notation of the triangle, verify from three equations of the form $a = b\cos C + c\cos B$, or three of the form $2bc\cos A = b^2 + c^2 - a^2$, that

$$\partial(A, B, C)/\partial(a, b, c) \equiv 0.$$

(8) If $\partial(f,g)/\partial(x,y)$ vanishes at (a,b) but not identically, show that, in general, the curves $f(x,y) = f(a,b)$, $g(x,y) = g(a,b)$ have a common tangent at (a,b).

(9) If $\partial(f,g,h)/\partial(x,y,z)$ vanishes at (a,b,c) but not identically, show that, in general, the surfaces

$$f(x,y,z) = f(a,b,c), \quad g(x,y,z) = g(a,b,c), \quad h(x,y,z) = h(a,b,c)$$

have a common tangent line at (a,b,c).

(10) If $u = f(x,y,z)$ and $v = g(x,y,z)$, where f and g have continuous partial derivatives in the neighbourhood N of a point (a,b,c), prove that

(i) necessary conditions that u and v be connected by a relation $F(u,v) = 0$, the function F being differentiable, are $J_1 \equiv 0$, $J_2 \equiv 0$, $J_3 \equiv 0$, where

$$J_1 = \partial(u,v)/\partial(y,z), \quad J_2 = \partial(u,v)/\partial(z,x), \quad J_3 = \partial(u,v)/\partial(x,y);$$

(ii) these conditions are also sufficient provided that one partial derivative is not zero at (a, b, c).

Examples 6 E

(1) If w_1, w_2 are two analytic functions of z, $= x+iy$, prove that the sum $w_1 + w_2$ and the product $w_1 w_2$ are analytic functions, and that so is the quotient w_1/w_2 except where $w_2 = 0$. Also, verify that the derivatives are given by the same rules as for functions of a real variable.

(2) If χ, $= \phi + i\psi$, is an analytic function of w, $= u + iv$, and if w is an analytic function of z, $= x + iy$, prove that χ is an analytic function of z and that $d\chi/dz$ is given by

$$\frac{d\chi}{dz} = \frac{d\chi}{dw}\frac{dw}{dz}.$$

(3) Prove that the function e^z defined by

$$w = e^z = e^{x+iy} = e^x \cdot e^{iy} = e^x(\cos y + i\sin y)$$

is analytic for all values of z, and that its derivative is given by the usual rule: $dw/dz = e^z$.

[*Proof.* If $w = e^x \cdot e^{iy}$, then $\partial w/\partial x = e^x \cdot e^{iy}$ and $-i\,\partial w/\partial y = e^x \cdot e^{iy}$. Hence $\partial w/\partial x = -i\,\partial w/\partial y$; thus, the Riemann–Cauchy conditions are satisfied and so w is an analytic function of z. Also, by (66), $dw/dz = \partial w/\partial x = e^x \cdot e^{iy} = e^{x+iy} = e^z$.

Otherwise. Since

$$\frac{\partial u}{\partial x} = e^x \cos y, \quad \frac{\partial u}{\partial y} = -e^x \sin y, \quad \frac{\partial v}{\partial x} = e^x \sin y, \quad \frac{\partial v}{\partial y} = e^x \cos y,$$

it follows that the Riemann–Cauchy conditions are satisfied, and, by (66),

$$\frac{dw}{dz} = \frac{\partial w}{\partial x} = \frac{\partial u}{\partial x} + i\frac{\partial v}{\partial x} = e^x(\cos y + i\sin y) = e^z.]$$

(4) Prove that the function z^n defined by

$$w = z^n = r^n e^{in\theta} = r^n(\cos n\theta + i\sin n\theta)$$

is an analytic function of z (except at $z = 0$ when n is not a positive integer); where n is real, $x = r\cos\theta$, $y = r\sin\theta$, $r = \sqrt{(x^2+y^2)}$, $\theta = \arg z = \tan^{-1}(y/x)$, $-\pi < \theta < \pi$, and θ lies in the same quadrant as the point z.

[*Proof.* Since $x = r \cos \theta$, $y = r \sin \theta$, $J = \partial(x,y)/\partial(r,\theta) = r$, we find, by §6.15,

$$\frac{\partial w}{\partial x} = \frac{\partial(w,y)}{\partial(x,y)} = \frac{1}{J}\frac{\partial(w,y)}{\partial(r,\theta)} = nr^{n-1}e^{i(n-1)\theta},$$

$$\frac{\partial w}{\partial y} = \frac{\partial(x,w)}{\partial(x,y)} = \frac{1}{J}\frac{\partial(x,w)}{\partial(r,\theta)} = inr^{n-1}e^{i(n-1)\theta},$$

from which follows

$$\frac{\partial w}{\partial x} = -i\frac{\partial w}{\partial y} = nr^{n-1}e^{i(n-1)\theta} = nz^{n-1}.$$

Thus the Riemann–Cauchy conditions are satisfied, and, further,

$$dw/dz = nz^{n-1}.]$$

(5) Prove that the function $\log z$ defined by

$$\log z = \log r + i\theta,$$

where $r = |z|$, $\theta = \arg z$ $(-\pi < \theta < \pi)$, is analytic except at $z = 0$, and that $d(\log z)/dz = 1/z$.

(6) Deduce from Examples 1, 2 that the following are analytic functions of z, $= x + iy$, except possibly at isolated points:

(i) e^{cz}, where $c = a + ib = $ constant.

[Put $\chi = e^w$, $w = cz$. Then χ is an analytic function of w, and w is an analytic function of z. Hence χ is an analytic function of z, by Example 2. Further,

$$\frac{d\chi}{dw} = e^w, \quad \frac{dw}{dz} = c; \quad \frac{d\chi}{dz} = \frac{d\chi}{dw}\frac{dw}{dz} = e^w . c = ce^{cz}.]$$

(ii) $\cos z = (e^{iz} + e^{-iz})/2$, $\quad \sin z = (e^{iz} - e^{-iz})/2i$.

(iii) $\sec z = 1/\cos z$, $\quad \tan z = \sin z/\cos z$, $\quad \cot z = 1/\tan z$.

(iv) $\cosh z = \frac{1}{2}(e^z + e^{-z})$, $\quad \sinh z = \frac{1}{2}(e^z - e^{-z})$.

(v) $\operatorname{sech} z = 1/\cosh z$, $\quad \tanh z = \sinh z/\cosh z$, $\quad \coth z = 1/\tanh z$.

(vi) $\log (z - c)$, and $\log \{(z-c)/(z+c)\}$, where $c = a + ib = $ constant.

(7) Separate the real and imaginary parts of the functions given in the last example, and in each case verify that the Riemann–Cauchy equations are satisfied.

(8) In the following cases, given that $u + iv$ is an analytic function of $x + iy$, find v:

(i) $u = x^2 - y^2 - 3$,

(ii) $u = \tan^{-1}(y/x)$,

(iii) $u = e^{-2xy}\sin(x^2 - y^2)$,

(iv) $u = \{x + \sqrt{(x^2 + y^2)}\}^{\frac{1}{2}}$,

(v) $u = \dfrac{x^3 + xy^2 + y}{x^2 + y^2}$,

(vi) $u = \dfrac{\cos x \cosh y}{\cosh 2y + \cos 2x}$,

(vii) $u = \tan^{-1}(\cos x/\sinh y)$.

(9) If $x = r\cos\theta$, $y = r\sin\theta$, show that from the Riemann–Cauchy conditions (64) follow (except at $r = 0$) the equations

$$\frac{\partial u}{\partial r} = \frac{1}{r}\frac{\partial v}{\partial \theta}, \quad \frac{1}{r}\frac{\partial u}{\partial \theta} = -\frac{\partial v}{\partial r},$$

which are called the Riemann–Cauchy conditions in polar coordinates.

(10) In the following cases, express the conjugate functions u and v in terms of r and θ, and verify that the equations of the last example are satisfied:

(i) z,   (ii) z^3,   (iii) $1/z$,   (iv) $\log z$,

(v) e^z,   (vi) $\sin z$,   (vii) z^n.

7

SUCCESSIVE PARTIAL DIFFERENTIATION

7.1 Second partial derivatives

The usual notation for the second partial derivatives of $f(x,y)$ is indicated by

$$\frac{\partial}{\partial x}\frac{\partial f}{\partial x} = \frac{\partial^2 f}{\partial x^2} = f''_{xx} = f_{xx} = f_{11},$$

$$\frac{\partial}{\partial x}\frac{\partial f}{\partial y} = \frac{\partial^2 f}{\partial x\,\partial y} = f''_{xy} = f_{xy} = f_{12},$$

$$\frac{\partial}{\partial y}\frac{\partial f}{\partial x} = \frac{\partial^2 f}{\partial y\,\partial x} = f''_{yx} = f_{yx} = f_{21},$$

$$\frac{\partial}{\partial y}\frac{\partial f}{\partial y} = \frac{\partial^2 f}{\partial y^2} = f''_{yy} = f_{yy} = f_{22}.$$

The conditions that hold good in most ordinary applications are sufficient to ensure that $f_{yx} \equiv f_{xy}$, so that usually there are only three distinct second derivatives f_{xx}, f_{xy}, f_{yy} to be taken into account. In particular, this is so when f_{yx} and f_{xy} are continuous functions of x, y (§7.2, Theorem 1).

When $f(x,y)$ is denoted by z, the following abbreviations are sometimes used:

$$p = \frac{\partial z}{\partial x}, \quad q = \frac{\partial z}{\partial y}, \quad r = \frac{\partial^2 z}{\partial x^2}, \quad s = \frac{\partial^2 z}{\partial x\,\partial y}, \quad t = \frac{\partial^2 z}{\partial y^2}. \tag{1}$$

Note that, provided that the limit exists,

$$f_{yx} = \lim_{k \to 0} \frac{f_x(x, y+k) - f_x(x, y)}{k} \tag{2}$$

so that the existence of f_{yx} at any point (x, y) implies the existence of $f_x(x, y+k)$ as a function of k near $k = 0$ and its continuity at $k = 0$.

7.2 Changing the order of differentiation

The following three theorems are concerned with sufficient conditions that the two 'mixed' partial derivatives $\partial^2 f/\partial x\,\partial y$, $\partial^2 f/\partial y\,\partial x$, which differ only in the order of the independent variables x, y, should have the same value at a point P.

In the first theorem both $\partial^2 f/\partial x\,\partial y$ and $\partial^2 f/\partial y\,\partial x$ are assumed to be continuous functions of position at P, in the second theorem only one of them is assumed to be continuous at P, and in the third neither is assumed to be continuous at P.

We shall take P to be the origin: this involves only an apparent loss of generality.

By putting $x = 0$, $y = 0$ in (2), we have formally

$$f_{yx}(0,0) = \lim_{k \to 0}\{f_x(0,k) - f_x(0,0)\}/k$$

$$= \lim_{k \to 0} \frac{1}{k}\left\{\lim_{h \to 0}\frac{f(h,k) - f(0,k)}{h} - \lim_{h \to 0}\frac{f(h,0) - f(0,0)}{h}\right\}$$

$$= \lim_{k \to 0} \frac{1}{k}\left\{\lim_{h \to 0}\frac{\Delta(h,k)}{h}\right\},$$

where

$$\Delta(h,k) = f(h,k) - f(0,k)$$
$$- f(h,0) + f(0,0). \quad (3)$$

It is therefore natural to consider this function $\Delta(h,k)$ in proving the following theorems.

Theorem 1. Sufficient conditions for $f_{xy}(0,0)$ and $f_{yx}(0,0)$ to be equal are that $f_{xy}(x,y)$ and $f_{yx}(x,y)$ should be continuous functions of position at $(0,0)$.

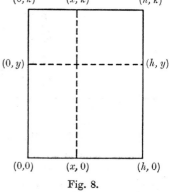

Fig. 8.

Proof. Let $f_{yx}(x,y)$ be continuous at $(0,0)$. Then there exists a circular region R, centre $(0,0)$, in which $f_{yx}(x,y)$ exists. Let (h,k) be an interior point of R (see Fig. 8). Put

$$\phi(x) = f(x,k) - f(x,0), \quad (4)$$

then, from (3), $$\Delta(h,k) = \phi(h) - \phi(0). \quad (5)$$

Since $f_{yx}(x,y)$ exists in R, therefore $f_x(x,y)$ exists in R. It follows that $\phi'(x)$ exists and hence also that $\phi(x)$ is continuous in the closed interval $[0,h]$ of the single variable x. Consequently, by the mean-value theorem, at least one number θ exists ($0 < \theta < 1$) such that

$$\Delta(h,k) = h\phi'(\theta h) = h\{f_x(\theta h, k) - f_x(\theta h, 0)\}. \quad (6)$$

By a further application of the mean-value theorem, since $f_{yx}(\theta h, y)$ exists in the closed interval $[0,k]$ of the single variable y, at least one number θ' exists ($0 < \theta' < 1$) such that

$$\Delta(h,k) = hk f_{yx}(\theta h, \theta' k) \quad (7)$$

and since $f_{yx}(x,y)$ is continuous at $(0,0)$ we can therefore put, if $hk \neq 0$,

$$\Delta(h,k)/(hk) = f_{yx}(0,0) + \alpha, \tag{8}$$

where $\alpha \to 0$ when $(h,k) \to (0,0)$ in any manner.

By similar reasoning, after putting $\Delta(h,k) = \psi(k) - \psi(0)$, with $\psi(y) = f(h,y) - f(0,y)$, we could show that, if $f_{xy}(x,y)$ is also continuous at $(0,0)$, then we can also put

$$\Delta(h,k)/(hk) = f_{xy}(0,0) + \beta, \tag{9}$$

where $\beta \to 0$ when $(h,k) \to (0,0)$ in any manner.

By (8) and (9), provided that both f_{yx} and f_{xy} are continuous at $(0,0)$ and that $hk \neq 0$,

$$f_{xy}(0,0) + \beta = f_{yx}(0,0) + \alpha$$

from which the truth of the theorem follows when $(h,k) \to (0,0)$.

Theorem 2. Sufficient conditions for $f_{xy}(0,0)$ and $f_{yx}(0,0)$ to be equal are that $f_{yx}(x,y)$ should be continuous at $(0,0)$ and that $f_y(x,y)$ should exist in the neighbourhood of $(0,0)$. [Schwarz]?

Proof. Let the circular region R be now one in which f_{yx} and f_y both exist.

As in the proof of Theorem 1, since f_{yx} is continuous at $(0,0)$, equation (8) holds good, and it follows that, when $(h,k) \to (0,0)$ in any manner, $\lim \{\Delta(h,k)/(hk)\}$ exists and is equal to $f_{yx}(0,0)$. In particular, the limit exists when first $k \to 0$, then $h \to 0$, and therefore

$$\lim_{h\to 0} \lim_{k\to 0} \{\Delta(h,k)/(hk)\} = f_{yx}(0,0). \tag{10}$$

But the L.H.S. may be written in the form

$$\lim_{h\to 0} \frac{1}{h}\left[\lim_{k\to 0}\left\{\frac{f(h,k)-f(h,0)}{k} - \frac{f(0,k)-f(0,0)}{k}\right\}\right]$$

that is, since f_y exists in R,

$$\lim_{h\to 0} \frac{1}{h}[f_y(h,0) - f_y(0,0)] \tag{11}$$

and this, by definition, is $f_{xy}(0,0)$. Thus $f_{xy}(0,0) = f_{yx}(0,0)$.

Note 1. The need for the condition that f_y should exist in R may be shown by an example: thus if $f(x,y) = x^3y^2 + g(y)$, then $f_{yx} = 6x^2y$, which is continuous everywhere, but f_y and f_{xy} do not exist in case $g(y)$ is not differentiable. In such a case, it is still true that

$$\lim_{h\to 0} \lim_{k\to 0} \Delta(h,k)/(hk) = f_{yx}(0,0),$$

but the L.H.S. is not equal to $f_{xy}(0,0)$, which has no meaning.

Note 2. Theorem 1 is a corollary of Theorem 2. For if f_{xy} is continuous at $(0,0)$, then f_y exists in the neighbourhood of $(0,0)$. Nevertheless, Theorem 1 is given as a separate theorem because it serves for most applications.

Theorem 3. Sufficient conditions for $f_{xy}(0,0)$ and $f_{yx}(0,0)$ to be equal are that $f_x(x,y)$ and $f_y(x,y)$ should both be differentiable at $(0,0)$. [Young]?

Proof. If f_x and f_y are differentiable at $(0,0)$, they must both exist near $(0,0)$. Let the circular region R be now one in which both exist.

Since f_x exists in R, we can put, by (6), if $h \neq 0$,

$$\Delta(h,k)/h = f_x(\theta h, k) - f_x(\theta h, 0) \quad (0 < \theta < 1). \tag{12}$$

Since f_x is differentiable at $(0,0)$, we can put, by §5.12,

$$f_x(\theta h, k) - f_x(0,0) = \theta h f_{xx}(0,0) + k f_{yx}(0,0) + \alpha \theta h + \beta k,$$

$$f_x(\theta h, 0) - f_x(0,0) = \theta h f_{xx}(0,0) + \gamma \theta h,$$

where α, β both vanish with h, k and γ vanishes with h. Substituting in (12) and dividing by k, we get, if $k \neq 0$,

$$\Delta(h,k)/(hk) = f_{yx}(0,0) + (\alpha - \gamma)\theta h/k + \beta.$$

Now put $k = h$ and let $h \to 0$: we get

$$\lim \Delta(h,h)/h^2 = f_{yx}(0,0).$$

Since f_y is also differentiable at $(0,0)$, the same limit is also equal to $f_{xy}(0,0)$. Hence the theorem.

Generalization. The above theorems can be generalized. Thus, by Theorem 1, at a point P,

$$\frac{\partial}{\partial x}\left(\frac{\partial}{\partial x}\frac{\partial f}{\partial y}\right) = \frac{\partial}{\partial x}\left(\frac{\partial}{\partial y}\frac{\partial f}{\partial x}\right) = \left(\frac{\partial}{\partial x}\frac{\partial}{\partial y}\right)\frac{\partial f}{\partial x} = \left(\frac{\partial}{\partial y}\frac{\partial}{\partial x}\right)\frac{\partial f}{\partial x}$$

that is, $f_{xxy} = f_{xyx} = f_{yxx}$, provided that each of these partial derivatives of the third order exists and is a continuous function of position at P. Similarly, $f_{xyy} = f_{yxy} = f_{yyx}$, provided that each of these partial derivatives is continuous at P. Consequently, the function $f(x,y)$ has, in all, four distinct partial derivatives of the third order $f_{xxx}, f_{xxy}, f_{xyy}, f_{yyy}$, provided that every mixed third-order derivative is continuous whatever the order in which the differentiations wo x and y are carried out.

In general, a mixed partial derivative of $f(x,y)$ of order n, the result of r differentiations wo x and s differentiations wo y $(r+s = n)$, is independent of the order in which the differentiations are performed, provided that the result is continuous, whatever the order. For we can effect any change from one order to another by a succession of reversals in the order of two consecutive differentiations, one wo x, the other wo y, and each such reversal will not affect the result, by Theorem 1.

Similar remarks apply to the mixed partial derivatives of a function of three or more independent variables.

Example

Consider at all points of the xy plane the second derivatives f_{xy} and f_{yx} of the function defined by

$$f(x, y) = x^2 \tan^{-1}(y/x) \quad (x \neq 0), \quad f(0, y) = 0.$$

If $x \neq 0$, so that the point (x, y) is not on the axis of y, we find by the ordinary rules of differentiation

$$f_x = 2x \tan^{-1}\frac{y}{x} - \frac{x^2 y}{x^2 + y^2}, \quad f_y = \frac{x^3}{x^2 + y^2},$$

$$f_{xy} = f_{yx} = x^2(x^2 + 3y^2)/(x^2 + y^2)^2.$$

If $x = 0$, so that the point (x, y) now lies on the axis of y, we use the fundamental definitions of §5.8 and find

$$f_x(0, y) = \lim \{f(h, y) - f(0, y)\}/h = \lim \{h \tan^{-1}(y/h)\} = 0,$$

and hence

$$f_{yx}(0, y) = \lim \{f_x(0, y + k) - f_x(0, y)\}/k = \lim \{0 - 0\}/k = 0$$

for all values of y; while

$$f_y(0, y) = \lim \{f(0, y + k) - f(0, y)\}/k = \lim \{0 - 0\}/k = 0,$$

and hence

$$f_{xy}(0, y) = \lim \{f_y(h, y) - f_y(0, y)\}/h = \lim \left\{\frac{h^2}{h^2 + y^2} - 0\right\}$$

from which follows

$$f_{xy}(0, y) = 0 \quad (y \neq 0), \quad f_{xy}(0, 0) = 1.$$

We thus see that $f_{xy} = f_{yx}$ at all points of the xy plane except $(0, 0)$, where $f_{xy} = 1 \neq f_{yx} = 0$.

From Theorem 1 we infer that $x^2(x^2 + 3y^2)/(x^2 + y^2)^2$ cannot be continuous at $(0, 0)$, as may be easily verified.

7.3   Total or exact differentials

We have seen that the total differential du of a differentiable function $u(x, y)$ of two independent variables x and y is defined by

$$du = p\,dx + q\,dy,$$

where $p = \partial u/\partial x$, $q = \partial u/\partial y$. Also, a sufficient condition for a function $u(x, y)$ to be differentiable, and therefore for du to exist, throughout a closed region R, is that $\partial u/\partial x$ and $\partial u/\partial y$ should be continuous functions of position (§5.15). The question arises: given a differential expression $p\,dx + q\,dy$, where p and q are functions of two variables x and y, under what conditions can we put

$$p\,dx + q\,dy = du,$$

where u is a function of x and y and the variables x and y are independent; in other words, under what conditions will there exist a function $u(x, y)$ of which $p\,dx + q\,dy$ is the total differential? For instance, we can put $x\,dy + y\,dx = du$, where $u = xy$; but we cannot put

$$x\,dy - y\,dx = du$$

if x and y are independent. If the function u exists, we say that

$$p\,dx + q\,dy$$

is a *total* or *perfect* (*exact*, *complete*) differential.

7.4

Theorem. Let p and q be functions of x and y with continuous first partial derivatives.

A necessary condition that $p\,dx + q\,dy$ should be a total, or exact, differential is

$$\partial p/\partial y = \partial q/\partial x. \tag{i}$$

Sufficient conditions are that (i) should hold good and that

$$(\partial/\partial y)\int p\,dx \text{ should exist.} \tag{ii}$$

Proof. Condition (i) *is necessary.* Suppose that a function u exists such that $du = p\,dx + q\,dy$ and therefore $\partial u/\partial x = p$, $\partial u/\partial y = q$. Then $\partial^2 u/\partial y\,\partial x = \partial p/\partial y$ which is continuous (given) and $\partial^2 u/\partial x\,\partial y = \partial q/\partial x$ which is also continuous (given). Hence, by §7.2, Theorem 1,

$$\partial^2 u/\partial y\,\partial x = \partial^2 u/\partial x\,\partial y$$

since both are continuous. Consequently, $\partial p/\partial y = \partial q/\partial x$; condition (i) is thus necessarily satisfied if u exists.

Conditions (i) *and* (ii) *are sufficient.* Let (i) and (ii) both hold good. Put $v = \int p\,dx$, the integration being carried out with y constant. Then $\partial v/\partial x = p$ and so $(\partial/\partial y)(\partial v/\partial x) = \partial p/\partial y$ which is continuous (given). Thus $\partial^2 v/\partial y\,\partial x$ is continuous; also $\partial v/\partial y$ exists, by (ii); hence, by §7.2, Theorem 2, $(\partial/\partial x)(\partial v/\partial y) = (\partial/\partial y)(\partial v/\partial x)$; that is, using (i), $(\partial/\partial x)(\partial v/\partial y) = \partial p/\partial y = \partial q/\partial x$, from which follows

$$(\partial/\partial x)(\partial v/\partial y - q) = 0.$$

This shows that $\partial v/\partial y - q$ is a function of y only; we can therefore put

$$q = \partial v/\partial y + \phi(y),$$

where $\phi(y)$ is known, since q and v are known. Hence

$$p\,dx + q\,dy = (\partial v/\partial x)\,dx + (\partial v/\partial y)\,dy + \phi(y)\,dy$$
$$= dv + \phi(y)\,dy;$$

thus $p\,dx + q\,dy$ is a perfect differential du, where

$$u = v + \int \phi(y)\,dy = \int p\,dx + \int \phi(y)\,dy.$$

Example

Show that, if x, y, and $x - y$ are all positive, a function u of the two independent variables x and y exists, and find u, such that

$$du = \frac{x^2\,dy - y^2\,dx}{x^2 y - xy^2}.$$

Here $\qquad p = -y^2/(x^2 y - xy^2) = 1/x - 1/(x - y),$

$$q = x^2/(x^2 y - xy^2) = 1/y + 1/(x - y),$$

from which the necessary condition $\partial p/\partial y = \partial q/\partial x$ is satisfied.

Putting $v = \int p\,dx = \log x - \log(x - y)$, we see that $\partial v/\partial y$ exists and that $\partial v/\partial y = 1/(x - y)$. Hence follows $q = 1/y + \partial v/\partial y$ and

$$p\,dx + q\,dy = \frac{\partial v}{\partial x}\,dx + \left(\frac{\partial v}{\partial y} + \frac{1}{y}\right) dy = du,$$

where $u = v + \log y + C = \log\{xy/(x - y)\} + C$, and C denotes an arbitrary constant.

7.5   Integrating factors

If a given differential expression $p\,dx + q\,dy$ is not an exact differential but becomes exact when multiplied by a factor $\lambda = \lambda(x, y)$, so that we can put $\qquad du = \lambda(p\,dx + q\,dy),$

where $u = u(x,y)$ is a differentiable function of the independent variables x and y, then λ is called an *integrating factor* (I.F.) of $p\,dx + q\,dy$.

Theorem 1. If λ is an I.F. of $p\,dx + q\,dy$, then $\lambda f(u)$ is also an I.F. where $f(u)$ is any integrable function of u.

Proof. We can put $\lambda(p\,dx + q\,dy) = du$ and therefore

$$\lambda f(u)\,(p\,dx + q\,dy) = f(u)\,du = dv,$$

where v denotes the integral of $f(u)\,du$; thus $\lambda f(u)$ is an I.F. It follows that if there is one I.F., there is an infinity of I.F.'s.

Theorem 2. If λ is one I.F. of $p\,dx + q\,dy$ and if $\lambda(p\,dx + q\,dy) = du$ then every I.F. is of the form $\lambda f(u)$.

Proof. Let λ, μ be two distinct I.F.'s and put

$$du = \lambda(p\,dx + q\,dy), \quad dv = \mu(p\,dx + q\,dy). \tag{i}$$

Then
$$\partial u/\partial x = \lambda p, \quad \partial u/\partial y = \lambda q, \quad \partial v/\partial x = \mu p, \quad \partial v/\partial y = \mu q. \tag{ii}$$

We verify at once that $\partial(u,v)/\partial(x,y) \equiv 0$, and assuming that p, q are not both zero, so that at least one of the four partial derivatives in (ii) is not zero, it follows (§ 6.19, Theorem 2) that we can put $v = \psi(u)$. We then have, from (i), $\mu/\lambda = dv/du = \psi'(u)$

and therefore $\mu = \lambda\psi'(u)$, so that μ is of the form $\lambda f(u)$.

Example

One I.F. of $x\,dy - y\,dx$ is $\lambda = 1/x^2$ and we have

$$(x\,dy - y\,dx)/x^2 = du, \quad \text{where} \quad u = y/x.$$

Another I.F. is $1/y^2$, or λ/u^2; another is $1/(xy)$, or λ/u.

7.6

If the differential equation $p\,dx + q\,dy = 0$ can be solved and the solution expressed in the form $u(x,y) = A$, where the function $u(x,y)$ is differentiable and A is an arbitrary constant, then, along every curve of the system of curves that represents the solution, both

$$(\partial u/\partial x)\,dx + (\partial u/\partial y)\,dy = 0 \quad \text{and} \quad p\,dx + q\,dy = 0$$

hold good, and hence also $(\partial u/\partial x)/p = (\partial u/\partial y)/q$. We can therefore put $\partial u/\partial x = \lambda p, \partial u/\partial y = \lambda q$ and hence $du = \lambda(p\,dx + q\,dy)$. Thus, if the differential equation $p\,dx + q\,dy = 0$ can be solved in the form $u(x,y) = A$, then $\lambda = (\partial u/\partial x)/p = (\partial u/\partial y)/q$ is an I.F. of $p\,dx + q\,dy$.

Conversely, if an integrating factor λ of the differential expression $p\,dx + q\,dy$ can be found, then the differential equation $p\,dx + q\,dy = 0$

can at once be solved, for if $\lambda(p\,dx + q\,dy) = du$, we can replace the differential equation $p\,dx + q\,dy = 0$ by $\lambda(p\,dx + q\,dy) = 0$, that is, $du = 0$, from which $u = A$ follows.

Example

The general solution of the equation $dy + ky\,dx = 0$ can be written $y\,e^{kx} = A$, from which we have

$$d(y\,e^{kx}) = e^{kx}(dy + ky\,dx) = 0.$$

It follows that $\lambda = e^{kx}$ is an I.F. of the differential $dy + ky\,dx$, if k is a constant.

7.7

It is known* that if $F(x, y)$ and $(\partial/\partial y)\,F(x, y)$ are continuous in a region R, then the differential equation $dy/dx = F(x, y)$ has a solution representable graphically by a one-parameter family of curves, no two of which have a common point in R.

It follows that if p and q satisfy the conditions that p/q and $(\partial/\partial y)$ (p/q) are continuous in a region R, then the differential equation

$$dy/dx = -p/q, \quad \text{or} \quad p\,dx + q\,dy = 0$$

has such a solution. Consequently, when p and q satisfy these conditions, which apply in the more important applications, the differential equation $p\,dx + q\,dy = 0$ is known to have a solution, and the solution will provide a theoretically general means of finding an I.F. of $p\,dx + q\,dy$. We say 'theoretically' because there are only a few types of differential equation which can be solved in finite form.

A second theoretically available method of finding an I.F. is to note that, by §7.4, if $\lambda(p\,dx + q\,dy)$ is exact, then λ must satisfy the partial differential equation

$$(\partial/\partial y)\,(\lambda p) = (\partial/\partial x)\,(\lambda q)$$

or

$$p\,\partial\lambda/\partial y - q\,\partial\lambda/\partial x + \lambda(\partial p/\partial y - \partial q/\partial x) = 0;$$

but again, attempts to solve this equation in finite form will only succeed in particular cases.

7.8 Inexact differentials

If $p\,dx + q\,dy$ is not an exact differential, so that we cannot put $du = p\,dx + q\,dy$ in the sense that there exists a function u such that $p\,dx + q\,dy$ is its total or exact differential when x and y are independent

* For example, J. C. Burkill, *The Theory of Ordinary Differential Equations*, § 3.

variables, nevertheless we can put $du = p\,dx + q\,dy$ if x and y are not independent. Thus, if y is a continuous function of x with a continuous derivative dy/dx, then we can put $du = p\,dx + q\,dy$ in the sense that

$$du/dx = p + q\,dy/dx$$

so that du is the differential of the function of the single variable x given by the indefinite integral

$$u = \int \left(p + q\frac{dy}{dx} \right) dx.$$

Or, again, if x and y are functions of a single variable t, such that dx/dt and dy/dt are continuous, we can put $du = p\,dx + q\,dy$, where u is the function of the single variable t given by

$$u = \int \left(p\frac{dx}{dt} + q\frac{dy}{dt} \right) dt.$$

7.9   Successive differentiation of functions of functions

To save repetition we shall suppose that all the functions and their derivatives that occur in the following articles are continuous. To illustrate methods, it will usually be sufficient not to go beyond second derivatives.

We first recall from §5.22 that, if z is a function of x, y and x, y are functions of a single variable t:

$$z = \phi(x,y), \quad x = f(t), \quad y = g(t), \tag{13}$$

so that z is a function of t through x, y, then

$$\frac{dz}{dt} = \frac{\partial z}{\partial x}\frac{dx}{dt} + \frac{\partial z}{\partial y}\frac{dy}{dt} = \frac{dx}{dt}\frac{\partial z}{\partial x} + \frac{dy}{dt}\frac{\partial z}{\partial y}. \tag{14}$$

If z is explicitly a function of t as well as of x, y, so that

$$z = \phi(x,y,t), \quad x = f(t), \quad y = g(t), \tag{15}$$

then $\qquad\qquad \dfrac{dz}{dt} = \dfrac{\partial z}{\partial x}\dfrac{dx}{dt} + \dfrac{\partial z}{\partial y}\dfrac{dy}{dt} + \dfrac{\partial z}{\partial t}. \tag{16}$

7.10

Let z be a function of x, y and let x, y be functions of t, as in (13). To find d^2z/dt^2.

We find dz/dt from (14); then, since $\partial z/\partial x$ and $\partial z/\partial y$ are functions of x and y, whereas dx/dt and dy/dt are functions of t, we see from (16) that

$$\frac{d}{dt}\left(\frac{dz}{dt}\right) = \left(\frac{dx}{dt}\frac{\partial}{\partial x} + \frac{dy}{dt}\frac{\partial}{\partial y} + \frac{\partial}{\partial t}\right)\left(\frac{dx}{dt}\frac{\partial z}{\partial x} + \frac{dy}{dt}\frac{\partial z}{\partial y}\right)$$

$$= \frac{dx}{dt}\left(\frac{dx}{dt}\frac{\partial^2 z}{\partial x^2} + \frac{dy}{dt}\frac{\partial^2 z}{\partial x\,\partial y}\right)$$

$$+ \frac{dy}{dt}\left(\frac{dx}{dt}\frac{\partial^2 z}{\partial x\,\partial y} + \frac{dy}{dt}\frac{\partial^2 z}{\partial y^2}\right) + \frac{d^2 x}{dt^2}\frac{\partial z}{\partial x} + \frac{d^2 y}{dt^2}\frac{\partial z}{\partial y},$$

that is,
$$\frac{d^2 z}{dt^2} = \dot{x}^2\frac{\partial^2 z}{\partial x^2} + 2\dot{x}\dot{y}\frac{\partial^2 z}{\partial x\,\partial y} + \dot{y}^2\frac{\partial^2 z}{\partial y^2} + \ddot{x}\frac{\partial z}{\partial x} + \ddot{y}\frac{\partial z}{\partial y} \qquad (17)$$

where dots denote differentiation wo t.

Example

$$z = f(x, y), \quad x = a + ht, \quad y = b + kt,$$

where a, b, h, k are constants.

Here x and y are linear functions of t, and $\dot{x} = h$, $\ddot{x} = 0$, $\dot{y} = k$, $\ddot{y} = 0$; hence, by (17),

$$\frac{d^2 z}{dt^2} = h^2\frac{\partial^2 z}{\partial x^2} + 2hk\frac{\partial^2 z}{\partial x\,\partial y} + k^2\frac{\partial^2 z}{\partial y^2}$$

which is often conveniently written symbolically in the form

$$\frac{d^2 z}{dt^2} = \left(h\frac{\partial}{\partial x} + k\frac{\partial}{\partial y}\right)^2 z,$$

where, after squaring, $(h\,\partial/\partial x)(k\,\partial/\partial y)$ is to be replaced by $hk\,\partial^2/\partial x\,\partial y$.

It follows by induction, without difficulty, that the nth derivative may be expressed symbolically by

$$\frac{d^n z}{dt^n} = \left(h\frac{\partial}{\partial x} + k\frac{\partial}{\partial y}\right)^n z, \qquad (18)$$

where, after expanding the operator by the binomial theorem, $(h\,\partial/\partial x)^{n-r}(k\,\partial/\partial y)^r$ is to be replaced by $h^{n-r}k^r\,\partial^n/\partial x^{n-r}\partial y^r$.

7.11

Let T be a function of u, v and let u, v be functions of x, y, z, so that T is a function of x, y, z through u, v: say

$$T = \phi(u, v), \quad u = f(x, y, z), \quad v = g(x, y, z). \qquad (19)$$

To find $\partial^2 T/\partial x^2$, $\partial^2 T/\partial x\,\partial y$, ...

By the same kind of reasoning as in §7.10,

$$\frac{\partial}{\partial x}\left(\frac{\partial T}{\partial x}\right) = \left(\frac{\partial u}{\partial x}\frac{\partial}{\partial u}+\frac{\partial v}{\partial x}\frac{\partial}{\partial v}+\frac{\partial}{\partial x}\right)\left(\frac{\partial u}{\partial x}\frac{\partial T}{\partial u}+\frac{\partial v}{\partial x}\frac{\partial T}{\partial v}\right)$$

which gives

$$\frac{\partial^2 T}{\partial x^2} = u_1^2\frac{\partial^2 T}{\partial u^2} + 2u_1 v_1\frac{\partial^2 T}{\partial u\,\partial v} + v_1^2\frac{\partial^2 T}{\partial v^2} + u_{11}\frac{\partial T}{\partial u} + v_{11}\frac{\partial T}{\partial v}, \qquad (20)$$

where $u_1 = \partial u/\partial x$, $u_{11} = \partial^2 u/\partial x^2, \dots$.   Again,

$$\frac{\partial}{\partial y}\left(\frac{\partial T}{\partial x}\right) = \left(\frac{\partial u}{\partial y}\frac{\partial}{\partial u}+\frac{\partial v}{\partial y}\frac{\partial}{\partial v}+\frac{\partial}{\partial y}\right)\left(\frac{\partial u}{\partial x}\frac{\partial T}{\partial u}+\frac{\partial v}{\partial x}\frac{\partial T}{\partial v}\right),$$

which gives

$$\frac{\partial^2 T}{\partial x\,\partial y} = u_1 u_2\frac{\partial^2 T}{\partial u^2} + (u_1 v_2 + u_2 v_1)\frac{\partial^2 T}{\partial u\,\partial v} + v_1 v_2\frac{\partial^2 T}{\partial v^2} + u_{12}\frac{\partial T}{\partial u} + v_{12}\frac{\partial T}{\partial v}. \quad (21)$$

Similar expressions give $\partial^2 T/\partial x\,\partial z$, $\partial^2 T/\partial y^2, \dots$.

7.12

Let T be a function of u, v and let u, v be defined inversely as functions of x, y by equations which give x, y explicitly in terms of u, v: say

$$T = \phi(u, v), \quad x = f(u, v), \quad y = g(u, v), \quad \partial(x,y)/\partial(u,v) \neq 0. \quad (22)$$

To find expressions for $\partial^2 T/\partial x^2$, $\partial^2 T/\partial y^2$, $\partial^2 T/\partial x\,\partial y$.

As in §6.15, we have

$$\frac{\partial T}{\partial x} = \frac{\partial(T,y)}{\partial(x,y)} = \frac{1}{J}\frac{\partial(T,y)}{\partial(u,v)} = \left(\frac{y_2}{J}\frac{\partial}{\partial u} - \frac{y_1}{J}\frac{\partial}{\partial v}\right)T,$$

$$\frac{\partial T}{\partial y} = \frac{\partial(x,T)}{\partial(x,y)} = \frac{1}{J}\frac{\partial(x,T)}{\partial(u,v)} = \left(-\frac{x_2}{J}\frac{\partial}{\partial u} + \frac{x_1}{J}\frac{\partial}{\partial v}\right)T,$$

where $J = \partial(x,y)/\partial(u,v)$, $x_1 = \partial x/\partial u, \dots$.   Hence

$$\frac{\partial^2 T}{\partial x^2} = \left(\frac{y_2}{J}\frac{\partial}{\partial u} - \frac{y_1}{J}\frac{\partial}{\partial v}\right)\left(\frac{y_2}{J}\frac{\partial T}{\partial u} - \frac{y_1}{J}\frac{\partial T}{\partial v}\right), \qquad (23)$$

$$\frac{\partial^2 T}{\partial y^2} = \left(\frac{x_2}{J}\frac{\partial}{\partial u} - \frac{x_1}{J}\frac{\partial}{\partial v}\right)\left(\frac{x_2}{J}\frac{\partial T}{\partial u} - \frac{x_1}{J}\frac{\partial T}{\partial v}\right), \qquad (24)$$

$$\frac{\partial^2 T}{\partial x\,\partial y} = \left(\frac{y_2}{J}\frac{\partial}{\partial u} - \frac{y_1}{J}\frac{\partial}{\partial v}\right)\left(-\frac{x_2}{J}\frac{\partial T}{\partial u} + \frac{x_1}{J}\frac{\partial T}{\partial v}\right). \qquad (25)$$

Example

$$T = \phi(r, \theta), \quad x = r \cos \theta, \quad y = r \sin \theta, \quad J = \partial(x, y)/\partial(r, \theta) = r \neq 0.$$

Putting $T_1 = \partial T/\partial r$, $T_2 = \partial T/\partial \theta$, ... we find

$$\frac{\partial^2 T}{\partial x^2} = \left(\cos \theta \frac{\partial}{\partial r} - \frac{\sin \theta}{r} \frac{\partial}{\partial \theta}\right)\left(\cos \theta \frac{\partial T}{\partial r} - \frac{\sin \theta}{r} \frac{\partial T}{\partial \theta}\right)$$

$$= \cos^2 \theta\, T_{11} - 2 \sin \theta \cos \theta \left(\frac{T_{12}}{r} - \frac{T_2}{r^2}\right) + \sin^2 \theta \left(\frac{T_1}{r} + \frac{T_{22}}{r^2}\right), \quad (26)$$

$$\frac{\partial^2 T}{\partial x \partial y} = \left(\cos \theta \frac{\partial}{\partial r} - \frac{\sin \theta}{r} \frac{\partial}{\partial \theta}\right)\left(\sin \theta \frac{\partial T}{\partial r} + \frac{\cos \theta}{r} \frac{\partial T}{\partial \theta}\right)$$

$$= \sin \theta \cos \theta \left(T_{11} - \frac{T_1}{r} - \frac{T_{22}}{r^2}\right) + (\cos^2 \theta - \sin^2 \theta)\left(\frac{T_{12}}{r} - \frac{T_2}{r^2}\right), \quad (27)$$

and similarly,

$$\frac{\partial^2 T}{\partial y^2} = \sin^2 \theta\, T_{11} + 2 \sin \theta \cos \theta \left(\frac{T_{12}}{r} - \frac{T_2}{r^2}\right) + \cos^2 \theta \left(\frac{T_1}{r} + \frac{T_{22}}{r^2}\right). \quad (28)$$

By adding (26) and (28),

$$\frac{\partial^2 T}{\partial x^2} + \frac{\partial^2 T}{\partial y^2} = \frac{\partial^2 T}{\partial r^2} + \frac{1}{r}\frac{\partial T}{\partial r} + \frac{1}{r^2}\frac{\partial^2 T}{\partial \theta^2}. \quad (29)$$

7.13 Successive differentiation of implicit functions

Let y be defined implicitly (§ 6.3) as a function of x by the equation

$$F(x, y) = 0, \quad \partial F/\partial y \neq 0. \quad (30)$$

Assuming, for sufficiency, that the successive partial derivatives of the function F are continuous, we find by repeated differentiations wo x

$$F_1 + F_2 \frac{dy}{dx} = 0, \quad (31)$$

$$F_{11} + 2F_{12}\frac{dy}{dx} + F_{22}\left(\frac{dy}{dx}\right)^2 + F_2\frac{d^2y}{dx^2} = 0, \quad (32)$$

....

where $F_1 = \partial F/\partial x$, $F_2 = \partial F/\partial y$, ...

These are simple equations in dy/dx, d^2y/dx^2, ... and since $F_2 \neq 0$, equation (31) gives dy/dx in terms of x, y; then, from (32), d^2y/dx^2 can be expressed in terms of x, y; then d^3y/dx^3, and so on.

7.14

Let u, v be defined implicitly (§6.10) as functions of the independent variables x, y by the equations

$$F(x, y, u, v) = 0, \quad G(x, y, u, v) = 0, \quad \partial(F, G)/\partial(u, v) \neq 0. \quad (33)$$

The first partial derivatives $\partial u/\partial x$, $\partial v/\partial x$ are determined in terms of x, y, u, v by the pair of equations

$$\left.\begin{aligned}
\frac{\partial F}{\partial u}\frac{\partial u}{\partial x} + \frac{\partial F}{\partial v}\frac{\partial v}{\partial x} + \frac{\partial F}{\partial x} = 0, \\
\frac{\partial G}{\partial u}\frac{\partial u}{\partial x} + \frac{\partial G}{\partial v}\frac{\partial v}{\partial x} + \frac{\partial G}{\partial x} = 0,
\end{aligned}\right\} \quad (34)$$

and $\partial u/\partial y$, $\partial v/\partial y$ by the pair of equations

$$\left.\begin{aligned}
\frac{\partial F}{\partial u}\frac{\partial u}{\partial y} + \frac{\partial F}{\partial v}\frac{\partial v}{\partial y} + \frac{\partial F}{\partial y} = 0, \\
\frac{\partial G}{\partial u}\frac{\partial u}{\partial y} + \frac{\partial G}{\partial v}\frac{\partial v}{\partial y} + \frac{\partial G}{\partial y} = 0.
\end{aligned}\right\} \quad (35)$$

By differentiating the pair (34) partially wo x, we obtain two equations that determine $\partial^2 u/\partial x^2$, $\partial^2 v/\partial x^2$. By differentiating (35) partially wo y, we obtain two equations that determine $\partial^2 u/\partial y^2$, $\partial^2 v/\partial y^2$. By differentiating the pair (34) wo y, or the pair (35) wo x, we obtain two equations that determine $\partial^2 u/\partial x\,\partial y$, $\partial^2 v/\partial x\,\partial y$.

Further differentiation would lead to equations that determine third and higher partial derivatives of u and v.

7.15   Change of variables by substitutions

When the independent variables in such an expression (the *Laplacian* of T) as

$$\nabla^2 T \equiv \frac{\partial^2 T}{\partial x^2} + \frac{\partial^2 T}{\partial y^2} + \frac{\partial^2 T}{\partial z^2} \quad (36)$$

are changed to u, v, w by means of a set of equations such as

$$x = f(u, v, w), \quad y = g(u, v, w), \quad z = h(u, v, w) \quad (37)$$

or inversely by such a set as

$$u = \phi(x, y, z), \quad v = \psi(x, y, z), \quad w = \chi(x, y, z) \quad (38)$$

the set of equations is called a *substitution* or a *transformation*. A substitution may also be in the form of a set of equations that define the relations between the old and new variables implicitly.

7.16   Orthogonal substitutions

A substitution is said to be *orthogonal* if the identities

$$
\left.
\begin{aligned}
v_1 w_1 + v_2 w_2 + v_3 w_3 &\equiv 0, \\
w_1 u_1 + w_2 u_2 + w_3 u_3 &\equiv 0, \\
u_1 v_1 + u_2 v_2 + u_3 v_3 &\equiv 0,
\end{aligned}
\right\}
\tag{39}
$$

are satisfied, where $u_1 = \partial u/\partial x, \ldots$. Geometrically, these identities express that the normals to the three surfaces $u = $ const., $v = $ const., $w = $ const., that pass through any point (x, y, z) are mutually perpendicular at that point. The triply infinite system of surfaces obtained by giving all possible values to u, v, w is then called a *triply orthogonal system*.

The conditions of orthogonality can also be expressed by the identities

$$
\left.
\begin{aligned}
x_2 x_3 + y_2 y_3 + z_2 z_3 &\equiv 0, \\
x_3 x_1 + y_3 y_1 + z_3 z_1 &\equiv 0, \\
x_1 x_2 + y_1 y_2 + z_1 z_2 &\equiv 0,
\end{aligned}
\right\}
\tag{40}
$$

where $x_1 = \partial x/\partial u, \ldots$. Geometrically, these identities express that the curves of intersection of the surfaces $u = $ const., $v = $ const., $w = $ const. in pairs are mutually perpendicular at a common point.

For example, the equations

$$
x = r \sin \cos \phi, \quad y = r \sin \theta \sin \phi, \quad z = r \cos \theta
$$

express the usual relations between Cartesian and polar coordinates in three dimensions. The surfaces $r = $ const. are concentric spheres with their common centre at the origin, the surfaces $\theta = $ const. are the right circular cones with their common apex at the origin and their common axis the axis of z, the surfaces $\phi = $ const. are planes having the axis of z as a common line of intersection. These three families of surfaces intersect mutually at right angles, that is, the transformation is orthogonal, as may be verified by constructing the set of identities that correspond to (40).

7.17

In two dimensions, a substitution or transformation such as

$$
x = f(u, v), \quad y = g(u, v)
\tag{41}
$$

or

$$
u = \phi(x, y), \quad v = \psi(x, y)
\tag{42}
$$

is said to be orthogonal if $u_1 v_1 + u_2 v_2 \equiv 0$, where $u_1 = \partial u/\partial x, \ldots$ or, what comes to the same thing, if

$$x_1 x_2 + y_1 y_2 \equiv 0, \quad \text{where} \quad x_1 = \partial x/\partial u, \ldots.$$

A class of two-dimensional orthogonal substitutions of special interest is that in which u and v satisfy the two conditions

$$u_1 = v_2, \quad u_2 = -v_1; \tag{43}$$

these are the Riemann–Cauchy conditions (§6.24), showing that $u + iv$ is an analytic function of $x + iy$. That such a substitution is orthogonal is an immediate consequence of (43), from which follows $u_1 v_1 + u_2 v_2 \equiv 0$. The Jacobian of the substitution can be put in various forms, thus

$$J = \frac{\partial(u,v)}{\partial(x,y)} = \begin{vmatrix} u_1 & u_2 \\ v_1 & v_2 \end{vmatrix} = u_1^2 + v_1^2 = u_2^2 + v_2^2 = u_1^2 + u_2^2 = v_1^2 + v_2^2. \tag{44}$$

7.18   Linear substitutions

A linear substitution in three dimensions is one of the form

$$\begin{aligned} x &= l_1 u + l_2 v + l_3 w, \\ y &= m_1 u + m_2 v + m_3 w, \\ z &= n_1 u + n_2 v + n_3 w, \end{aligned} \right\} \tag{45}$$

where $l_1, \ldots$ are constants. The substitution is orthogonal if

$$\begin{aligned} l_2 l_3 + m_2 m_3 + n_2 n_3 &= 0, \\ l_3 l_1 + m_3 m_1 + n_3 n_1 &= 0, \\ l_1 l_2 + m_1 m_2 + n_1 n_2 &= 0. \end{aligned} \right\} \tag{46}$$

If (l_r, m_r, n_r), $(r = 1, 2, 3)$, are the direction cosines of three mutually perpendicular directions, such a substitution is equivalent to a rotation of rectangular coordinate axes about some axis through the origin.

In two dimensions a linear substitution is of the form

$$x = l_1 u + l_2 v, \quad y = m_1 u + m_2 v, \tag{47}$$

where $l_1, \ldots$ are constants. It is orthogonal if $l_1 l_2 + m_1 m_2 = 0$. In particular, a linear substitution of the form

$$x = u \cos \alpha - v \sin \alpha, \quad y = u \sin \alpha + v \cos \alpha \tag{48}$$

is equivalent to a rotation of rectangular axes about the origin.

7.19   Examples of substitutions

To change the independent variables in the equation

$$\frac{\partial^2 z}{\partial t^2} = c^2 \frac{\partial^2 z}{\partial x^2}$$

from x, t to u, v by means of the linear substitution

$$u = x + ct, \quad v = x - ct,$$

we have

$$\frac{\partial z}{\partial x} = \frac{\partial z}{\partial u}\frac{\partial u}{\partial x} + \frac{\partial z}{\partial v}\frac{\partial v}{\partial x} = \left(\frac{\partial}{\partial u} + \frac{\partial}{\partial v}\right) z,$$

$$\frac{\partial z}{\partial t} = \frac{\partial z}{\partial u}\frac{\partial u}{\partial t} + \frac{\partial z}{\partial v}\frac{\partial v}{\partial t} = c\left(\frac{\partial}{\partial u} - \frac{\partial}{\partial v}\right) z.$$

The equation therefore becomes

$$c^2 \left(\frac{\partial}{\partial u} - \frac{\partial}{\partial v}\right)^2 z = c^2 \left(\frac{\partial}{\partial u} + \frac{\partial}{\partial v}\right)^2 z$$

which, if $c \neq 0$, reduces to $\partial^2 z / \partial u\, \partial v = 0$.

7.20

To transform the expression (the *Laplacian* of T in two dimensions)

$$\nabla_1^2 T \equiv \frac{\partial^2 T}{\partial x^2} + \frac{\partial^2 T}{\partial y^2} \tag{49}$$

by a substitution of the form $u = \phi(x, y)$, $v = \psi(x, y)$.

Put $a = u_1^2 + u_2^2, \quad h = u_1 v_1 + u_2 v_2, \quad b = v_1^2 + v_2^2,$

where $u_1 = \partial u / \partial x, \dots$. Then, using $\partial^2 T / \partial x^2$ from (20) and adding a similar expression for $\partial^2 T / \partial y^2$, we get

$$\nabla_1^2 T = a \frac{\partial^2 T}{\partial u^2} + 2h \frac{\partial^2 T}{\partial u\, \partial v} + b \frac{\partial^2 T}{\partial v^2} + \nabla_1^2 u \frac{\partial T}{\partial u} + \nabla_1^2 v \frac{\partial T}{\partial v}. \tag{50}$$

Corollary 1. If the substitution is orthogonal, then $h = 0$, and

$$\nabla_1^2 T = a \frac{\partial^2 T}{\partial u^2} + b \frac{\partial^2 T}{\partial v^2} + \nabla_1^2 u \frac{\partial T}{\partial u} + \nabla_1^2 v \frac{\partial T}{\partial v}. \tag{51}$$

Corollary 2. If u, v are a conjugate pair of harmonic functions of x, y (see p. 216) so that $u_1 = v_2$, $u_2 = -v_1$, that is, if $w, = u + iv$, is an analytic function of $z, = x + iy$, then

$$h = 0, \quad a = b = \frac{\partial(u, v)}{\partial(x, y)} = \left|\frac{dw}{dz}\right|^2 = \frac{\partial w}{\partial x}\frac{\partial \overline{w}}{\partial x}, \tag{52}$$

where $\overline{w} = u - iv$, and

$$\nabla_1^2 T = \frac{\partial(u, v)}{\partial(x, y)}\left(\frac{\partial^2 T}{\partial u^2} + \frac{\partial^2 T}{\partial v^2}\right) = \frac{\partial w}{\partial x}\frac{\partial \overline{w}}{\partial x}\left(\frac{\partial^2 T}{\partial u^2} + \frac{\partial^2 T}{\partial v^2}\right). \tag{53}$$

7.21

To transform $\nabla_1^2 T$, given by (49), by a substitution of the form

$$x = f(u,v), \quad y = g(u,v), \quad J = \partial(x,y)/\partial(u,v) \neq 0.$$

Put $A = x_1^2 + y_1^2, \quad H = x_1 x_2 + y_1 y_2, \quad B = x_2^2 + y_2^2,$

where $x_1 = \partial x/\partial u, \ldots.$ As in §6.15, we have, using $y_{12} = y_{21}$ and $x_{12} = x_{21}$,

$$\frac{\partial T}{\partial x} = \frac{1}{J}\left(y_2 \frac{\partial T}{\partial u} - y_1 \frac{\partial T}{\partial v}\right) \tag{54}$$

$$= \frac{1}{J}\left\{\frac{\partial}{\partial u}(y_2 T) - \frac{\partial}{\partial v}(y_1 T)\right\}, \tag{55}$$

$$\frac{\partial T}{\partial y} = \frac{1}{J}\left(x_1 \frac{\partial T}{\partial v} - x_2 \frac{\partial T}{\partial u}\right) \tag{56}$$

$$= \frac{1}{J}\left\{\frac{\partial}{\partial v}(x_1 T) - \frac{\partial}{\partial u}(x_2 T)\right\}. \tag{57}$$

Replacing T in (55) by $\partial T/\partial x$, and in (57) by $\partial T/\partial y$, and adding, we get

$$\nabla_1^2 T = \frac{1}{J}\left\{\frac{\partial}{\partial u}\left(y_2 \frac{\partial T}{\partial x} - x_2 \frac{\partial T}{\partial y}\right) + \frac{\partial}{\partial v}\left(x_1 \frac{\partial T}{\partial y} - y_1 \frac{\partial T}{\partial x}\right)\right\}.$$

Then, substituting for $\partial T/\partial x$ from (54) and for $\partial T/\partial y$ from (56), we find

$$\nabla_1^2 T = \frac{1}{J}\left\{\frac{\partial}{\partial u}\left(\frac{BT_1 - HT_2}{J}\right) + \frac{\partial}{\partial v}\left(\frac{AT_2 - HT_1}{J}\right)\right\}. \tag{58}$$

Note that

$$J^2 = \begin{vmatrix} x_1^2 + y_1^2 & x_1 x_2 + y_1 y_2 \\ x_1 x_2 + y_1 y_2 & x_2^2 + y_2^2 \end{vmatrix} = \begin{vmatrix} A & H \\ H & B \end{vmatrix}. \tag{59}$$

Corollary 1. If the substitution is orthogonal, $H = 0$, $J^2 = AB$, and

$$\nabla_1^2 T = \frac{1}{J}\left\{\frac{\partial}{\partial u}\left(\frac{J}{A}\frac{\partial T}{\partial u}\right) + \frac{\partial}{\partial v}\left(\frac{J}{B}\frac{\partial T}{\partial v}\right)\right\}, \tag{60}$$

where J may be replaced by $|J| = \sqrt{(AB)}$, since the R.H.S. is unaltered if we replace J by $-J$.

Corollary 2. If x, y are a conjugate pair of functions of u, v, so that $x_1 = y_2$, $x_2 = -y_1$, or $z_1 = x + iy$, is an analytic function of $w_1 = u + iv$, then

$$H = 0, A = B = J = \frac{\partial(x,y)}{\partial(u,v)} = \left|\frac{dz}{dw}\right|^2 = \frac{\partial z}{\partial u}\frac{\partial \bar{z}}{\partial u},$$

where $\bar{z} = x - iy$, and then ∇_1^2 takes the form

$$\nabla_1^2 T = \frac{1}{J}\left(\frac{\partial^2 T}{\partial u^2} + \frac{\partial^2 T}{\partial v^2}\right). \tag{61}$$

Examples

(1) Transform $\nabla_1^2 T$ to polar coordinates r, θ where

$$x = r \cos \theta, \quad y = r \sin \theta.$$

Here $A = 1, H = 0, B = r^2$,

$$\nabla_1^2 T = \frac{1}{r} \left\{ \frac{\partial}{\partial r} \left(r \frac{\partial T}{\partial r} \right) + \frac{\partial}{\partial \theta} \left(\frac{1}{r} \frac{\partial T}{\partial \theta} \right) \right\} \tag{62}$$

which is another form of (29).

(2) Transform $\nabla_1^2 T$ by the substitution

$$x = c \cosh \xi \cos \eta, \quad y = c \sinh \xi \sin \eta.$$

Here $x_1 = y_2, x_2 = -y_1$, where $x_1 = \partial x/\partial \xi, \ldots$; so that $x + iy$ is an analytic function of $\xi + i\eta$. In fact

$$z = x + iy = c \cosh (\xi + i\eta),$$

$$\frac{\partial z}{\partial \xi} \frac{\partial \bar{z}}{\partial \xi} = c \cosh (\xi + i\eta) c \cosh (\xi - i\eta) = \tfrac{1}{2} c^2 (\cosh 2\xi + \cos 2\eta),$$

$$\nabla_1^2 T = \frac{2}{c^2 (\cosh 2\xi + \cos 2\eta)} \left(\frac{\partial^2 T}{\partial \xi^2} + \frac{\partial^2 T}{\partial \eta^2} \right). \tag{63}$$

(3) Transform $\nabla_1^2 T$ to confocal coordinates λ, μ, defined by the equations

$$\left. \begin{aligned} \frac{x^2}{a^2 + \lambda} + \frac{y^2}{b^2 + \lambda} = 1 \quad (-b^2 < \lambda < \infty), \\[2mm] \frac{x^2}{a^2 + \mu} + \frac{y^2}{b^2 + \mu} = 1 \quad (-a^2 < \mu < -b^2). \end{aligned} \right\} \tag{64}$$

Solving for x^2, y^2, we find

$$x^2 = \frac{(a^2 + \lambda)(a^2 + \mu)}{a^2 - b^2}, \quad y^2 = -\frac{(b^2 + \lambda)(b^2 + \mu)}{a^2 - b^2}, \tag{65}$$

and hence

$$\left. \begin{aligned} 2\,\partial x/\partial \lambda = x/(a^2 + \lambda), \quad 2\partial x/\partial \mu = x/(a^2 + \mu), \\[1mm] 2\,\partial y/\partial \lambda = y/(b^2 + \lambda), \quad 2\,\partial y/\partial \mu = y/(b^2 + \mu). \end{aligned} \right\} \tag{66}$$

From (66) and (65), putting $x_1 = \partial x/\partial \lambda, \ldots$ we now find

$$A = x_1^2 + y_1^2 = (\lambda - \mu)/4\Lambda, \quad H = x_1 x_2 + y_1 y_2 = 0,$$

$$B = x_2^2 + y_2^2 = (\lambda - \mu)/4\mathrm{M},$$

where $\Lambda = (a^2 + \lambda)(b^2 + \lambda)$, $M = -(a^2 + \mu)(b^2 + \mu)$. Also,

$$J = \begin{vmatrix} x_1 & x_2 \\ y_1 & y_2 \end{vmatrix} = -\frac{\lambda - \mu}{4xy(a^2 - b^2)} = \mp \frac{\lambda - \mu}{4\sqrt{(\Lambda M)}}$$

according as $xy > 0$ or $xy < 0$. After substituting in (60), we find

$$\nabla_1^2 T = \frac{4}{\lambda - \mu}\left\{\Lambda^{\frac{1}{2}}\frac{\partial}{\partial \lambda}\left(\Lambda^{\frac{1}{2}}\frac{\partial T}{\partial \lambda}\right) + M^{\frac{1}{2}}\frac{\partial}{\partial \mu}\left(M^{\frac{1}{2}}\frac{\partial T}{\partial \mu}\right)\right\}. \qquad (67)$$

7.22

Transform the Laplacian

$$\nabla^2 T \equiv \frac{\partial^2 T}{\partial x^2} + \frac{\partial^2 T}{\partial y^2} + \frac{\partial^2 T}{\partial z^2} \qquad (68)$$

(i) to cylindrical coordinates z, ρ, ϕ;

(ii) to spherical polar coordinates r, θ, ϕ.

(a) Since $x = \rho \cos \phi$, $y = \rho \sin \phi$, by (62) we have

$$\nabla^2 T = \frac{\partial^2 T}{\partial \rho^2} + \frac{1}{\rho}\frac{\partial T}{\partial \rho} + \frac{1}{\rho^2}\frac{\partial^2 T}{\partial \phi^2} + \frac{\partial^2 T}{\partial z^2}. \qquad (69)$$

(b) Now make the substitution $z = r \cos \theta$, $\rho = r \sin \theta$. Then, again by (62) we have

$$\frac{\partial^2 T}{\partial z^2} + \frac{\partial^2 T}{\partial \rho^2} = \frac{\partial^2 T}{\partial r^2} + \frac{1}{r}\frac{\partial T}{\partial r} + \frac{1}{r^2}\frac{\partial^2 T}{\partial \theta^2}.$$

Also, $\dfrac{\partial T}{\partial \rho} = \dfrac{\partial(z, T)}{\partial(z, \rho)} = \dfrac{\partial(z, T)}{\partial(r, \theta)}\bigg/\dfrac{\partial(z, \rho)}{\partial(r, \theta)} = \sin\theta\dfrac{\partial T}{\partial r} + \dfrac{\cos\theta}{r}\dfrac{\partial T}{\partial \theta}.$

Hence, after a little reduction,

$$\nabla^2 T = \frac{1}{r^2}\frac{\partial}{\partial r}\left(r^2\frac{\partial T}{\partial r}\right) + \frac{1}{r^2}\frac{\partial^2 T}{\partial \theta^2} + \frac{\cot\theta}{r^2}\frac{\partial T}{\partial \theta} + \frac{1}{r^2 \sin^2\theta}\frac{\partial^2 T}{\partial \phi^2}. \qquad (70)$$

7.23

Transform $\nabla^2 T$ by the substitution

$$x = f(u, v, w), \quad y = g(u, v, w), \quad z = h(u, v, w). \qquad (71)$$

Let $J \neq 0$ be the determinant of the substitution and let $\text{adj}\, J$ be its adjugate, so that

$$J = \begin{vmatrix} x_1 & x_2 & x_3 \\ y_1 & y_2 & y_3 \\ z_1 & z_2 & z_3 \end{vmatrix}, \quad \text{adj}\, J = \begin{vmatrix} X_1 & X_2 & X_3 \\ Y_1 & Y_2 & Y_3 \\ Z_1 & Z_2 & Z_3 \end{vmatrix}, \qquad (72)$$

where $x_1 = \partial x/\partial u, \ldots$; $X_1 = y_2 z_3 - y_3 z_2, \ldots$. Then, by §6.16 (49) and since $\partial X_1/\partial u + \partial X_2/\partial v + \partial X_3/\partial w \equiv 0$,

$$\frac{\partial T}{\partial x} = \frac{1}{J}\left(X_1 \frac{\partial T}{\partial u} + X_2 \frac{\partial T}{\partial v} + X_3 \frac{\partial T}{\partial w}\right) \tag{73}$$

$$= \frac{1}{J}\left\{\frac{\partial}{\partial u}(X_1 T) + \frac{\partial}{\partial v}(X_2 T) + \frac{\partial}{\partial w}(X_3 T)\right\}. \tag{74}$$

Replacing T by $\partial T/\partial x$ gives

$$\frac{\partial^2 T}{\partial x^2} = \frac{1}{J}\left\{\frac{\partial}{\partial u}\left(X_1 \frac{\partial T}{\partial x}\right) + \frac{\partial}{\partial v}\left(X_2 \frac{\partial T}{\partial x}\right) + \frac{\partial}{\partial w}\left(X_3 \frac{\partial T}{\partial x}\right)\right\}.$$

From this and two similar expressions, by addition,

$$\nabla^2 T = \frac{1}{J}\Sigma \frac{\partial}{\partial u}\left(X_1 \frac{\partial T}{\partial x} + Y_1 \frac{\partial T}{\partial y} + Z_1 \frac{\partial T}{\partial z}\right) \tag{75}$$

and, by using (73) and two similar equations,

$$\nabla^2 T = \frac{1}{J}\Sigma \frac{\partial}{\partial u}\left(\frac{\Sigma X_1^2}{J}\frac{\partial T}{\partial u} + \frac{\Sigma X_1 X_2}{J}\frac{\partial T}{\partial v} + \frac{\Sigma X_1 X_3}{J}\frac{\partial T}{\partial w}\right).$$

To calculate $\Sigma X_1^2, \Sigma X_1 X_2, \ldots$ put

$$\begin{aligned}
A &= x_1^2 + y_1^2 + z_1^2, & F &= x_2 x_3 + y_2 y_3 + z_2 z_3, \\
B &= x_2^2 + y_2^2 + z_2^2, & G &= x_3 x_1 + y_3 y_1 + z_3 z_1, \\
C &= x_3^2 + y_3^2 + z_3^2, & H &= x_1 x_2 + y_1 y_2 + z_1 z_2;
\end{aligned} \tag{76}$$

then we find the two sets of three equations typified by

$$\Sigma X_1^2 = \Sigma \begin{vmatrix} y_2 & y_3 \\ z_2 & z_3 \end{vmatrix}^2 = BC - F^2, \tag{77}$$

$$\Sigma X_2 X_3 = \Sigma \begin{vmatrix} y_3 & y_1 \\ z_3 & z_1 \end{vmatrix} \cdot \begin{vmatrix} y_1 & y_2 \\ z_1 & z_2 \end{vmatrix} = GH - AF. \tag{78}$$

Consequently, $J\nabla^2 T =$

$$\frac{\partial}{\partial u}\frac{1}{J}\begin{vmatrix} T_u & T_v & T_w \\ H & B & F \\ G & F & C \end{vmatrix} + \frac{\partial}{\partial v}\frac{1}{J}\begin{vmatrix} A & H & G \\ T_u & T_v & T_w \\ G & F & C \end{vmatrix} + \frac{\partial}{\partial w}\frac{1}{J}\begin{vmatrix} A & H & G \\ H & B & F \\ T_u & T_v & T_w \end{vmatrix}. \tag{79}$$

Note that, by the rule for multiplication of determinants,

$$J^2 = \begin{vmatrix} A & H & G \\ H & B & F \\ G & F & C \end{vmatrix}. \tag{80}$$

Corollary. Orthogonal substitution. If the substitution is orthogonal, then $F = G = H = 0$, $J^2 = ABC$, and

$$\nabla^2 T = \frac{1}{J}\left\{\frac{\partial}{\partial u}\left(\frac{J}{A}\frac{\partial T}{\partial u}\right) + \frac{\partial}{\partial v}\left(\frac{J}{B}\frac{\partial T}{\partial v}\right) + \frac{\partial}{\partial w}\left(\frac{J}{C}\frac{\partial T}{\partial w}\right)\right\}, \quad (81)$$

where J may be replaced by $|J| = (ABC)^{\frac{1}{2}}$, since the R.H.S. is unaltered if J is replaced by $-J$.

Examples

(1) Transform $\nabla^2 T$ to spherical polar coordinates r, θ, ϕ, where

$$x = r\sin\theta\cos\phi, \quad y = r\sin\theta\sin\phi, \quad z = r\cos\theta.$$

Here $A = 1$, $B = r^2$, $C = r^2\sin^2\theta$, $F = G = H = 0$, $J = r^2\sin\theta$. By substitution in (81),

$$\nabla^2 T = \frac{1}{r^2\sin\theta}\left\{\frac{\partial}{\partial r}\left(r^2\sin\theta\frac{\partial T}{\partial r}\right) + \frac{\partial}{\partial\theta}\left(\sin\theta\frac{\partial T}{\partial\theta}\right) + \frac{\partial}{\partial\phi}\left(\frac{1}{\sin\theta}\frac{\partial T}{\partial\phi}\right)\right\}$$

$$= \frac{1}{r^2}\frac{\partial}{\partial r}\left(r^2\frac{\partial T}{\partial r}\right) + \frac{1}{r^2\sin\theta}\frac{\partial}{\partial\theta}\left(\sin\theta\frac{\partial T}{\partial\theta}\right) + \frac{1}{r^2\sin^2\theta}\frac{\partial^2 T}{\partial\phi^2}$$

which is seen at once to agree with (70).

(2) Transform the equation $\nabla^2 T = 0$ to confocal coordinates λ, μ, ν defined by the equations

$$\left.\begin{array}{l} \dfrac{x^2}{a^2+\lambda}+\dfrac{y^2}{b^2+\lambda}+\dfrac{z^2}{c^2+\lambda} = 1 \quad (-c^2 < \lambda < \infty), \\[2mm] \dfrac{x^2}{a^2+\mu}+\dfrac{y^2}{b^2+\mu}+\dfrac{z^2}{c^2+\mu} = 1 \quad (-b^2 < \mu < -c^2), \\[2mm] \dfrac{x^2}{a^2+\nu}+\dfrac{y^2}{b^2+\nu}+\dfrac{z^2}{c^2+\nu} = 1 \quad (-a^2 < \nu < -b^2). \end{array}\right\} \quad (82)$$

Solving for x^2, y^2, z^2, we find

$$\left.\begin{array}{l} x^2 = \dfrac{(a^2+\lambda)(a^2+\mu)(a^2+\nu)}{(a^2-b^2)(a^2-c^2)}, \\[3mm] y^2 = \dfrac{(b^2+\lambda)(b^2+\mu)(b^2+\nu)}{(b^2-c^2)(b^2-a^2)}, \quad z^2 = \dfrac{(c^2+\lambda)(c^2+\mu)(c^2+\nu)}{(c^2-a^2)(c^2-b^2)} \end{array}\right\} \quad (83)$$

Hence the partial derivatives of x are given by

$$2\frac{\partial x}{\partial\lambda} = \frac{x}{a^2+\lambda}, \quad 2\frac{\partial x}{\partial\mu} = \frac{x}{a^2+\mu}, \quad 2\frac{\partial x}{\partial\nu} = \frac{x}{a^2+\nu} \quad (84)$$

and there are six similar expressions for the derivatives of y and z. We now find, with the aid of (83), that $F = G = H = 0$, so that the transformation is orthogonal, and that the values of A, B, C are respectively

$$(\lambda - \mu)(\lambda - \nu)/4\Lambda, \quad (\lambda - \mu)(\mu - \nu)/4\mathrm{M}, \quad (\lambda - \nu)(\mu - \nu)/4\mathrm{N}, \quad (85)$$

where $$\Lambda = (a^2 + \lambda)(b^2 + \lambda)(c^2 + \lambda),$$

$$\mathrm{M} = -(a^2 + \mu)(b^2 + \mu)(c^2 + \mu), \quad \mathrm{N} = (a^2 + \nu)(b^2 + \nu)(c^2 + \nu).$$

Also we find

$$J = \frac{\partial(x, y, z)}{\partial(\lambda, \mu, \nu)} = -\frac{\Pi(\mu - \nu)}{8xyz\,\Pi\,(b^2 - c^2)} = \mp \frac{(\lambda - \mu)(\lambda - \nu)(\mu - \nu)}{8\sqrt{(\Lambda\mathrm{MN})}}$$

according as $xyz > 0$ or $xyz < 0$. By substitution in (81), the equation $\nabla^2 T = 0$ reduces to

$$\left\{ (\mu - \nu)\left(\Lambda^{\frac{1}{2}}\frac{\partial}{\partial\lambda}\right)^2 + (\lambda - \nu)\left(\mathrm{M}^{\frac{1}{2}}\frac{\partial}{\partial\mu}\right)^2 + (\lambda - \mu)\left(\mathrm{N}^{\frac{1}{2}}\frac{\partial}{\partial\nu}\right)^2 \right\} T = 0. \quad (86)$$

7.24   Orientation

Let OA, OB, OC be drawn parallel to three given directions not all parallel to the same plane. Let α, β, γ be the angles BOC, COA, AOB, each between 0 and π. Also, let OA' be drawn perpendicular to the plane BOC, the angle AOA' being acute.

The given directions in the order OA, OB, OC are said to form a system with a right-handed (left-handed) *orientation* if a right-handed (left-handed) rotation α about the axis OA' will bring OB into coincidence with OC.

Two ordered pairs of directions in a plane are said to have the same orientation if they both form right-handed or both form left-handed systems with any third direction not parallel to the plane.

7.25

Let (u, v) be Cartesian coordinates in a uv plane, and (x, y) rectangular Cartesian coordinates in an xy plane.

A substitution $x = f(u, v)$, $y = g(u, v)$ in general transforms a region of the uv plane into a corresponding region of the xy plane. In particular, an infinitesimal parallelogram $pq_1p'q_2$, of which the four corners are (u, v), $(u + du, v)$, $(u + du, v + dv)$, $(u, v + dv)$ respectively, is transformed into a corresponding infinitesimal parallelogram

14-2

$PQ_1P'Q_2$. Let $ds = PP'$, and let ω be the angle Q_1PQ_2. If $x_1 = \partial x/\partial u$, $x_2 = \partial x/\partial v, \ldots$, then

$$ds^2 = (x_1\,du + x_2\,dv)^2 + (y_1\,du + y_2\,dv)^2$$

$$= A\,du^2 + B\,dv^2 + 2H\,du\,dv \tag{87}$$

$$= H_1^2\,du^2 + H_2^2\,dv^2 + 2H_1H_2\cos\omega\,du\,dv, \tag{88}$$

where

$$A = H_1^2 = x_1^2 + y_1^2, \quad B = H_2^2 = x_2^2 + y_2^2, \quad H = H_1H_2\cos\omega = x_1x_2 + y_1y_2.$$

The infinitesimal ds is called *the line element* of the transformation. In particular, $ds_1 = PQ_1 = H_1\,du$, $ds_2 = PQ_2 = H_2\,dv$.

Further, if $d\sigma\,(>0)$ is the area of the parallelogram $PQ_1P'Q_2$, then

$$d\sigma = \pm \begin{vmatrix} 1 & x & y \\ 1 & x+x_1\,du & y+y_1\,du \\ 1 & x+x_2\,dv & y+y_2\,dv \end{vmatrix} = \pm \begin{vmatrix} 1 & x & y \\ 0 & x_1\,du & y_1\,du \\ 0 & x_2\,dv & y_2\,dv \end{vmatrix}$$

that is,

$$d\sigma = \pm \frac{\partial(x,y)}{\partial(u,v)}\,du\,dv = \pm J\,du\,dv. \tag{89}$$

It follows that, if $du > 0$, $dv > 0$, and $J > 0$, then $d\sigma = J\,du\,dv$ and PQ_1, PQ_2 have the same orientation as the coordinate axes Ox, Oy; but if $J < 0$, then $d\sigma = -J\,du\,dv$ and the orientation of PQ_1, PQ_2 is contrary to that of the axes Ox, Oy.

7.26   The Hessian

Definition. If T is a function of x, y, z, the Hessian of T wo x, y, z is defined by

$$H(T)_{xyz} = \frac{\partial(T_x, T_y, T_z)}{\partial(x,y,z)} = \begin{vmatrix} T_{xx} & T_{xy} & T_{xz} \\ T_{xy} & T_{yy} & T_{yz} \\ T_{xz} & T_{yz} & T_{zz} \end{vmatrix}. \tag{90}$$

Theorem. If T is a function of u, v, w and u, v, w are linear functions of x, y, z so that T is a function of x, y, z through a linear substitution, then

$$H(T)_{xyz} = J^2 . H(T)_{uvw} \tag{91}$$

where $J = \partial(u,v,w)/\partial(x,y,z)$.

Proof. Let the substitution be

$$u = l_1x + m_1y + n_1z, \quad J = \begin{vmatrix} l_1 & m_1 & n_1 \\ l_2 & m_2 & n_2 \\ l_3 & m_3 & n_3 \end{vmatrix}; \tag{92}$$

$$v = l_2x + m_2y + n_2z,$$

$$w = l_3x + m_3y + n_3z,$$

then
$$T_x = l_1 T_u + l_2 T_v + l_3 T_w$$
$$T_y = m_1 T_u + m_2 T_v + m_3 T_w$$
$$T_z = n_1 T_u + n_2 T_v + n_3 T_w$$
(93)

and hence

$$\frac{\partial(T_x, T_y, T_z)}{\partial(x, y, z)} = \frac{\partial(T_x, T_y, T_z)}{\partial(T_u, T_v, T_w)} \frac{\partial(T_u, T_v, T_w)}{\partial(u, v, w)} \frac{\partial(u, v, w)}{\partial(x, y, z)},$$

that is
$$H(T)_{xyz} = J' . H(T)_{uvw} . J = J^2 . H(T)_{uvw}$$

since J' (the transpose of J) $= J$.

7.27 Taylor's theorem for two independent variables

Taylor's formula (§4.3) for $f(x)$ to base a expresses $f(x)$ as a sum of powers of $x - a$ and a 'remainder'. One way of expressing the formula is

$$f(x) = f(a) + \sum_{s=1}^{n-1} \frac{h^s}{s!} f^{(s)}(a) + R_n$$
(94)

where $R_n = (1/n!)\, h^n f^{(n)}(\xi)$, $h = x - a$, $\xi = a + \theta h$, and θ is some number between 0 and 1, depending on x, a and n.

The corresponding Taylor formula for a function $f(x, y)$ to base (a, b) expresses $f(x, y)$ as a sum of powers and products of powers of $x - a$ and $y - b$ and a remainder. The formula is

$$f(x, y) = f(a, b) + \sum_{s=1}^{n-1} \frac{1}{s!} (hf_1 + kf_2)^s (a, b) + R_n$$
(95)

where
$$R_n = (1/n!)\, (hf_1 + kf_2)^n (\xi, \eta),$$
(96)

$h = x - a$, $k = y - b$, $\xi = a + \theta h$, $\eta = b + \theta k$, and θ is some number, depending on x, y, a, b and n, between 0 and 1; while $(hf_1 + kf_2)^s (a, b)$ denotes $(h\,\partial/\partial x + k\,\partial/\partial y)^s f(x, y)$ with x, y replaced by a, b after differentiation.

Proof of (95). We assume that $f(x, y)$ and its partial derivatives up to the nth order are continuous throughout some circular region, centre (a, b). Let (x, y) be another point of this region. Then, on the straight line
$$x = a + ht, \quad y = b + kt \quad (0 \leqslant t \leqslant 1)$$
(97)

$f(x, y)$ is a function of the single variable t of which the sth derivative is, by (18),

$$\left(\frac{d}{dt}\right)^s f(x, y) = \left(h\frac{\partial}{\partial x} + k\frac{\partial}{\partial y}\right)^s f(x, y) \quad (s = 1, 2, ..., n)$$
(98)

and when $t = 0$ this becomes $(hf_1 + kf_2)^s (a, b)$ in accordance with the notation just explained. Consequently, by Taylor's theorem for a single variable, the expansion of $f(x, y)$ in powers of t is given by

$$f(x, y) = f(a, b) + \sum_{s=1}^{n-1} \frac{t^s}{s!} (hf_1 + kf_2)^s (a, b) + R_n \qquad (99)$$

where $R_n = (t^n/n!) (hf_1 + kf_2)^n (\xi, \eta), \xi = a + h\theta t, \eta = b + k\theta t$, and θ is some number between 0 and 1. From (99) follows (95) when $t = 1$.

7.28

Taylor's formula (95) becomes Maclaurin's formula for a function of two variables when we put $a = 0$, $b = 0$ and therefore $h = x$, $k = y$; this gives

$$f(x, y) = f(0, 0) + \sum_{s=1}^{n-1} \frac{1}{s!} (xf_1 + yf_2)^s (0, 0) + R_n \qquad (100)$$

where $\qquad R_n = (1/n!) (xf_1 + yf_2)^n (\theta x, \theta y) \quad (0 < \theta < 1). \qquad (101)$

7.29 Three or more variables

A formula similar to (95) holds good for three or more variables; thus, for three variables, inside a sphere, centre (a, b, c), within which all the partial derivatives of $f(x, y, z)$ up to the nth order are continuous, the formula can be expressed by

$$f(x, y, z) = f(a, b, c) + \sum_{s=1}^{n-1} \frac{1}{s!} (hf_1 + kf_2 + lf_3)^s (a, b, c) + R_n$$

where $\qquad R_n = (1/n!) (hf_1 + kf_2 + lf_3)^n (\xi, \eta, \zeta),$

$h = x - a, k = y - b, l = z - c, \ \xi = a + \theta h, \ \eta = b + \theta k, \ \zeta = c + \theta l$ and θ is some number between 0 and 1.

Examples 7 A

In many of the following examples, it may be assumed that sufficient conditions are satisfied for the required results to hold good:

(1) $\qquad f(x, y) = x^3 \sin (y/x) \quad (x \neq 0), \quad f(0, y) = 0.$

Verify that $f_{xy} = f_{yx}$ at all points.

(2) $\qquad f(x, y) = xy \dfrac{x^2 - y^2}{x^2 + y^2} \quad (x^2 + y^2 \neq 0), \quad f(0, 0) = 0.$

Show that $\qquad f_{xy}(0, 0) = 1, \quad f_{yx}(0, 0) = -1.$

(3) Verify that $\partial^2 z/\partial y \, \partial x = \partial^2 z/\partial x \, \partial y$ in the following cases:

$\qquad$ (i) $z = \dfrac{x}{y}$, $\quad$ (ii) $z = \dfrac{f(xy)}{x}$, $\quad$ (iii) $z = xf\left(\dfrac{y}{x}\right).$

(4) Verify that the given functions satisfy the given partial differential equations:

(i) $y = e^{-ax} \cos(\omega t - ax)$, $\qquad \dfrac{\partial^2 y}{\partial x^2} = \dfrac{2a^2}{\omega} \dfrac{\partial y}{\partial t}$;

(ii) $V = \dfrac{1}{\sqrt{t}} \exp\left(-\dfrac{x^2}{4kt}\right)$, $\qquad \dfrac{\partial V}{\partial t} = k \dfrac{\partial^2 V}{\partial x^2}$;

(iii) $u = \dfrac{2x + (1+x^2)y}{1 + x^2 + 2xy}$, $\qquad \dfrac{\partial^2 u}{\partial x\, \partial y} = -\dfrac{4(1-x^2)u}{(1+x^2+2xy)^2}$;

(iv) $u = \dfrac{1}{z} \exp\left(-\dfrac{x^2+y^2}{4z}\right)$, $\qquad \dfrac{\partial u}{\partial z} = \dfrac{\partial^2 u}{\partial x^2} + \dfrac{\partial^2 u}{\partial y^2}$.

(5) Verify the given results, in which p, q, r, s, t have the meanings of §7.1, and f, g denote arbitrary functions:

(i) $z = f(x - cy) + g(x + cy)$, $\qquad c^2 r = t$;

(ii) $z = f(x - cy) + xg(x - cy)$, $\qquad c^2 r + 2cs + t = 0$;

(iii) $z = g\{x + f(y)\}$, $\qquad ps = qr$;

(iv) $z = x^n f(y/x) + x^{-n} g(y/x)$, $\qquad x^2 r + 2xys + y^2 t + px + qy = n^2 z$.

(6) If $z = x^m y^n$, where $m + n = 1$, verify that

$$\frac{\partial^2 z}{\partial x^2} \frac{\partial^2 z}{\partial y^2} - \left(\frac{\partial^2 z}{\partial x\, \partial y}\right)^2 = 0 \qquad (xy \neq 0).$$

(7) If $u = f(x, y)$, verify that

$$\frac{\partial}{\partial x}\left\{g(u)\frac{\partial u}{\partial y}\right\} = \frac{\partial}{\partial y}\left\{g(u)\frac{\partial u}{\partial x}\right\}.$$

(8) If f, g are functions of x, y, z, show that

$$\frac{\partial}{\partial x}\frac{\partial(f, g)}{\partial(y, z)} + \frac{\partial}{\partial y}\frac{\partial(f, g)}{\partial(z, x)} + \frac{\partial}{\partial z}\frac{\partial(f, g)}{\partial(x, y)} = 0.$$

(9) If $x = f(u, v, w)$, $y = g(u, v, w)$, $z = h(u, v, w)$,

$$J = \begin{vmatrix} x_1 & x_2 & x_3 \\ y_1 & y_2 & y_3 \\ z_1 & z_2 & z_3 \end{vmatrix}, \qquad \operatorname{adj} J = \begin{vmatrix} X_1 & X_2 & X_3 \\ Y_1 & Y_2 & Y_3 \\ Z_1 & Z_2 & Z_3 \end{vmatrix},$$

where $x_1 = \partial x/\partial u, \ldots$ and $\operatorname{adj} J$ denotes the adjugate of J, show that

$$\frac{\partial X_1}{\partial u} + \frac{\partial X_2}{\partial v} + \frac{\partial X_3}{\partial w} = \frac{\partial Y_1}{\partial u} + \frac{\partial Y_2}{\partial v} + \frac{\partial Y_3}{\partial w} = \frac{\partial Z_1}{\partial u} + \frac{\partial Z_2}{\partial v} + \frac{\partial Z_3}{\partial w} = 0.$$

(10) If $u = a \cos\theta + b \sin\theta \cos\phi$, show that:

(i) $(\partial u/\partial\theta)^2 + \operatorname{cosec}^2\theta \, (\partial u/\partial\phi)^2 = a^2 + b^2 - u^2$,

(ii) $\partial^2 u/\partial\theta^2 + \cot\theta \, (\partial u/\partial\theta) + \operatorname{cosec}^2\theta \, (\partial^2 u/\partial\phi^2) = -2u$.

(11) In the usual notation of a triangle, show that

$$\left(\frac{\partial^2\Delta}{\partial A\, \partial B}\right)_c = \frac{2\Delta}{\sin^2 C}, \qquad \left(\frac{\partial^2\Delta}{\partial A\, \partial b}\right)_c = \frac{c^2}{b}, \qquad \left(\frac{\partial^2\Delta}{\partial a\, \partial b}\right)_c = \frac{R^2 \cos C}{\Delta}.$$

(12) *Harmonic functions.* A solution of the partial differential equation (Laplace's equation in two dimensions)

$$\partial^2 T/\partial x^2 + \partial^2 T/\partial y^2 = 0$$

is called a *harmonic* function of x and y.

Verify that the following are harmonic functions:

(i) $x^3 - 3xy^2$,   (ii) $x/(x^2 + y^2)$,   (iii) $\log(x^2 + y^2)$,

(iv) $\tan^{-1}(y/x)$,   (v) $e^{-2xy}\cos(x^2 - y^2)$.

(13) If $u + iv$ is an analytic function of $x + iy$ show, from the Riemann–Cauchy equations (§6.24), that u and v are harmonic functions of x and y.

(14) If u and v satisfy the Riemann–Cauchy equations in polar coordinates (Examples 6E, 9), show that each satisfies the equation (the polar form of Laplace's equation in two dimensions)

$$\frac{\partial^2 T}{\partial r^2} + \frac{1}{r}\frac{\partial T}{\partial r} + \frac{1}{r^2}\frac{\partial^2 T}{\partial \theta^2} = 0.$$

Verify that this equation is satisfied by

(i) $T = r^n \cos n\theta$,   (ii) $T = \cos(m\log r)\cosh m\theta$.

(15) If the equation in the last example is satisfied by $T = r^n f(\theta)$, prove that $f(\theta) = A\cos n\theta + B\sin n\theta$, where A, B are constants. Deduce that there are only two independent homogeneous harmonic functions of order n.

(16) If u, v are functions of x, y and if $T(u, v)\,du\,dv \equiv F(x, y)\,dx\,dy$ identically when the L.H.S. is expressed in terms of x, y, dx, dy, show that:

(i) either u is a function of x only and v is a function of y only, or u is a function of y only and v is a function of x only;

(ii) $\dfrac{1}{T}\dfrac{\partial^2}{\partial u\,\partial v}(\log T) = \dfrac{1}{F}\dfrac{\partial^2}{\partial x\,\partial y}(\log F).$

Examples 7 B

(1) Show that the following differential expressions are exact, and find the function of which each is the exact differential:

(i) $(2xy - y^2 + 6x)\,dx + (x^2 - 2xy + 10y)\,dy$,

(ii) $(x\,dy - y\,dx)/(x^2 - y^2)$,   (iii) $(x\,dy - y\,dx)/(x^2 + xy + y^2)$,

(iv) $e^{3x}\,dy + 3e^{3x}y\,dx$,   (v) $(x^3\,dy - y^3\,dx)/(x^3 y - xy^3)$,

(vi) $r^{-n}\cos n\theta\,d\theta - r^{-n-1}\sin n\theta\,dr$,   (vii) $\{dy + f(y/x)\,dx\}/\{y + xf(y/x)\}$.

(2) Show that a function u exists such that

$$du = (3x^2 + 2y^2 - 3y)\,dx + (4xy - 3x - 3y^2 + 1)\,dy.$$

Find u, given that $u = 4$ when $x = 2$, $y = 1$.

(3) If $v = \phi(x - y)$, show that a function u exists such that

$$du = \{\partial v/\partial y + 2v + f(x)\}\,dx + \{\partial v/\partial x - 2v + g(y)\}\,dy$$

where ϕ, f, g denote arbitrary functions.

(4) Show that r^{n-1} is an integrating factor of $\cos n\theta\, dr - r\sin n\theta\, d\theta$.

(5) Find u, given that

$$du = \mu\{(2x-3)\,y\,dx + (2x^2 - 6x + 9y)\,dy\}$$

and that μ is a function of y only.

(6) Find u, given that

$$du = \mu\{(3x^2y - 4y + 1)\,dx + (x^3 - 2x)\,dy\}$$

and that μ is of the form $xf(x^2y)$. Also find an integrating factor which is a function of x only.

(7) Show that every integrating factor of $y\,dx + dy$ is expressible in the form $(1/y)f(ye^x)$.

(8) Solve the equation $x\,dy + (x - 2y)\,dx = 0$. Hence or otherwise show that $1/x^3$ and $1/(x^2 - xy)$ are integrating factors.

(9) Find all the integrating factors of

$$(2y + 5\cos x)\,dx - dy.$$

(10) Find all the integrating factors of

$$(5y - 6x)\,dx + (y - 4x)\,dy.$$

(11) Show that $\lambda = 1/(xy)^n$ and $\mu = 1/(x^ny - xy^n)$ are integrating factors of $x^n\,dy - y^n\,dx$. If $du = \lambda(x^n\,dy - y^n\,dx)$ and $dv = \mu(x^n\,dy - y^n\,dx)$, find the relation between u and v.

(12) Show that $\lambda = 1/(\sin x \sinh y)$ and $\mu = 1/(\cosh y - \cos x)$ are integrating factors of $\sinh y\,dx - \sin x\,dy$. If

$$du = \frac{\sinh y\,dx - \sin x\,dy}{\sin x \sinh y} \quad\text{and}\quad dv = \frac{\sinh y\,dx - \sin x\,dy}{\cosh y - \cos x},$$

show that $e^u = \tan \tfrac{1}{2}v$.

(13) Show that $\lambda = r$ and $\mu = 1/(r\cos\theta)$ are integrating factors of

$$2\cos\theta\, dr - r\sin\theta\, d\theta.$$

If

$$du = r(2\cos\theta\, dr - r\sin\theta\, d\theta) \quad\text{and}\quad dv = (2\cos\theta\, dr - r\sin\theta\, d\theta)/(r\cos\theta),$$

show that $u = e^v$.

(14) Show that $P\,dx + Q\,dy + R\,dz$, where P, Q, R and their first partial derivatives are continuous, cannot be a total differential unless P, Q, R satisfy the conditions

$$\frac{\partial R}{\partial y} = \frac{\partial Q}{\partial z}, \quad \frac{\partial P}{\partial z} = \frac{\partial R}{\partial x}, \quad \frac{\partial Q}{\partial x} = \frac{\partial P}{\partial y}.$$

(15) If P, Q, R are functions of x, y, z and if z is a differentiable function of x and y, show that a necessary condition that $P\,dx + Q\,dy + R\,dz$ should be the total differential of a function of the two independent variables x and y is

$$\frac{\partial P}{\partial y} + \frac{\partial R}{\partial y}\frac{\partial z}{\partial x} = \frac{\partial Q}{\partial x} + \frac{\partial R}{\partial x}\frac{\partial z}{\partial y}.$$

Examples 7C

(1) (i) If $V = \log r$, where $r^2 = x^2 + y^2$, show that $\nabla_1^2 V = 0$.

(ii) If V is a function of r only, $r^2 = x^2 + y^2$, and if $\nabla_1^2 V = 0$, show that $V = A + B \log r$, where A and B are arbitrary constants.

(2) (i) If $V = 1/r$, where $r^2 = x^2 + y^2 + z^2$, show that $\nabla^2 V = 0$.

(ii) If V is a function of r only, $r^2 = x^2 + y^2 + z^2$, and if $\nabla^2 V = 0$, show that $V = A + B/r$, where A and B are arbitrary constants.

(3) If $V = f(u)$, where $u = x^3 y - xy^3$, show that $\nabla_1^2 V = (x^2 + y^2)^3 f''(u)$.

(4) If $rV = f(r + ct)$, verify that

$$\frac{\partial V}{\partial r} + \frac{V}{r} = \frac{1}{c}\frac{\partial V}{\partial t}, \quad \frac{\partial^2 V}{\partial r^2} + \frac{2}{r}\frac{\partial V}{\partial r} = \frac{1}{c^2}\frac{\partial^2 V}{\partial t^2}.$$

(5) If $V = f(u)$ and $u = x/\sqrt{(x^2 + y^2)}$, verify that

$$(x^2 + y^2)\left(\frac{\partial^2 V}{\partial x^2} + \frac{\partial^2 V}{\partial y^2}\right) = (1 - u^2)f''(u) - uf'(u).$$

(6) If $V = \sqrt{(x + r)}$, where $r = \sqrt{(x^2 + y^2)}$, show that

(i) $x\dfrac{\partial V}{\partial x} + y\dfrac{\partial V}{\partial y} = \tfrac{1}{2}V$, (ii) $\left(\dfrac{\partial V}{\partial x}\right)^2 + \left(\dfrac{\partial V}{\partial y}\right)^2 = \dfrac{1}{2r}$, (iii) $\nabla_1^2 V = 0$.

(7) If $r = \sqrt{(x^2 + y^2 + z^2)}$, verify that $\nabla^2 V = 0$ when

(i) $V = x/r^3$,   (ii) $V = xy/r^5$,   (iii) $V = xyz/r^7$,

(iv) $V = \log(z + r)$,   (v) $V = r - z\log(z + r)$.

(8) If ϕ is a homogeneous function of x and y of degree n, show that

$$x^2\frac{\partial^2 \phi}{\partial x^2} + 2xy\frac{\partial^2 \phi}{\partial x\,\partial y} + y^2\frac{\partial^2 \phi}{\partial y^2} = n(n-1)\,\phi.$$

(9) If $\quad \phi = (ax^2 + 2hxy + by^2)f\left(\dfrac{y}{x}\right) + \dfrac{1}{Ax + By}\,g\left(\dfrac{y}{x}\right),$

verify that $\quad x^2\dfrac{\partial^2 \phi}{\partial x^2} + 2xy\dfrac{\partial^2 \phi}{\partial x\,\partial y} + y^2\dfrac{\partial^2 \phi}{\partial y^2} = 2\phi.$

(10) If u and v are harmonic functions of x and y, show that

$$\nabla_1^2(uv) = 2\left(\frac{\partial u}{\partial x}\frac{\partial v}{\partial x} + \frac{\partial u}{\partial y}\frac{\partial v}{\partial y}\right).$$

If $z = u\log r$, where $u = x^3 - 3xy^2$ and $r = \sqrt{(x^2 + y^2)}$, verify that $\nabla_1^2 z = 6u/r^2$.

(11) If u, v are a conjugate pair of harmonic functions, show that the product uv is harmonic. Also show that

$$\nabla_1^2 f(uv) = (u^2 + v^2)f''(uv)\,\partial(u, v)/\partial(x, y).$$

(12) If u is homogeneous of degree n in x and y and if $\nabla_1^2 u = 0$, show that $\nabla_1^2(u/r^{2n}) = 0$, where $r^2 = x^2 + y^2$.

(13) If V is homogeneous of degree n in x, y, z and if $\nabla^2 V = 0$, show that $\nabla^2(V/r^{2n+1}) = 0$, where $r^2 = x^2 + y^2 + z^2$. Deduce that

$$\nabla^2\{x\sqrt{(x^2+y^2+z^2)}/(x^2+y^2)\} = 0.$$

(14) If w is an analytic function of z, $= x + iy$, show that

$$\nabla_1^2\{\nabla_1^2(z\bar{w} + \bar{z}w)\} = 0.$$

(15) If z is a function of t determined by the equations

$$z = \phi(x, y, t), \quad x = f(t), \quad y = g(t),$$

show that d^2z/dt^2 is given by the formula

$$d^2z/dt^2 = \dot{x}^2\phi_{11} + 2\dot{x}\dot{y}\phi_{12} + \dot{y}^2\phi_{22} + 2\dot{x}\phi_{13} + 2\dot{y}\,\phi_{23} + \phi_{33} + \ddot{x}\phi_1 + \ddot{y}\phi_2$$

where $\dot{x} = dx/dt$, $\phi_1 = \partial\phi/\partial x$, etc.

Verify this formula when $z = (x^2 - y^2)\,t^3$, $x = t$, $y = t^2$.

Examples 7 D

(1) If $r^2 = x^2 + y^2$ and $\tan\theta = y/x$, show that

(i) $\dfrac{\partial^2 r}{\partial x^2}\dfrac{\partial^2 r}{\partial y^2} = \left(\dfrac{\partial^2 r}{\partial x\,\partial y}\right)^2$,   (ii) $2\dfrac{\partial^2 r}{\partial x\,\partial y} = r\tan 2\theta\,\dfrac{\partial^2\theta}{\partial x\,\partial y}$,

(iii) $\dfrac{\partial^2\theta}{\partial x\,\partial y} = \dfrac{\partial^2}{\partial x^2}(\log r) = -\dfrac{\partial^2}{\partial y^2}(\log r) = -\dfrac{1}{r^2}\cos 2\theta$.

(2) If $0 < e < 1$, show that the equation $\mu = \phi - e\sin\phi$ defines ϕ implicitly as a differentiable function of μ and e, and that

$$\frac{\partial^2\phi}{\partial e^2} + \frac{1}{e}\frac{\partial\phi}{\partial e} + \frac{1-e^2}{e^2}\frac{\partial^2\phi}{\partial\mu^2} = 0.$$

(3) Show that the equation $u = x\cos u + y\sin u$ defines u implicitly as a function of x and y inside the circle $x^2 + y^2 = 1$, and that $\nabla_1^2 u = -u/v^3$, where $v = 1 + x\sin u - y\cos u$.

(4) If u, v are defined as functions of the independent variables x, y by the equations $u^2 + v^2 + x^2 + y^2 = a^2, \quad uv + xy = b^2,$

show that $(u^2 - v^2)^3\,\partial^2 u/\partial x^2 = 2b^2 v(3u^2 + v^2) - a^2 u(u^2 + 3v^2).$

(5) If the equation $F(x, y, z) = 0$, in which $F(x, y, z)$ is homogeneous in x, y, z, defines z in terms of the independent variables x and y, show that

$$x^2\frac{\partial^2 z}{\partial x^2} = -xy\frac{\partial^2 z}{\partial x\,\partial y} = y^2\frac{\partial^2 z}{\partial y^2}.$$

(6) Given that the equation $F(x, y, z) = 0$ defines z as a function of the independent variables x and y, show that

$$\partial^2 z/\partial x^2 = -(F_{33}F_1^2 - 2F_{13}F_1F_3 + F_{11}F_3^2)/F_3^3$$

where $F_1 = \partial F/\partial x$, $F_{11} = \partial^2 F/\partial x^2$, etc.

Find a similar expression for $\partial^2 z/\partial x\,\partial y$.

(7) If $F(x, y) = 0$, show that y''' is given by

$$F_{111} + 3F_{112}y' + 3F_{122}y'^2 + F_{222}y'^3 + 3(F_{12} + F_{22}y')y'' + F_2y''' = 0,$$

where $y' = dy/dx$, $F_1 = \partial F/\partial x$, etc.

(8) Given that the equations $F(x, t) = 0$, $G(y, t) = 0$ define y as a function of x, find an expression for d^2y/dx^2.

Examples 7 E

(1) In this example, it is tacitly assumed that T is a function of u, v in the first place, and hence can be regarded as a function of x, y through u, v. Similar remarks apply to later examples.

If $u = ax + by$, $v = bx - ay$, show that

$$\nabla_1^2 T = (a^2 + b^2)(\partial^2 T/\partial u^2 + \partial^2 T/\partial v^2),$$

$$\frac{\partial^2 T}{\partial x \, \partial y} = ab\left(\frac{\partial^2 T}{\partial u^2} - \frac{\partial^2 T}{\partial v^2}\right) + (b^2 - a^2)\frac{\partial^2 T}{\partial u \, \partial v}.$$

(2) If $u = x + y$, $v = xy$, show that

$$\frac{\partial^2 T}{\partial x^2} - 2\frac{\partial^2 T}{\partial x \, \partial y} + \frac{\partial^2 T}{\partial y^2} = (u^2 - 4v)\frac{\partial^2 T}{\partial v^2} - 2\frac{\partial T}{\partial v}.$$

(3) If $u = xy$, $v = x/y$, show that

$$x\frac{\partial T}{\partial x} = u\frac{\partial T}{\partial u} + v\frac{\partial T}{\partial v}, \quad y\frac{\partial T}{\partial y} = u\frac{\partial T}{\partial u} - v\frac{\partial T}{\partial v},$$

$$x^2\frac{\partial^2 T}{\partial x^2} + 2xy\frac{\partial^2 T}{\partial x \, \partial y} + y^2\frac{\partial^2 T}{\partial y^2} = 4u^2\frac{\partial^2 T}{\partial u^2} + 2u\frac{\partial T}{\partial u}.$$

(4) If $x = r\cos\theta$, $y = r\sin\theta$, show that

$$T_{xx} = L\cos^2\theta - 2M\sin\theta\cos\theta + N\sin^2\theta,$$

$$T_{yy} = L\sin^2\theta + 2M\sin\theta\cos\theta + N\cos^2\theta,$$

$$T_{xy} = (L - N)\sin\theta\cos\theta + M(\cos^2\theta - \sin^2\theta),$$

where $\quad L = \dfrac{\partial^2 T}{\partial r^2}, \quad M = \dfrac{1}{r}\dfrac{\partial^2 T}{\partial r \, \partial\theta} - \dfrac{1}{r^2}\dfrac{\partial T}{\partial\theta}, \quad N = \dfrac{1}{r}\dfrac{\partial T}{\partial r} + \dfrac{1}{r^2}\dfrac{\partial^2 T}{\partial\theta^2}.$

(5) If $x + iy = e^{u+i\theta} = re^{i\theta}$, show that

$$\nabla_1^2 T = e^{-2u}\left(\frac{\partial^2 T}{\partial u^2} + \frac{\partial^2 T}{\partial\theta^2}\right) = \frac{\partial^2 T}{\partial r^2} + \frac{1}{r}\frac{\partial T}{\partial r} + \frac{1}{r^2}\frac{\partial^2 T}{\partial\theta^2}.$$

(6) If $u = e^x(x\cos y - y\sin y)$, $v = e^x(x\sin y + y\cos y)$, prove that

$$\frac{\partial^2 T}{\partial u^2} + \frac{\partial^2 T}{\partial v^2} = \left(\frac{\partial^2 T}{\partial x^2} + \frac{\partial^2 T}{\partial y^2}\right)e^{-2x}\{(x+1)^2 + y^2\}^{-1}.$$

Also prove that $\quad \dfrac{\partial u}{\partial x} = e^{2x}\{(x+1)^2 + y^2\}\dfrac{\partial x}{\partial u}.$

(7) If $x = f(u) \cos nv$, $y = f(u) \sin nv$, show that

$$\nabla_1^2 T = \frac{1}{f'^2}\left\{\frac{\partial^2 T}{\partial u^2} + \left(\frac{f'}{f} - \frac{f''}{f'}\right)\frac{\partial T}{\partial u} + \frac{f'^2}{n^2 f^2}\frac{\partial^2 T}{\partial v^2}\right\}.$$

In particular, if $f(u) = u^n$ show that

$$\nabla_1^2 T = \frac{1}{n^2 u^{2n-2}}\left(\frac{\partial^2 T}{\partial u^2} + \frac{1}{u}\frac{\partial T}{\partial u} + \frac{1}{u^2}\frac{\partial^2 T}{\partial v^2}\right).$$

(8) If $\xi = x^2 - y^2$, $\eta = 2xy$, show that

$$(x^2 + y^2)\left(\frac{\partial^2 T}{\partial x^2} + \frac{\partial^2 T}{\partial y^2}\right) = 4(\xi^2 + \eta^2)\left(\frac{\partial^2 T}{\partial \xi^2} + \frac{\partial^2 T}{\partial \eta^2}\right).$$

Deduce that the most general function of xy satisfying the equation $\nabla_1^2 T = 0$ is $Axy + B$, where A, B are constants, and find the most general function of $xy(x^2 - y^2)$ satisfying the same equation.

(9) If $x = r \cos \theta$, $y = r \sin \theta$, $\xi = r^n \cos n\theta$, $\eta = r^n \sin n\theta$, show that

$$n^2(\xi^2 + \eta^2)\left(\frac{\partial^2 T}{\partial \xi^2} + \frac{\partial^2 T}{\partial \eta^2}\right) = (x^2 + y^2)\left(\frac{\partial^2 T}{\partial x^2} + \frac{\partial^2 T}{\partial y^2}\right).$$

(10) Show that the substitution given by $xu = yv = x^2 + y^2$ is orthogonal and that

$$\nabla_1^2 T = \frac{(u^2 + v^2)^2}{u^4 v^4}\left\{\left(u^2\frac{\partial}{\partial u}\right)^2 + \left(v^2\frac{\partial}{\partial v}\right)^2\right\}T.$$

(11) If $x = (1/u)\operatorname{sech} v$, $y = (1/u)\tanh v$, show that

$$\nabla_1^2 T = u^3\frac{\partial}{\partial u}\left(u\frac{\partial T}{\partial u}\right) + u^2 \cosh v \frac{\partial}{\partial v}\left(\cosh v\frac{\partial T}{\partial v}\right).$$

(12) If $x = r \cos \theta$, $y = r \sin \theta$, $u = r^n \cos n\theta$, $v = r^n \sin n\theta$, verify that $u_1 = v_2$, $u_2 = -v_1$, $\nabla_1^2 u = 0$, $\nabla_1^2 v = 0$, where $u_1 = \partial u/\partial x$, ... Also verify that

$$\frac{\partial^{p+q} u}{\partial x^p \partial y^q} = n(n-1)(n-2)...(n-p-q+1)r^{n-p-q}\cos\{(n-p-q)\theta + \tfrac{1}{2}q\pi\}.$$

(13) If $r^2 = x_1^2 + x_2^2 + ... + x_n^2$ and if T is a function of r only, show that

$$\frac{\partial^2 T}{\partial x_1^2} + \frac{\partial^2 T}{\partial x_2^2} + ... + \frac{\partial^2 T}{\partial x_n^2} = \frac{1}{r^{n-1}}\frac{d}{dr}\left(r^{n-1}\frac{dT}{dr}\right).$$

(14) If the equation $\nabla^2 T = 0$ is satisfied by $T = f(x, y, z)$, show that it is also satisfied by

$$T = \frac{1}{r}f\left(\frac{x}{r^2}, \frac{y}{r^2}, \frac{z}{r^2}\right) \quad (r^2 = x^2 + y^2 + z^2).$$

(15) If $x = vw$, $y = wu$, $z = uv$, show that

$$\Sigma x^2\frac{\partial^2 T}{\partial x^2} + \Sigma yz\frac{\partial^2 T}{\partial y \partial z} = \tfrac{1}{2}\Sigma u^2\frac{\partial^2 T}{\partial u^2}.$$

(16) Show that both of the expressions

$$\left(\frac{\partial T}{\partial x}\right)^2 + \left(\frac{\partial T}{\partial y}\right)^2 + \left(\frac{\partial T}{\partial z}\right)^2, \quad \frac{\partial^2 T}{\partial x^2} + \frac{\partial^2 T}{\partial y^2} + \frac{\partial^2 T}{\partial z^2}$$

are unaltered in form (are *absolute invariants*) under any rotation of rectangular axes.

(17) Show that under the substitution

$$u = \phi(x,y,z), \quad v = \psi(x,y,z), \quad w = \chi(x,y,z),$$

$$\nabla^2 T = (\Sigma a \, \partial^2/\partial u^2 + 2\Sigma f \, \partial^2/\partial v \, \partial w) \, T + \nabla^2 u \frac{\partial T}{\partial u} + \nabla^2 v \frac{\partial T}{\partial v} + \nabla^2 w \frac{\partial T}{\partial w},$$

where $a = u_1^2 + u_2^2 + u_3^2$, $f = v_1 w_1 + v_2 w_2 + v_3 w_3, \ldots,$ $u_1 = \partial u/\partial x, \ldots.$

(18) Show that the substitution $xu = yv = zw = x^2 + y^2 + z^2$ is orthogonal and that

$$\nabla^2 T = R^4 \left\{ \left(\frac{u^2}{R} \frac{\partial}{\partial u} \right)^2 + \left(\frac{v^2}{R} \frac{\partial}{\partial v} \right)^2 + \left(\frac{w^2}{R} \frac{\partial}{\partial w} \right)^2 \right\} T,$$

where $R = u^{-2} + v^{-2} + w^{-2}$.

Examples 7 F

(1) In the notation of §7.25, for the following transformations obtain the given results, assuming that $du \, dv > 0$:

(i) $x = e^u \cos v$, $y = e^u \sin v$:
$$\cos \omega = 0, \quad d\sigma = e^{2u} \, du \, dv.$$

(ii) $x = au + bv$, $y = cu + dv$:
$$\cos \omega = \frac{ab + cd}{\sqrt{(a^2 + c^2)} \sqrt{(b^2 + d^2)}}, \quad d\sigma = |ad - bc| \, du \, dv.$$

(iii) $x = u^2 v/(1 + u^2)$, $y = uv/(1 + u^2)$:
$$\cos \omega = \frac{1}{\sqrt{(1 + u^2)}}, \quad d\sigma = \frac{u^2 v}{(1 + u^2)^2} du \, dv \quad (v > 0).$$

(2) If $x = f(\xi, \eta), y = g(\xi, \eta)$ and

$$dx^2 + dy^2 = \lambda(d\xi^2 + d\eta^2)$$

show that
$$\left(\frac{\partial T}{\partial x} \right)^2 + \left(\frac{\partial T}{\partial y} \right)^2 = \frac{1}{\lambda} \left\{ \left(\frac{\partial T}{\partial \xi} \right)^2 + \left(\frac{\partial T}{\partial \eta} \right)^2 \right\},$$

$$\frac{\partial^2 T}{\partial x^2} + \frac{\partial^2 T}{\partial y^2} = \frac{1}{\lambda} \left(\frac{\partial^2 T}{\partial \xi^2} + \frac{\partial^2 T}{\partial \eta^2} \right).$$

(3) If $x = f(u,v), y = g(u,v)$ and

$$dx^2 + dy^2 = E \, du^2 + 2F \, du \, dv + G \, dv^2$$

show that, if ϕ is a function of u, v and $\phi_1 = \partial\phi/\partial u$, $\phi_2 = \partial\phi/\partial v$,

$$\left(\frac{\partial\phi}{\partial x} \right)^2 + \left(\frac{\partial\phi}{\partial y} \right)^2 = \frac{G\phi_1^2 - 2F\phi_1\phi_2 + E\phi_2^2}{EG - F^2},$$

$$\frac{\partial^2\phi}{\partial x^2} + \frac{\partial^2\phi}{\partial y^2} = \frac{1}{\sqrt{(EG - F^2)}} \left\{ \frac{\partial}{\partial u} \frac{G\phi_1 - F\phi_2}{\sqrt{(EG - F^2)}} + \frac{\partial}{\partial v} \frac{E\phi_2 - F\phi_1}{\sqrt{(EG - F^2)}} \right\}.$$

(4) If $u = axy + bx + cy + d$, $v = pxy + qx + ry + s$, prove that

$$\frac{\partial}{\partial y}\frac{u}{v} = \frac{f(x)}{v^2}, \quad \frac{\partial}{\partial x}\frac{u}{v} = \frac{g(y)}{v^2},$$

where $f(x)$, $g(y)$ denote quadratic functions of x, y respectively. Hence prove that u/v is a function of a product XY, with X, Y certain functions of x, y respectively.

(5) Let (u, v, w), (x, y, z) be two sets of Cartesian coordinates of which the second is rectangular.

A substitution $x = f(u, v, w)$, $y = g(u, v, w)$, $z = h(u, v, w)$ in general transforms an infinitesimal parallelepiped of which two opposite corners are $p(u, v, w)$ and $p'(u+du, v+dv, w+dw)$ and three concurrent edges are pq_1, pq_2, pq_3 parallel to the u, v, w axes respectively, into an infinitesimal parallelepiped of which the two corresponding corners will be denoted by P, P' and the corresponding edges by PQ_1, PQ_2, PQ_3. Let $ds = PP'$, and let α, β, γ be the angles of the corresponding faces at P. Show that

$$ds^2 = A\,du^2 + B\,dv^2 + C\,dw^2 + 2F\,dv\,dw + 2G\,dw\,du + 2H\,du\,dv$$
$$= H_1^2\,du^2 + H_2^2\,dv^2 + H_3^2\,dw^2 + 2\Sigma H_2 H_3 \cos \alpha\, dv\,dw$$

where, with $x_1 = \partial x/\partial u, \ldots,$

$$A = H_1^2 = x_1^2 + y_1^2 + z_1^2, \ldots, \quad F = H_2 H_3 \cos \alpha = x_2 x_3 + y_2 y_3 + z_2 z_3, \ldots.$$

The infinitesimal ds is called *the line element* of the transformation. Let $ds_1 = PQ_1, ds_2 = PQ_2, ds_3 = PQ_3$, and let $d\tau\,(>0)$ be the volume of the parallelepiped PP'. Show that

$$ds_1 = H_1 du, \quad ds_2 = H_2 dv, \quad ds_3 = H_3 dw, \quad d\tau = |J|\,du\,dv\,dw,$$
$$J^2 = H_1^2 H_2^2 H_3^2 (1 - \cos^2 \alpha - \cos^2 \beta - \cos^2 \gamma + 2 \cos \alpha \cos \beta \cos \gamma),$$

where $J = \partial(x, y, z)/\partial(u, v, w)$.

Also show that PQ_1, PQ_2, PQ_3 have the same orientation as the coordinate axes Ox, Oy, Oz if $J > 0$, but the contrary if $J < 0$.

(6) *Matrix notation.* Some of the results of this and the preceding chapter can be compactly expressed in the notation of matrices. Let T be a function of $y_1, y_2, \ldots, y_n$ and let $y_1, y_2, \ldots, y_n$ be functions of $x_1, x_2, \ldots, x_n$. Prove that, in the notation of matrices,

$$[\partial T/\partial x] = [\partial T/\partial y][\partial y/\partial x]$$

or

$$[\partial T/\partial x]' = [\partial y/\partial x]'[\partial T/\partial y]'$$

where $[\partial T/\partial x]$, $[\partial T/\partial x]'$ denote a row matrix and its transpose, and $[\partial y/\partial x]$ denotes the $n \times n$ matrix of which $\partial y_r/\partial x_s$ is the (r, s)th element.

(7) Let T be a function of $y_1, y_2, \ldots, y_n$ and let $y_1, y_2, \ldots, y_n$ be functions of $x_1, x_2, \ldots, x_n$ defined inversely by equations giving $x_1, x_2, \ldots, x_n$ explicitly in terms of $y_1, y_2, \ldots, y_n$. Prove that, in the notation of matrices,

$$[\partial T/\partial x] = [\partial T/\partial y][\partial x/\partial y]^{-1}$$
$$[\partial y/\partial x] = [\partial x/\partial y]^{-1}.$$

(8) If the n functions $y_1, y_2, \ldots, y_n$ of the m variables $x_1, x_2, \ldots, x_m$ are defined implicitly by the equations

$$F_r(x_1, x_2, \ldots, x_m, y_1, y_2, \ldots, y_n) = 0 \quad (r = 1, 2, \ldots, n)$$

prove that the partial derivatives $\partial y_r/\partial x_s$ are given by the matrix equation

$$[\partial y/\partial x] = -[\partial F/\partial y]^{-1}[\partial F/\partial x]$$

the orders of the matrices from left to right being $n \times m$, $n \times n$, $n \times m$.

(9) If T is a function of u, v, and u, v are each functions of x, y, show that

$$\begin{bmatrix} T_{11} & T_{12} \\ T_{12} & T_{22} \end{bmatrix} = \begin{bmatrix} u_1 & u_2 \\ v_1 & v_2 \end{bmatrix}' \begin{bmatrix} T_{uu} & T_{uv} \\ T_{uv} & T_{vv} \end{bmatrix} \begin{bmatrix} u_1 & u_2 \\ v_1 & v_2 \end{bmatrix}$$

$$+ T_u \begin{bmatrix} u_{11} & u_{12} \\ u_{12} & u_{22} \end{bmatrix} + T_v \begin{bmatrix} v_{11} & v_{12} \\ v_{12} & v_{22} \end{bmatrix},$$

where $T_1 = \partial T/\partial x$, $T_2 = \partial T/\partial y$, ... Generalize this.

(10) If $T = f(u,v)$, $u = ax+by$, $v = cx+dy$, show that

$$\begin{bmatrix} T_{11} & T_{12} \\ T_{12} & T_{22} \end{bmatrix} = \begin{bmatrix} a & c \\ b & d \end{bmatrix} \begin{bmatrix} T_{uu} & T_{uv} \\ T_{uv} & T_{vv} \end{bmatrix} \begin{bmatrix} a & b \\ c & d \end{bmatrix}.$$

Generalize this.

Examples 7G

(1) Show that Maclaurin's formula (100), applied to the function $x^3 - 3xy^2$,

(i) for $n = 1$ gives $x^3 - 3xy^2 = 3\theta^2(x^3 - 3xy^2)$, $\theta = 1/\sqrt{3}$,

(ii) for $n = 2$ gives $x^3 - 3xy^2 = 3\theta(x^3 - 3xy^2)$, $\theta = \frac{1}{3}$.

(2) Show that Maclaurin's formula (100), applied to the function

$$y^2 - 3x^2y + 2x^4, \text{ for } n = 2 \text{ gives } 12\theta^2 x^2 - 9\theta y + 3y - 2x^2 = 0,$$

and verify that this equation, as a quadratic equation in θ, has a root between 0 and $\frac{1}{2}$.

(3) By any method, obtain the first few terms given in the following expansions:

(i) $\sin x \cosh y = x - \frac{1}{6}(x^3 - 3xy^2) + \frac{1}{120}(x^5 - 10x^3y^2 + 5xy^4) - \dots.$

(ii) $e^{-2xy}\cos(x^2 - y^2) = 1 - 2xy - \frac{1}{2}(x^4 - 6x^2y^2 + y^4) + \dots.$

(iii) If $f(x,y) = x^3 - 3xy^2$, then

$$f(2+h, 1+k) = 2 + 9h - 12k + 6(h^2 - hk - k^2) + h^3 - 3hk^2.$$

(iv) If $f(x,y) = x/(x^2 + y^2)$, then

$$f(1+h, 1+k) = \tfrac{1}{2} - \tfrac{1}{2}k - \tfrac{1}{4}(h^2 - 2hk - k^2) - \dots.$$

(4) If the function $\phi(x,y)$, defined implicitly by the equation $y = \phi - x\sin\phi$, is expanded in powers and products of powers of x and y, show that, as far as the terms of degree four,

$$\phi = y + xy + x^2y + x^3y - \tfrac{1}{6}xy^3 \dots.$$

(5) *Euler's theorem on homogeneous functions* (§ 5.29) may be generalized by means of Taylor's theorem.

Let $f(x, y, z)$ be homogeneous of degree m, and let all the partial derivatives of $f(x, y, z)$ up to the nth order be continuous. Then, if $r < n$,

$$\left(x\frac{\partial}{\partial x} + y\frac{\partial}{\partial y} + z\frac{\partial}{\partial z}\right)^r f(x, y, z) = m(m-1)\ldots(m-r+1)f(x, y, z). \qquad (102)$$

Proof. Since $f(x, y, z)$ is homogeneous of order m,

$$f(a+ta, b+tb, c+tc) = (1+t)^m f(a, b, c).$$

The L.H.S. can be expanded in ascending powers of t by Taylor's theorem and the R.H.S. by the binomial theorem, and by equating coefficients of t^r in the two expansions, we obtain (102), with a, b, c in place of x, y, z.

(6) *A meaning for* $\nabla_1^2 f(x, y)$. Let R be a circular region of radius r, centre $P_0(a, b)$. Let $f(x, y)$ have continuous partial derivatives up to the third order (a sufficient condition) in R. Let $P(x, y)$ be any point on the circumference of R. Then, by (95), we can put

$$f(x, y) = f_0 + hf_1 + kf_2 + \tfrac{1}{2}(h^2 f_{11} + 2hk f_{12} + k^2 f_{22}) + \rho_3,$$

where $f_0, f_1, f_2, f_{11}, f_{12}, f_{22}$ refer to P_0 and

$$\rho_3 = \tfrac{1}{6}(h^3 f_{111} + 3h^2 k f_{112} + 3hk^2 f_{122} + k^3 f_{222})\,(\xi, \eta)$$

refers to some point (ξ, η) on the radius $P_0 P$.

Now put $h = r\cos\theta$, $k = r\sin\theta$, so that

$$x = a + h = a + r\cos\theta, \quad y = b + k = b + r\sin\theta;$$

then $f(x, y) = f_0 + r(f_1 \cos\theta + f_2 \sin\theta)$

$$+ \tfrac{1}{2}r^2(f_{11}\cos^2\theta + 2f_{12}\cos\theta\sin\theta + f_{22}\sin^2\theta) + \tfrac{1}{6}r^3\Theta,$$

where Θ is a bounded function of θ, since the third partial derivatives are continuous in R.

The average value $\bar{f}$ of $f(x, y)$ on the circumference of R is given by (see §9.14)

$$\bar{f} = \frac{1}{2\pi r}\int_0^{2\pi} f(x, y)\, r\, d\theta;$$

consequently, if $-K < \Theta < K$, where K is independent of θ,

$$2\pi r\bar{f} = 2\pi r f_0 + \tfrac{1}{2}r^3(\pi f_{11} + \pi f_{22}) + \sigma_3$$

where

$$|\sigma_3| < \int_0^{2\pi} \tfrac{1}{6}r^3 K . r\, d\theta = \tfrac{1}{6}(2\pi r^4 K),$$

and hence

$$\bar{f} = f_0 + \tfrac{1}{4}r^2\nabla_1^2 f + \tfrac{1}{6}r^3\tau_3$$

where $|\tau_3| < K$. It follows that $\tfrac{1}{4}r^2\nabla_1^2 f$ is an approximation to the excess of $\bar{f}$, the average value of f on the circumference of R, over f_0 the value at the centre P_0, since, if $\nabla_1^2 f \neq 0$, the term $\tfrac{1}{6}r^3\tau_3$ is as small as we please compared with the term $\tfrac{1}{4}r^2\nabla_1^2 f$ when r is small enough.

This leads to the following definitions of subharmonic and superharmonic functions:

Definitions. (i) If $\nabla_1^2 f > 0$ at P_0, so that $f_0 < \bar{f}$, then $f(x, y)$ is said to be *subharmonic* at P_0.

(ii) If $\nabla_1^2 f < 0$ at P_0, so that $f_0 > \bar{f}$, then $f(x, y)$ is said to be *superharmonic* at P_0.

Example. If $f(x, y) = (h^2 - x^2)(k^2 - y^2)$, we find

$$\nabla_1^2 f(x, y) = 2(x^2 + y^2 - h^2 - k^2);$$

by definition, therefore, the function $f(x, y)$ is superharmonic at every interior point of the circle $x^2 + y^2 = h^2 + k^2$, or of the rectangle in which $x^2 \leqslant h^2, y^2 \leqslant k^2$.

(7) *Taylor's theorem for a harmonic function.*

Theorem. If the function $f(x, y)$ of §7.27 is harmonic, Taylor's theorem (95) can be expressed in the form

$$f(x, y) = f(a, b) + \sum_{s=1}^{n-1} \frac{r^s}{s!} (A_s \cos s\theta + B_s \sin s\theta) + R_n$$

where $A_s = \partial^s f / \partial x^s$ and $B_s = \partial^s f / \partial x^{s-1} \partial y$ at the point (a, b), $x = a + r \cos \theta$, $y = b + r \sin \theta$, and R_n is of the form $r^n \Theta$, where Θ is a function of θ which is bounded in the neighbourhood of (a, b).

Proof. Put $f_{s,0} = \partial^s f / \partial x^s$, $f_{s-1,1} = \partial^s f / \partial x^{s-1} \partial y$, etc. Then, since f is harmonic, $f_{2,0} + f_{0,2} \equiv 0$; therefore $f_{2,0} \equiv -f_{0,2}$ and by repeated differentiation

$$f_{s,0} \equiv -f_{s-2,2} \equiv f_{s-4,4} \equiv \cdots,$$

$$f_{s-1,1} \equiv -f_{s-3,3} \equiv f_{s-5,5} \equiv \cdots.$$

The factor $(hf_1 + kf_2)^s (a, b)$ in (95), when expanded, reads

$$h^s f_{s,0} + {}^sC_1 h^{s-1} k f_{s-1,1} + {}^sC_2 h^{s-2} k^2 f_{s-2,2} + \cdots$$

which can now be written

$$f_{s,0}(h^s - {}^sC_2 h^{s-2} k^2 + \cdots) + f_{s-1,1}({}^sC_1 h^{s-1} k - \cdots)$$

which becomes, when we put $h = x - a = r \cos \theta$, $k = y - b = r \sin \theta$,

$$A_s r^s (\cos^s \theta - {}^sC_2 \cos^{s-2} \theta \sin^2 \theta + \cdots) + B_s r^s ({}^sC_1 \cos^{s-1} \theta \sin \theta - \cdots)$$

$$= r^s (A_s \cos s\theta + B_s \sin s\theta)$$

by de Moivre's theorem. The proof can now be completed by showing that R_n is of the form $r^n \Theta$, where Θ is a bounded function of θ if n is fixed.

(8) *Interpolation, two independent variables.* Obtain the linear interpolation formula

$$f(a + \lambda h, b + \mu k) \doteqdot f(a, b) + \lambda \Delta_1 + \mu \Delta_2$$

and the quadratic interpolation formula

$$f(a + \lambda h, b + \mu k) \doteqdot f(a, b) + \lambda \Delta_1 + \mu \Delta_2 + \tfrac{1}{2}\{\lambda(\lambda - 1)\Delta_{11} + 2\lambda\mu\Delta_{12} + \mu(\mu - 1)\Delta_{22}\}$$

where

$$\Delta_1 = f(a + h, b) - f(a, b), \quad \Delta_2 = f(a, b + k) - f(a, b),$$

$$\Delta_{11} = f(a + 2h, b) - 2f(a + h, b) + f(a, b),$$

$$\Delta_{12} = f(a + h, b + k) - f(a + h, b) - f(a, b + k) + f(a, b),$$

$$\Delta_{22} = f(a, b + 2k) - 2f(a, b + k) + f(a, b).$$

8

MAXIMA AND MINIMA

8.1 Maxima and minima of a function of one variable

Let $x = a$ be an interior point of an interval in which a function $f(x)$ is defined.

Definition. We say that $f(a)$ is a *minimum* of $f(x)$, or that $f(x)$ has a *minimum* $f(a)$ at $x = a$, if number δ exists such that $f(a) < f(x)$ for all values of x except $x = a$ in the subinterval $a - \delta < x < a + \delta$; or briefly, such that

$$f(a) < f(x), \quad 0 < (x-a)^2 < \delta^2. \tag{1}$$

If there are values of $f(x)$ less than or equal to $f(a)$ at other points of the original interval, it may be convenient to call $f(a)$ a *local minimum*.

If a is an end point of a closed interval, we consider only values of x on the side of $x = a$ which belongs to the interval.

Examples

(1) If $f(x) = x^{\frac{2}{3}}$ then $f(0) = 0$ and $0 < f(x)$ if $x \neq 0$. Hence $f(x)$ has the minimum value 0 at $x = 0$.

(2) If $f(x) = e^{-1/x^2}$ when $x \neq 0$ and $f(0) = 0$, then $f(x)$ has a minimum 0 at $x = 0$, for the same reason as in the last example.

(3) The function $f(x) = |\sin x|$ has a minimum at $x = 0$ and at $x = k\pi$, where k is any integer.

A *maximum* of $f(x)$ is defined in a similar way: thus, $f(a)$ is a maximum of $f(x)$ if a number δ exists such that

$$f(a) > f(x), \quad 0 < (x-a)^2 < \delta^2. \tag{2}$$

8.2

The above definitions of maxima and minima apply whether $f(x)$ is differentiable or not. Suppose now that $f(x)$ is differentiable at $x = a$, so that $f'(a)$ exists.

If $f'(a) = 0$ the function $f(x)$ is said to be *stationary* at $x = a$, and $f(a)$ is called a *stationary value* of $f(x)$. On the curve $y = f(x)$ the point where $x = a$ is then called a *stationary point*. At a stationary point the curve has a tangent parallel to the axis of x.

8.3

If $f(x)$ is differentiable at an interior point $x = a$, a *necessary* condition that $f(a)$ should be a maximum or a minimum is $f'(a) = 0$.

For, by §2.11, III, Corollary 1, if $f'(a) \neq 0$ the function $f(x)$ is either strictly increasing or decreasing at $x = a$, contrary to the definition of a maximum or minimum, by which $f(x) - f(a)$ has the same sign whether $x > a$ or $x < a$.

It follows that the maxima and minima of $f(x)$ are stationary values of $f(x)$. Consequently, if $f(x)$ is differentiable the values of x for which $f(x)$ has maxima or minima must be sought among the solutions of the equation $f'(x) = 0$.

8.4

If there is a value of n such that $f^{(r)}(a) = 0$ for $r = 1, 2, \ldots, n-1$, but $f^{(n)}(a) \neq 0$, then a necessary and sufficient condition for $f(a)$ to be a minimum or a maximum is that n be even.

For, by Taylor's formula, we can then put

$$f(x) - f(a) = (x-a)^n \{f^{(n)}(a) + \eta\}/n! \tag{3}$$

where $\eta \to 0$ when $x \to a$. It follows that, for small enough values of $x - a$:

(i) if n is even, the difference $f(x) - f(a)$ has the same sign as $f^{(n)}(a)$ whether $x - a > 0$ or $x - a < 0$, and hence that $f(a)$ is a minimum if $f^{(n)}(a) > 0$, but a maximum if $f^{(n)}(a) < 0$;

(ii) if n is odd, the difference $f(x) - f(a)$ changes sign when $x - a$ changes sign, in which case $f(a)$ is neither a minimum nor a maximum.

Note that for the function $f(x) = e^{-1/x^2}$ of §8.1, Example 2, $f^{(n)}(0) = 0$ for every value of n (cf. Examples 3B, 30), but $f(0)$ is a minimum, showing that it is not *necessary* that there should be any value of n for which $f^{(n)}(a) \neq 0$.

8.5   Maxima and minima of a function of two variables

Let (a, b) be an interior point of a region in which a function $f(x, y)$ is defined.

Definition. We say that $f(a, b)$ is a *minimum* of $f(x, y)$ if a number δ exists such that

$$f(a, b) < f(x, y), \quad 0 < (x-a)^2 + (y-b)^2 < \delta^2, \tag{4}$$

that is, such that $f(a, b) < f(x, y)$ at every point (x, y) inside the circle $(x-a)^2 + (y-b)^2 = \delta^2$, except the centre.

If (a, b) is a point on the boundary of the region we consider only neighbouring points belonging to the region and its boundary.

A *maximum* is defined in a similar way: thus, $f(a, b)$ is a maximum of $f(x, y)$ if a number δ exists such that

$$f(a, b) > f(x, y), \quad 0 < (x-a)^2 + (y-b)^2 < \delta^2. \tag{5}$$

Examples

(1) The function $f(x, y) = x^2 - 2xy^2 + 2y^4$ has a minimum 0 at $(0, 0)$, since the value of the function is 0 at the point $(0, 0)$, and

$$f(x, y) = (x - y^2)^2 + y^4 > 0$$

at every other point.

(2) The function $f(x, y) = (x + y)^2$ has the value 0 at $(0, 0)$ and has a positive value at every point not on the line $x + y = 0$. But $f(0, 0)$ is not a minimum because within every circle, centre $(0, 0)$, however small, $f(x, y) = 0$ at every point on the line $x + y = 0$.

(3) The function $z = f(x, y) = (y - x^2)(y - 3x^2)$ is such that if $y = mx$, then $z = m^2x^2 - 4mx^3 + 3x^4$, which shows that z has a minimum 0 at $x = 0$ for every finite value of m: that is, z is a minimum at $x = 0$ along every straight line $y = mx$; it is also a minimum on the line $x = 0$, since $z = y^2$ on $x = 0$. Nevertheless, $f(0, 0)$ is not a minimum of $f(x, y)$; for since

$$f(x, y) = y^2 - 4yx^2 + 3x^4 = (y - 2x^2)^2 - x^4$$

it follows that $f(x, y) < 0$ at all points except $(0, 0)$ on the parabola $y = 2x^2$; consequently, every circle, centre $(0, 0)$, contains points where $f(x, y) < f(0, 0)$ (see Examples 8A, 10).

(4) Consider the function $f(x, y) = x^2 - y^2$ in the circular region $x^2 + y^2 \leqslant 1$.

In accordance with the definition, $f(1, 0)$ and $f(-1, 0)$ are maxima, $f(0, 1)$ and $f(0, -1)$ are minima. For at any point $x = r\cos\theta, y = r\sin\theta$, we have $f(x, y) = r^2\cos 2\theta < 1$ at all points in the region except the points $(1, 0)$ and $(-1, 0)$, where $r = 1, \theta = 0$ and $r = 1, \theta = \pi$ respectively. At these points $f(x, y) = 1$. Consequently, $f(1, 0)$ and $f(-1, 0)$ are local maxima. Similarly, $f(0, 1)$ and $f(0, -1)$ are local minima.

8.6

We assume now that $f(x, y)$ is differentiable.

A point on the surface $z = f(x, y)$ at which the partial derivatives $\partial z/\partial x$, $\partial z/\partial y$ both vanish, is called a *stationary point*.

At such a point there is a tangent plane which is parallel to the xy plane, and the function $f(x,y)$ is said to have a stationary value.

In order to locate the stationary points we seek the values of x and y that satisfy the simultaneous equations

$$\partial z/\partial x = 0, \quad \partial z/\partial y = 0. \tag{6}$$

If $f(x,y)$ is defined over a region R in the xy plane, any interior point of R where $f(x,y)$ has a maximum or minimum must be a stationary point. For at such a point, $f(x,y)$ must evidently be stationary with respect to each variable separately, and hence $f_x = 0, f_y = 0$. Further, a maximum or minimum on the boundary of R may be a stationary point, but there may be on the boundary maximum or minimum points which are not stationary points (§ 8.5, Example 4).

8.7

Next, let $f(x,y)$ be differentiable and let (a,b) be a point where $f(x,y)$ is stationary, so that $f_x = 0$ and $f_y = 0$ at (a,b). Also suppose that there is a neighbourhood of (a,b) within which all the second-order partial derivatives of $f(x,y)$ exist and are continuous, and hence in which $f_{yx} \equiv f_{xy}$.

Then, by Taylor's formula (§ 7.27) to base (a,b), we can put

$$f(x,y) - f(a,b) = \tfrac{1}{2}\{h^2 f_{11}(\xi,\eta) + 2hk f_{12}(\xi,\eta) + k^2 f_{22}(\xi,\eta)\}, \tag{7}$$

where $h = x - a, k = y - b, \xi = a + \theta h, \eta = b + \theta k$, and θ is some number between 0 and 1, depending on h and k.

In order that $f(a,b)$ may be a minimum (maximum) of $f(x,y)$ it is necessary and sufficient that there should exist a circle, centre (a,b), radius $\delta > 0$, within which the R.H.S. of (7) is positive (negative) at all points except the centre.

As a preliminary to the general case (see § 8.11) we begin by considering the simplest case, in which $z = f(x,y) = ax^2 + 2hxy + by^2$, where a, h, b are constants. Here a, b are not to be confused with the coordinates (a,b) in (7).

8.8   Sign of a quadratic form

The expression
$$f(x,y) = ax^2 + 2hxy + by^2, \tag{8}$$

where a, h, b are constants and x, y variables, is called a *quadratic form* in x and y. It is said to be *positive-definite* if its value is positive for every pair of values (x,y) except only $(0,0)$. It is said to be *negative-definite* if the form $-f(x,y)$ is positive-definite.

Theorem. If a, h, b, x, y are real, necessary and sufficient conditions for $f(x, y)$ to be positive-definite are

$$a > 0, \quad ab - h^2 > 0. \tag{9}$$

The conditions are necessary: that is, given that $f(x, y)$ is positive-definite, the conditions (9) necessarily follow.

Proof. Put $x = 1$, $y = 0$. Since $f(x, y)$ is positive-definite (given), therefore $f(1, 0) > 0$. But $f(1, 0) = a$, therefore $a > 0$.

Since $a > 0$, therefore $a \neq 0$ and $f(x, y)$ can be expressed in the form

$$f(x, y) = a(x + hy/a)^2 + (ab - h^2) y^2/a. \tag{10}$$

Put $y = 1$, $x = -h/a$. Since $f(x, y)$ is positive-definite, therefore $f(-h/a, 1) > 0$; that is, $(ab - h^2)/a > 0$ and hence $ab - h^2 > 0$, since $a > 0$.

The conditions are sufficient: that is, if the conditions are satisfied, it must follow that $f(x, y)$ is positive-definite.

Proof. Since $a > 0$ (given), $f(x, y)$ can be expressed in the form (10). Since $a > 0$ and $ab - h^2 > 0$, each of the two terms on the R.H.S. of (10) must be either positive or zero, and they can both be zero only if $x + hy/a = 0$ and $y = 0$, which requires both $x = 0$ and $y = 0$. Hence $f(x, y)$ is positive-definite.

Corollary. Necessary and sufficient conditions that $f(x, y)$ should be negative-definite are

$$a < 0, \quad ab - h^2 > 0, \tag{11}$$

since $-f(x, y)$ will then be positive-definite.

8.9

Next, consider the cases (i) $ab - h^2 = 0$, (ii) $ab - h^2 < 0$.

(i) $ab - h^2 = 0$. Either a and b are both zero, in which case $h = 0$ too, and $f(x, y) \equiv 0$; or a and b are not both zero, in which case suppose a not zero: then, by (10),

$$f(x, y) = a(x + hy/a)^2, \tag{12}$$

which shows that $f(x, y) = 0$ when (x, y) is any point on the straight line $x + hy/a = 0$, and that $f(x, y)$ has the sign of a when (x, y) is any point off the line.

(ii) $ab - h^2 < 0$. The function can now be expressed as the product of distinct real linear factors, thus

$$f(x, y) = (l_1 x + m_1 y)(l_2 x + m_2 y) \tag{13}$$

from which it follows that $f(x, y) = 0$ when (x, y) is any point on either of the two straight lines $l_1 x + m_1 y = 0$, $l_2 x + m_2 y = 0$; that $f(x, y) > 0$

when (x, y) lies inside one of the angles between these lines, and that $f(x, y) < 0$ when (x, y) lies in the supplementary angle. The sign of $f(x, y)$ changes when the point (x, y) crosses one of these lines.

8.10   The surface $z = ax^2 + 2hxy + by^2$

Returning to the case called the simplest case at the end of §8.7, we deduce from the algebra of §§8.8 and 8.9 that on the surface

$$z = f(x, y) = ax^2 + 2hxy + by^2 \tag{14}$$

the origin is a minimum or a maximum point if $ab - h^2 > 0$. This inequality being satisfied, the origin is a minimum point if $a > 0$, a maximum point if $a < 0$. (The surface is an elliptic paraboloid.)

If $ab - h^2 = 0$ and a, b, h are not all zero, the origin is neither a maximum nor a minimum point, by the definition of §8.5, because $z = 0$ at every point on a certain straight line through the origin and so we cannot draw round the origin any circle within which $z > 0$ or $z < 0$ at every point except the origin. Every point on this straight line is a stationary point, and z has the same sign at all points off the line. (The surface is a parabolic cylinder.)

When $ab - h^2 < 0$ the origin is called a *saddle point*. Near the origin the shape of the surface resembles that of a saddle or a pass between hills. (The surface is a hyperbolic paraboloid.)

8.11   The general case

We now consider the general case (7) of §8.7 in which $f(x, y)$ is stationary at the point (a, b) and the partial derivatives of the second order are continuous in the neighbourhood of (a, b). In particular, these derivatives being continuous at (a, b), we can put

$$f_{11}(\xi, \eta) = r + \alpha, \quad f_{12}(\xi, \eta) = s + \beta, \quad f_{22}(\xi, \eta) = t + \gamma, \tag{15}$$

where $r = f_{11}(a, b)$, $s = f_{12}(a, b)$, $t = f_{22}(a, b)$ and α, β, γ all tend to zero as $(h, k) \to (0, 0)$ in any manner. Then (7) will read

$$\delta z = \tfrac{1}{2}\{h^2(r + \alpha) + 2hk(s + \beta) + k^2(t + \gamma)\}, \tag{16}$$

where $\delta z = f(x, y) - f(a, b) = f(a + h, b + k) - f(a, b).$

In particular, at the point at which

$$h = \rho \cos \psi, \quad k = \rho \sin \psi \quad (-\pi < \psi \leqslant \pi),$$

$$\delta z = \tfrac{1}{2}\rho^2\{F(\psi) + \zeta\}, \tag{17}$$

where $$F(\psi) = r \cos^2 \psi + 2s \cos \psi \sin \psi + t \sin^2 \psi, \tag{18}$$

$$\zeta = \alpha \cos^2 \psi + 2\beta \cos \psi \sin \psi + \gamma \sin^2 \psi. \tag{19}$$

Here $F(\psi)$ is a function of the one variable ψ; it is continuous and therefore attains its upper and lower bounds (see §1.22, V; also Examples 8 A, 1). We can now prove the following theorem:

8.12

Theorem. If $rh^2 + 2shk + tk^2$ is a positive-definite form, then $f(a, b)$ is a minimum of $f(x, y)$.

Proof. In (18), $F(\psi)$ will now be a positive-definite form in $\cos\psi$ and $\sin\psi$, and since $\cos\psi$ and $\sin\psi$ cannot vanish together, the lower bound of $F(\psi)$ will be positive: let L be this lower bound.

Put $\epsilon = \frac{1}{6}L > 0$. Since α, β, γ all tend to zero as $(h, k) \rightarrow (0, 0)$, there exists a number $\delta > 0$ such that α, β, γ are numerically less than ϵ inside a circle, radius δ, centre (a, b). It follows from (19) that inside this circle

$$|\zeta| \leqslant |\alpha \cos^2\psi + \beta \sin 2\psi + \gamma \sin^2\psi|$$

$$\leqslant |\alpha| + |\beta| + |\gamma| < 3\epsilon = \tfrac{1}{2}L$$

and hence from (17) that inside this circle

$$\delta z > \tfrac{1}{2}\rho^2(L - \tfrac{1}{2}L) = \tfrac{1}{4}\rho^2 L > 0 \quad (\rho \neq 0),$$

and hence further that $f(x, y) > f(a, b)$ except at (a, b); consequently, $f(a, b)$ is a minimum of $f(x, y)$.

Corollary. If $rh^2 + 2shk + tk^2$ is negative-definite, then $f(a, b)$ will be a maximum of $f(x, y)$.

8.13

Suppose next that $rh^2 + 2shk + tk^2$ is not a definite form. There are two cases (i) $rt - s^2 = 0$, (ii) $rt - s^2 < 0$.

Since $F(\psi)$ is independent of ρ, it follows from (17) that δz has the sign of $F(\psi)$ if $F(\psi) \neq 0$, when ρ is small enough, $\rho \neq 0$; but has the sign of ζ if $F(\psi) = 0$.

(i) $rt - s^2 = 0$. Either r, s, t are all zero and then the sign of δz is unknown for all values of ψ, being the sign of the unknown ζ; or r, s, t are not all zero, and then the sign of δz is unknown for values of ψ that satisfy the equation $F(\psi) = 0$.

The case $rt - s^2 = 0$ is therefore called *the doubtful case*: the sign of δz cannot be decided by the values at (a, b) of the second derivatives only.

(ii) $rt - s^2 < 0$. Since $\rho^2 F(\psi) = rh^2 + 2shk + tk^2$, the sign of $F(\psi)$ is the same as that of the quadratic form $rh^2 + 2shk + tk^2$ when $F(\psi) \neq 0$. This form is no longer definite, but is positive for some values of (h, k), negative for others (§8.9, (ii)). Also, the sign of δz is the same as that of $F(\psi)$ when $F(\psi) \neq 0$ and ρ is small enough. The point where $x = a$, $y = b$ is now called a *saddle point* (end of §8.10).

8.14 Summary

If $x = a$, $y = b$ satisfy the equations $\partial z / \partial x = 0$, $\partial z / \partial y = 0$, then the point at which $x = a$, $y = b$ on the surface $z = f(x, y)$ is a stationary point. It is further

(i) a minimum point if $rt - s^2 > 0$, $r > 0$;
(ii) a maximum point if $rt - s^2 > 0$, $r < 0$;
(iii) a saddle point if $rt - s^2 < 0$.

The case $rt - s^2 = 0$ requires more information than is provided by the second derivatives alone.

8.15

In the following examples p, q, r, s, t may be used to denote

$$\partial z / \partial x, \partial z / \partial y, \partial^2 z / \partial x^2, \partial^2 z / \partial x \, \partial y, \partial^2 z / \partial y^2$$

respectively.

Examples

(1) $$z = xy^2 (3x + 6y - 2).$$

By partial differentiation

$$p = 2y^2 (3x + 3y - 1), \quad q = 2xy(3x + 9y - 2),$$

$$r = 6y^2, \quad s = 6y^2 + 4y(3x + 3y - 1), \quad t = 18xy + 2x(3x + 9y - 2).$$

The equations ($p = 0$, $q = 0$) which give the stationary points are

$$y^2(3x + 3y - 1) = 0, \quad xy(3x + 9y - 2) = 0,$$

and these equations are satisfied by

(i) $x = h$, $y = 0$, where h is arbitrary.

The xy plane is the tangent plane at every point on the axis of x, but no point of it is a maximum or minimum point, since the whole of the axis of x lies in the surface.

(ii) $3x + 3y - 1 = 0$, $x = 0$, and hence $x = 0$, $y = \frac{1}{3}$, $z = 0$. At this point $r = \frac{2}{3}$, $s = \frac{2}{3}$, $t = 0$, $rt - s^2 = -\frac{4}{9}$. Since $rt - s^2 < 0$, the point is a saddle point.

(iii) $3x + 3y - 1 = 0$, $3x + 9y - 2 = 0$; $x = \frac{1}{6}$, $y = \frac{1}{6}$, $z = -\frac{1}{432}$.

At this point $r = \frac{1}{6}$, $s = \frac{1}{6}$, $t = \frac{1}{2}$, $rt - s^2 = \frac{1}{18}$.
Since $rt - s^2 > 0$ and $r > 0$, the point is a minimum point.

(2) $$z = \{(x-3)^2 + 2y^2 + 6\}/(x+y)^2.$$

Write this equation in the form

$$(x+y)^2 z = (x-3)^2 + 2y^2 + 6. \qquad (a)$$

Then by partial differentiation

$$(x+y)^2 p + 2(x+y)z = 2(x-3), \quad (x+y)^2 q + 2(x+y)z = 4y, \qquad (b)$$

$$(x+y)^2 r + 4(x+y)p + 2z = 2, \qquad (c)$$

$$(x+y)^2 s + 2(x+y)(p+q) + 2z = 0, \qquad (d)$$

$$(x+y)^2 t + 4(x+y)q + 2z = 4. \qquad (e)$$

Putting $p = 0$, $q = 0$ in (b), we get the two equations

$$(x+y)z = x-3, \quad (x+y)z = 2y$$

which, with (a), give the values of x, y, z at the stationary points. They give only $x = 5$, $y = 1$, $z = \frac{1}{3}$. Then (c), (d), (e) give $36r = \frac{4}{3}$, $36s = -\frac{2}{3}$, $36t = \frac{10}{3}$, from which follows $rt - s^2 > 0$. Since also $r > 0$, we infer that $(5, 1, \frac{1}{3})$ is a minimum point.

(3) $$z = x^2 - x^3 - 2xy^2 + y^4.$$

We find $$p = 2(x - y^2) - 3x^2, \quad q = -4y(x - y^2)$$

and the equations $p = 0$, $q = 0$ give (i) $x = 0$, $y = 0$; (ii) $x = \frac{2}{3}$, $y = 0$.

(i) At $x = 0$, $y = 0$, we find $r = 2$, $s = 0$, $t = 0$, $rt - s^2 = 0$. This is therefore an example of the doubtful case. But, since z can at once be expressed in the form $z = (x - y^2)^2 - x^3$, we see that $z \leqslant 0$ on the parabola $y^2 = x$, whereas $z \geqslant 0$ on the axis $x = 0$. Consequently, $(0, 0, 0)$ is neither a maximum nor a minimum point.

(ii) At $x = \frac{2}{3}$, $y = 0$, we find $r = -2$, $s = 0$, $t = -\frac{8}{3}$, $rt - s^2 > 0$. The point $(\frac{2}{3}, 0, \frac{4}{27})$ is therefore a maximum point.

8.16 Quadratic forms in more than two variables

Before continuing the discussion of maxima and minima, we shall verify a few theorems concerning quadratic forms. The theorems are enunciated in the notation of determinants, as this notation is natural for more general cases (see e.g. Ferrar, *Finite Matrices*).

8.17　Three variables

Theorem. Necessary and sufficient conditions that the quadratic form

$$f(x, y, z) = ax^2 + by^2 + cz^2 + 2fyz + 2gzx + 2hxy \tag{20}$$

should be positive-definite, i.e. should be positive for every set of values (x, y, z) except only $(0, 0, 0)$, are

$$D_1 > 0, \quad D_2 > 0, \quad D_3 > 0, \tag{21}$$

where
$$D_1 = a, \quad D_2 = \begin{vmatrix} a & h \\ h & b \end{vmatrix}, \quad D_3 = \begin{vmatrix} a & h & g \\ h & b & f \\ g & f & c \end{vmatrix}. \tag{22}$$

Proof. Put $x = 1$, $y = 0$, $z = 0$. Since $f(x, y, z)$ is positive-definite, therefore $f(1, 0, 0) > 0$. But $f(1, 0, 0) = a$; hence $a > 0$.

Since $a > 0$, therefore $a \neq 0$ and by completing the square of the terms in x, we can put $f(x, y, z)$ in the form

$$f(x, y, z) \equiv a(x + \alpha y + \beta z)^2 + b_1 y^2 + 2f_1 yz + c_1 z^2. \tag{23}$$

Now put $z = 0$, $y = 1$, $x = -\alpha$. Since $f(x, y, z)$ is positive-definite, therefore $f(-\alpha, 1, 0) > 0$. Hence $b_1 > 0$ and we can put

$$f(x, y, z) \equiv a(x + \alpha y + \beta z)^2 + b_1(y + \gamma z)^2 + c_2 z^2 \tag{24}$$

and by putting $z = 1$, $y = -\gamma$, $x = \alpha\gamma - \beta$, we see that $c_2 > 0$. Consequently, in order that the quadratic form $f(x, y, z)$ should be positive-definite, a *necessary* condition is that it should be expressible in the form (24) with

$$a > 0, \quad b_1 > 0, \quad c_2 > 0. \tag{25}$$

This condition is also *sufficient*, as we see from (24) and (25), which show that $f(x, y, z)$ can only be zero if $x + \alpha y + \beta z = 0$, $y + \gamma z = 0$, $z = 0$, and therefore $z = 0$, $y = 0, x = 0$.

It remains to show that conditions (21) are equivalent to (25). Now, by equating coefficients of like terms in the identity (24), we find

$$a = a, \qquad b = a\alpha^2 + b_1, \quad c = a\beta^2 + b_1\gamma^2 + c_2,$$

$$f = a\alpha\beta + b_1\gamma, \quad g = a\beta, \qquad h = a\alpha$$

and hence, by applying the properties of determinants,

$$\begin{vmatrix} a & h \\ h & b \end{vmatrix} = ab_1, \quad \begin{vmatrix} a & h & g \\ h & b & f \\ g & f & c \end{vmatrix} = ab_1 c_2.$$

It follows that conditions (21) and (25) are equivalent.

Corollary. Necessary and sufficient conditions that $f(x, y, z)$ be negative-definite are

$$D_1 < 0, \quad D_2 > 0, \quad D_3 < 0, \tag{26}$$

since $-f(x, y, z)$ will then be positive-definite.

8.18 Generalization

The theorem of §8.17 can be extended to a quadratic form in n variables. Thus, if

$$f(x_1, x_2, \ldots, x_n) = a_{11}x_1^2 + a_{22}x_2^2 + \ldots + 2a_{12}x_1x_2 + \ldots \tag{27}$$

it can be proved that necessary and sufficient conditions for this form to be positive-definite are

$$D_1 > 0, \quad D_2 > 0, \quad D_3 > 0, \ldots, D_n > 0, \tag{28}$$

where $D_r = |a_{pq}|$ denotes the symmetric determinant of order r of which a_{pq} is the element belonging to the pth row and qth column, and $p = 1, 2, \ldots, r; q = 1, 2, \ldots, r$.

8.19 Maxima and minima. Three independent variables

Let $f(x, y, z)$ be a function of three independent variables x, y, z throughout a region R, including the boundary.

Definition. We say that $f(a, b, c)$ is a *minimum* of $f(x, y, z)$ if a number δ exists such that

$$f(a, b, c) < f(x, y, z), \quad 0 < (x-a)^2 + (y-b)^2 + (z-c)^2 < \delta^2 \tag{29}$$

that is, such that $f(a, b, c) < f(x, y, z)$ at every point (x, y, z) inside the sphere $(x-a)^2 + (y-b)^2 + (z-c)^2 = \delta^2$, except the centre.

For a maximum and for a point on the boundary, the same kind of modifications as in §8.5 must be made.

8.20

Suppose now that $f(x, y, z)$ has continuous partial derivatives of the second order throughout R, and that (a, b, c) is an interior point of R. Let suffixes 1, 2, 3 indicate partial derivatives wo x, y, z respectively.

Theorem. Necessary conditions that $f(a, b, c)$ should be a minimum (maximum) of $f(x, y, z)$ are that $x = a, y = b, z = c$ should be a solution of the simultaneous equations

$$f_x(x, y, z) = 0, \quad f_y(x, y, z) = 0, \quad f_z(x, y, z) = 0. \tag{30}$$

These conditions being satisfied, sufficient conditions are
(i) for a minimum,

$$D_1 > 0, \quad D_2 > 0, \quad D_3 > 0; \tag{31}$$

(ii) for a maximum,

$$D_1 < 0, \quad D_2 > 0, \quad D_3 < 0; \tag{32}$$

where
$$D_1 = f_{11}, \quad D_2 = \begin{vmatrix} f_{11} & f_{12} \\ f_{21} & f_{22} \end{vmatrix}, \quad D_3 = \begin{vmatrix} f_{11} & f_{12} & f_{13} \\ f_{21} & f_{22} & f_{23} \\ f_{31} & f_{32} & f_{33} \end{vmatrix} \tag{33}$$

the second derivatives in (33) being evaluated at (a, b, c).

Proof. Conditions (30) are evidently necessary because $f(x, y, z)$ must be a minimum (maximum) in each variable separately.

That conditions (31) for a minimum or (32) for a maximum are then sufficient follows by applying to the quadratic form

$$h^2 f_{11} + k^2 f_{22} + l^2 f_{33} + 2kl f_{23} + 2lh f_{31} + 2hk f_{12} \tag{34}$$

with the aid of the theorem of §8.17, the same kind of reasoning as that in §§8.11, 8.12, replacing $\cos\psi$, $\sin\psi$ by a set of direction cosines.

8.21　Sign of a conditioned quadratic form

Theorem. The necessary and sufficient condition that the form

$$f_2 = f(x, y) = ax^2 + 2hxy + by^2 \tag{35}$$

should be positive for all not-both-zero values of x and y that satisfy the linear condition

$$lx + my = 0 \tag{36}$$

is
$$\begin{vmatrix} a & h & l \\ h & b & m \\ l & m & 0 \end{vmatrix} < 0. \tag{37}$$

Proof. Suppose $lm \neq 0$, and let f_2 be divided by $lx + my$. If $2(px + qy)$ is the quotient and $b_1 y^2$ the remainder, then, identically in x and y,

$$ax^2 + 2hxy + by^2 \equiv 2(lx + my)(px + qy) + b_1 y^2 \tag{38}$$

and hence, by putting $x = -my/l$ $(l \neq 0)$, we see that for all x, y that satisfy (36)

$$ax^2 + 2hxy + by^2 \equiv b_1 y^2 \tag{39}$$

is an identity in y. It follows that the required condition is $b_1 > 0$.

To find b_1 in terms of a, h, b, l, m, we have, by equating coefficients of like terms in the identity (38),

$$a = 2lp, \quad h = lq + mp, \quad b = 2mq + b_1$$

and hence, by the elementary properties of determinants,

$$\begin{vmatrix} a & h & l \\ h & b & m \\ l & m & 0 \end{vmatrix} = -l^2 b_1$$

from which and $b_1 > 0$ follows (37).

8.22

Theorem. Necessary and sufficient conditions that the form

$$f_3 = f(x, y, z) = ax^2 + by^2 + cz^2 + 2fyz + 2gzx + 2hxy \qquad (40)$$

should be positive for all not-all-zero values of x, y, z that satisfy the linear condition

$$lx + my + nz = 0 \quad (l \neq 0), \qquad (41)$$

are

$$\begin{vmatrix} a & h & l \\ h & b & m \\ l & m & 0 \end{vmatrix} < 0, \qquad \begin{vmatrix} a & h & g & l \\ h & b & f & m \\ g & f & c & n \\ l & m & n & 0 \end{vmatrix} < 0. \qquad (42)$$

Proof. Suppose $l \neq 0$ and let f_3 be divided by $lx + my + nz$. Let $2(px + qy + rz)$ be the quotient and $b_1 y^2 + 2f_1 yz + c_1 z^2$ the remainder. Then, identically in x, y, z,

$$f(x, y, z) \equiv 2(lx + my + nz)(px + qy + rz) + b_1 y^2 + 2f_1 yz + c_1 z^2 \qquad (43)$$

and hence, putting $x = -(my + nz)/l$, we see that for all x, y, z that satisfy (41)

$$f(x, y, z) \equiv b_1 y^2 + 2f_1 yz + c_1 z^2 \qquad (44)$$

is an identity in y and z. It follows (§8.8) that the required conditions are

$$b_1 > 0, \quad b_1 c_1 - f_1^2 > 0. \qquad (45)$$

To express these conditions in terms of the coefficients in (40) and (41), we equate coefficients of like terms in the identity (43), obtaining

$$a = 2lp, \qquad b = 2mq + b_1, \qquad c = 2nr + c_1,$$

$$f = mr + nq + f_1, \qquad g = lr + np, \qquad h = lq + mp.$$

Substituting these values of $a, b, \ldots$ in the determinants in (42), and expanding them by the elementary properties of determinants, we find that conditions (42) are equivalent to (45).

8.23

Theorem. The necessary and sufficient condition that the form

$$f_3 = f(x, y, z) = ax^2 + by^2 + cz^2 + 2fyz + 2gzx + 2hxy \qquad (46)$$

should be positive for all not-all-zero values of x, y, z that satisfy the two linear conditions

$$lx + my + nz = 0, \quad l'x + m'y + n'z = 0 \tag{47}$$

is

$$\begin{vmatrix} a & h & g & l & l' \\ h & b & f & m & m' \\ g & f & c & n & n' \\ l & m & n & 0 & 0 \\ l' & m' & n' & 0 & 0 \end{vmatrix} > 0. \tag{48}$$

Proof. We assume that none of $lm' - l'm, mn' - m'n, nl' - n'l$ is zero. If, for instance, $lm' - l'm$ were zero, and $mn' - m'n$ not zero, then conditions (47) would be equivalent to $lx + my = 0$, $z = 0$, and with $z = 0$ the form f_3 would reduce to the form f_2 of §8.21.

Then l, m, l', m' cannot all be zero, since $lm' - l'm \neq 0$; suppose $l \neq 0$. Then we can put, identically in x, y, z,

$$f(x, y, z) \equiv 2(lx + my + nz)(px + qy + rz)$$
$$+ 2(l'x + m'y + n'z)(q'y + r'z) + c_2 z^2 \tag{49}$$

where, by equating coefficients of like terms,

$$a = 2lp, \quad b = 2(mq + m'q'), \quad c = 2(nr + n'r') + c_2, \quad \text{(i), (ii), (iii)}$$
$$f = mr + m'r' + nq + n'q', \tag{iv}$$
$$g = np + lr + l'r', \tag{v}$$
$$h = lq + l'q' + mp. \tag{vi}$$

For these equations determine the six unknowns p, q, r, q', r', c_2; thus (i) gives p, since $l \neq 0$; then (ii), (vi) give q, q', since $lm' - l'm \neq 0$; then (iv), (v) give r, r', since $lm' - l'm \neq 0$; and (iii) gives c_2.

It then follows from (49) that the necessary and sufficient condition that f_3 be positive for all not-all-zero x, y, z satisfying (47) is $c_2 > 0$. But by substituting the values of $a, b, c, \ldots$ given by (i), (ii), (iii), $\ldots$ in the determinant on the L.H.S. of (48), and expanding the resulting determinant, we find the expansion to be $(lm' - l'm)^2 c_2$. Hence and from $c_2 > 0$ follows (48).

8.24   Conditioned maxima and minima

We now consider a few cases of maxima and minima of a function of two or more variables when the variables are subject to one or more conditions. The simplest case is that of a function $f(x, y)$ of two variables connected by an equation $\phi(x, y) = 0$.

We assume that (a, b) is an interior point of a region in which the functions f and ϕ are defined, and that both functions have continuous partial derivatives of the second order throughout the region.

Let $x = a + h$, $y = b + k$; and put

$$\delta f = f(x, y) - f(a, b) = f(a + h, b + k) - f(a, b). \tag{50}$$

Since $\phi(x, y) = 0$,

$$0 = \phi(x, y) - \phi(a, b) = \phi(a + h, b + k) - \phi(a, b). \tag{51}$$

By Taylor's theorem, there is a neighbourhood of (a, b) in which we can put, as convenient,

$$\delta f = h(f_1 + \xi) + k(f_2 + \eta), \tag{52}$$

$$0 = h(\phi_1 + \xi_1) + k(\phi_2 + \eta_1), \tag{53}$$

or $\quad \delta f = hf_1 + kf_2 + \tfrac{1}{2}\{h^2(f_{11} + \alpha) + 2hk(f_{12} + \beta) + k^2(f_{22} + \gamma)\}, \tag{54}$

$$0 = h\phi_1 + k\phi_2 + \tfrac{1}{2}\{h^2(\phi_{11} + \alpha_1) + 2hk(\phi_{12} + \beta_1) + k^2(\phi_{22} + \gamma_1)\}, \tag{55}$$

where $f_1, \phi_1, f_{11}, \phi_{11}, \ldots$ denote partial derivatives evaluated at (a, b), and $\xi, \eta, \ldots, \alpha, \beta, \gamma, \ldots$ are functions of h, k which all vanish as $(h, k) \rightarrow (0, 0)$.

8.25

Theorem. A necessary condition that $f(a, b)$ should be a minimum of $f(x, y)$ subject to the condition $\phi(x, y) = 0$ is that $x = a$, $y = b$ should be a solution of the simultaneous equations

$$f_x/\phi_x = f_y/\phi_y, \quad \phi(x, y) = 0. \tag{56}$$

This necessary condition being satisfied, if we put

$$f_1/\phi_1 = f_2/\phi_2 = \lambda, \tag{57}$$

a sufficient condition is

$$\begin{vmatrix} f_{11} - \lambda\phi_{11} & f_{12} - \lambda\phi_{12} & \phi_1 \\ f_{12} - \lambda\phi_{12} & f_{22} - \lambda\phi_{22} & \phi_2 \\ \phi_1 & \phi_2 & 0 \end{vmatrix} < 0, \tag{58}$$

where $f_1, \phi_1, \ldots$ denote the values of the partial derivatives $f_x, \phi_x, \ldots$ at (a, b).

Proof. We assume that neither ϕ_x nor ϕ_y is identically zero. If, for instance, $\phi_y \equiv 0$, then ϕ would be a function of x only; if $\phi_x \equiv 0$, then ϕ would be a function of y only.

Suppose $\phi_1 = \phi_x(a, b) \neq 0$ and put $\lambda = f_1/\phi_1$. Then, after multiplying (53) by λ and subtracting from (52), we see that for all h, k that satisfy (53), δf can be put in the form

$$\delta f = h(\xi - \lambda \xi_1) + k(f_2 - \lambda \phi_2 + \eta - \lambda \eta_1)$$
$$= h\xi_2 + k(f_2 - \lambda \phi_2 + \eta_2), \qquad (59)$$

where ξ_2, η_2 vanish with (h, k). It follows that when h, k are small enough (in particular, when $h = 0$ and k is small enough) the sign of δf changes with the sign of k unless $f_2 - \lambda \phi_2 = 0$. Hence, in order that $\delta f > 0$ for all small enough h, k, a necessary condition is $f_2 - \lambda \phi_2 = 0$. Consequently, with the convention that $f_2 = 0$ in case $\phi_2 = 0$, we must have

$$f_2/\phi_2 = \lambda = f_1/\phi_1 \qquad (60)$$

which establishes (57).

It remains to prove that (58) is sufficient for a minimum.

By multiplying (55) by λ and subtracting from (54), having regard to (60), we can now put δf in the form

$$\delta f = \tfrac{1}{2} \{ h^2(f_{11} - \lambda \phi_{11} + \alpha_2) + 2hk(f_{12} - \lambda \phi_{12} + \beta_2) + k^2(f_{22} - \lambda \phi_{22} + \gamma_2) \} \qquad (61)$$

and from (53) we can put

$$h = -k(\phi_2/\phi_1 + \zeta), \qquad (62)$$

where α_2, β_2, γ_2 and ζ all vanish with (h, k). Substituting this value of h in (61), we see that δf can be put in the form (cf. §8.21)

$$\delta f = k^2(b_1 + \epsilon), \qquad (63)$$

where ϵ vanishes with (h, k). It follows that the sufficient condition for $\delta f > 0$ is $b_1 > 0$. Here b_1 is independent of h, k and is evidently unaffected by $\alpha_2, \beta_2, \gamma_2$ and ζ. We conclude that $\delta f > 0$ provided that the form

$$h^2(f_{11} - \lambda \phi_{11}) + 2hk(f_{12} - \lambda \phi_{12}) + k^2(f_{22} - \lambda \phi_{22}) \qquad (64)$$

is positive under the linear condition

$$h\phi_1 + k\phi_2 = 0 \qquad (65)$$

from which follows (58), by the theorem of §8.21.

Corollary. If $f(a, b)$ is a maximum of $f(x, y)$ under the condition $\phi(x, y) = 0$, the sign of inequality in (58) must be reversed.

8.26   Lagrange's multiplier

It may be noted that if we put

$$F(x, y, \lambda) = f(x, y) - \lambda \phi(x, y), \qquad (66)$$

then the equations that determine the possible values of a, b are the same as those we obtain by treating $F(x, y, \lambda)$ as a function of three independent variables x, y, λ. Thus, corresponding to (30), we find from (66)

$$F_x \equiv f_x - \lambda\phi_x = 0, \quad F_y \equiv f_y - \lambda\phi_y = 0, \quad F_\lambda \equiv -\phi(x,y) = 0. \quad (67)$$

Moreover, the determinant in (58) is the same as D_3 in (33) when f is replaced by F in (33).

Used in this way, λ is called a *Lagrange's multiplier*.

Example

Find the maximum and minimum values of the function

$$f = y^2 - 8x + 17$$

subject to the condition

$$\phi \equiv x^2 + y^2 - 9 = 0.$$

Putting $\quad F(x, y, \lambda) = y^2 - 8x + 17 - \lambda(x^2 + y^2 - 9)$,

the equations $F_x = 0, F_y = 0, F_\lambda = 0$ are

$$-8 - 2\lambda x = 0, \quad 2(1 - \lambda)y = 0, \quad -(x^2 + y^2 - 9) = 0$$

which have the solutions

    (i) $x = -4, y$ imaginary, $\lambda = 1$;

    (ii) $x = 3, y = 0, \lambda = -\frac{4}{3}$;   (iii) $x = -3, y = 0, \lambda = \frac{4}{3}$.

Then, by putting

$$D = \begin{vmatrix} F_{xx} & F_{xy} & \phi_x \\ F_{xy} & F_{yy} & \phi_y \\ \phi_x & \phi_y & 0 \end{vmatrix} = \begin{vmatrix} -2\lambda & 0 & 2x \\ 0 & 2(1-\lambda) & 2y \\ 2x & 2y & 0 \end{vmatrix} = 8\lambda y^2 - 8(1-\lambda)x^2$$

we see that $D < 0$ in case (ii), $D > 0$ in case (iii). Hence, by the theorem of §8.25, f has a minimum value (-7) for $x = 3, y = 0$;

a maximum value (41) for $x = -3, y = 0$.

Note that eliminating y and treating f as a function of x only does not help. This gives $f = 26 - 8x - x^2 = 42 - (x+4)^2$ which, as a function of x in an interval including $x = -4$, has a maximum at $x = -4$, but this value of x lies outside the range $(-3 \leqslant x \leqslant 3)$ to which x is limited by the condition $x^2 + y^2 = 9$.

8.27   Three variables, one condition

We assume that $f(x, y, z)$ and $\phi(x, y, z)$ are functions having continuous partial derivatives of the second order in a region of which (a, b, c) is an interior point.

Theorem. A necessary condition that $f(a, b, c)$ should be a minimum of $f(x, y, z)$ subject to the condition $\phi(x, y, z) = 0$ is that $x = a$, $y = b$, $z = c$ should be a solution of the simultaneous equations

$$f_x/\phi_x = f_y/\phi_y = f_z/\phi_z, \quad \phi(x, y, z) = 0. \tag{68}$$

Such a solution having been found, put

$$f_1/\phi_1 = f_2/\phi_2 = f_3/\phi_3 = \lambda, \tag{69}$$

where $f_1, \phi_1, \ldots$ denote values at (a, b, c). Also put

$$F(x, y, z, \lambda) = f(x, y, z) - \lambda\phi(x, y, z), \tag{70}$$

$$D = \begin{vmatrix} F_{xx} & F_{xy} & F_{xz} & \phi_x \\ F_{xy} & F_{yy} & F_{yz} & \phi_y \\ F_{xz} & F_{yz} & F_{zz} & \phi_z \\ \phi_x & \phi_y & \phi_z & 0 \end{vmatrix}. \tag{71}$$

Then sufficient conditions for $f(a, b, c)$ to be a minimum are that D should be negative and that one of the first three principal minors should also be negative when $x = a$, $y = b$, $z = c$.

To begin the proof, we suppose that $\phi_1 \neq 0$; then continue on the lines of §8.25 making use of §8.22.

Corollary. Sufficient conditions for a maximum are that D should be negative and one of its first three principal minors positive.

Note that the equations satisfied by a, b, c are obtainable from (70) by equating to zero the partial derivatives of F wo x, y, z, λ (cf. §8.26).

Example

Prove that the least value of the sum of the squares of the perpendiculars to the sides of a triangle from any point is

$$4S^2/(a^2 + b^2 + c^2),$$

where S is the area of the triangle and a, b, c are the lengths of its sides.

Let x, y, z be the lengths of the perpendiculars from any point P, and put

$$f(x, y, z) = x^2 + y^2 + z^2.$$

Evidently when $f(x, y, z)$ is a minimum, P will be in the plane of and inside the triangle, and then $ax + by + cz = 2S$; so put

$$\phi(x, y, z) = ax + by + cz - 2S = 0$$

and consider the function

$$F(x, y, z, \lambda) = x^2 + y^2 + z^2 - \lambda(ax + by + cz - 2S).$$

The equations to be satisfied at a point where f is a minimum (or maximum) are $F_x = 0$, $F_y = 0$, $F_z = 0$, $F_\lambda = 0$; that is, $2x - \lambda a = 0$, $2y - \lambda b = 0$, $2z - \lambda c = 0$, $ax + by + cz - 2S = 0$, from which we find

$$\frac{x}{a} = \frac{y}{b} = \frac{z}{c} = \frac{2S}{a^2 + b^2 + c^2}$$

and at this point the value of $f(x, y, z)$ is $4S^2/(a^2 + b^2 + c^2)$.

The determinant D of (71) becomes

$$D = \begin{vmatrix} 2 & 0 & 0 & a \\ 0 & 2 & 0 & b \\ 0 & 0 & 2 & c \\ a & b & c & 0 \end{vmatrix} = -4(a^2 + b^2 + c^2) < 0.$$

Since $D < 0$, the value of f just found is either a minimum or a maximum. Since the first three principal minors are all negative, the value is actually a minimum.

Note. This is a case in which it is practically evident that the value found is a minimum without examining the determinant D. There are other cases in which it is intuitive that a known stationary point is a maximum or minimum or neither, e.g. when the function concerned is a simple one or when there is a simple physical meaning. None the less, as we have seen at the end of the example in §8.26, what appears at first sight to be an obvious step may be misleading.

8.28 Three variables, two conditions

The following theorem will be stated without proof. The proof would follow the same lines as the proofs of the preceding theorems, the condition for sufficiency depending on the theorem of §8.23.

For the general case of a function of n variables with m conditions ($m < n$), see e.g. Hancock, *Theory of Maxima and Minima*, Dover Publications.

Theorem. A necessary condition that $f(a, b, c)$ should be a minimum (maximum) of $f(x, y, z)$, subject to the two conditions

$$\phi(x, y, z) = 0, \quad \psi(x, y, z) = 0,$$

is that a, b, c should be values of x, y, z in a solution of the five simultaneous equations

$$\left. \begin{array}{l} f_x - \lambda \phi_x - \mu \psi_x = 0, \ f_y - \lambda \phi_y - \mu \psi_y = 0, \ f_z - \lambda \phi_z - \mu \psi_z = 0, \\ \phi(x, y, z) = 0, \quad \psi(x, y, z) = 0. \end{array} \right\} \quad (72)$$

[It is usually convenient to introduce Lagrangian multipliers λ, μ and to put

$$F(x,y,z,\lambda,\mu) \equiv f(x,y,z) - \lambda\phi(x,y,z) - \mu\psi(x,y,z). \qquad (73)$$

Equations (72) can then be written down by partial differentiation of the function F wo x, y, z, λ, μ; thus

$$F_x = 0, \quad F_y = 0, \quad F_z = 0, \quad F_\lambda = 0, \quad F_\mu = 0.] \qquad (74)$$

A sufficient condition is obtained by substituting $x = a$, $y = b$, $z = c$ and the corresponding values of λ, μ in the determinant

$$\begin{vmatrix} F_{xx} & F_{xy} & F_{xz} & \phi_x & \psi_x \\ F_{xy} & F_{yy} & F_{yz} & \phi_y & \psi_y \\ F_{xz} & F_{yz} & F_{zz} & \phi_z & \psi_z \\ \phi_x & \phi_y & \phi_z & 0 & 0 \\ \psi_x & \psi_y & \psi_z & 0 & 0 \end{vmatrix}, \qquad (75)$$

where F denotes the function defined by (73). If D denotes the value of the determinant after the substitution, then a sufficient condition for the corresponding value of f to be a minimum is $D > 0$, a maximum is $D < 0$.

Examples 8 A

(1) Show that the function

$$a \cos^2\theta + 2h \cos\theta \sin\theta + b \sin^2\theta$$

oscillates between the maximum and minimum values

$$\tfrac{1}{2}(a+b) \pm \tfrac{1}{2}\sqrt{\{(a-b)^2 + 4h^2\}}$$

both of which are positive if $a > 0$ and $ab - h^2 > 0$.

(2) Show that the function $(\tan\theta)^{\sin\theta - \cos\theta}$ has a minimum at $\theta = \tfrac{1}{4}\pi$.

(3) Show that at $\theta = 0$ the function $a \cos\theta + \theta \sin\theta$ has a minimum if $a < 2$, a maximum if $a \geqslant 2$.

(4) Show that the function $\operatorname{sech} x - \tanh x$ has a maximum value $\sqrt{2}$. Sketch the graph of the function.

(5) Sketch the graphs of the functions

$$\text{(i) } \sin(e^x), \quad \text{(ii) } \sin(\cosh x), \quad \text{(iii) } \cosh(\sin x).$$

State their maxima and minima.

(6) Show that the function
(i) $\cosh(\sinh x) + \cos(\sin x)$ has a minimum at $x = 0$,
(ii) $\cosh(\sin x) + \cos(\sinh x)$ has a maximum at $x = 0$.

(7) Show that $3a \sin\theta + \sin 3\theta$ is an odd function of θ, periodic with period 2π. Show that there are three turning values between 0 and π if $-1 < a < 3$. Sketch the graphs for $a = -2$, $a = 1$, $a = 4$.

(8) Find the maxima and minima of the function

$$\frac{\sin\theta}{1} + \frac{\sin 3\theta}{3} + \frac{\sin 5\theta}{5}$$

between $\theta = 0$ and $\theta = \pi$, and the corresponding values of θ.

(9) Show that the values of θ for the maxima and minima of the function

$$\frac{\sin\theta}{1} + \frac{\sin 3\theta}{3} + \frac{\sin 5\theta}{5} + \dots + \frac{\sin(2n-1)\theta}{2n-1}$$

in the interval from $\theta = 0$ to $\theta = \pi$ divide the interval into equal parts.

(10) Prove that, along the curve of intersection of the surface

$$z = (y - x^2)(y - 3x^2)$$

by any plane $y = mx$ through the axis of z, the point $(0, 0, 0)$ is a minimum point; but that along the curve of intersection by the cylinder $y = \lambda x^2 (1 < \lambda < 3)$, the point $(0, 0, 0)$ is a maximum point.

(11) If $f(\theta) = \sin^3\theta/(\theta^3\cos\theta)$ when $\theta \neq 0$, and $f(0) = 1$, show that $f(0)$ is a minimum.

(12) If $f''(x)$ is one-signed in the interval $[a, b]$, prove that the area between the curve $y = f(x)$ and the tangent at $x = \xi$ $(a \leqslant \xi \leqslant b)$, from $x = a$ to $x = b$, is a minimum when $\xi = \frac{1}{2}(a + b)$.

Examples 8 B

Examples 4–9 may be deferred if the reader is unfamiliar with the operation of differentiating under the integral sign (§ 9.27).

(1) Examine whether each of the following functions has a minimum or maximum at $(0, 0)$, or neither:

(i) xy,   (ii) $x^2 + xy + y^2$,   (iii) $x^2 + 2xy + y^2$,

(iv) $x^2 - y^2$,   (v) $2x^2 - 5xy + 2y^2$,   (vi) $3x^2 - 2xy + y^2$,

(vii) $x^6 + y^2$,   (viii) $x^6 + (x - y)^3$,   (ix) $(y + 2x)^2 + (x + y)^5$.

(2) Show that the stationary values of the function

$$(a'x^2 + 2h'xy + b'y^2)/(ax^2 + 2hxy + by^2)$$

are the roots of the quadratic equation

$$(ab - h^2)\lambda^2 - (ab' + a'b - 2hh')\lambda + a'b' - h'^2 = 0.$$

(3) If (x_r, y_r) $(r = 1, 2, \dots, n)$ are known pairs of values of x and y, show that the values of a and b which minimize the sum

$$(y_1 - ax_1 - b)^2 + (y_2 - ax_2 - b)^2 + \dots + (y_n - ax_n - b)^2$$

are given by

$$a/(\Sigma x_r y_r - n\bar{x}\bar{y}) = b/(\bar{y}\Sigma x_r^2 - \bar{x}\Sigma x_r y_r) = 1/(\Sigma x_r^2 - n\bar{x}^2),$$

where

$$\bar{x} = (\Sigma x_r)/n, \quad \bar{y} = (\Sigma y_r)/n.$$

(4) Show that $\frac{1}{180}$ is the minimum value of the integral

$$\int_0^1 (x^2 - ax - b)^2 \, dx.$$

(5) Show that the integral

$$\int_0^\pi (x - a\sin x - b\sin 2x)^2\,dx$$

has the minimum value $\pi(\frac{1}{3}\pi^2 - \frac{5}{2})$ when $a = 2$, $b = -1$.

(6) Show that the integral

$$\int_0^x (1 + at + bt^2)^2\,dt \quad (x > 0),$$

as a function a and b, has the minimum value $\frac{1}{9}x$ when $a = -4/x$, $b = 10/(3x^2)$.

(7) By finding the minimum value of the integral

$$\int_0^x (1 + t)\left(\frac{1}{1+t} - a - bt\right)^2\,dt \quad (x > 0),$$

considered as a function of a and b, show that

$$\log(1 + x) > (x + \tfrac{1}{2}x^2)/(1 + x + \tfrac{1}{6}x^2) \quad (x > 0).$$

(8) If λ, u, v, w are functions of t, and $\lambda > 0$, show that the values of a and b which minimize the integral

$$\int_0^x \lambda(u + av + bw)^2\,dt \quad (x > 0),$$

considered as a function of a and b, are given by

$$\int_0^x \lambda v(u + av + bw)\,dt = 0, \quad \int_0^x \lambda w(u + av + bw)\,dt = 0$$

and that the minimum value is then equal to

$$\int_0^x \lambda u(u + av + bw)\,dt.$$

(9) If u, v are fixed and positive, show that $\frac{1}{7}uv$ is the minimum value of the double integral

$$\int_{x=0}^{x=u} \int_{y=0}^{y=v} (1 + ax + by)^2\,dx\,dy.$$

Examples 8C

In examples 1–15, find the coordinates of the stationary points, and state whether each point is a maximum or minimum or neither.

(1) $z = 2x^3 + 6xy^2 - 9x^2 + 9y^2 + 10$.　　(2) $z = 1 - 3x^2 - 4xy - y^2 - x^3$.

(3) $z = 4y^2 - x^4 - 6x^2y^2 - y^4$.　　(4) $z = 2x^4 + 12x^2y^2 + 2y^4 - x^2 - 3y^2$.

(5) $z = 6ax^3y^2 - 3x^4y^2 - 2x^3y^3$.　　(6) $z = (ax + by + c)/(1 + x^2 + y^2)$.

(7) $z = (y^2 - 4ax)(x^2 - 4ay)$.　　(8) $z = (x + y)^2 \exp(1 - x^2 - y^2)$.

(9) $z = xy\log(x + y)$.　　(10) $z = 9y^2 - 6xy + x^2 + x^3$.

(11) $z = (x - y)^2(x^2 + y^2 - 3)$.　　(12) $z = x^2 + 3xy + y^2 + x^3 + y^3$.

(13) $z = x^2 + y^2 + (ax + by + c)^2$.　　(14) $z = (x + ay + b)^2 + 2(y + c)^2 + 3$.

(15) $z = ax^2 + 2hxy + by^2 + 2gx + 2fy + c$ $(ab - h^2 > 0)$.

(16) Show that the surface $z = 2a^2x^2 - x^4 - y^2$ resembles two hills with a pass between them. Sketch a few typical contour lines.

(17) Show that the surface $z = \sin^2 x - y^2$ resembles a chain of hills with passes between them. Sketch a few contour lines.

(18) Show that the surface $z = \cos^2 x \cos^2 y$ resembles a doubly infinite series of hills uniformly distributed over the xy plane, with their peaks above the points of intersection of the straight lines $x = m\pi, y = n\pi$, where m, n are integers.

(19) If $\qquad z = (3 + 8\cos x)\cos y + 10\sin x \sin y$,

show that z has a maximum value when x and y are acute angles such that $\cos x = \sin y = \frac{2}{3}$.

(20) Find the stationary points on the surface

$$z = a\sin x \cos y + b\cos x \sin y,$$

where a and b are constants, $a > b > 0$, $0 \leqslant x \leqslant \pi$, $0 \leqslant y \leqslant \pi$.
Examine the nature of the stationary points.

(21) Show that the function $(4ax - y^2)\exp(x + y + 1)$ has the maximum value $-4ae^{-a}$ if $a < 0$.

(22) Show that each of the functions $u = \cos x \cosh y$, $v = \sin x \sinh y$ has saddle points at $x = s\pi$, $y = 0$, where s is an integer.

(23) If $M = |\sin(x + iy)|$, show that M has a minimum at $x = \frac{1}{2}s\pi$, $y = 0$ if s is even, but a saddle point if s is odd.

(24) If $u + iv = (x + iy)^n$, show that u and v have neither maxima nor minima inside any circle $x^2 + y^2 = a^2$. Find their maxima and minima on the circumference, n being an integer.

(25) If u is a harmonic function of x and y, and v its conjugate, show that where $u_x = 0$, $u_y = 0$, and $u_{xx} \neq 0$, both u and v have saddle points.

(26) Prove that, if $f(x, y)$ is subharmonic (superharmonic) in a region R, it cannot have a maximum (minimum) at an interior point of R. [See the definitions in Examples 7 G, 6.]

(27) Prove that, if $f(x, y)$ is harmonic in a region R, it cannot have a maximum or a minimum at an interior point of R.

[*Proof.* In the notation of Examples 7 G, 7, let m be the least integer for which A_m and B_m are not both zero at an interior point (a, b). Then, approximately, at a neighbouring point (x, y)

$$f(x, y) = f(a, b) + (r^m/m!)(A_m \cos m\theta + B_m \sin m\theta)$$

$$= f(a, b) + (r^m/m!)\sqrt{(A_m^2 + B_m^2)}\cos(m\theta - \alpha),$$

where $\tan \alpha = B_m/A_m$. It follows that in any direction θ from the point (a, b) in which $\cos(m\theta - \alpha) > 0$, the function $f(x, y)$ increases, while if $\cos(m\theta - \alpha) < 0$, the function decreases. Consequently, $f(a, b)$ cannot be either a maximum or a minimum.]

(28) Show that the integral

$$\int_0^\pi x \sin x \left(\frac{1}{x} - a - bx\right)^2 dx$$

has a minimum for $a = \frac{1}{2}\pi$, $b = -\frac{1}{2}$. Deduce that

$$\int_0^\pi \frac{\sin x}{x} dx > \frac{\pi}{2}.$$

(29) Show that on the surface

$$z = (1 + x^2 + y^2 - x - xy)/(1 + 2x^2 + 3y^2)$$

there is a maximum, a minimum, and a saddle point, and find their coordinates.

Examples 8 D

(1) A, B, C are three fixed points and P is any point in space. Prove that the sum $AP^2 + BP^2 + CP^2$ is a minimum when P coincides with the centroid of the points A, B, C.

Prove that the same is true for any number of fixed points $A, B, C, \ldots$.

(2) If $f = x^3 - 3xy^2 + 6y$ and $\phi \equiv 3x^2y - y^3 - 2x = 0$, prove that f is a maximum or minimum when $x = y = \pm 1$.

(3) Prove that the area enclosed by a convex four-bar linkage is a maximum when the linkage is cyclic.

(4) If $\tan x + \tan y = 2$, prove that $\sin^n x + \sin^n y$, where n is a positive integer, has a maximum at $x = y = \frac{1}{4}\pi$ when $n < 4$ but a minimum when $n > 4$.

(5) If $r^2 = x^2 + y^2$ and $ax^2 + 2hxy + by^2 = 1$, find the equation satisfied by the maximum and minimum values of r^2 when $ab - h^2 > 0$. Examine also the case $ab - h^2 < 0$.

(6) Prove that the maximum value of $|\sin z|$ where $z = x + iy$, $x^2 + y^2 = a^2$ and $a > 0$, is $\sinh a$, and that the minimum value is $|\sin a|$.

(7) If the volume of a rectangular parallelepiped is fixed, prove that its surface area is a minimum when it is a cube. Also prove that the sum of the lengths of its edges is then a minimum.

(8) Prove that the maximum area of a triangle inscribed in a circle of radius a is $(3\sqrt{3}) a^2/4$. Find the maximum area of a triangle inscribed in an ellipse of semi-axes a, b.

(9) If x, y, z are positive and $x^2/a^2 + y^2/b^2 + z^2/c^2 = 1$, show that the maximum value of $x + y + z$ is $\sqrt{(a^2 + b^2 + c^2)}$.

(10) If $a^2/x + b^2/y + c^2/z = 1$, show that $x + y + z$ has a minimum value $(a + b + c)^2$, if a, b, c are positive.

(11) Consider the maximum and/or minimum values of $x + y + z$,
  (i) subject to the condition $yz + zx + xy = 3a^2$,
  (ii) subject to the condition $xyz = a^3$,
  (iii) subject to both conditions.

(12) Consider the maximum or minimum values of
$$(x-2)^2+(y-4)^2+(z-4)^2$$
subject to the condition $x^2+y^2+z^2=1$.

(13) Show that the function $f=xyz$, subject to the condition
$$x+y+z=3a>0,$$
has a maximum a^3, but no other maximum or minimum.

(14) Show that the function $f=x^2+y^2+z^2$, subject to the condition
$$lx+my+nz=p,$$
has a minimum value $p^2/(l^2+m^2+n^2)$.

(15) Consider the maximum and minimum values of the function $x^2+y^2+z^2$, subject to the condition
$$ax^2+by^2+cz^2+2fyz+2gzx+2hxy=1$$
with
$$a>0, \quad ab-h^2>0, \quad \begin{vmatrix} a & h & g \\ h & b & f \\ g & f & c \end{vmatrix} > 0.$$

(16) Consider the maximum and minimum values of the function $x^2+y^2+z^2$, subject to the two conditions:
$$x^2/a^2+y^2/b^2+z^2/c^2=1, \quad lx+my+nz=0.$$

(17) Show that the stationary values of
$$f=ax_1x_2+bx_1y_2+cx_2y_1+dy_1y_2$$
where x_1,x_2,y_1,y_2 are connected by the two equations
$$x_1^2+y_1^2=1, \quad x_2^2+y_2^2=1,$$
satisfy the equation
$$\begin{vmatrix} -f & 0 & a & b \\ 0 & -f & c & d \\ a & c & -f & 0 \\ b & d & 0 & -f \end{vmatrix} = 0.$$

(18) If r is the distance between two points P and Q situated respectively on the curves
$$y^2=4ax, \quad (x-3a)^2+y^2=12a^2,$$
find the positions of P and Q when r is a maximum or a minimum. Interpret the results geometrically.

9

INTEGRATION

9.1 The definite integral

Let the function $f(x)$ be bounded over the closed finite interval $[a,b], a < b$.

Given $\delta > 0$, let the interval be divided into n subintervals each of length not greater than δ, so that $n\delta \geqslant b-a$. Let the points of division be $x_1, x_2, ..., x_{n-1}$, so that if we put $x_0 = a$, $x_n = b$, then

$$a = x_0 < x_1 < x_2 < \ ... \ < x_{n-1} < x_n = b. \tag{1}$$

Such a division of an interval into subintervals will be called a *dissection* $D(\delta)$, or briefly D, of the interval.

Let δ_r be the length of the rth subinterval $[x_{r-1}, x_r]$ and let ξ_r be any value of x in this subinterval, so that

$$0 < \delta_r = x_r - x_{r-1} \leqslant \delta, \quad x_{r-1} \leqslant \xi_r \leqslant x_r \quad (r = 1, 2, ..., n). \tag{2}$$

Also, put
$$\sigma(D) = \sum_{r=1}^{n} f(\xi_r)\, \delta_r. \tag{3}$$

Definition. The definite integral of $f(x)\, dx$ over the interval $[a, b]$ is defined by

$$\int_a^b f(x)\, dx = \lim_{\delta \to 0} \sum_{r=1}^{n} f(\xi_r)\, \delta_r = \lim_{\delta \to 0} \sigma(D) \tag{4}$$

provided that the limit exists.

The function $f(x)$ is then said to be *integrable* over the interval (or 'integrable in the sense of Riemann' or 'Riemann-integrable' if there is risk of confusion with other definitions of integration.)

Note that $n \to \infty$ as $\delta \to 0$, since $n\delta \geqslant b-a$; that the points of division may be any $n-1$ distinct points subject only to $0 < \delta_r \leqslant \delta$; and that ξ_r may be any point in $[x_{r-1}, x_r]$.

The variable x is called the *variable of integration*. The value of the integral is independent of x: any other letter could be used; the variable of integration is therefore sometimes called a *dummy* variable. The interval from a to b is called the *range of integration*; a and b are called the *limits* of integration, the lower and upper limit respectively. The function $f(x)$ is called the *integrand*.

Definition. If $a > b$ we define the integral by

$$\int_a^b f(x)\,dx = -\int_b^a f(x)\,dx. \tag{5}$$

One consequence of this definition is (put $b = a$)

$$\int_a^a f(x)\,dx = 0. \tag{6}$$

9.2   Properties of the integral

The following theorems follow at once from the definition.

Theorem 1. If $f(x)$ is integrable and A is constant, then $Af(x)$ is integrable and

$$\int_a^b Af(x)\,dx = A\int_a^b f(x)\,dx. \tag{7}$$

Proof. For every dissection $D(\delta)$, since A is independent of $D(\delta)$,

$$\Sigma Af(\xi_r)\,\delta_r = A\Sigma f(\xi_r)\,\delta_r$$

from which and (4) follows (7) when $\delta \to 0$, since $f(x)$ is integrable.

Theorem 2. If $f(x)$ and $g(x)$ are integrable, then their sum is integrable and

$$\int_a^b \{f(x)+g(x)\}\,dx = \int_a^b f(x)\,dx + \int_a^b g(x)\,dx. \tag{8}$$

Proof. In every dissection $D(\delta)$

$$\Sigma\{f(\xi_r)+g(\xi_r)\}\,\delta_r = \Sigma f(\xi_r)\,\delta_r + \Sigma g(\xi_r)\,\delta_r$$

from which follows (8), when $\delta \to 0$, since $f(x)$ and $g(x)$ are integrable.

Corollary. It follows by induction that the sum of a finite number of integrable functions is integrable, and that the integral of their sum is equal to the sum of their integrals.

Theorem 3. If (i) $f(x)$ is integrable in $[a,b]$, (ii) $f(x)$ is bounded, so that we may put $m \leqslant f(x) \leqslant M$, where m, M denote lower and upper bounds of $f(x)$, (iii) $a \leqslant b$; then

$$m(b-a) \leqslant \int_a^b f(x)\,dx \leqslant M(b-a). \tag{9}$$

Proof. In every dissection $D(\delta)$, since $0 < \delta_r$,

$$m\delta_r \leqslant f(\xi_r)\,\delta_r \leqslant M\delta_r$$

and by summing from $r = 1$ to $r = n$, since m and M are independent of $D(\delta)$,

$$m(b-a) \leqslant \Sigma f(\xi_r)\,\delta_r \leqslant M(b-a)$$

from which and (4) follows (9) when $\delta \to 0$, since $f(x)$ is integrable.

Corollary 1. If $f(x) \geqslant 0$, then

$$\int_a^b f(x)\, dx \geqslant 0. \qquad (10)$$

This follows from (9) since, if $f(x) \geqslant 0$, we can put $m = 0$.

Corollary 2. If $f(x)$ and $g(x)$ are integrable and $f(x) \geqslant g(x)$, then

$$\int_a^b f(x)\, dx \geqslant \int_a^b g(x)\, dx.$$

This follows from Corollary 1 by considering that $f(x) - g(x) \geqslant 0$ and using (8).

Theorem 4. If $f(x)$ is integrable in $[a, b]$ and if there exists a differentiable function $\phi(x)$ whose derivative is $f(x)$ throughout $[a, b]$, then, in a familiar notation,

$$\int_a^b f(x)\, dx = [\phi(x)]_a^b = \phi(b) - \phi(a). \qquad (11)$$

This theorem is sometimes called *the fundamental theorem of the integral calculus*.

Proof. Let $D(\delta)$ be a dissection of $[a, b]$. Since $\phi(x)$ is differentiable, it is continuous throughout $[a, b]$. We may therefore apply the mean value theorem ($\S 2.14$) to each subinterval. The rth subinterval gives

$$\phi(x_r) - \phi(x_{r-1}) = \phi'(\xi_r)\,(x_r - x_{r-1}) = f(\xi_r)\,\delta_r,$$

where ξ_r belongs to the subinterval. If we now sum from $r = 1$ to $r = n$, we get

$$\phi(b) - \phi(a) = \Sigma f(\xi_r)\,\delta_r$$

which gives (11) when $\delta \to 0$, since $f(x)$ is integrable and, by definition, the limit of the sum on the right is the integral on the L.H.S. of (11).

The reader will no doubt, in a first course on Integral Calculus, have used these theorems frequently to evaluate definite integrals, usually when $f(x)$ was continuous. But they assume that $f(x)$ and $g(x)$ are integrable, and from our present standpoint of more careful reasoning, before applying them we ought to prove that integrable functions exist, and in particular that continuous functions are integrable. To this end, a useful condition of integrability is obtained by considering the *upper sum* and the *lower sum* which will now be defined.

9.3   The upper and lower sums

Definition. Let m_r be the greatest lower bound (g.l.b.) and M_r the least upper bound (l.u.b.) of $f(x)$ in the rth subinterval of a dissection $D(\delta)$ of the interval $[a, b]$, $a < b$.

The *lower sum* $s(D)$ and the *upper sum* $S(D)$ for the function $f(x)$ over the interval $[a, b]$, due to the dissection $D(\delta)$, are defined by

$$s(D) = \Sigma m_r \delta_r, \quad S(D) = \Sigma M_r \delta_r. \qquad (12)$$

Evidently, $s(D) \leqslant S(D)$, since $m_r \leqslant M_r$ $(r = 1, 2, \ldots, n)$.

Moreover, if m is the g.l.b. and M the l.u.b. of $f(x)$ over the whole interval $[a, b]$, then

$$m(b-a) \leqslant s(D) \leqslant S(D) \leqslant M(b-a) \qquad (13)$$

since $m \leqslant m_r$ and $M_r \leqslant M$ $(r = 1, 2, \ldots, n)$. The sign of equality is possible throughout (13) only if $f(x)$ is constant.

9.4   A necessary and sufficient condition of integrability

Our aim will now be to show that a necessary and sufficient condition for $f(x)$ to be integrable over $[a, b]$ is that the difference between the upper and lower sums should tend to zero as $\delta \to 0$; that is,

$$S(D) - s(D) \to 0 \quad \text{as} \quad \delta \to 0. \qquad (14)$$

We first prove that this condition is *necessary*: that is, given that the integral exists, the condition must be satisfied.

Proof. Let I denote the integral, which exists (given). Then, by (4), whatever the choice of $\xi_1, \xi_2, \ldots$, when $\delta \to 0$

$$\lim \sigma(D) = I. \qquad (15)$$

Put $\epsilon = \delta/(b-a)$. By §1.16, we can choose the ξ's so that

$$f(\xi_r) > M_r - \epsilon \ (r = 1, 2, \ldots, n)$$

and hence $M_r < f(\xi_r) + \epsilon$, and by multiplying by $\delta_r > 0$,

$$M_r \delta_r < f(\xi_r) \delta_r + \epsilon \delta_r.$$

By summing with respect to r and using σ_1 to refer to this choice of the ξ's we find

$$S(D) < \sigma_1(D) + \epsilon \Sigma \delta_r = \sigma_1(D) + \epsilon(b-a),$$

that is

$$S(D) < \sigma_1(D) + \delta. \qquad (16)$$

Again by §1.16, we can also choose the ξ's so that $f(\xi_r) < m_r + \epsilon$ and hence $m_r > f(\xi_r) - \epsilon$, and by multiplying by $\delta_r > 0$ and summing, we find

$$s(D) > \sigma_2(D) - \delta, \qquad (17)$$

where σ_2 refers to this new choice of the ξ's.

From (16) and (17), and since $S(D) \geqslant s(D)$,

$$0 \leqslant S(D) - s(D) < \sigma_1(D) - \sigma_2(D) + 2\delta$$

and hence, by (15), when $\delta \to 0$,

$$0 \leqslant \lim \{S(D) - s(D)\} \leqslant I - I + 0 = 0,$$

from which follows (14).

It remains to prove that condition (14) is *sufficient* to ensure that $f(x)$ be integrable; that is, to show that if the condition is satisfied, then the limit in (4) exists. This requires further properties of the upper and lower sums.

9.5

The upper and lower sums: further discussion.

Definition. Given a dissection D of an interval, if we form a new dissection by adding more points of division to those of D, the new dissection is called a *refinement* of D.

A refinement of D which consists in adding only one more point of division will be called a *simple refinement* of D.

Theorem 1. The lower sum $s(D)$ cannot be diminished and the upper sum $S(D)$ cannot be increased by any refinement of D. That is, if D_1 is any refinement of D, then

$$s(D) \leqslant s(D_1) \leqslant S(D_1) \leqslant S(D). \tag{18}$$

Proof. Evidently any refinement of D can be arrived at by a succession of simple refinements. It will therefore be sufficient to suppose D_1 to be a simple refinement of D.

Let D_1 be a simple refinement of D due to one additional point of division x_r' dividing the rth subinterval into two parts. Let the g.l.b. of $f(x)$ be m_r' in the first of these parts of length δ_r', and m_r'' in the second part of length δ_r''.

By §9.3, from the rth subinterval of D the contribution to $s(D)$ is $m_r \delta_r$, while that to $s(D_1)$ is $m_r' \delta_r' + m_r'' \delta_r''$. But $m_r' \geqslant m_r$ and $m_r'' \geqslant m_r$; hence

$$m_r' \delta_r' + m_r'' \delta_r'' \geqslant m_r(\delta_r' + \delta_r'') = m_r \delta_r. \tag{19}$$

Since the contributions from the other subintervals of D are unchanged, it follows that $s(D_1) \geqslant s(D)$, proving that the lower sum is not diminished.

Similarly, it may be proved that the upper sum is not increased.

Theorem 2. Every lower sum is less than every upper sum (unless $f(x) = \text{const.}$).

Let D, D' be any two dissections. We shall prove that $s(D) < S(D')$, except when $f(x) = \text{const.}$ in which case $s(D) = S(D')$.

Proof. Let D_1 be the dissection whose points of division consist of all those belonging to D or D'. Then D_1 is a refinement of D and also of D'. Hence, by (18),

$$s(D) \leqslant s(D_1) \quad \text{and} \quad S(D_1) \leqslant S(D'); \tag{20}$$

but $s(D_1) \leqslant S(D_1)$ by (13), the sign of equality being possible only when $f(x) = \text{const.}$; it follows that $s(D) \leqslant S(D')$.

9.6

From (13) it follows that for all dissections D the lower and upper sums are bounded. Let j be the l.u.b. of $s(D)$ and J the g.l.b. of $S(D)$.

By means of § 9.5, Theorem 2, we can prove that $j \leqslant J$.

Proof. Suppose $j > J$ possible, and put $j - J = 2\epsilon > 0$.

By the definitions of j and J it follows (§1.16) that dissections D and D' exist such that

$$s(D) > j - \epsilon, \quad S(D') < J + \epsilon \tag{21}$$

and therefore such that

$$s(D) - S(D') > j - J - 2\epsilon = 2\epsilon - 2\epsilon = 0$$

contrary to Theorem 2. Thus $j > J$ is impossible.

9.7

We can now complete the proof begun in §9.4, by showing that the condition $S(D) - s(D) \to 0$ is sufficient to ensure that the limit of the sum $\sigma(D)$ should exist when $\delta \to 0$.

First, observe that $\sigma(D)$ lies between $s(D)$ and $S(D)$. For, by definition of m_r and M_r, $m_r \leqslant f(\xi_r) \leqslant M_r$, and by multiplying by $\delta_r > 0$ and summing from $r = 1$ to $r = n$,

$$s(D) \leqslant \sigma(D) \leqslant S(D). \tag{22}$$

Completion of proof. Putting s for $s(D)$ and S for $S(D)$, we have identically
$$S - s \equiv (S - J) + (J - j) + (j - s). \tag{23}$$

Now let $\delta \to 0$ and suppose that $S - s \to 0$. Then, since each of the three terms on the R.H.S. of (23) is positive or zero (§9.6), it follows that $S - J \to 0$, $j - s \to 0$, and $J - j = 0$ since J and j are independent of δ. Hence $S \to J$, $s \to j$, and $J = j$.

From (22) when $\delta \to 0$ now follows

$$j = \lim \sigma(D) = J; \tag{24}$$

thus, if $S - s \to 0$, then $\lim \sigma(D)$ exists, that is, $f(x)$ is integrable, and

$$j = \int_a^b f(x)\,dx = J. \tag{25}$$

This completes the proof.

9.8

As applications of the condition of integrability just discussed, two further properties of the definite integral will be proved.

Theorem 1. If $a \leqslant a' < b' \leqslant b$ and $f(x)$ is integrable over $[a, b]$, then $f(x)$ is integrable over $[a', b']$.

Proof. Let $D(\delta)$ be any dissection of $[a, b]$ in which $x = a'$, $x = b'$ are two of the points of division. Let the upper and lower sums of $f(x)$ be S, s over $[a, b]$ and S', s' over $[a', b']$.

Then $S - s = \Sigma (M_r - m_r) \delta_r$ is a sum of non-negative terms which include all the terms belonging to $S' - s'$ and more. Consequently,

$$0 \leqslant S' - s' \leqslant S - s.$$

But $S - s \to 0$ since $f(x)$ is integrable over $[a, b]$, by §9.4. It follows that $S' - s' \to 0$ and hence that $f(x)$ is integrable over $[a', b']$, by §9.7.

Theorem 2. If $a < c < b$ and $f(x)$ is integrable over $[a, b]$, then

$$\int_a^b f(x)\,dx = \int_a^c f(x)\,dx + \int_c^b f(x)\,dx. \tag{26}$$

Proof. Let $D(\delta)$ be any dissection of $[a, b]$ which has $x = c$ for one of its points of division. In the notation of §9.1 let $\sigma(D)$, $\sigma_1(D)$, $\sigma_2(D)$ refer to $[a, b]$, $[a, c]$, $[c, b]$ respectively. Then

$$\sigma(D) = \sigma_1(D) + \sigma_2(D). \tag{27}$$

But it is given that $f(x)$ is integrable over $[a, b]$ and by the last theorem it is therefore integrable over $[a, c]$ and $[c, b]$. Hence $\sigma(D)$, $\sigma_1(D)$, $\sigma_2(D)$ all tend to limits when $\delta \to 0$, and from (27) follows

$$\lim \sigma(D) = \lim \sigma_1(D) + \lim \sigma_2(D)$$

which is another form of (26).

Evidently, (26) could be generalized in the form

$$\int_a^b f(x)\,dx = \int_a^{c_1} f(x)\,dx + \int_{c_1}^{c_2} f(x)\,dx + \ldots + \int_{c_{n-1}}^b f(x)\,dx, \tag{28}$$

where $a \leqslant c_1 \leqslant c_2 \ldots \leqslant c_{n-1} \leqslant b$.

Moreover, recalling definitions (5) and (6) of §9.1, we could deduce that (26) holds good for any limits a, b, c, provided that $f(x)$ is integrable over any range involved. A corresponding statement applies to (28).

9.9 Integrable functions

The condition of integrability will now be used to prove that certain types of functions are integrable.

Theorem 1. *A continuous function is integrable.* More precisely, if $f(x)$ is continuous in the closed interval $[a, b]$, then $f(x)$ is integrable over the interval.

Proof. Suppose $\epsilon > 0$ given, and $a < b$. Put $\epsilon_1 = \epsilon/(b-a)$.

Since $f(x)$ is continuous in the interval $[a, b]$, by §1.20 a number δ_0 exists such that, for any dissection $D(\delta)$, $\delta < \delta_0$, the spread of $f(x)$ in every subinterval is less than ϵ_1. Thus δ_0 exists such that, for all $\delta < \delta_0$,

$$S - s = \Sigma(M_r - m_r)\,\delta_r < \epsilon_1 \Sigma \delta_r = \epsilon_1(b-a) = \epsilon.$$

It follows that $S - s \to 0$, and therefore $f(x)$ is integrable.

Theorem 2. *A piecemeal continuous function is integrable.* That is, if $f(x)$ is bounded in the interval $[a, b]$ and continuous except at a finite number of points, then $f(x)$ is integrable over the interval.

Proof. Suppose first that there is only one point of discontinuity, at $x = c$.

In any dissection $D(\delta)$ of the interval $[a, b]$ the point $x = c$ will belong to one subinterval or possibly to two; in either case the contribution to $S - s$ cannot exceed $2M\delta - 2m\delta$, where m, M are the closest bounds of $f(x)$ in $[a, b]$.

Now, given $\epsilon > 0$, let $\epsilon_1 = \epsilon/(b-a)$. Then, as in Theorem 1, δ_0 exists such that, for all $\delta < \delta_0$, the spread of $f(x)$ in every other subinterval, to the left and right of $x = c$, is less than ϵ_1. Consequently, the contributions from all these subintervals will be less than

$$\epsilon_1(b-a) = \epsilon.$$

Hence, for every $\delta < \delta_0$,

$$S - s < \epsilon + 2M\delta - 2m\delta$$

and if we take δ less than both δ_0 and ϵ,

$$S - s < (1 + 2M - 2m)\,\epsilon.$$

It follows that $S - s \to 0$, since M and m are independent of δ.

Evidently, if there were a finite number k of points of discontinuity it could be proved that δ_0 exists such that, for all δ less than both δ_0 and ϵ,

$$S - s < (1 + 2kM - 2km)\,\epsilon$$

and hence that $S - s \to 0$.

Theorem 3. *A bounded monotonic function is integrable*. That is, if $f(x)$ is bounded and monotonic in the interval $[a, b]$, then $f(x)$ is integrable over the interval.

Proof. It will be sufficient to suppose $f(x)$ to be increasing.

As always, for any dissection $D(\delta)$ of the interval, since $\delta_r \leqslant \delta$,

$$S - s = \Sigma(M_r - m_r)\,\delta_r \leqslant \delta\Sigma(M_r - m_r).$$

But, in this case, $m_r = f(x_{r-1})$, $M_r = f(x_r)$ and

$$\Sigma(M_r - m_r) = \Sigma\{f(x_r) - f(x_{r-1})\} = f(b) - f(a);$$

consequently,

$$S - s \leqslant \delta\{f(b) - f(a)\} \to 0 \quad \text{when} \quad \delta \to 0.$$

Theorem 4. *A bounded piecemeal monotonic function is integrable*. That is, if the interval $[a, b]$ can be divided into a finite number of parts in each of which $f(x)$ is bounded and monotonic, then $f(x)$ is integrable over the interval.

Proof. The proof follows the same lines as that of Theorem 2, the first step being to suppose there are only two parts, in one of which $f(x)$ is increasing, in the other decreasing.

9.10

A few examples will now be given of integrable functions of functions.

Let $f(x)$ be integrable over $[a, b]$. Let the closest bounds of $f(x)$ in a dissection $D(\delta)$ be m, M in $[a, b]$ and m_r, M_r in $[x_{r-1}, x_r]$.

Let the corresponding bounds of $|f(x)|$ be k, K and k_r, K_r.

Evidently, the spread of $|f(x)|$ in any interval cannot exceed that of $f(x)$, so that

$$K - k \leqslant M - m, \quad K_r - k_r \leqslant M_r - m_r. \tag{29}$$

Theorem 1. (i) The function $|f(x)|$ is integrable;

$$\text{(ii)} \quad \left| \int_a^b f(x)\,dx \right| \leqslant \int_a^b |f(x)|\,dx, \quad \text{if} \quad a < b. \tag{30}$$

Proof of (i). Let the upper and lower sums in a dissection $D(\delta)$ be S and s for $f(x)$, S_1 and s_1 for $|f(x)|$.

Since $f(x)$ is integrable, therefore $S - s \to 0$, by §9.4. Hence and by (29),

$$S_1 - s_1 = \Sigma\,(K_r - k_r)\,\delta_r \leqslant \Sigma\,(M_r - m_r)\,\delta_r = S - s \to 0; \tag{31}$$

consequently, $S_1 - s_1 \to 0$, which proves (i), by §9.7.

Proof of (ii). Since the modulus of a sum of terms cannot exceed the sum of their moduli, therefore, if $0 < \delta_r$,

$$\left| \Sigma f(\xi_r)\, \delta_r \right| \leqslant \Sigma |f(\xi_r)|\, \delta_r \tag{32}$$

which becomes (30) when $\delta \to 0$, since $f(x)$ is integrable (given) and $|f(x)|$ is integrable, by (i).

Corollary. If $-K < f(x) < K$ in $[a, b]$, then

$$\left| \int_a^b f(x)\, dx \right| < K|b-a|. \tag{33}$$

Note. The converse of (i) is not necessarily true; in other words, if $|f(x)|$ is integrable it does not necessarily follow that $f(x)$ is integrable. For instance, in the interval $[a, b]$, $a < b$, let $f(x) = c > 0$ if x is rational, $f(x) = -c$ if x is irrational. Then, for all x, $|f(x)| = c$ and is integrable, the integral from a to b being $c(b-a)$. But the closest bounds in any interval $[x_{r-1}, x_r]$ are $m_r = -c$ and $M_r = c$; consequently, for any dissection $D(\delta)$ of the interval $[a, b]$,

$$S - s = \Sigma(M_r - m_r)\,\delta_r = 2c\,\Sigma\delta_r = 2c(b-a) \neq 0$$

and so $f(x)$ is not integrable, by §9.4.

Theorem 2. The function $f^2(x)$, i.e. $\{f(x)\}^2$, is integrable.

Proof. The closest bounds of $f^2(x)$ are k^2, K^2 in $[a, b]$ and k_r^2, K_r^2 in $[x_{r-1}, x_r]$.

Let S_2 and s_2 be the upper and lower sums for $f^2(x)$ in a dissection $D(\delta)$. Then $S_2 - s_2 = \Sigma(K_r^2 - k_r^2)\,\delta_r$. But

$$K_r^2 - k_r^2 = (K_r + k_r)\,(K_r - k_r) \leqslant 2K(K_r - k_r)$$

and hence $\quad S_2 - s_2 \leqslant 2K\Sigma(K_r - k_r)\delta_r = 2K(S_1 - s_1),$

S_1, s_1 being as in Theorem 1. The theorem follows, because $S_1 - s_1 \to 0$ by Theorem 1 and therefore $S_2 - s_2 \to 0$.

Theorem 3. The product of two integrable functions is integrable.

Proof. Let $f = f(x)$, $g = g(x)$ be two functions, each integrable in $[a, b]$. From Theorem 2 and (7), (8) and the identity

$$2fg \equiv (f+g)^2 - f^2 - g^2,$$

it follows that fg is integrable.

Corollary. If m and n are positive integers, $f^m g^n$ is integrable.

Proof. By putting $g = f^2$ in fg, we see that f^3 is integrable, and by induction that f^m is integrable. Similarly, g^n is integrable. Since f^m and g^n are integrable, their product is integrable.

Theorem 4. The function $1/f(x)$ is integrable if $f(x)$ is one-signed and $1/f(x)$ is bounded.

Proof. Suppose $f(x) > 0$. Then the closest bounds of $1/f(x)$ are $1/m, 1/M$ in $[a, b]$ and $1/m_r, 1/M_r$ in $[x_{r-1}, x_r]$.

Let S_3 and s_3 be the upper and lower sums for $1/f(x)$ in a dissection $D(\delta)$. Then, since

$$\frac{1}{m_r} - \frac{1}{M_r} = \frac{M_r - m_r}{m_r M_r} \leqslant \frac{M_r - m_r}{m_r^2} \leqslant \frac{M_r - m_r}{m^2}$$

it follows, after multiplying by δ_r and summing, that

$$S_3 - s_3 \leqslant (S - s)/m^2.$$

But $S - s \to 0$ since $f(x)$ is integrable (§ 9.4), and m is independent of δ; it follows that $S_3 - s_3 \to 0$ and that $1/f(x)$ is therefore integrable (§ 9.7).

Example

If $f(0) = 1$ and $f(x) = x$ for $x > 0$, the function $f(x)$ is integrable over the interval $[0, 1]$; also $f(x)$ is one-signed; but $1/f(x)$ is not integrable (the function $1/f(x)$ is not bounded).

Theorem 5. Bounded rational functions of integrable functions are integrable.

This follows by generalization from the last two theorems, and the first two theorems of § 9.2.

9.11   A mean value theorem for integrals

Let $f(x)$ be bounded and integrable, and let m, M be the closest bounds of $f(x)$ in $[a, b]$.

Theorem. There exists a number $\mu(m \leqslant \mu \leqslant M)$ such that

$$\int_a^b f(x)\, dx = \mu(b - a). \tag{34}$$

This follows at once from (9).

Corollary 1. If $f(x)$ is continuous at a, we can put

$$\int_a^b f(x)\, dx = \{f(a) + \eta\}(b - a), \tag{35}$$

where $\eta \to 0$ when $b \to a$.

This follows since, if $f(x)$ is continuous at a, both m and M, and therefore also μ, must tend to $f(a)$ as $b \to a$.

Corollary 2. If $f(x)$ is continuous over the interval $[a, b]$, there exists a number ξ, belonging to the interval, such that

$$\int_a^b f(x)\, dx = f(\xi)\, (b - a). \tag{36}$$

This follows from §1.22, V and VII.

9.12 Inequalities

Theorem. In the interval $[a, b]$, $a < b$, if (i) $f(x)$ is continuous, (ii) $f(x) \geqslant 0$, and (iii) $f(x)$ is not zero identically, then

$$\int_a^b f(x)\, dx > 0.$$

Proof. By (ii) and (iii), there must be an interior point c where $f(c) > 0$. Therefore, by §1.22, II, there must be an interval

$$[c - h, c + h], \quad 0 < h,$$

throughout which $f(x) > 0$. Hence, by (36),

$$\int_{c-h}^{c+h} f(x)\, dx = f(\xi) \cdot 2h > 0,$$

since ξ belongs to this interval. Further, by (28),

$$\int_a^b f(x)\, dx = \left(\int_a^{c-h} + \int_{c-h}^{c+h} + \int_{c+h}^b \right) f(x)\, dx.$$

Since $f(x) \geqslant 0$, the first and third of the integrals on the R.H.S. cannot be negative, by (10); and the second integral is positive (just proved); consequently, the integral on the L.H.S. must be positive, as was to be proved.

Corollary. If

(i) $f(x)$ is continuous,   (ii) $f(x) \geqslant 0$,   (iii) $\int_a^b f(x)\, dx = 0$,

then $f(x)$ must be identically zero.

Note. Here $f(x)$ is continuous, whereas $f(x)$ in (10) was not necessarily continuous. Thus, in (10) we could put $f(x) = 0$ for $0 \leqslant x < 1$, $f(1) = 5$, $f(x) = 0$ for $1 < x \leqslant 2$; then the integral from 0 to 2 would be zero, although $f(x) \geqslant 0$ and $f(x)$ is not identically zero.

9.13   Schwarz's inequality

Theorem. Let $f(x)$ and $g(x)$ be continuous in $[a, b]$, $a < b$. Then

$$\int_a^b f^2(x)\,dx \int_a^b g^2(x)\,dx \geqslant \left\{ \int_a^b f(x)\,g(x)\,dx \right\}^2 \qquad (37)$$

according as $f(x)$, $g(x)$ are linearly independent or not (§3.12).

Proof. Let λ, μ be any real constants. Then by (7), (8), we can put

$$\int_a^b \{\lambda f(x) + \mu g(x)\}^2\,dx = A\lambda^2 + 2B\lambda\mu + C\mu^2, \qquad (38)$$

where A, B, C denote respectively the integrals of f^2, fg, g^2, which are known to exist (§9.10).

First, suppose that $f(x)$, $g(x)$ are linearly independent; then we have to prove that $AC > B^2$, or $AC - B^2 > 0$.

Now since f and g are linearly independent, $\lambda f + \mu g$ cannot vanish identically unless λ, μ are both zero. Also $(\lambda f + \mu g)^2 \geqslant 0$. It follows from the theorem of §9.12 that the L.H.S. of (38) must be positive for all not-both-zero values of λ, μ. Hence the quadratic form in λ, μ on the R.H.S. must be positive-definite, and (see §8.8) one necessary condition for this is $AC - B^2 > 0$, which was to be proved.

Secondly, let $f(x)$, $g(x)$ be linearly dependent; then we have to prove that $AC = B^2$, or $AC - B^2 = 0$.

Since f and g are now linearly dependent, constants l, m, not both zero, exist such that $lf + mg \equiv 0$, and therefore such that

$$\int_a^b (lf + mg) f\,dx = 0, \quad \int_a^b (lf + mg) g\,dx = 0,$$

that is, $lA + mB = 0, \quad lB + mC = 0.$

It follows, from the algebra of homogeneous linear equations, since l, m are not both zero, that $AC - B^2 = 0$.

9.14   The average value of a function over an interval

Definition. The number μ given by (34), viz.

$$\mu = \frac{1}{b-a} \int_a^b f(x)\,dx \qquad (39)$$

is defined to be the *average value* or *mean value* of $f(x)$ with respect to x over the interval $[a, b]$.

Example

The average value of $\sin^2 x$ over the interval $[0, \frac{1}{2}\pi]$ is

$$\frac{1}{\frac{1}{2}\pi} \int_0^{\frac{1}{2}\pi} \sin^2 x \, dx = \frac{1}{2}. \tag{40}$$

It is, moreover, evident from the way in which $\sin^2 x$ varies that the average value of $\sin^2 x$ is $\frac{1}{2}$ over any interval $[\frac{1}{2}p\pi, \frac{1}{2}q\pi]$, where p and q are integers. The same is also evidently true of $\cos^2 x$.

By memorizing this particular average value, we can at once evaluate an integral that often occurs in practice, the integral of $\sin^2 x$ from $\frac{1}{2}p\pi$ to $\frac{1}{2}q\pi$; thus, from (34), remembering that here $\mu = \frac{1}{2}$,

$$\int_{\frac{1}{2}p\pi}^{\frac{1}{2}q\pi} \sin^2 x \, dx = \frac{1}{2}(\tfrac{1}{2}q\pi - \tfrac{1}{2}p\pi) = (q-p)\tfrac{1}{4}\pi. \tag{41}$$

We say 'with respect to x' in the definition because in practice a variable may often be regarded as a function of one of several independent variables, and the average value will depend upon the independent variable concerned. Thus, in the rectilinear motion of a point, we may regard the velocity as a function of time or of displacement; again, we may regard the length of a chord of a circle as a function of its distance from the centre or as a function of the angle subtended at the centre.

9.15   The root-mean-square value of a function

Another kind of average of a function is the *root-mean-square* (R.M.S.) value. This is defined to be the square root of the average square of the function; thus, if κ is the R.M.S. value of the integrable function $f(x)$ over $[a, b]$, then

$$\kappa^2 = \frac{1}{b-a} \int_a^b f^2(x) \, dx. \tag{42}$$

By putting $g(x) = 1$ in Schwarz's inequality (37), we get

$$\int_a^b f^2(x) \, dx . (b-a) \geqslant \left\{ \int_a^b f(x) \, dx \right\}^2$$

and by dividing by $(b-a)^2$ follows $\kappa^2 \geqslant \mu^2$, with equality only when $f(x)$ is constant. Thus, unless $f(x)$ is constant, its average square exceeds the square of its average, over any interval.

9.16 The definite integral as a function of its limits

Let $f(x)$ be integrable in $[a, b]$ and put

$$F(x) = \int_a^x f(t)\, dt. \tag{43}$$

The function $F(x)$ exists for any x in $[a, b]$ by §9.8, Theorem 1.

Theorem 1. The function $F(x)$ is continuous in $[a, b]$.

Proof. By (26) and (5),

$$F(x+h) - F(x) = \int_x^{x+h} f(t)\, dt. \tag{44}$$

Now since $f(x)$ is bounded, we may suppose $-K < f(x) < K$ where K is independent of x. Then, by (33), follows

$$|F(x+h) - F(x)| < K|h|$$

and hence $F(x+h) - F(x) \to 0$ when $h \to 0$; therefore $F(x)$ is continuous.

Theorem 2. The function $F(x)$ is differentiable at any point where $f(x)$ is continuous, and $F'(x) = f(x)$ at any such point.

Proof. By (35) and (44), if $f(x)$ is continuous at x, we can put

$$F(x+h) - F(x) = \int_x^{x+h} f(t)\, dt = \{f(x) + \eta\}h$$

where $\eta \to 0$ when $h \to 0$. It follows by §2.2 that $F(x)$ is differentiable at x and that $F'(x) = f(x)$, that is,

$$\frac{d}{dx} \int_a^x f(t)\, dt = f(x). \tag{45}$$

Corollary. Provided that $f(x)$ is continuous at x,

$$\frac{d}{dx} \int_x^b f(t)\, dt = -f(x). \tag{46}$$

Theorem 3. If $f(x)$ is continuous, and if $\phi(x)$ is any function such that $\phi'(x) = f(x)$ throughout $[a, b]$, then

$$\int_a^b f(x)\, dx = [\phi(x)]_a^b = \phi(b) - \phi(a). \tag{47}$$

Such a function $\phi(x)$ is called an *indefinite integral* of $f(x)$.

Proof. The theorem follows at once from the more general Theorem 4 of §9.2, since a continuous function is integrable.

Alternatively, $F'(x) = f(x)$ by Theorem 2, and if also $\phi'(x) = f(x)$

then
$$(d/dx)\{F(x) - \phi(x)\} = F'(x) - \phi'(x) = f(x) - f(x) = 0$$

throughout the interval $[a, b]$, and hence, by §2.16,

$$F(x) - \phi(x) = C,$$

where C is independent of x. Putting $x = a$, $x = b$ in turn,

$$F(a) - \phi(a) = C, \quad F(b) - \phi(b) = C,$$

and by subtraction, since $F(a) = 0$, by (6), we obtain (47) in the form

$$F(b) - F(a) = F(b) = \phi(b) - \phi(a).$$

Note. It may be emphasized that here and in (11) the function $\phi(x)$ is differentiable, and therefore necessarily continuous, throughout $[a, b]$. For example, we can put

$$\int_0^{\sqrt{3}} \frac{2\,dx}{1+x^2} = [2\tan^{-1}x]_0^{\sqrt{3}} \quad \text{but not} \quad \left[\tan^{-1}\frac{2x}{1-x^2}\right]_0^{\sqrt{3}}$$

since the function $\tan^{-1}\{2x/(1-x^2)\}$ is not continuous at $x = 1$ (between 0 and $\sqrt{3}$) if by $\tan^{-1}$ is meant the principal value of the inverse tangent.

9.17 Evaluation of definite integrals

The following methods of evaluating or attempting to evaluate definite integrals may be noted here:

(i) Direct application of the definition (§9.1).

(ii) Application of the fundamental theorem, summarized in (11) and (47), when the indefinite integral is known.

(iii) Integration by parts.

(iv) Integration by substitution, or change of variable of integration, by which an unfamiliar integral can sometimes be converted into another integral of well-known or 'standard' form.

(v) Consideration of the equation or equations connecting two or more related integrals.

(vi) Differentiation with respect to a parameter in the integrand.

(vii) Methods of approximation which can be applied when the integral cannot be evaluated in finite form, or which may be more practical than analytical methods when numerical results are required.

9.18   Integration by direct application of the definition

We divide the interval of integration into n subintervals, and choose a convenient ξ in each subinterval (§9.1).

Examples

(1) Evaluate $\int_0^{\frac{1}{2}\pi} \cos x \, dx$ directly from the definition of a definite integral.

Divide the interval $[0, \frac{1}{2}\pi]$ into n equal subintervals each of length $\frac{1}{2}\pi/n$. Put $\theta = \frac{1}{2}\pi/n$. Then the rth subinterval is from $(r-1)\theta$ to $r\theta$. Put $\xi_r = (r-1)\theta$, the value of x at the beginning of the rth subinterval. Then, by definition (§9.1),

$$\int_0^{\frac{1}{2}\pi} \cos x \, dx = \lim_{n\to\infty} \theta\{1 + \cos\theta + \cos 2\theta + \ldots + \cos(n-1)\theta\}.$$

Putting $C_n = 1 + \cos\theta + \cos 2\theta + \ldots + \cos(n-1)\theta$, we find by trigonometry, after multiplying by $2\sin\frac{1}{2}\theta$,

$$2C_n \sin\tfrac{1}{2}\theta = \sin\tfrac{1}{2}\theta + \sin(n-\tfrac{1}{2})\theta = \sin\tfrac{1}{2}\theta + \cos\tfrac{1}{2}\theta$$

since $n\theta = \frac{1}{2}\pi$; and hence

$$\theta C_n = (\sin\tfrac{1}{2}\theta + \cos\tfrac{1}{2}\theta)(\tfrac{1}{2}\theta/\sin\tfrac{1}{2}\theta)$$

from which follows

$$\int_0^{\frac{1}{2}\pi} \cos x \, dx = \lim_{n\to\infty} (\theta C_n) = \lim_{\theta\to 0} (\theta C_n) = (0+1) \times 1 = 1.$$

(2) When the value of the definite integral is known (e.g. via the indefinite integral) the definition (§9.1) provides a means of evaluating the limit of the sum on the R.H.S. of (4).

Thus, if we put $f(x) = 1/x$, $a = 1$, $b = 2$, and divide the interval from 1 to 2 into n equal subintervals, each of length $1/n$, taking ξ_r now to be the value at the end of its subinterval, so that $\xi_r = r/n$, we have, by (4),

$$\int_1^2 \frac{1}{x} \, dx = \lim_{n\to\infty} \frac{1}{n}\left(\frac{1}{1+1/n} + \frac{1}{1+2/n} + \ldots + \frac{1}{1+n/n}\right)$$

from which, assuming that the value of the integral on the L.H.S. is known to be $\log 2$, we deduce that

$$\lim_{n\to\infty}\left(\frac{1}{n+1} + \frac{1}{n+2} + \ldots + \frac{1}{2n}\right) = \log 2.$$

9.19   Integration by the fundamental theorem

Examples

(1) Since $\sin x$ is a continuous function whose derivative is $\cos x$, it follows from (11) or (47) that, for all values of a and b,

$$\int_a^b \cos x \, dx = [\sin x]_a^b = \sin b - \sin a.$$

It follows that the numerical value of this integral cannot exceed 2.

(2) A second example will illustrate again the significance of the *Note* at the end of §9.16.

Let $1(x)$ denote the unit function defined by

$$1(x) = 0 \quad (x \leqslant 0), \quad 1(x) = 1 \quad (0 < x).$$

Then in (47) we can put $f(x) = 2x \, 1(x)$, $\phi(x) = x^2 \, 1(x)$, and get

$$\int_a^b 2x \, 1(x) \, dx = [x^2 \, 1(x)]_a^b = b^2 \, 1(b) - a^2 \, 1(a)$$

for all values of a and b, because $\phi(x)$ is differentiable and

$$\phi'(x) = 2x \, 1(x) \quad \text{for all } x.$$

But we cannot put

$$\int_a^b 2x \, 1(x) \, dx = [(x^2 + C) \, 1(x)]_a^b$$

with $a < 0 < b$ and $C \neq 0$, because $(x^2 + C) \, 1(x)$ is not continuous (and hence not differentiable) at $x = 0$ if $C \neq 0$. This may be verified by expressing the integral from a to b as the sum of the integrals from a to 0 and 0 to b.

9.20   Integration by parts

Theorem. Let $f = f(x)$, $g = g(x)$ be differentiable and therefore continuous functions whose derivatives $f'(x)$, $g'(x)$ are integrable. Then

$$\int_a^b (fg' + f'g) \, dx = [fg]_a^b. \tag{48}$$

Proof. By §9.10, the products fg', $f'g$ are integrable and so, by §9.2, is the sum $fg' + f'g$.

Also, the product fg is a continuous function whose derivative is $fg' + f'g$. Hence, by (11), follows (48).

It may also be written

$$\int_a^b f\,dg = [fg]_a^b - \int_a^b g\,df,$$

(49)

where the limits a, b refer to x.

9.21　Integration by substitution

Theorem. Let $x = \phi(t)$ be a continuous function of t with a continuous derivative $\phi'(t)$ from $t = \alpha$ to $t = \beta$.

Let $a = \phi(\alpha)$, $b = \phi(\beta)$, and let $f(x)$ be a continuous function of x from $x = a$ to $x = b$. Then

$$\int_a^b f(x)\,dx = \int_\alpha^\beta f\{\phi(t)\}\,\phi'(t)\,dt.$$

(50)

Proof. Put $v = \phi(u)$ and

$$F(u) = \int_a^v f(x)\,dx, \quad G(u) = \int_\alpha^u f\{\phi(t)\}\,\phi'(t)\,dt.$$

Then, by (45) and §2.20 (39), the integrands of $F(u)$ and $G(u)$ being continuous,

$$\frac{dF}{du} = \frac{dF}{dv}\frac{dv}{du} = f(v)\,\phi'(u), \quad \frac{dG}{du} = f\{\phi(u)\}\,\phi'(u);$$

thus $dF/du = dG/du$. Also $F = G = 0$ when $u = \alpha$. It follows that $F(u) = G(u)$ and in particular that $F(\beta) = G(\beta)$, which is equivalent to (50).

Example

Consider the integral

$$I = \int_{-1}^2 x^2\,dx = 3.$$

To evaluate this integral we could not (even if we wished!) make the substitution $x = \sqrt{t^3}$ because this would imply $t > 0$ and $x > 0$ and so could not cover the part of the range from $x = -1$ to $x = 0$. We could, however, put

$$I = I_1 + I_2 = \int_{-1}^0 x^2\,dx + \int_0^2 x^2\,dx$$

and make the substitutions:

in　I_1: $x = -\sqrt{t^3}$,　$dx/dt = -\tfrac{3}{2}\sqrt{t}$,　$t = 1$　to　0,

in　I_2: $x = \sqrt{t^3}$,　$dx/dt = \tfrac{3}{2}\sqrt{t}$,　$t = 0$　to　$\sqrt[3]{4}$,

leading to $I = \tfrac{1}{3} + \tfrac{8}{3} = 3$.

9.22

In much of what follows we shall assume that the integrands with which we are concerned are integrable.

We shall also assume that the reader is familiar with the 'graphical meaning' or 'representation' of the integral (4) as the 'area' under the graph of $y = f(x)$ from $x = a$ to $x = b$. From a strict mathematical point of view, we first define the integral and then *define* the 'area' to be the value of the integral. Nevertheless, the physical meaning of an area is often helpful in considering the integral which it 'represents.' It must be kept in mind that, over any range in which $f(x)$ is one-signed from a to b, the integral has the same sign as $(b-a) f(x)$.

9.23   Related definite integrals

A relation between two or more definite integrals may enable us to evaluate one or more of them, possibly even when the indefinite integrals are not expressible in a finite number of terms. For instance, it is evident from the graphical representation, or by making the substitution $\theta = \frac{1}{2}\pi - x$ in either, that the two integrals

$$I = \int_0^{\frac{1}{2}\pi} \sin^2 \theta \, d\theta, \quad J = \int_0^{\frac{1}{2}\pi} \cos^2 \theta \, d\theta$$

are equal, i.e. $I = J$. Also, by adding,

$$I + J = \int_0^{\frac{1}{2}\pi} (\sin^2 \theta + \cos^2 \theta) \, d\theta = \int_0^{\frac{1}{2}\pi} d\theta = \frac{1}{2}\pi,$$

and hence $I = J = \frac{1}{4}\pi$. Thus we are able to evaluate both definite integrals without finding either of the indefinite integrals, though, in this case, that would have been easy.

In the following theorem, the limits of integration are equal in magnitude but opposite in sign:

Theorem.

$$\int_{-a}^a f(x) \, dx = \int_0^a \{f(x) + f(-x)\} \, dx. \tag{51}$$

Proof. By (26) we can put

$$\int_{-a}^a f(x) \, dx = \left(\int_{-a}^0 + \int_0^a \right) f(x) \, dx = \left(\int_0^a + \int_{-a}^0 \right) f(x) \, dx.$$

In the last integral make the substitution $x = -t$, $dx = -dt$. Then, using (5), we get

$$\int_{-a}^a f(x) \, dx = \int_0^a f(x) \, dx + \int_0^a f(-t) \, dt$$

which proves (51) since t is a dummy variable (§9.1).

Corollary. If $f(x)$ is even, i.e. if $f(-x) = f(x)$, then

$$\int_{-a}^{a} f(x)\,dx = 2\int_{0}^{a} f(x)\,dx. \tag{52}$$

If $f(x)$ is odd, i.e. if $f(-x) = -f(x)$, then

$$\int_{-a}^{a} f(x)\,dx = 0. \tag{53}$$

In (53), the value of the integral is found to be zero whether the indefinite integral can be obtained or not. For instance,

$$\int_{-a}^{a} x^3 f(\cos x)\,dx = 0$$

since the integrand is odd, whether the indefinite integral of $x^3 f(\cos x)$ can be expressed in terms of elementary functions or not.

9.24

Another useful relation between two definite integrals is provided by the theorem:

Theorem.
$$\int_{a}^{b} f(x)\,dx = \int_{a}^{b} f(a+b-x)\,dx. \tag{54}$$

Proof. The theorem is proved by making the substitution

$$x = a+b-t, \quad dx = -dt,$$

in either integral.

Corollary.
$$\int_{0}^{a} f(x)\,dx = \int_{0}^{a} f(a-x)\,dx. \tag{55}$$

Examples

(1) By (54), the two integrals

$$I = \int_{\alpha}^{\pi-\alpha} x \sin x \cos^2 x\,dx, \quad J = \int_{\alpha}^{\pi-\alpha} (\pi-x)\sin x \cos^2 x\,dx$$

are equal, since $\sin(\pi-x)\cos^2(\pi-x) = \sin x \cos^2 x$. By addition,

$$I+J = 2I = \int_{\alpha}^{\pi-\alpha} \pi \sin x \cos^2 x\,dx = \pi[-\tfrac{1}{3}\cos^3 x]_{\alpha}^{\pi-\alpha}$$

and hence $\quad I = J = \tfrac{1}{2}\pi(\tfrac{1}{3}\cos^3\alpha + \tfrac{1}{3}\cos^3\alpha) = \tfrac{1}{3}\pi\cos^3\alpha.$

(2) By (55),
$$\int_{0}^{\frac{1}{2}\pi} f(\sin\theta)\,d\theta = \int_{0}^{\frac{1}{2}\pi} f(\cos\theta)\,d\theta. \tag{56}$$

9.25 Reduction formulae

The reader will be familiar with examples of reduction formulae, whereby the evaluation of an integral I_n, involving a constant n, can be facilitated by relating it to a similar integral in which the constant n is reduced. The reduction formula in the following example finds frequent application:

Example

If p and q are integers, and $n \geq 2$,

$$\int_{\frac{1}{2}p\pi}^{\frac{1}{2}q\pi} \cos^n \theta \, d\theta = \frac{n-1}{n} \int_{\frac{1}{2}p\pi}^{\frac{1}{2}q\pi} \cos^{n-2} \theta \, d\theta. \tag{57}$$

(i) *First method of proof*: by integration by parts. By (49),

$$\int_\alpha^\beta \cos^n \theta \, d\theta = \int_\alpha^\beta \cos^{n-1}\theta \, d(\sin\theta)$$
$$= [\cos^{n-1}\theta \sin\theta]_\alpha^\beta - \int_\alpha^\beta \sin\theta \, d(\cos^{n-1}\theta).$$

Now if $\alpha = \frac{1}{2}p\pi$, $\beta = \frac{1}{2}q\pi$, either $\sin\theta$ or $\cos\theta$ is zero at each limit, and hence
$$\int_{\frac{1}{2}p\pi}^{\frac{1}{2}q\pi} \cos^n \theta \, d\theta = (n-1) \int_{\frac{1}{2}p\pi}^{\frac{1}{2}q\pi} \cos^{n-2}\theta \sin^2\theta \, d\theta.$$

Denote the integral on the L.H.S. by I_n. Then, after replacing $\sin^2\theta$ by $1 - \cos^2\theta$ on the R.H.S. we get

$$I_n = (n-1)(I_{n-2} - I_n)$$

which is equivalent to (57).

(ii) *Second method*: by differentiating a suitable function of θ, here the function $\cos^n \theta \sin\theta$. Thus,

$$(d/d\theta)(\cos^n\theta \sin\theta) = \cos^{n+1}\theta - n\cos^{n-1}\theta \sin^2\theta$$
$$= (n+1)\cos^{n+1}\theta - n\cos^{n-1}\theta,$$

from which, by integration from $\frac{1}{2}p\pi$ to $\frac{1}{2}q\pi$, follows

$$0 = (n+1)I_{n+1} - nI_{n-1}$$

and hence (57) on replacing n by $n-1$.

By repeated application of (57), if n is a positive integer we can express I_n in terms of I_1 if n is odd, or I_0 if n is even, and hence evaluate I_n without further integration. Thus

$$I_n = \frac{n-1}{n} I_{n-2} = \frac{n-1}{n}\frac{n-3}{n-2} I_{n-4} = \dots \tag{58}$$

and since I_1 and I_0 are given by

$$I_1 = \sin \tfrac{1}{2}q\pi - \sin \tfrac{1}{2}p\pi, \quad I_0 = \tfrac{1}{2}q\pi - \tfrac{1}{2}p\pi,$$

therefore, if n is odd,

$$I_n = \frac{(n-1)(n-3)\ldots4.2}{n(n-2)\ldots5.3}(\sin \tfrac{1}{2}q\pi - \sin \tfrac{1}{2}p\pi), \tag{59}$$

if n is even, $\quad I_n = \dfrac{(n-1)(n-3)\ldots3.1}{n(n-2)\ldots4.2}(\tfrac{1}{2}q\pi - \tfrac{1}{2}p\pi).$ $\tag{60}$

In the same kind of way it can be proved that

$$\int_{\frac{1}{2}p\pi}^{\frac{1}{2}q\pi} \sin^n \theta \, d\theta = \frac{n-1}{n}\int_{\frac{1}{2}p\pi}^{\frac{1}{2}q\pi} \sin^{n-2}\theta \, d\theta \tag{61}$$

and that, if $\quad I_{m,n} = \displaystyle\int_{\frac{1}{2}p\pi}^{\frac{1}{2}q\pi} \sin^m \theta \cos^n \theta \, d\theta,$

then $\quad I_{m,n} = \dfrac{m-1}{m+n}I_{m-2,n} = \dfrac{n-1}{m+n}I_{m,n-2}.$ $\tag{62}$

9.26　Differentiation of definite integrals

The definite integral $\displaystyle\int_a^b f(x,y)\,dx$ depends on a, b and y, but not on x which is a dummy variable (§9.1). Suppose that a and b are fixed; then we may put

$$G(y) = \int_a^b f(x,y)\,dx \tag{63}$$

and consider the integral as a function of y, which is called a *parameter* in the integrand.

We first note that $G(y)$ may not even be a continuous function of y. For instance, if

$$f(x,y) = 3x^2 + g(y)$$

then, by integration,

$$G(y) = b^3 - a^3 + g(y)(b-a),$$

where $g(y)$ denotes any function of y whatever.

Suppose now that (x,y) belongs to the rectangle

$$R\,(a \leqslant x \leqslant b, \quad \alpha \leqslant y \leqslant \beta).$$

Theorem 1. If $f(x,y)$ is a continuous function of position in R, then $G(y)$ is continuous in the interval $[\alpha, \beta]$.

Proof. Let y_0 be any point in $[\alpha, \beta]$. Then

$$G(y_0+k)-G(y_0) = \int_a^b \{f(x,y_0+k)-f(x,y_0)\}\,dx. \tag{64}$$

Now, by §5.5, II, since $f(x,y)$ is a continuous function of position in R: given $\epsilon > 0$, a number δ exists such that the spread of $f(x,y)$ is less than ϵ in every circle of radius δ with its centre in R. Hence, if $|k| < \delta$,

$$|f(x,y_0+k)-f(x,y_0)| < \epsilon.$$

It follows from (64), using (30), that

$$|G(y_0+k)-G(y_0)| < \int_a^b \epsilon\,dx = \epsilon(b-a)$$

and hence, $b-a$ being fixed, that $G(y)$ is continuous at y_0, and since y_0 is any point in the interval $[\alpha, \beta]$, that $G(y)$ is continuous in this interval (§1.20, I).

Examples

(1) Let $f(x,y) = \cos xy$. Then $f(x,y)$ is continuous over all the xy plane, and, in particular, in the rectangle $(0 \leqslant x \leqslant 1, -K \leqslant y \leqslant K)$ for any positive K. By the theorem, if we put

$$G(y) = \int_0^1 \cos xy\,dx,$$

then $G(y)$ is a continuous function of y in the interval $(-K \leqslant y \leqslant K)$.

To verify this, we find $G(y) = (\sin y)/y\ (y \neq 0), G(0) = 1$; from which we see that, when $y \to 0$, $G(y) \to G(0)$, showing that $G(y)$ is continuous at $y = 0$. It is evidently continuous for every other value of y.

(2) Evaluate the integral

$$G(y) = \int_0^\alpha \frac{1-y\cos x}{1-2y\cos x+y^2}\,dx \quad (0 \leqslant y \leqslant 1, 0 < \alpha < \pi).$$

Show that $G(y)$ is not a continuous function of y at $y = 1$.

When $y = 0$, $G(0) = \alpha$. When $y = 1$, $G(1) = \frac{1}{2}\alpha$. When $0 < y < 1$,

$$G(y) = \int_0^\alpha \frac{1}{2}\left(1 + \frac{1-y^2}{1-2y\cos x+y^2}\right)dx$$

and after putting $t = \tan \frac{1}{2}x$, we find

$$G(y) = \frac{1}{2}\alpha + \tan^{-1}\left(\frac{1+y}{1-y}\tan \frac{1}{2}\alpha\right).$$

It follows that when $y \to 1$, $\lim G(y) = \frac{1}{2}\alpha + \frac{1}{2}\pi$. This limit is not equal to $G(1) = \frac{1}{2}\alpha$, and so $G(y)$ is not a continuous function of y at $y = 1$.

We infer from the above theorem that the integrand in $G(y)$ cannot be a continuous function of position throughout the rectangle

$$R \; (0 \leqslant x \leqslant \alpha, \quad 0 \leqslant y \leqslant 1).$$

In fact, if we put $x = h$, $y = 1 - k$ in the corner $(0, 1)$ of R, we find approximately

$$\frac{1 - y \cos x}{1 - 2y \cos x + y^2} = \frac{\frac{1}{2}h^2 + k}{h^2 + k^2}$$

which has the limit $\frac{1}{2}$ when first $k \to 0$, then $h \to 0$; but tends to ∞ when first $h \to 0$, then $k \to 0$.

9.27

Theorem 2. If $f(x, y)$ and $f_y(x, y)$ are continuous functions of position in R, then $G(y)$ is differentiable in $[\alpha, \beta]$ and

$$G'(y) = \int_a^b f_y(x, y) \, dx, \tag{65}$$

where $f_y(x, y)$ denotes the partial derivative of $f(x, y)$ wo y. Also, $G'(y)$ is continuous in $[\alpha, \beta]$.

Proof. Since $f(x, y)$ is continuous and $f_y(x, y)$ exists, it follows from (64) and §2.14 (22) that

$$\frac{G(y_0 + k) - G(y_0)}{k} = \int_a^b f_y(x, y_0 + \theta k) \, dx, \tag{66}$$

where $0 < \theta < 1$, and θ depends on x, y_0 and k. Now the integral in (66), by adding and subtracting $f_y(x, y_0)$ in its integrand, can be expressed in the form

$$\int_a^b \{ f_y(x, y_0 + \theta k) - f_y(x, y_0) \} \, dx + \int_a^b f_y(x, y_0) \, dx. \tag{67}$$

By §5.5, II, because $f_y(x, y)$ is a continuous function of position in R, given $\epsilon > 0$, a number δ exists such that for all x, y_0 and for all $|k| < \delta$ the integrand of the first integral in (67) is numerically less than ϵ, and hence such that the integral itself is numerically less than $\epsilon(b - a)$. It follows that this integral tends to zero when $k \to 0$. Hence, by (66) and (67),

$$\lim \frac{G(y_0 + k) - G(y_0)}{k} = \int_a^b f_y(x, y_0) \, dx. \tag{68}$$

Thus $G'(y_0)$ exists and is equal to the integral on the R.H.S. of (68). Consequently, for every value y_0 of y in $[\alpha, \beta]$, the integral $G(y)$ is differentiable and $G'(y)$ is given by (65). Moreover, $G'(y)$ is continuous, by Theorem 1.

9.28

Two ways in which it may be possible to apply (65), provided that the conditions of Theorem 2 hold good, are:

(i) By evaluating $G(y)$ and then finding the integral on the R.H.S. of (65) by differentiating $G(y)$.

(ii) By evaluating the integral on the R.H.S. and so obtaining a differential equation satisfied by $G(y)$, by solving which $G(y)$ may be found.

Examples

(1) Since $\cos xy$ and $(\partial/\partial y) \cos xy$, $= -x \sin xy$, are continuous at every point (x, y), we can differentiate the equation of §9.26, Example 1, differentiating the R.H.S. under the integral sign and obtaining

$$G'(y) = \int_0^1 -x \sin xy \, dx.$$

The new integral on the R.H.S. can therefore be evaluated for all values of y by finding $G'(y)$.

But $G(y) = (\sin y)/y$, $y \neq 0$; $G(0) = 1$; from which we find

$$G'(y) = (y \cos y - \sin y)/y^2 \quad (y \neq 0), \quad G'(0) = 0;$$

these equations therefore give the value of the new integral.

In this case, the new integral could be easily evaluated directly.

It may be verified that the integral is a continuous function of y for all values of y, including $y = 0$.

(2) Consider the integral

$$G = G(y) = \int_0^{\frac{1}{2}\pi} \log (1 + y \cos^2 x) \, dx.$$

The integrand and its partial derivative wo y are continuous functions of position in any region in which $y > -1$. If $y > -1$ we may therefore apply (65), obtaining

$$\frac{dG}{dy} = \int_0^{\frac{1}{2}\pi} \frac{\cos^2 x \, dx}{1 + y \cos^2 x}.$$

This integral can be evaluated, e.g. by means of the substitution $t = \tan x$. After the integration we find

$$\frac{dG}{dy} = \frac{\pi}{2y}\left\{1 - \frac{1}{\sqrt{(1+y)}}\right\} = \frac{\pi}{1 + \sqrt{(1+y)}}\frac{1}{2\sqrt{(1+y)}}.$$

The general solution of this differential equation is

$$G = \pi \log\{1 + \sqrt{(1+y)}\} + C,$$

where C is independent of y. Since $G = 0$ when $y = 0$, therefore $C = -\pi\log 2$ and finally

$$G = \pi[\log\{1 + \sqrt{(1+y)}\} - \log 2]. \tag{69}$$

Assuming that G exists and is continuous at $y = -1$, we deduce by putting $y = -1$ that (see §10.3, Example)

$$\int_0^{\frac{1}{2}\pi} \log(\sin x)\,dx = -\tfrac{1}{2}\pi\log 2. \tag{70}$$

9.29

Let the limits of integration a, b be variable, and put

$$G = G(a, b, y) = \int_a^b f(x, y)\,dx. \tag{71}$$

Then, if $f(x, y)$ and $f_y(x, y)$ are continuous functions of position in R, the integral G is, by §5.25, a differentiable function of a, b and y, regarded as independent variables, having the continuous partial derivatives

$$\frac{\partial G}{\partial a} = -f(a, y), \quad \frac{\partial G}{\partial b} = f(b, y), \quad \frac{\partial G}{\partial y} = \int_a^b f_y(x, y)\,dx \tag{72}$$

by (45), (46) and (65). It follows by §5.27 (31) that, if a and b are differentiable functions of y, then G will be a differentiable function of the single variable y, the total derivative dG/dy being given by

$$\frac{dG}{dy} = -f(a, y)\frac{da}{dy} + f(b, y)\frac{db}{dy} + \int_a^b f_y(x, y)\,dx. \tag{73}$$

Example

Verify that
$$y = \frac{1}{\omega}\int_0^t f(x)\sin\omega(t-x)\,dx$$

is the solution of the differential equation

$$d^2y/dt^2 + \omega^2 y = f(t)$$

which satisfies the conditions : $y = dy/dt = 0$ at $t = 0$.

Here y is a function of t, with $a = 0$ and $b = t$, and therefore $da/dt = 0$, $db/dt = 1$. Also, the value of the integrand at $x = t$, the upper limit of integration, is

$$f(t)\sin\omega(t-t) = f(t)\sin 0 = 0.$$

Hence, by (73), with y in place of G and t in place of y,

$$\frac{dy}{dt} = 0+0+\frac{1}{\omega}\int_0^t f(x)\,.\,\omega\cos\omega(t-x)\,dx.$$

By applying (73) again, and putting $\cos\omega\,(t-t) = \cos 0 = 1$,

$$\frac{d^2y}{dt^2} = 0+f(t)-\frac{1}{\omega}\int_0^t f(x)\,.\,\omega^2\sin\omega(t-x)\,dx$$

$$= f(t) - \omega^2 y.$$

Thus the differential equation is satisfied, and both y and dy/dt vanish at $t = 0$, by (6).

9.30   The first mean value theorem for integrals

The following theorem, more general than that of §9.2, Theorem 3 or §9.11, is known as the *first mean value theorem for integrals*:

Theorem. In the interval $[a, b]$, $a < b$, let $f(x)$, $g(x)$ be bounded and integrable and $g(x) \geqslant 0$. Also, let m, M be the closest bounds of $f(x)$. Then

$$m\int_a^b g(x)\,dx \leqslant \int_a^b f(x)\,g(x)\,dx \leqslant M\int_a^b g(x)\,dx. \tag{74}$$

Proof. The proof follows the same lines as that of §9.2, Theorem 3.

If $g(x) \leqslant 0$, instead of $g(x) \geqslant 0$, the theorem holds good with the signs of inequality reversed.

Corollary 1. If $g(x)$ does not change sign in $[a, b]$, a number μ exists $(m \leqslant \mu \leqslant M)$ such that

$$\int_a^b f(x)\,g(x)\,dx = \mu\int_a^b g(x)\,dx. \tag{75}$$

Corollary 2. If $f(x)$ is monotonic, the number μ lies between $f(a)$ and $f(b)$.

Corollary 3. If $f(x)$ is continuous at $x = a$, we can put

$$\int_a^b f(x)\,g(x)\,dx = \{f(a)+\eta\}\int_a^b g(x)\,dx, \tag{76}$$

where $\eta \to 0$ when $b \to a$.

Corollary 4. If $f(x)$ is continuous throughout $[a, b]$,   a number ξ exists $(a \leqslant \xi \leqslant b)$ such that

$$\int_a^b f(x) g(x) \, dx = f(\xi) \int_a^b g(x) \, dx. \tag{77}$$

Example

Prove that, if $a > 0$,

$$\lim_{n \to \infty} \int_{1/(n+a)}^{1/n} \frac{\sin (x^2)}{x^4} \, dx = a.$$

Put $f(x) = (\sin x^2)/x^2$, $g(x) = 1/x^2$. Then, if $n > 0$, both $f(x)$ and $g(x)$ are continuous and therefore integrable over the range of integration, and $g(x) > 0$. Hence, by Corollary 4, a number ξ exists between $1/(n+a)$ and $1/n$ such that

$$\int_{1/(n+a)}^{1/n} \frac{\sin (x^2)}{x^2} \frac{1}{x^2} \, dx = \frac{\sin (\xi^2)}{\xi^2} \int_{1/(n+a)}^{1/n} \frac{dx}{x^2}$$

from which the required result follows, since $\xi \to 0$ when $n \to \infty$ and therefore $(\sin \xi^2)/\xi^2 \to 1$, while the value of the last integral is a.

9.31   The second mean value theorem for integrals

The following theorem is known as the *second mean value theorem for integrals*. Conditions (i) and (ii) are *sufficient* conditions for it to be true.

Theorem. (i) Let $f(x)$ be differentiable and monotonic, so that $f'(x)$ does not change sign; and let $f'(x)$ be integrable.

(ii) Let $g(x)$ be continuous.

Then a number ξ exists $(a \leqslant \xi \leqslant b)$ such that

$$\int_a^b f(x) g(x) \, dx = f(a) \int_a^\xi g(x) \, dx + f(b) \int_\xi^b g(x) \, dx. \tag{78}$$

Proof. Put $G(x) = \int_a^x g(t) \, dt$;

then, since $g(x)$ is continuous, $G'(x) = g(x)$, by §9.16, Theorem 2. Hence, by (48), since $f'(x)$ and $G'(x)$ are integrable, and $G(a) = 0$,

$$\int_a^b (fG' + f'G) \, dx = [fG]_a^b = f(b) \, G(b).$$

But, by (77), since $f'(x)$ is integrable and of constant sign and $G(x)$

is continuous by §9.16, Theorem 1, a number ξ exists $(a \leqslant \xi \leqslant b)$ such that

$$\int_a^b G(x) f'(x) \, dx = G(\xi) \int_a^b f'(x) \, dx = G(\xi) \{f(b) - f(a)\},$$

by (11). Hence $\int_a^b fG' \, dx + G(\xi) \{f(b) - f(a)\} = f(b) G(b)$

and therefore, since $G' = g$,

$$\int_a^b fg \, dx = f(a) G(\xi) + f(b) \{G(b) - G(\xi)\} \tag{79}$$

which is equivalent to (78).

Corollary 1. *Bonnet's form of the second mean-value theorem.*

If $f(x)$ is positive monotonic decreasing, $f'(x)$ integrable, and $g(x)$ continuous, then a number X exists $(a \leqslant X \leqslant b)$ such that

$$\int_a^b f(x) g(x) \, dx = f(a) \int_a^X g(x) \, dx. \tag{80}$$

Proof. Under these conditions, $f(a) > f(b) \geqslant 0$.

(i) Suppose $f(b) = 0$; then $X = \xi$ from (78); thus X exists.

(ii) Suppose $f(b) > 0$; then (79) can be expressed in the form

$$\int_a^b f(x) g(x) \, dx = m_1 G(\xi) + m_2 G(b) = (m_1 + m_2) \mu$$

where μ lies between $G(\xi)$ and $G(b)$, since $m_1 = f(a) - f(b)$ and $m_2 = f(b)$ are both positive. Also, since $G(x)$ is continuous, X exists $(\xi \leqslant X \leqslant b)$ such that $\mu = G(X)$. This proves (80) since $m_1 + m_2 = f(a)$.

Corollary 2. If the monotonic function $f(x)$ is positive increasing, then a number X exists $(a \leqslant X \leqslant b)$ such that

$$\int_a^b f(x) g(x) \, dx = f(b) \int_X^b g(x) \, dx. \tag{81}$$

The proof follows the same lines as that of Corollary 1.

Example

If $f(x)$ is piecemeal differentiable and has only a finite number of maxima and minima in $[0, \pi]$, and if $f'(x)$ is integrable, prove that the integrals

$$\int_0^\pi f(x) \sin nx \, dx, \quad \int_0^\pi f(x) \cos nx \, dx$$

tend to zero as $n \to \infty$.

Since maxima and minima and points of discontinuity are finite in number, the interval $[0, \pi]$ can be divided into a finite number of subintervals in each of which the conditions of the theorem are satisfied. Then by (78), for each subinterval $[a, b]$ a number ξ exists $(a \leqslant \xi \leqslant b)$ such that

$$\int_a^b f(x) \sin nx\, dx = f(a) \int_a^\xi \sin nx\, dx + f(b) \int_\xi^b \sin nx\, dx.$$

Now the integral of $\sin nx\, dx$ over any interval cannot exceed $2/n$ in numerical value $(n > 0)$; consequently

$$\left| \int_a^b f(x) \sin nx\, dx \right| \leqslant |f(a)| \frac{2}{n} + |f(b)| \frac{2}{n}$$

and hence, if the number of subintervals is p,

$$\left| \int_0^\pi f(x) \sin nx\, dx \right| \leqslant (|f(a)| + |f(b)|) \, 2p/n$$

which tends to zero when $n \to \infty$.

This proves the result for the first of the given integrals. The second can be treated in the same way.

9.32 Approximate integration

When it is necessary to evaluate a definite integral which is not susceptible to accurate methods, a method of approximation must be used. This will usually depend upon replacing the intractable integrand by one approximating to it over the range of integration, but having an easily found indefinite integral.

Examples

(1) To find upper and lower bounds to the integral

$$I = \int_0^1 \frac{x\, dx}{\sqrt{(1 + x^3)}}.$$

(i) If $0 < x < 1$, then $1 + x^2 > 1 + x^3 > 1 + x^4$, and hence

$$\int_0^1 \frac{x\, dx}{\sqrt{(1 + x^2)}} < I < \int_0^1 \frac{x\, dx}{\sqrt{(1 + x^4)}}$$

which gives $\sqrt{2} - 1 < I < \tfrac{1}{2} \log(1 + \sqrt{2})$

and hence $0 \cdot 414 < I < 0 \cdot 441.$

(ii) An upper bound could be found by using Schwarz's inequality. In (37) put $f(x) = x/\sqrt{(1+x^3)}$, $g(x) = 1$; then follows

$$I^2 < \int_0^1 \frac{x^2\, dx}{1+x^3} = \tfrac{1}{3}\log 2$$

and hence $I < 0\cdot481$.

(2) With the help of tables of elliptic functions, it can be proved that

$$I = \int_0^\pi \sqrt{(\sin x)}\, dx \doteqdot 2\cdot3962,$$

but in the absence of some knowledge of elliptic functions, we can only approximate to the value of I. Simpson's rule (§ 9.33) might be used. We notice also that:

(i) Schwarz's inequality gives, on putting $f(x) = \sqrt{(\sin x)}$, $g(x) = 1$,

$$I^2 < \int_0^\pi \sin x\, dx \int_0^\pi dx = 2\pi$$

and therefore $I < \sqrt{(2\pi)} < 2\cdot51$.

(ii) Seeing that $y = \sqrt{(\sin x)}$ from $x = 0$ to $x = \pi$ resembles the upper half of an ellipse of semi-axes $\tfrac{1}{2}\pi$ and 1, we have, by the formula for the area of an ellipse,

$$I \doteqdot \tfrac{1}{2}\pi \cdot \tfrac{1}{2}\pi \cdot 1 = \tfrac{1}{4}\pi^2 \doteqdot 2\cdot47.$$

9.33   Simpson's rule

It is easy to prove that the average value $\bar{y}$ of the cubic function

$$y = f(x) = Ax^3 + Bx^2 + Cx + D$$

over any interval $[a, b]$ is given by

$$\bar{y} = \tfrac{1}{6}(y_1 + 4y_2 + y_3), \tag{82}$$

where $y_1 = f(a)$, $y_2 = f\{\tfrac{1}{2}(a+b)\}$, $y_3 = f(b)$. Consequently, if $f(x)$ is a cubic function,

$$\int_a^b f(x)\, dx = \frac{b-a}{6}\left\{f(a) + 4f\left(\frac{a+b}{2}\right) + f(b)\right\}; \tag{83}$$

in particular,

$$\int_{-h}^h f(x)\, dx = \tfrac{1}{3}h\{f(-h) + 4f(0) + f(h)\}. \tag{84}$$

Formula (82) or (83) may be called *Simpson's rule*.

One of many methods (see e.g. C. A. Stewart, *Advanced Calculus*, p. 164) of approximating to the value of a numerical definite integral depends on Simpson's rule. For instance, if we apply (83) to any ordinary function, not a cubic, we shall effectively be using a cubic passing through the points where $x = a$, $x = \frac{1}{2}(a+b)$, $x = b$ as an approximating function, and we shall expect the resulting approximation to the integral to be good if the range of integration is small enough.

To improve the accuracy, we divide the interval $[a, b]$ into an even number $2n$ of subintervals each of width h, the points of division being at

$$x = a, \quad x = a+h, \quad x = a+2h, \ldots, \quad x = a+2nh = b$$

and the corresponding values of the function $y = f(x)$ being

$$y = y_1, \quad y = y_2, \quad y = y_3, \ldots, \quad y = y_{2n+1}.$$

Then, by applying Simpson's rule to the subintervals taken in pairs, we obtain as an approximation to the integral

$$2h\{\tfrac{1}{6}(y_1 + 4y_2 + y_3) + \tfrac{1}{6}(y_3 + 4y_4 + y_5) + \tfrac{1}{6}(y_5 + 4y_6 + y_7) + \ldots\}$$
$$= \tfrac{1}{3}h\{(y_1 + y_{2n+1}) + 4(y_2 + \ldots + y_{2n}) + 2(y_3 + \ldots + y_{2n-1})\}. \quad (85)$$

9.34   Estimation of the error in applying Simpson's rule

Consider formula (84) when the integrand $f(x)$ is not a cubic function. Put

$$F(x) = \int_{-x}^{x} f(t)\,dt - \tfrac{1}{3}x\{f(-x) + 4f(0) + f(x)\}; \quad (86)$$

then $F(h)$ will be the error in (84), and so we seek an upper bound to $|F(h)|$. We shall prove that

$$|F(h)| < Mh^5/90, \quad (87)$$

where M is the greatest numerical value of $f^{\mathrm{iv}}(x)$ in $[-h, h]$, assuming that $f^{\mathrm{iv}}(x)$ exists.

Proof. Using (73) to obtain $F'(x)$, we find

$$F'(x) = \tfrac{2}{3}\{f(x) + f(-x)\} - \tfrac{4}{3}f(0) - \tfrac{1}{3}x\{f'(x) - f'(-x)\},$$
$$F''(x) = \tfrac{1}{3}\{f'(x) - f'(-x)\} - \tfrac{1}{3}x\{f''(x) + f''(-x)\},$$
$$F'''(x) = -\tfrac{1}{3}x\{f'''(x) - f'''(-x)\}.$$

Now put $G(x) = x^5$. Then $F(x)$, $G(x)$ and their first two derivatives all vanish at $x = 0$. Consequently, by an extension of the mean-value theorem (Examples 4F, 13) a number ξ exists such that

$$\frac{F(h)}{h^5} = \frac{F'''(\xi)}{G'''(\xi)} = \frac{F'''(\xi)}{60\xi^2} = -\frac{1}{180}\frac{f'''(\xi) - f'''(-\xi)}{\xi}$$

where $0 < \xi < h$ and we have supposed $0 < h$. By applying the ordinary mean-value theorem to the last numerator, we see further that a number ξ_1 exists such that

$$\frac{F(h)}{h^5} = -\frac{1}{90} f^{\mathrm{iv}}(\xi_1) \quad (-\xi < \xi_1 < \xi),$$

from which (87) follows.

9.35 Repeated integrals

An integral of the form

$$\int_\alpha^\beta dy \int_a^b f(x,y)\, dx, \tag{88}$$

which indicates that we first integrate wo x and then wo y, is called a *repeated* integral.

Theorem. Let $f(x,y)$ be a continuous function of position in the rectangle $(a \leqslant x \leqslant b, \alpha \leqslant y \leqslant \beta)$. Then

$$\int_a^b dx \int_\alpha^\beta f(x,y)\, dy = \int_\alpha^\beta dy \int_a^b f(x,y)\, dx. \tag{89}$$

In words, the order of integration may be reversed; i.e. we may integrate first wo x, or first wo y, without affecting the final result.

Proof. Put

$$G(\eta) = \int_a^b dx \int_\alpha^\eta f(x,y)\, dy, \tag{90}$$

where $\alpha \leqslant \eta \leqslant \beta$. Then, by (65) and (45),

$$\frac{dG(\eta)}{d\eta} = \int_a^b dx\, \frac{\partial}{\partial \eta} \int_\alpha^\eta f(x,y)\, dy = \int_a^b f(x,\eta)\, dx$$

and hence, by integration from $\eta = \alpha$ to $\eta = \beta$

$$G(\beta) - G(\alpha) = \int_\alpha^\beta d\eta \int_a^b f(x,\eta)\, dx$$

which, since $G(\alpha) = 0$, is the same as (89), with the dummy variable y replaced by η.

Examples 9 A

(1) If a and b are given by the equations

$$\int_0^1 x(x^2+ax+b)\,dx = 0, \qquad \int_0^1 x^2(x^2+ax+b)\,dx = 0,$$

show that

(i) $\displaystyle\int_0^1 (x^2+ax+b)^2\,dx = \tfrac{1}{100},$ (ii) $\displaystyle\int_0^1 (x^2+ax+b)^3\,dx = \tfrac{13}{7000}.$

(2) If $\displaystyle\int_0^1 x^n f(x)\,dx = 0$ for $n = 1, 2, 3,$ where $f(x) = x^3+px^2+qx+r$, find p, q, r and show that

$$\int_0^1 x\{f(x)\}^2\,dx = \frac{1}{2^3 . 5^2 . 7^2}.$$

(3) The area bounded by the curve $y = a\cos(x^2/a^2)$ from $x = 0$ to $x = a\sqrt{(\tfrac{1}{2}\pi)}$ and the coordinate axes, makes one complete revolution about the axis of y. Prove that the volume generated is πa^3.

(4) Let y be a function of x which steadily decreases from $y = b$ to $y = 0$ as x increases from $x = 0$ to $x = a$. Prove that

$$\int_0^a 2xy\,dx = \int_0^b x^2\,dy.$$

The area in the first quadrant bounded by the coordinate axes and the curve $y = (\sin\tfrac{1}{2}\pi x)/x$, from $x = 0$ to $x = 2$, makes one complete rotation about the axis of y. Find the volume swept out.

(5) Find the area of the portion of the circle

$$(x - 2)^2 + (y - 3)^2 = 5^2$$

in each quadrant of the coordinate axes.

(6) A volume V of water is poured into a spherical bowl of radius R, and the depth of water at the deepest point is found to be h. Find R.

(7) A triangle of height h is divided into two parts by a line parallel to the base at a distance x from it. The triangle is then rotated about the base. Show that the ratio of the volumes generated by the two parts is

$$\frac{(h-x)^2\,(h+2x)}{x^2(3h-2x)}.$$

If they are equal, find x.

(8) Sketch the curve $y = \operatorname{sech} x - \tanh x$. Prove that the area in the first quadrant bounded by the curve and the coordinate axes is $\tfrac{1}{4}(\pi - \log 4)$.

(9) Evaluate $\displaystyle\int_{-1}^{-2} \frac{dx}{x\sqrt{(7x^2 - 18x)}}$ by putting $x = 1/t$.

(10) Evaluate

$$\int_1^4 (x^2 - 6x + 8)\,dx \quad \text{by putting} \quad t = x^2 - 6x + 8.$$

(11) If n and s are positive integers and

$$C_n = \int_0^{2\pi} x^n \cos sx\, dx, \quad S_n = \int_0^{2\pi} x^n \sin sx\, dx,$$

show that $\qquad sC_n = -nS_{n-1}, \quad sS_n = -(2\pi)^n + nC_{n-1}.$

(12) If $\qquad I_n = \int_0^{\frac{1}{2}\pi} \cos a\theta \cos^n \theta\, d\theta \quad (n > 1),$

show that $\qquad (n^2 - a^2) I_n = n(n-1) I_{n-2}.$

Deduce that, if $n > 2s - 1$ (s a positive integer),

$$\int_0^{\frac{1}{2}\pi} \cos n\theta \cos^{n-2s} \theta\, d\theta = 0.$$

(13) Find all the possible values of the integrals

$\qquad$ (i) $\displaystyle\int_0^{\pi} \cos mx \cos nx\, dx,$ $\quad$ (ii) $\displaystyle\int_0^{\pi} \cos lx \cos mx \cos nx\, dx,$

where l, m, n may each be any positive integer or zero.

(14) If α is a root of the equation $x = \tan x$, show that

$$\int_0^1 \sin^2 \alpha t\, dt = \frac{\alpha^2}{2(1+\alpha^2)}, \quad \int_0^1 \cos^2 \alpha t\, dt = \frac{2+\alpha^2}{2(1+\alpha^2)}.$$

If α, β are two distinct positive roots of the same equation, show that

$$\int_0^1 \sin \alpha t \sin \beta t\, dt = 0, \quad \int_0^1 \cos \alpha t \cos \beta t\, dt = \cos \alpha \cos \beta.$$

(15) Evaluate the sum

$$C_n = \sum_{s=1}^n \exp(sx/n) \cos(sy/n).$$

Find the limit of C_n/n as $n \to \infty$, and compare its value with that of the integral

$$\int_0^1 \exp(tx) \cos(ty)\, dt.$$

Examples 9 B

(1) If $n > -1$, prove that

$$\int_0^a x(a-x)^n\, dx = \frac{a^{n+2}}{(n+1)(n+2)}.$$

(2) Show that $\qquad \displaystyle\int_0^a e^x \{3(a-x)^2 - (a-x)^3\}\, dx = a^3.$

(3) Show that $\qquad \displaystyle\int_0^a \cos mx \sin m(a-x)\, dx = \tfrac{1}{2} a \sin ma.$

(4) Prove that $\qquad \displaystyle\int_0^{\pi} \frac{x \sin x\, dx}{1+\cos^2 x} = \frac{\pi^2}{4}.$

(5) If l, m, n are integers and $l+m+n$ is even, show that

$$\int_\alpha^{\pi-\alpha} \sin lx \sin mx \sin nx\, dx = 0.$$

(6) Prove that

$$\int_\alpha^{\frac{1}{2}\pi-\alpha} \frac{d\theta}{1+\tan^n \theta} = \tfrac{1}{4}\pi - \alpha \quad \text{(independent of } n).$$

(7) Show that, if n is a positive integer,

$$\int_0^\pi \theta \sin^{2n} \theta\, d\theta = \int_0^\pi \theta \cos^{2n} \theta\, d\theta.$$

(8) Evaluate

$$\int_0^\pi |1+2\cos x|\, dx.$$

(9) If $n+1 > 0$, show that

$$\int_0^x t^{n-r}|t|^r dt = \frac{x^{n-r+1}|x|^r}{n+1}.$$

(10) If n is an integer, positive or negative, prove that

$$\text{(i)} \int_0^{n\pi} |\sin x|\, dx = 2n, \quad \text{(ii)} \int_0^{n\pi} x|\sin x|\, dx = n^2\pi.$$

(11) With $1(x)$ defined as in § 9.19, Example 2, show that

$$\int_a^b (e^x-1)1(x)\, dx = [(e^x-1-x)1(x)]_a^b \quad \text{for all} \quad a, b;$$
$$\neq [(e^x-x)1(x)]_a^b \quad \text{if} \quad a < 0 < b.$$

(12) If $[x]$ denotes the greatest integer not greater than x, show that

$$\int_0^x (t-[t])^2\, dt = \tfrac{1}{3}[x] + \tfrac{1}{3}(x-[x])^3.$$

(13) Let $f(x)$ be integrable and $(d/dx)\,\phi(x) = f(x)$. If n is a positive integer and $[x]$ denotes the greatest integer not greater than x, show that

$$\int_0^n f(x)\,(x-[x])\, dx = \sum_{r=1}^n \phi(r) - \int_0^n \phi(x)\, dx.$$

(14) Show that the function $\sqrt{f(x)}$ is integrable over any finite range in which $f(x)$ is positive and integrable.

(15) Show that the function defined by $f(x) = \sin(1/x)$, $x \neq 0$, $f(0) = k$ is integrable over any interval.

[For if the interval does not include $x = 0$, the function is continuous. If $x = 0$ is included, the function is bounded and has only one point of discontinuity (§ 9.9, Theorem 2).]

(16) If $f(x) = 2x\sin(1/x) - \cos(1/x)$, $x \neq 0$, and $f(0) = k$, show that $f(x)$ is integrable over any finite range and that, if n is an integer, the value of the integral from 0 to $1/n\pi$ is zero.

Examples 9 C

(1) Using the inequalities $x^4 < x^3 < x^2$ when $0 < x < 1$, show that

$$\tfrac{1}{3} < \int_0^1 \frac{x\,dx}{\sqrt{(x^3 + 2x^2 + 1)}} < \tfrac{1}{2}\log 2.$$

(2) If $\quad I = \int_0^1 \sqrt{(3x^2 + 4)}\,\sqrt{(3x^2 + 9)}\,dx,\quad J = \int_0^1 (3x^2 + 6)\,dx,$

show that $I > J$. Also, by using Schwarz's inequality, show that $I^2 < 50$. Deduce that $7 < I < 7\!\cdot\!08$.

(3) By expanding the integrand by the binomial theorem, show that the value of the integral

$$\int_{-1}^1 \sqrt{(25 + x^3)}\,dx$$

is less than 10 by about 0·0003.

(4) By considering the minimum value of the integral

$$\int_0^x (1 + t)\left(\frac{1}{1+t} - a\right)^2 dt \quad (0 < x),$$

as a function of a, prove that

$$\log(1 + x) > x/(1 + \tfrac{1}{2}x) \quad (0 < x).$$

Deduce that, if $n > 0$, $\quad \left(1 + \dfrac{1}{n}\right)^{n+\frac{1}{2}} > e.$

(5) By considering that the integral

$$\int_a^b \{(\lambda f_1 + g_1)^2 + (\lambda f_2 + g_2)^2\}\,dx \quad (a < b)$$

is necessarily positive or zero, show that

$$\int_a^b (f_1^2 + f_2^2)\,dx \int_a^b (g_1^2 + g_2^2)\,dx \geqslant \left\{\int_a^b (f_1 g_1 + f_2 g_2)\,dx\right\}^2$$

where $f_1, g_1, \ldots$ are integrable functions of x. Generalize this.

(6) Let $f = f(x)$, $g = g(x)$, $h = h(x)$ be integrable functions, and let A, B, C, F, G, H denote respectively the integrals from a to b of f^2, g^2, h^2, gh, hf, fg. Show that the minimum value of the integral

$$\int_a^b (\lambda f + \mu g + h)^2\,dx \quad (a < b),$$

regarded as a function of λ and μ, is $\Delta/(AB - H^2)$, where

$$\Delta = \begin{vmatrix} A & H & G \\ H & B & F \\ G & F & C \end{vmatrix}.$$

(7) Let $f_1, f_2, \ldots, f_n$ be continuous functions of x in $[a, b]$, and let I_{rs} denote the integral of $f_r f_s$ over $[a, b]$, $a < b$. Also, let Δ be the determinant of which I_{rs} is the r, s^{th} element.

Prove that $\Delta \geqslant 0$ according as $f_1, f_2, \ldots, f_n$ are linearly independent or not. [See §9.13].

(8) Prove that the average value of the cubic function

$$f(x) = Ax^3 + Bx^2 + Cx + D$$

between $x = -h$ and $x = +h$ can be expressed in the form $\frac{1}{2}\{f(-\alpha h) + f(\alpha h)\}$ where $\alpha = 1/\sqrt{3}$.

Show that the error in the approximation

$$\int_{-h}^{h} f(x)\,dx = h\{f(-h/\sqrt{3}) + f(h/\sqrt{3})\},$$

when $f(x)$ is not a cubic, is about $h^5 f^{iv}(0)/135$ if h is small enough.

(9) Prove that the average value of the cubic function

$$f(x) = Ax^3 + Bx^2 + Cx + D$$

between $x = -h$ and $x = +h$ can, for any value of α, be expressed in the form

$$\{f(-\alpha h) + (6\alpha^2 - 2)f(0) + f(\alpha h)\}/6\alpha^2.$$

Show that in general, when $f(x)$ is not a cubic, the defect in the approximation

$$\int_{-h}^{h} f(x)\,dx = \frac{h}{3\alpha^2}\{f(-\alpha h) + (6\alpha^2 - 2)f(0) + f(\alpha h)\}$$

when $\alpha^2 = \frac{3}{5}$ is about $h^7 f^{vi}(0)/15{,}750$, if h is small enough.

Show that this approximation, applied to the integral

$$\frac{\pi}{3\sqrt{3}} = \int_{-1}^{1} \frac{dx}{3 + x^2}$$

gives $\pi = 3{\cdot}143$ approximately.

(10) Prove that the average value of the quintic

$$f(x) = Ax^5 + Bx^4 + Cx^3 + Dx^2 + Ex + F$$

between $x = -h$ and $x = +h$ can be expressed in the form

$$(5y_1 + 8y_2 + 5y_3)/18$$

where $y_1 = f(-\alpha h)$, $y_2 = f(0)$, $y_3 = f(\alpha h)$, and $\alpha = \sqrt{(\frac{3}{5})}$.

(11) *Wallis's formula.* Evaluate the integrals in the inequalities

$$\int_0^{\frac{1}{2}\pi} \sin^{2n+1}\theta\,d\theta < \int_0^{\frac{1}{2}\pi} \sin^{2n}\theta\,d\theta < \int_0^{\frac{1}{2}\pi} \sin^{2n-1}\theta\,d\theta$$

and deduce that

$$1 < \left\{\frac{1.3.5....(2n-1)}{2.4.6...(2n)}\right\}^2 (2n+1)\tfrac{1}{2}\pi < 1 + \frac{1}{2n}$$

and hence that, when $n \to \infty$,

$$\lim\left\{\frac{2.4.6...(2n)}{1.3.5....(2n-1)}\right\}^2 \frac{1}{2n+1} = \frac{\pi}{2}. \tag{91}$$

This is called *Wallis's formula.* After taking square roots, it can be expressed in the form

$$\lim \frac{(2^n n!)^2}{(2n)!\sqrt{(2n+1)}} = \sqrt{(\tfrac{1}{2}\pi)}. \tag{92}$$

(12) *Stirling's formula.* The factorial $n!$ can be expressed asymptotically by the formula

$$n! = n^n e^{-n} \sqrt{(2\pi n)} (1 + \epsilon_n) \tag{93}$$

where $\epsilon_n \to 0$ when $n \to \infty$. This is known as *Stirling's formula*. To verify it, put

$$\phi(n) = n!/\{n^n e^{-n} \sqrt{(2\pi n)}\}.$$

Then, by simplifying the ratio $\phi(n+1)/\phi(n)$ and using Example 4, we find

$$\frac{\phi(n+1)}{\phi(n)} = e \Big/ \left(1 + \frac{1}{n}\right)^{n+\frac{1}{2}} < 1$$

from which it follows that $\phi(n)$ decreases monotonically as n increases and, being positive, must therefore tend to a limit. Hence we can put $\phi(n) = C + \epsilon_n$ and so

$$n! = n^n e^{-n} \sqrt{(2\pi n)} (C + \epsilon_n) \tag{94}$$

where C is independent of n and $\epsilon_n \to 0$ when $n \to \infty$.

To prove that $C = 1$, we can use Wallis's formula. Substituting for $n!$ and $(2n)!$ from (94) in (92), we find

$$\lim \frac{2^{2n} n^{2n} e^{-2n} 2\pi n (C + \epsilon_n)^2}{(2n)^{2n} e^{-2n} \sqrt{(4\pi n)} (C + \epsilon_{2n})} \frac{1}{\sqrt{(2n+1)}} = \sqrt{(\tfrac{1}{2}\pi)}$$

which reduces to $\sqrt{\pi} C / \sqrt{2} = \sqrt{(\tfrac{1}{2}\pi)}$, or $C = 1$.

Even for $n = 10$, the approximation given by putting $\epsilon_n = 0$ in (93) is within 1% of the true value; thus

$$10! = 3{,}628{,}800, \quad 10^{10} e^{-10} \sqrt{(20\pi)} \doteq 3{,}598{,}699.$$

Examples 9D

(1) Show that
$$\lim_{\lambda \to 0} \lambda \int_{\lambda^2}^{\lambda} \frac{\cos x \, dx}{x^{\frac{3}{2}}} = 2.$$

(2) If $f(x)$ is integrable, and continuous at $x = 0$, show that
$$\lim_{n \to \infty} \int_{1/(n+a)^2}^{1/n^2} \frac{f(x) \, dx}{x^{\frac{3}{2}}} = 2af(0) \quad (0 < a).$$

(3) Show that, if $f(x)$ is continuous,
$$\lim_{a \to 0} \int_0^{\sqrt{a}} \frac{af(x) \, dx}{a^2 + x^2} = \lim_{a \to 0} \int_0^1 \frac{af(x) \, dx}{a^2 + x^2} = \tfrac{1}{2}\pi f(0).$$

(4) If $0 < r < \tfrac{1}{2}m - 1$, where r and m are integers, and if
$$I_r = (-)^r \int_{r\pi/m}^{(r+1)\pi/m} \frac{\sin mx}{\sin x} dx$$
prove that $\qquad I_{r-1} > (2/m) \operatorname{cosec}(r\pi/m) > I_r.$

(5) If m is an integer (zero excluded) prove that
$$\lim_{a \to \infty} \int_{-a}^{a} \frac{\sin^2 \pi x}{x^2 - m^2} dx = 0.$$

Deduce that $\qquad \displaystyle\int_0^\infty \frac{\sin^2 \pi x}{x^2 - m^2} dx = 0.$

(6) If $f(x)$ is bounded and integrable over $[0, \pi]$, prove that, as $n \to \infty$ through integral values,

$$\lim \int_0^\pi f(x) \left| \sin nx \right| dx = \frac{2}{\pi} \int_0^\pi f(x)\, dx.$$

(7) Show that a number ξ exists $(0 < \xi < \pi)$ such that

$$\int_0^\pi (2\pi - x) \sin x\, dx = 2\pi \int_0^\xi \sin x\, dx + \pi \int_\xi^\pi \sin x\, dx$$

and find ξ. Also show that X exists $(0 < X < \pi)$ such that

$$\int_0^\pi (2\pi - x) \sin x\, dx = 2\pi \int_0^X \sin x\, dx.$$

(8) If $\sin x$ in the last example is replaced by $\cos x$, show that $\xi = \sin^{-1}(2/\pi)$, $X = \sin^{-1}(1/\pi)$.

(9) If $0 < a < b$ and $0 \leqslant n$, prove that

$$\left| \int_a^b \frac{\cos x}{x^n} dx \right| \leqslant \frac{2}{a^n}.$$

(10) If $ab > 0$, prove that

$$\int_a^b \frac{\sin nx}{x} dx \to 0 \quad (n \to \infty).$$

(11) If $f(x)$ is monotonic and differentiable, and $ab > 0$, prove that

$$\int_a^b f(x) \frac{\sin nx}{x} dx \to 0 \quad (n \to \infty).$$

(12) Show by considering separately the intervals $[0, 1/n]$ and $[1/n, \pi]$ that, if $1 \leqslant n$,

$$\int_0^\pi \frac{\sin nx}{\sqrt{x}} dx < \frac{3}{\sqrt{n}}.$$

Also show that the integral is positive.

(13) If $0 < \delta < 2/\pi$, prove that

$$\left| \int_\delta^{2/\pi} \sin \frac{1}{x} dx \right| \leqslant \frac{4}{\pi^2}.$$

(14) Prove that $$\left| \int_0^\pi \sin x \cos (x^2)\, dx \right| \leqslant \tfrac{1}{2}.$$

(15) If $f(x) \geqslant 0$ and $xf'(x) + f(x) \geqslant 0$ for $0 < a \leqslant x \leqslant b$, prove that

$$\left| \int_a^b f(x) \cos (\log x)\, dx \right| \leqslant 2bf(b).$$

(16) Prove that, when $a \to 0$,

$$\lim \int_a^1 \left(\frac{e^x - e^{3x}}{x} + \frac{2e^{2x+1}}{2x+1} \right) dx = \log 3.$$

Examples 9 E

(1) Given $\displaystyle\int_0^x \frac{dt}{(a^2-t^2)^{\frac12}} = \sin^{-1}\frac{x}{a}$, find $\displaystyle\int_0^x \frac{dt}{(a^2-t^2)^{\frac32}}$.

(2) Verify that, for every value of x,

$$\frac{d}{dx}\int_0^x |\cos t|\,dt = |\cos x|.$$

(3) Given $\displaystyle I = \int_0^a \frac{dx}{a^2+x^2}$, find $\dfrac{dI}{da}$ in two ways.

(4) Verify that, if $f'(x)$ is continuous,

$$\frac{d}{da}\int_0^a f(x)\,dx = \frac{d}{da}\int_0^a f(a-x)\,dx.$$

(5) If $0 < y < \pi$, and

$$G(y) = \int_0^\pi f(x,y)\,dx, \quad f(x,y) = \begin{cases} \sin x + \cos x & (x \leqslant y), \\ \cos(\pi-x) & (y < x), \end{cases}$$

find $G'(y)$ in two ways.

(6) Verify that $\displaystyle y = \int_0^x f(u)\sinh(x-u)\,du$

satisfies the equation $d^2y/dx^2 - y = f(x)$. Find the general solution of the equation $d^2y/dx^2 - y = \operatorname{sech} x$.

(7) Verify that, if $f(t)$ is continuous for $0 < t$, and $\beta \neq \alpha$, then

$$y = -\frac{1}{\alpha-\beta}\int_0^t \{e^{-\alpha(t-u)} - e^{-\beta(t-u)}\} f(u)\,du$$

satisfies the equation
$$d^2y/dt^2 + (\alpha+\beta)\,dy/dt + \alpha\beta y = f(t)$$
and the conditions
$$y = 0 \quad\text{and}\quad dy/dt = 0 \quad\text{at}\quad t = 0.$$

Evaluate y for $0 < t < 1$ and for $1 < t$ when

$$\alpha = 1, \quad \beta = 2 \quad\text{and}\quad f(t) = \mathbf{1}(t) - \mathbf{1}(t-1)$$

where $\mathbf{1}$ denotes Heaviside's unit function.

(8) Solve the equation
$$d^2y/dx^2 - dy/dx = e^{2x}.e^{e^x}.$$

(9) Verify that, if $f(t)$ is continuous for $0 < t$, then

$$y = \int_1^t (t-u)\,e^{\alpha(t-u)} f(u)\,du \quad (0 < t)$$

satisfies the equation $\quad d^2y/dt^2 - 2\alpha\,dy/dt + \alpha^2 y = f(t)$

and the conditions $\quad y = 0 \quad\text{and}\quad dy/dt = 0 \quad\text{at}\quad t = 1$.
Solve the equation
$$d^2y/dt^2 - 2\,dy/dt + y = e^t/t \quad (0 < t).$$

(10) Differentiate each of the following integral equations, and hence find the function $f(x)$ that satisfies it:

(i) $f(x) = 2 + \displaystyle\int_1^x \{f(t)\}^2 \, dt.$

(ii) $x^5 = \displaystyle\int_0^x (x+t)f(t) \, dt$, given that $f(x)$ is finite at $x = 0$.

(iii) $x^3 = \displaystyle\int_0^x (2x - 3u)f(u) \, du.$

(iv) $f'(x) = 3 - \displaystyle\int_1^x \dfrac{f(u)}{u^2} \, du$, given that $f(1) = 1$.

(11) If $f(x) = P \displaystyle\int_0^x f(t) \, dt$, where P is a polynomial in x, prove that

$$f(x) = AP \exp\left(\int P \, dx \right)$$

where A is an arbitrary constant.

(12) Given that for all positive values of x and y,

$$\int_x^{xy} f(t) \, dt = \phi(y),$$

where $\phi(y)$ is independent of x, find the most general forms of the functions $f(x)$ and $\phi(x)$.

(13) An ellipsoid of revolution is formed by rotating the ellipse $x^2/a^2 + y^2/b^2 = 1$ about the axis of x. An axial hole of length $2h$ is bored through it. Find h when the total surface area of the remaining part is a maximum.

(14) Prove that $\displaystyle\int_0^{\frac{1}{2}\pi} \dfrac{d\theta}{a^2 \cos^2 \theta + b^2 \sin^2 \theta} = \dfrac{\pi}{2ab}.$

Deduce the values of

$$\int_0^{\frac{1}{2}\pi} \dfrac{\cos^2 \theta \, d\theta}{(a^2 \cos^2 \theta + b^2 \sin^2 \theta)^2}, \qquad \int_0^{\frac{1}{2}\pi} \dfrac{\sin^2 \theta \, d\theta}{(a^2 \cos^2 \theta + b^2 \sin^2 \theta)^2}.$$

(15) If $2n\pi < \alpha < (2n+1)\pi$, and n is an integer, prove that

$$\int_0^\alpha \dfrac{d\theta}{1 - \cos \alpha \cos \theta} = (2n + \tfrac{1}{2})\,\pi/\sin \alpha.$$

Evaluate $\displaystyle\int_0^\alpha \dfrac{\cos \theta \, d\theta}{(1 - \cos \alpha \cos \theta)^2}, \qquad \int_0^\alpha \dfrac{d\theta}{(1 - \cos \alpha \cos \theta)^2}.$

(16) If $I_n(x) = \dfrac{1}{n! \, 2^n} \displaystyle\int_0^1 (1 - t^2)^n \cos xt \, dt$

and n is a positive integer, prove that

$$I_n = \left(-\dfrac{1}{x} \dfrac{d}{dx} \right)^n \int_0^1 \cos xt \, dt.$$

Find I_1, I_2, I_3.

(17) Prove that, if $0 < y < \pi$,

$$G(y) = \int_{-1}^{1} \frac{\sin y \, dx}{1 - 2x \cos y + x^2} = \frac{\pi}{2}.$$

Show that $G(y)$ is not continuous at $y = \pm n\pi$. Sketch the graph of $G(y)$ as a function of y.

(18) Show that $1/\sqrt{(1 - 2xy + y^2)}$ is not a continuous function of position at the points $(1, 1)$, $(-1, -1)$ in the xy plane.

Show also that the integral

$$G(y) = \int_{-1}^{1} \frac{dx}{\sqrt{(1 - 2xy + y^2)}}$$

is continuous for all y, but that $G'(y)$ is not continuous at $y = \pm 1$. Sketch the graphs of $G(y)$ and $G'(y)$.

Show also that, although $G'(1)$ does not exist, the integral

$$\int_{-1}^{1} \frac{\partial}{\partial y} \frac{1}{\sqrt{(1 - 2xy + y^2)}} \, dx$$

has the value -1 at $y = 1$, the mean of $G'(1 - 0)$ and $G'(1 + 0)$.

(19) Prove that, if $P(x)$ is a polynomial of degree r, and n is an integer greater than $r - 1$, and if $a < b < c$, then

$$\int_{a}^{b} \frac{P(x) \, dx}{(x - c)^{n+1}} = -\frac{1}{n!} \left(\frac{d}{dc} \right)^n \left\{ P(c) \log \frac{c - a}{c - b} \right\}.$$

(20) Show that, if $-\frac{1}{2}\pi < \alpha < \frac{1}{2}\pi$,

$$I(\alpha) = \int_{0}^{\pi} \frac{\log (1 + \sin \alpha \cos x)}{\cos x} \, dx = \pi\alpha$$

the integrand being defined at $x = \frac{1}{2}\pi$ so as to be continuous.
Sketch the graph of $I(\alpha)$.

(21) If $-1 < a < 1$, show that

$$\int_{0}^{\frac{1}{2}\pi} \log (1 + \cos \pi a \cos x) \sec x \, dx = \frac{1}{2}\pi^2(\frac{1}{4} - a^2).$$

(22) Show that $\displaystyle \lim_{a \to 0} \int_{0}^{1} \frac{a \, dx}{a^2 + x^2} \neq \int_{0}^{1} \lim_{a \to 0} \left(\frac{a}{a^2 + x^2} \right) dx.$

(23) Let $\displaystyle G(y) = \int_{0}^{1} \log (x^2 + y^2) \, dx.$

Calculate $G(y)$, and show that

$$\lim_{y \to 0} \frac{G(y) - G(0)}{y} \neq \int_{0}^{1} \lim_{y \to 0} \left\{ \frac{\log (x^2 + y^2) - \log (x^2)}{y} \right\} dx.$$

(24) Let $y = u(x)$, $y = v(x)$ be independent solutions of the linear equation

$$y'' + Py' + Qy = 0,$$

where P, Q denote functions of x, and let $W(x)$ denote the Wronskian of $u(x)$ and $v(x)$. Verify that the solution of the equation

$$y'' + Py' + Qy = f(x)$$

such that $y = y' = 0$ at $x = c$ can be expressed by

$$y = \int_c^x \frac{f(t)}{W(t)} \begin{vmatrix} u(t) & v(t) \\ u(x) & v(x) \end{vmatrix} dt.$$

(25) If $0 < \alpha$, $0 < \beta$, prove that

$$\int_0^{\frac{1}{2}\pi} \log\left(\alpha^2 \cos^2\theta + \beta^2 \sin^2\theta\right) d\theta = \pi \log \frac{\alpha+\beta}{2}.$$

(26) If $\qquad G(y) = \int_0^1 \frac{\log(1+xy)}{1+x^2} dx$, find $\dfrac{dG(y)}{dy}$

and hence find $G(1)$.

(27) Justify differentiating the integral

$$\int_0^{\frac{1}{2}\pi} \tan^{-1}(\alpha^2 \tan^2 x)\, dx$$

with respect to α under the sign of integration, and hence prove that the value of the integral is $\pi \tan^{-1}\{\alpha/(\alpha+\sqrt{2})\}$ $(0 < \alpha)$.

(28) If $\qquad G(y) = \int_0^{\frac{1}{2}\pi} \tan^{-1}(\sinh y \sin x)\, dx$ $\quad (0 < y)$,

prove that

$$G(y) = \int_0^y \frac{t\, dt}{\sinh t} = 2\left(\frac{\pi^2}{8} - \frac{1+y}{1^2}e^{-y} - \frac{1+3y}{3^2}e^{-3y} - \cdots\right).$$

10

INFINITE INTEGRALS

10.1 Infinite integrals

The definition of a definite integral in §9.1 required the integrand $f(x)$ to be bounded and the range of integration to be finite. When these conditions are not fulfilled, other definitions are necessary.

Definition. Let $f(x)$ be bounded and integrable for $x > a$. The definite integral of $f(x)\,dx$ from a to $+\infty$ is defined by

$$\int_a^\infty f(x)\,dx = \lim_{X\to\infty} \int_a^X f(x)\,dx \tag{1}$$

provided that the limit exists, in which case the integral on the left is called an *infinite integral of the first kind* and is said to *exist* or to *converge*.

If the integral on the R.H.S. of (1) does not tend to a limit, the infinite integral is said to diverge or not to exist or not to converge or to oscillate, whichever words may be appropriate.

[The word 'infinite' in 'infinite integral' is used with the same kind of meaning as in 'infinite series'.]

For example, when $X \to +\infty$, if $a > 0$,

(i) $\lim \displaystyle\int_a^X \frac{dx}{x^2} = \lim \left[-\frac{1}{x} \right]_a^X = \lim \left(-\frac{1}{X} + \frac{1}{a} \right) = \frac{1}{a}$;

(ii) $\displaystyle\int_a^X \frac{dx}{\sqrt{x}} = [2\sqrt{x}]_a^X = 2\sqrt{X} - 2\sqrt{a} \to +\infty$;

(iii) $\displaystyle\int_0^X \cos x\,dx = [\sin x]_0^X = \sin X$.

In case (i) the infinite integral exists or converges; in (ii) it diverges; in (iii) it oscillates between ± 1.

A similar definition applies when the lower limit of integration is $-\infty$: thus,

$$\int_{-\infty}^b f(x)\,dx = \lim_{X\to\infty} \int_{-X}^b f(x)\,dx. \tag{2}$$

The integral from $-\infty$ to $+\infty$ is defined by

$$\int_{-\infty}^\infty f(x)\,dx = \lim_{X\to\infty} \int_{-X}^a f(x)\,dx + \lim_{X\to\infty} \int_a^X f(x)\,dx, \tag{3}$$

where a is arbitrary, provided that both these limits exist, in which case we can put

$$\int_{-\infty}^{\infty} = \lim_{X \to \infty} \left(\int_{-X}^{a} + \int_{a}^{X} \right), \tag{4}$$

but not otherwise, as we see by putting, e.g. $f(x) = \sin x$.

10.2

Definition. If $f(x)$ becomes infinite or undefined at $x = b$, the definite integral of $f(x)\,dx$ from a to b $(a < b)$ is defined by

$$\int_{a}^{b} f(x)\,dx = \lim_{\delta \to 0} \int_{a}^{b-\delta} f(x)\,dx \tag{5}$$

provided that the limit exists, in which case the integral is called *an infinite integral of the second kind*, and is said to exist or to converge.

If $f(x)$ becomes infinite or undefined at $x = a$, a similar definition applies: thus,

$$\int_{a}^{b} f(x)\,dx = \lim_{\delta \to 0} \int_{a+\delta}^{b} f(x)\,dx \tag{6}$$

provided that the limit exists.

If there is a point $x = c$ between a and b at which $f(x)$ becomes infinite or undefined, then the integral from a to b, $a < b$, is defined by

$$\int_{a}^{b} = \lim_{\delta \to 0} \int_{a}^{c-\delta} + \lim_{\delta \to 0} \int_{c+\delta}^{b} \tag{7}$$

provided that both limits exist, in which case we can put

$$\int_{a}^{b} = \lim_{\delta \to 0} \left(\int_{a}^{c-\delta} + \int_{c+\delta}^{b} \right). \tag{8}$$

Example

Consider the integral $\int_{0}^{1} x^n\,dx$. If $\delta > 0$ and $n+1 \neq 0$,

$$\int_{\delta}^{1} x^n\,dx = \frac{1}{n+1} - \frac{\delta^{n+1}}{n+1},$$

which has the limit $1/(n+1)$ when $\delta \to 0$ if $n+1 > 0$. Hence

$$\int_{0}^{1} x^n\,dx = \frac{1}{n+1} \quad (-1 < n),$$

although, when $-1 < n < 0$, the integrand $x^n \to \infty$ as $x \to 0$.

Note 1. By a change in the variable of integration, an ordinary definite integral can be converted into an infinite integral; an infinite integral of the second kind can be converted into one of the first kind and vice versa. For example, by the substitution $x = 1/t$, we find

$$\int_0^1 x^n \, dx = \int_1^\infty \frac{dt}{t^{n+2}}.$$

The integral on the left is an ordinary integral if $0 \leqslant n$, a convergent infinite integral of the second kind if $-1 < n < 0$, divergent if $n \leqslant -1$. The integral on the right is an infinite integral of the first kind, convergent if $-1 < n$, divergent if $n \leqslant -1$.

Note 2. The constants in the bilinear relation

$$xy + Ax + By + C = 0$$

can be chosen so as to make any three values of y correspond to three given values of x. For instance, if we write this relation in the form

$$y = c\frac{x-a}{b-x}, \quad \frac{dy}{dx} = c\frac{b-a}{(b-x)^2},$$

then $y = 0$, $y = \infty$, $y = -c$ will correspond respectively to $x = a$, $x = b$, $x = \infty$; and if $a < b$ and $0 < c$, the variable y will increase steadily from $y = 0$ to $y = +\infty$ as x increases from a to b. If, however, $a < b$ and $c < 0$, then y will decrease from 0 to $-\infty$ as x increases from a to b.

10.3

While considering the above examples, it was easy to see whether the infinite integrals converged or not, because the indefinite integrals could be found and their behaviour observed. But the indefinite integrals cannot be found in many cases, e.g.

$$\int_{-\infty}^\infty \exp(-x^2) \, dx, \quad \int_a^\infty \sin(x^2) \, dx, \quad \int_0^\infty \frac{\log x \, dx}{a^2 + x^2} \qquad (9)$$

and we need some other means of deciding whether such integrals exist or not, and, if they do, of evaluating them. An algebraic step may be sufficient, as in the next example; other methods are indicated below.

Example

Prove that the integrals

$$I_1 = \int_0^{\frac{1}{2}\pi} \log \sin \theta \, d\theta, \quad I_2 = \int_0^{\frac{1}{2}\pi} \log \cos \theta \, d\theta$$

converge, and evaluate them.

Note that $\log \sin \theta \to -\infty$ as $\theta \to 0+$, and that $\log \cos \theta \to -\infty$ as $\theta \to \frac{1}{2}\pi - 0$.

If $0 < \theta < \pi$ we can put $\log \sin \theta = \log(\sin \theta / \theta) + \log \theta$ and hence

$$\int_0^{\frac{1}{2}\pi} \log \sin \theta \, d\theta = \int_0^{\frac{1}{2}\pi} \left(\log \frac{\sin \theta}{\theta} + \log \theta \right) d\theta.$$

The first of the two integrals on the R.H.S. is an ordinary integral of a continuous function over a finite range (taking $\sin \theta / \theta = 1$ at $\theta = 0$), and the second is easily shown to converge by finding its indefinite integral. Hence I_1 converges.

By making the substitution $\theta = \frac{1}{2}\pi - \theta'$ in I_2, we see that $I_2 = I_1$. It follows that

$$I_1 = I_2 = \tfrac{1}{2}(I_1 + I_2) = \frac{1}{2} \int_0^{\frac{1}{2}\pi} \log \left(\tfrac{1}{2} \sin 2\theta \right) d\theta.$$

Hence, by putting $2\theta = \phi + \frac{1}{2}\pi$,

$$I_1 + I_2 = \frac{1}{2} \int_{-\frac{1}{2}\pi}^{\frac{1}{2}\pi} (\log \cos \phi - \log 2) \, d\phi = \tfrac{1}{2}(2I_2 - \pi \log 2)$$

and therefore $\qquad\qquad I_1 = I_2 = -\tfrac{1}{2}\pi \log 2.$ \hfill (10)

10.4   Integrands which do not change sign

Suppose that $f(x) \geqslant 0$ for all $x \geqslant a$, so that the integral

$$F(X) = \int_a^X f(x) \, dx \quad (a < X)$$

increases monotonically as X increases.

Theorem. If $f(x) \geqslant 0$ for all $x \geqslant a$, then

$$\int_a^\infty f(x) \, dx \quad \text{converges if} \quad \int_a^X f(x) \, dx \quad \text{is bounded as} \quad X \to \infty.$$

Proof. Since $F(X)$ increases monotonically it must (§ 1.14) tend either to a limit or to $+\infty$ as $X \to +\infty$. If it is bounded, it must therefore tend to a limit.

In other words, if $f(x) \geqslant 0$ for all $x \geqslant a$, and if there exists a number K, independent of X, such that $F(X) < K$, then $F(X)$ converges.

Corollary 1. If $0 \leqslant f(x) \leqslant g(x)$ for all $x \geqslant a$ and if

$$G(X) = \int_a^X g(x) \, dx \quad \text{converges, then} \quad F(X) = \int_a^X f(x) \, dx \quad \text{converges.}$$

For in this case $\lim G(X)$, when $X \to \infty$, exists and we can put $K = \lim G(X)$.

Evidently, if $0 \leqslant g(x) \leqslant f(x)$ for all $x \geqslant a$ and if $G(X)$ diverges, then $F(X)$ diverges. For in this case $G(X)$, and hence $F(X)$, is unbounded.

Corollary 2. (i) If a positive constant C exists such that,

for all $x \geqslant a$, $0 \leqslant f(x) \leqslant C/x^s$ $(1 < s)$, then $F(X)$ converges.

(ii) If a positive constant C exists such that,

for all $x \geqslant a$, $C/x^s \leqslant f(x)$ $(s \leqslant 1)$, then $F(X)$ diverges.

Example

Show that if the constant p is positive, the following integrals converge, whatever the value of the constant n:

$$\text{(i)} \int_1^\infty x^n e^{-px} dx, \quad \text{(ii)} \int_1^\infty x^n e^{-px^2} dx.$$

(i) As in §2.19 (37), when $x \to +\infty$, $\lim (x^m e^{-px}) = 0$ for all m if $p > 0$; it follows that $x^s(x^n e^{-px}) \to 0$ for all s. Consequently, if C is any positive constant, there exists a value $x = a$ such that $x^s(x^n e^{-px}) < C$ for all $x > a$ and all s; and therefore such that $x^n e^{-px} < C/x^s$, with $1 < s$. Hence, by Corollary 2, integral (i) converges.

(ii) If $x > 1$, then $px^2 > px$ and so $x^n e^{-px^2} < x^n e^{-px}$. It follows from Corollary 1 that integral (ii) converges, since integral (i) converges.

10.5   The series-integral theorem (Maclaurin–Cauchy)

Let $f(x)$ be positive monotonic decreasing for all $x \geqslant 1$. Being monotonic, $f(x)$ is integrable (§9.9). Let

$$S_n = f(1) + f(2) + \ldots + f(n), \quad F_X = \int_1^X f(x) \, dx. \tag{11}$$

Theorem. The difference $S_n - F_n$ tends to a limit when $n \to \infty$, such that

$$0 \leqslant \lim (S_n - F_n) \leqslant f(1). \tag{12}$$

Proof. Since $f(x)$ is monotonic decreasing, if $1 \leqslant r \leqslant x \leqslant r+1$,

$$f(r) \geqslant f(x) \geqslant f(r+1)$$

and hence, by integration from $x = r$ to $x = r+1$,

$$f(r) \geqslant \int_r^{r+1} f(x) \, dx \geqslant f(r+1).$$

Putting $r = 1, 2, \ldots, n-1$ and adding, we get

$$S_n - f(n) \geqslant F_n \geqslant S_n - f(1)$$

from which and since $0 \leqslant f(n)$ follows

$$0 \leqslant f(n) \leqslant S_n - F_n \leqslant f(1); \tag{13}$$

thus $S_n - F_n$ is bounded below and above. Further,

$$S_{n+1} - F_{n+1} - (S_n - F_n) = S_{n+1} - S_n - (F_{n+1} - F_n)$$
$$= f(n+1) - (F_{n+1} - F_n)$$
$$= \int_n^{n+1} \{f(n+1) - f(x)\} \, dx \leqslant 0,$$

since $f(n+1)$ is constant and $f(n+1) \leqslant f(x)$, $n \leqslant x \leqslant n+1$. Consequently $S_n - F_n$ is a monotonic decreasing function of n and, being bounded below, must tend to a limit which, by (13), lies between 0 and $f(1)$.

Corollary. The series S_n and the integral F_n either both converge or both diverge.

For, since both S_n and F_n are monotonic increasing, each must either converge or diverge. Since $S_n - F_n$ converges, if either S_n or F_n converges, so must the other; if either diverges, so must the other.

Examples

(1) Putting $f(x) = 1/x$, we deduce from the theorem that

$$1 + \tfrac{1}{2} + \tfrac{1}{3} + \ldots + 1/n - \log n$$

tends to a limit between 0 and 1 when $n \to \infty$. Denoting the limit by γ, we can therefore put

$$1 + \tfrac{1}{2} + \tfrac{1}{3} + \ldots + 1/n - \log n = \gamma + \eta_n, \tag{14}$$

where γ is independent of n and $0 < \gamma < 1$, while $\eta_n \to 0$ when $n \to \infty$. The constant γ is called Euler's constant: its value $0 \cdot 57721566\ldots$ has been calculated to more than 200 places of decimals.

(2) Put $f(x) = 1/x^s$, $0 < s$ $(s \neq 1)$. Then $f(x)$ is monotonic decreasing and from (13) follows

$$0 < \frac{1}{n^s} \leqslant \frac{1}{1^s} + \frac{1}{2^s} + \ldots + \frac{1}{n^s} - \left(\frac{n^{1-s}}{1-s} - \frac{1}{1-s} \right) \leqslant 1.$$

By the theorem, the third term of these inequalities tends to a limit when $n \to \infty$. It follows that we can put

$$\frac{1}{1^s} + \frac{1}{2^s} + \ldots + \frac{1}{n^s} - \frac{n^{1-s}}{1-s} = \zeta(s) + \eta_n(s) \tag{15}$$

where $\zeta(s)$ is independent of n and $\eta_n(s) \to 0$ when $n \to \infty$. Then, by letting $n \to \infty$, we see that

$$0 \leqslant \zeta(s) + 1/(1-s) \leqslant 1$$

and hence that $\zeta(s)$ lies between $1/(s-1)$ and $s/(s-1)$.

In particular, if $s > 1$, $n^{1-s} \to 0$ and we can put

$$\frac{1}{1^s} + \frac{1}{2^s} + \ldots + \frac{1}{n^s} = \zeta(s) + \eta_n(s). \tag{16}$$

For instance, if $s = 2$,

$$\frac{1}{1^2} + \frac{1}{2^2} + \ldots + \frac{1}{n^2} = \zeta(2) + \eta_n(2), \tag{17}$$

where $\zeta(2)$ lies between 1 and 2. By other methods, it is known that $\zeta(2) = \frac{1}{6}\pi^2$.

The function $\zeta(s)$ when $s > 1$ is called Riemann's Zeta function.

10.6 Absolute and conditional convergence

Theorem. The infinite integral

$$\int_a^\infty f(x)\,dx \quad \text{converges if} \quad \int_a^\infty |f(x)|\,dx \quad \text{converges},$$

in which case the first integral is said to *converge absolutely* or to be *absolutely convergent*.

Proof. Since $0 \leqslant f(x) + |f(x)| \leqslant 2|f(x)|$, and since

$$\int_a^X 2|f(x)|\,dx \quad \text{converges, therefore} \quad \int_a^X \{f(x) + |f(x)|\}\,dx$$

converges, by §10.4, Corollary 1. Then, since identically

$$\int_a^X f(x)\,dx = \int_a^X \{f(x) + |f(x)|\}\,dx - \int_a^X |f(x)|\,dx$$

and both integrals on the R.H.S. converge, the theorem follows.

An analogous theorem holds good for infinite integrals of the second kind.

An infinite integral which converges but not absolutely is said to be *conditionally convergent*.

10.7

Theorem 1. If $1 < s$ and if $x^s f(x)$ is bounded as $x \to +\infty$, in particular if $x^s f(x)$ tends to a limit as $x \to +\infty$, then the integral

$$\int_a^\infty f(x)\,dx \quad (0 < a)$$

converges absolutely.

Proof. Suppose $-C < x^s f(x) < C, 0 < a \leqslant x$, where C is independent of x: then $|f(x)| < C/x^s$ and

$$\int_a^X |f(x)| \, dx < \int_a^X \frac{C}{x^s} dx = \frac{C}{s-1} \left(\frac{1}{a^{s-1}} - \frac{1}{X^{s-1}} \right) < \frac{C}{(s-1) \, a^{s-1}}$$

if $X > a$, since $s > 1$. Hence the integral on the left converges, by § 10.4. The theorem follows.

Theorem 2. Let $f(x)$ be undefined or infinite at $x = 0$.

If $s < 1$ and if $x^s f(x)$ is bounded, in particular if $x^s f(x)$ tends to a limit, as $x \to 0+$, then the integral

$$\int_0^a f(x) \, dx \quad (0 < a)$$

converges absolutely.

Proof. Suppose $-C < x^s f(x) < C, 0 < x \leqslant a$, where C is independent of x; then $|f(x)| < C/x^s$ and

$$\int_\delta^a |f(x)| \, dx < \int_\delta^a \frac{C}{x^s} dx = \frac{C}{1-s} (a^{1-s} - \delta^{1-s}) < \frac{Ca^{1-s}}{1-s}$$

since $s < 1$. Hence the integral on the left converges when $\delta \to 0$. The theorem follows.

Corollary. Let $f(x)$ be undefined or infinite at $x = a$.

If $s < 1$ and if $(x-a)^s f(x)$ is bounded, in particular if $(x-a)^s f(x)$ tends to a limit, as $x \to a+$, then the integral

$$\int_a^b f(x) \, dx \quad (a < b)$$

converges absolutely.

This follows from Theorem 2 by putting $x - a = \xi$, in which case $\xi \to 0+$ when $x \to a+$.

Example

Consider the convergence of the integrals

$$I_1 = \int_1^\infty \frac{x^{\alpha-1} \log x}{1+x} \, dx, \quad I_2 = \int_0^1 \frac{x^{\alpha-1} \log x}{1+x} \, dx.$$

Note that $\log x \to +\infty$ as $x \to +\infty$ in I_1, and that $\log x \to -\infty$ as $x \to 0$ in I_2.

Put $f(x) = x^{\alpha-1} \log x/(1+x)$. Then, after multiplying by x^s, we can put $x^s f(x)$ in the form

$$x^s f(x) = x^{s+\alpha-2} \log x/(1+x^{-1}).$$

It follows from §2.19 (34) that $x^s f(x)$ tends to a limit (zero) when $x \to \infty$, provided that $s + \alpha - 2 < 0$, and therefore, by Theorem 1, that I_1 will converge if s can be found to satisfy both of the inequalities $1 < s, s < 2 - \alpha$, and therefore if $1 < 2 - \alpha$, or $\alpha < 1$.

To consider I_2 we can put $x^s f(x)$ in the form

$$x^s f(x) = x^{s+\alpha-1} \log x / (1+x),$$

and it follows from §2.19 (35) that $x^s f(x)$ tends to a limit (zero) when $x \to 0+$, provided that $s + \alpha - 1 > 0$, and therefore, by Theorem 2, that I_2 will converge if s can be found to satisfy both of the inequalities $s < 1, 1 - \alpha < s$, and therefore if $1 - \alpha < 1$, or $0 < \alpha$.

A deduction which follows from these results for I_1 and I_2 is that the integral

$$\int_0^\infty \frac{x^{\alpha-1} \log x}{1+x} \, dx$$

will exist if $0 < \alpha < 1$. It may be verified that it will not exist at the lower limit if $\alpha \leqslant 0$ or at the upper limit if $1 \leqslant \alpha$.

10.8   A test for conditional convergence

Theorem. If $g(x) = x^s f(x) \; (0 < s)$ and $f(x)$ is continuous, and if, as $X \to \infty$, the integral

$$G(X) = \int_a^X g(x) \, dx \quad \text{is bounded, then} \quad \int_a^X f(x) \, dx \quad (0 < a)$$

converges.

[The condition $0 < a$ is imposed merely to avoid considering the lower limit $a = 0$. This restriction can be removed if the integral is known to converge when $a = 0$.]

Proof. Put $G(x) = \int_a^x g(t) \, dt$; then $dG(x) = g(x) \, dx$, since $g(x)$ is continuous, and by integrating by parts,

$$\int_a^X f(x) \, dx = \int_a^X \frac{g(x) \, dx}{x^s} = \int_a^X \frac{1}{x^s} \, dG(x) = \left[\frac{G(x)}{x^s} \right]_a^X + \int_a^X \frac{s G(x) \, dx}{x^{s+1}}.$$

Since $G(a) = 0$, the first term on the R.H.S. is equal to $G(X)/X^s$, which tends to zero when $X \to \infty$, because $G(X)$ is bounded. The second term on the R.H.S. is an absolutely convergent integral, by §10.7, Theorem 1. It follows that the integral on the L.H.S. converges, and moreover that its limiting value is the same as that of the integral on the R.H.S. (It does not follow that the integral on the L.H.S. converges absolutely.)

Example

Consider the convergence of the integral

$$I = \int_0^\infty \frac{\sin x}{x^s}\,dx. \tag{18}$$

Put $f(x) = (\sin x)/x^s$. Then $x^{s-1}f(x) \to 1$ when $x \to 0$, and it follows from §10.7, Theorem 2, that I converges at its lower limit if $s - 1 < 1$, or $s < 2$.

At its upper limit

(i) if $1 < s$, then I converges absolutely, by §10.7, Theorem 1, since $\sin x$ is bounded as $x \to \infty$;

(ii) if $0 < s \leqslant 1$, convergence follows from the present theorem, since

$$G(X) = \int_0^X \sin x\,dx = 1 - \cos X$$

is bounded as $X \to \infty$. In this case, convergence is not absolute; to prove this, let n be a positive integer and put

$$I_n = \int_0^{n\pi} \frac{|\sin x|}{x^s}\,dx \geqslant \int_0^{n\pi} \frac{|\sin x|}{x}\,dx$$

since $1/x^s \geqslant 1/x$ if $0 < s \leqslant 1$; hence

$$I_n \geqslant \int_0^\pi \frac{\sin x\,dx}{x} - \int_\pi^{2\pi} \frac{\sin x\,dx}{x} + \int_{2\pi}^{3\pi} \frac{\sin x\,dx}{x} - \dots$$

to n terms. Now $1/x \geqslant 1/\pi$ in the first integral on the R.H.S.; in the second $1/x \geqslant 1/2\pi$; in the third $1/x \geqslant 1/3\pi, \dots$. Consequently

$$I_n > \int_0^\pi \frac{\sin x\,dx}{\pi} - \int_\pi^{2\pi} \frac{\sin x\,dx}{2\pi} + \int_{2\pi}^{3\pi} \frac{\sin x\,dx}{3\pi} - \dots$$

$$= \frac{2}{\pi}\left(1 + \tfrac{1}{2} + \tfrac{1}{3} + \dots + \tfrac{1}{n}\right) = \frac{2}{\pi}(\log n + \gamma + \eta_n)$$

by (14). It follows that $I_n \to +\infty$ when $n \to \infty$, and hence that the original integral is not absolutely convergent.

10.9 Cauchy's principle of convergence

The following theorem, known as *Cauchy's general principle of convergence*, applies to infinite integrals and to all problems concerned with limits.

Theorem. Let $f(x)$ be defined for all $x > a$.

A necessary and sufficient condition that $f(x)$ should tend to a limit as $x \to +\infty$ is that, for every $\epsilon > 0$, a number $x_0(\epsilon)$, that is, a number x_0 depending on ϵ, should exist such that

$$|f(x_2) - f(x_1)| < \epsilon \qquad (19)$$

for all $x_1 > x_0$ and all $x_2 > x_0$.

The condition is necessary: that is, if $f(x)$ tends to a limit, then (19) must be satisfied.

Proof. Let $f(x) \to l$ when $x \to +\infty$. Then, by definition of a limit, for every $\epsilon > 0$ a number x_0 exists such that

$$|f(x) - l| < \tfrac{1}{2}\epsilon$$

for all $x > x_0$. Hence, for all $x_1 > x_0$ and $x_2 > x_0$,

$$|f(x_2) - f(x_1)| \equiv |f(x_2) - l - \{f(x_1) - l\}|$$
$$\leqslant |f(x_2) - l| + |f(x_1) - l| < \tfrac{1}{2}\epsilon + \tfrac{1}{2}\epsilon = \epsilon;$$

the condition (19) is therefore satisfied.

The condition is sufficient: that is, if (19) is satisfied, then $f(x)$ must tend to a limit.

Proof. First we note that, if the condition is satisfied, then $f(x)$ is bounded. For, since

$$f(x_2) \equiv f(x_2) - f(x_1) + f(x_1)$$

it follows from (19) that, if $x_1 > x_0$ and $x_2 > x_0$, then

$$|f(x_2)| \leqslant |f(x_2) - f(x_1)| + |f(x_1)| < \epsilon + |f(x_1)|$$

and by supposing x_1 fixed and x_2 variable, we see that $f(x_2)$ and therefore $f(x)$ is bounded.

Next, in any interval $x_0 < x < \infty$, the function $f(x)$, being bounded, has (*a*) a least upper bound which decreases monotonically to a limit J as $x_0 \to \infty$, (*b*) a greatest lower bound which increases monotonically to a limit j as $x_0 \to \infty$. To show that $f(x)$ tends to a limit, it must be proved that $j = J$, in which case it will follow that $\lim f(x)$ exists and in fact that $\lim f(x) = j = J$.

Now by definition of J and j, for every $\epsilon > 0$ and for every x_0, numbers $x_1 > x_0$ and $x_2 > x_0$ exist such that

$$f(x_1) > J - \epsilon, \quad f(x_2) < j + \epsilon \qquad (20)$$

and hence such that

$$f(x_1) - f(x_2) > J - j - 2\epsilon. \qquad (21)$$

Again by definition of J and j, the inequality $J < j$ is impossible, so if we prove that $J > j$ is also impossible it will follow that $J = j$.

Assume therefore that $J > j$ is possible and put $J - j = \eta > 0$. Also let $\epsilon = \frac{1}{3}\eta$. Then, by (21), for this particular $\epsilon > 0$,

$$f(x_1) - f(x_2) > \eta - \frac{2}{3}\eta = \frac{1}{3}\eta = \epsilon.$$

But this is contrary to (19), which holds good for every $\epsilon > 0$. Thus $J > j$ is not possible and $J = j$ follows.

Corollary. A necessary and sufficient condition for the convergence of the infinite integral (1) is that, for every $\epsilon > 0$, a number $X_0(\epsilon)$ should exist such that

$$\left| \int_{X_1}^{X_2} f(x)\,dx \right| < \epsilon \tag{22}$$

for all $X_1 > X_0$ and all $X_2 > X_0$.

Example

Consider again when $0 < s$ the integral (18), viz.

$$I = \int_0^\infty \frac{\sin x}{x^s}\,dx.$$

Suppose $0 < X_0 < X_1 < X_2$. Then, since $1/x^s$ is positive and monotonically decreasing, and $\sin x$ is continuous, it follows from Bonnet's form of the second mean value theorem (§9.31, Corollary 1) that a number X exists $(X_1 < X < X_2)$ such that

$$\int_{X_1}^{X_2} \frac{\sin x\,dx}{x^s} = \frac{1}{X_1^s} \int_{X_1}^{X} \sin x\,dx = \frac{1}{X_1^s}(\cos X_1 - \cos X). \tag{i}$$

Now suppose $\epsilon > 0$ given, and let X_0 be given by $X_0^s = 2/\epsilon$. Then

$$\left| \int_{X_1}^{X_2} \frac{\sin x\,dx}{x^s} \right| \leqslant \frac{1}{X_1^s}(1+1) < \frac{2}{X_0^s} = \epsilon$$

and it follows, by the corollary, that I converges.

10.10   Uniform convergence

The infinite integral

$$G(y) = \int_a^\infty f(x, y)\,dx, \tag{23}$$

if convergent, is a function of the parameter y. Whether the integral can be evaluated in finite terms or not, the questions may arise (cf. §9.26) as to whether it is a continuous function of y and whether $G'(y)$, if it exists, can be found by differentiating under the sign of integration to give correctly

$$\frac{d}{dy} G(y) = \int_a^\infty \frac{\partial}{\partial y} f(x, y)\,dx. \tag{24}$$

For example, on the first question, the integral

$$G(y) = \int_0^\infty y e^{-xy} dx$$

converges for all values of $y \geqslant 0$, but $G(y)$ is not continuous at $y = 0$. For $G(y) = 1$ if $y > 0$, while $G(0) = 0$.

Similar remarks apply to infinite integrals of the second kind.

To lay down sufficient conditions under which such integrals are continuous and differentiable as in (24) requires the notion of *uniform* convergence.

10.11   Uniform convergence of a function

Let the function $f(x, y)$ tend to a limit $g(y)$ when $x \to \infty$. Also let

$$f(x, y) - g(y) = \eta(x, y) \tag{25}$$

so that $\eta(x, y) \to 0$ when $x \to \infty$.

Definition. The function $f(x, y)$ is said to tend to the limit $g(y)$, or to converge to $g(y)$, *uniformly wo y in an interval* $\alpha \leqslant y \leqslant \beta$, provided that, for every $\epsilon > 0$, a number $x_0(\epsilon)$ exists, *independent of y*, such that for all $x > x_0$ and all y in $\alpha \leqslant y \leqslant \beta$,

$$|f(x, y) - g(y)| = |\eta(x, y)| < \epsilon, \tag{26}$$

in other words, such that (26) is satisfied throughout the rectangle $x_0 < x < \infty, \alpha \leqslant y \leqslant \beta$.

Examples

(1) Consider $f(x, y) = e^{-xy}$ when $x \to \infty$ and y belongs to the interval $0 \leqslant y \leqslant K$.

Here $g(y) = 0$ if $0 < y$, while $g(0) = 1$; so that $f(x, y)$ converges for all y in $0 \leqslant y \leqslant K$. But it does not converge uniformly wo y, for, if $0 < y$,

$$f(x, y) - g(y) = \eta(x, y) = e^{-xy} - 0 = e^{-xy}$$

and so $|\eta(x, y)| < \epsilon$ requires $e^{xy} > 1/\epsilon$, or $x > (1/y) \log (1/\epsilon)$ which $\to \infty$ when $y \to 0$. Accordingly, no x_0 exists that will serve for all values of y in the interval $0 \leqslant y \leqslant K$, and therefore, by definition, e^{-xy} is not *uniformly* convergent in this interval.

It is uniformly convergent in the interval $\alpha \leqslant y \leqslant K$, where $0 < \alpha$, for we could take $x_0 = (1/\alpha) \log (1/\epsilon)$ for every y in this interval.

(2) Consider $f(x, y) = xy/(1 + x^2 y^2)$ when $x \to +\infty$.

Evidently, for all values of y, $f(x, y) \to 0$, so that $g(y) = 0$ and $\eta(x, y) = f(x, y) - 0 = xy/(1 + x^2 y^2)$. Then we seek x_0 so that, if possible, $xy/(1+x^2y^2)$ may be numerically less than any given $\epsilon > 0$ for all $x > x_0$. Suppose $y \geqslant 0$.

If $y = 0$, plainly any x_0 will serve.

If $y \neq 0$, the fraction $xy/(1+x^2y^2)$ lies between $\pm \frac{1}{2}$ (when $xy = \pm 1$). Take $\epsilon = \frac{3}{10}$. In order that the fraction may be less than $\frac{3}{10}$ for all $x > x_0$, the least value of x_0 is given by $x_0 y/(1+x_0^2 y^2) = \frac{3}{10}$, which leads to $x_0 y = 3$, $x_0 = 3/y$. It is therefore impossible to find x_0 independent of y, even for $\epsilon = \frac{3}{10}$, in any interval of y that includes $y = 0$. It follows that $f(x, y)$ is non-uniformly convergent in any such interval. The same is evidently true also when $x \to -\infty$.

A similar definition applies when a function $f(x, y)$ converges to a limit $g(y)$ as x tends to a finite value $x = a$. In this case we say that the convergence is uniform wo y in an interval $\alpha \leqslant y \leqslant \beta$ provided that for every $\epsilon > 0$ a positive number δ exists, independent of y, such that, for all values of x satisfying $|x-a| < \delta$ and all y in the interval $\alpha \leqslant y \leqslant \beta$, the inequality (26) is satisfied.

(3) Consider $f(x, y) = (\sin xy)/x$ when $x \to 0$.

In this case, for all values of y,

$$g(y) = y, \quad f(x, y) - g(y) = \eta(x, y) = (\sin xy)/x - y.$$

Now we can put $\sin xy = xy - \frac{1}{6}(xy)^3 \cos(\theta xy)$, by Maclaurin's formula, for some value of θ between 0 and 1, and hence

$$|\eta(x, y)| = |\tfrac{1}{6}x^2 y^3 \cos(\theta xy)| \leqslant \tfrac{1}{6}|x^2 y^3|.$$

Let y belong to the interval $-K \leqslant y \leqslant K$. Then $\frac{1}{6}|x^2 y^3| < \epsilon$ for all $|x| < \delta$ if we put $\frac{1}{6}\delta^2 K^3 = \epsilon$. Since this gives $\delta = \sqrt{(6\epsilon/K^3)}$, independent of y, it follows from the definition that $(\sin xy)/x$ converges to the limit y, uniformly wo y when $x \to 0$.

10.12

Theorem. A necessary and sufficient condition that, when $x \to +\infty$, $f(x, y)$ should converge uniformly wo y in an interval $\alpha \leqslant y \leqslant \beta$ is that, for every $\epsilon > 0$, a number $x_0(\epsilon)$ should exist, independent of y, such that

$$|f(x_2, y) - f(x_1, y)| < \epsilon \tag{27}$$

for all $x_1 > x_0$ and all $x_2 > x_0$ (cf. §10.9).

Proof. The condition is necessary. For suppose that $f(x, y) \to g(y)$ uniformly wo y when $x \to \infty$. Then, for every $\epsilon > 0$, a number $x_0(\epsilon)$ exists, independent of y, such that

$$|f(x, y) - g(y)| < \tfrac{1}{2}\epsilon \quad \text{for all} \quad x > x_0.$$

Accordingly, if $x_1 > x_0$ and $x_2 > x_0$,

$$|f(x_2, y) - f(x_1, y)| \equiv |f(x_2, y) - g(y) - \{f(x_1, y) - g(y)\}|$$
$$\leqslant |f(x_2, y) - g(y)| + |f(x_1, y) - g(y)|$$
$$< \tfrac{1}{2}\epsilon + \tfrac{1}{2}\epsilon = \epsilon,$$

thus, condition (27) is satisfied.

The condition is sufficient. For let (27) be satisfied. Then $f(x, y)$ converges for every y in $[\alpha, \beta]$, by §10.9. Let it converge to $g(y)$. Then, keeping x_2 fixed in (27), let $x_1 \to \infty$; we thus obtain

$$|f(x_2, y) - g(y)| \leqslant \epsilon$$

which shows that convergence is uniform, since this inequality holds good for all $x_2 > x_0$ independent of y.

10.13

The next theorem shows that the uniform convergence of $f(x, y)$ to $g(y)$ is sufficient to ensure the continuity of $g(y)$. The converse is not true (see §10.11, Example 2).

Theorem. In the rectangle $a \leqslant x < \infty$, $\alpha \leqslant y \leqslant \beta$,

   (i) let $f(x, y)$ be a continuous function of y for every fixed x,

   (ii) let $f(x, y) \to g(y)$ when $x \to \infty$, uniformly wo y;

then $g(y)$ is a continuous function of y in the interval $\alpha \leqslant y \leqslant \beta$.

Proof. Let y_0 be an interior point of $[\alpha, \beta]$. Then we have to prove that, given any $\epsilon > 0$, a number $\delta(\epsilon)$ exists such that

$$|g(y) - g(y_0)| < \epsilon \quad \text{for all } y \text{ such that} \quad |y - y_0| < \delta. \tag{28}$$

In the notation of (25), we have

$$g(y) - g(y_0) \equiv f(x, y) - f(x, y_0) - \eta(x, y) + \eta(x, y_0)$$

and hence

$$|g(y) - g(y_0)| \leqslant |f(x, y) - f(x, y_0)| + |\eta(x, y)| + |\eta(x, y_0)|. \tag{29}$$

Now, by (ii), for every $\epsilon > 0$, a number x_0 exists, independent of y, such that $|\eta(x, y)| < \tfrac{1}{3}\epsilon$ for all $x > x_0$ and all y in $[\alpha, \beta]$. Let any $x > x_0$ be chosen. Then, by (i), a number δ exists such that

$$|f(x, y) - f(x, y_0)| < \tfrac{1}{3}\epsilon \quad \text{for all } y \text{ such that} \quad |y - y_0| < \delta.$$

It follows that, for this number δ, from (29),

$$|g(y) - g(y_0)| \leqslant \tfrac{1}{3}\epsilon + \tfrac{1}{3}\epsilon + \tfrac{1}{3}\epsilon = \epsilon$$

as was to be proved.

The necessary modifications in the proof when y_0 is an end point of the interval $[\alpha, \beta]$ can easily be made.

10.14 Uniform convergence of infinite integrals

We consider infinite integrals of the first kind, leaving the reader to supply any modifications necessary for infinite integrals of the second kind. Let

$$F(X,y) = \int_a^X f(x,y)\,dx, \quad G(y) = \int_a^\infty f(x,y)\,dx, \tag{30}$$

supposing that the integral $F(X,y)$ converges to $G(y)$ when $X \to \infty$. Put $F(X,y) - G(y) = \eta(X,y)$, that is

$$\int_a^X f(x,y)\,dx - \int_a^\infty f(x,y)\,dx = \eta(X,y) = -\int_X^\infty f(x,y)\,dx. \tag{31}$$

Definition. (cf. §10.11). The integral $F(X,y)$ is said to converge to $G(y)$ *uniformly* wo y in an interval $\alpha \leqslant y \leqslant \beta$ provided that, for every $\epsilon > 0$, a number $X_0(\epsilon)$ exists, *independent of y*, such that for all $X > X_0$ and all y in $\alpha \leqslant y \leqslant \beta$

$$|\eta(X,y)| = \left| \int_X^\infty f(x,y)\,dx \right| < \epsilon. \tag{32}$$

10.15

Theorem. A necessary and sufficient condition that the integral $F(X,y)$ should converge uniformly wo y in the interval $\alpha \leqslant y \leqslant \beta$ is that, for every $\epsilon > 0$, a number $X_0(\epsilon)$ should exist, independent of y, such that

$$|F(X_2,y) - F(X_1,y)| = \left| \int_{X_1}^{X_2} f(x,y)\,dx \right| < \epsilon \tag{33}$$

for all $X_1 > X_0$ and all $X_2 > X_0$.

The proof follows the same lines as that of §10.12.

10.16 Three theorems depending on uniform convergence

The following theorems are concerned with the infinite integral

$$G(y) = \int_a^\infty f(x,y)\,dx \tag{34}$$

as a function of the parameter y. Sufficient conditions are stated for (1) the continuity of $G(y)$, (2) the integration of $G(y)$ 'under the integral sign', (3) the differentiation of $G(y)$ 'under the integral sign'.

Theorem 1. (i) Let $f(x,y)$ be a continuous function of position in the rectangle $a \leqslant x < \infty$, $\alpha \leqslant y \leqslant \beta$.

(ii) Let the infinite integral $G(y)$ converge uniformly wo y in the interval $\alpha \leqslant y \leqslant \beta$.

Then $G(y)$ is a continuous function of y in this interval.

Proof. By (i) and §9.26, Theorem 1, the integral $F(X, y)$ in (30) is a continuous function of y for every fixed X.

Also $F(X, y) \to G(y)$ uniformly wo y. It follows from §10.13, after replacing $f(x, y)$ there by $F(X, y)$ and $g(y)$ by $G(y)$ that $G(y)$ is continuous.

Theorem 2. Under conditions (i) and (ii) of Theorem 1,

$$\int_{\alpha}^{\gamma} dy \int_{a}^{\infty} f(x, y)\, dx = \int_{a}^{\infty} dx \int_{\alpha}^{\gamma} f(x, y)\, dy \tag{35}$$

where $\alpha \leqslant \gamma \leqslant \beta$.

Proof. By definition of uniform convergence, for every $\epsilon > 0$ a number X_0 exists, independent of y, such that

$$|F(X, y) - G(y)| < \epsilon/(\beta - \alpha) \quad \text{for all} \quad X > X_0. \tag{36}$$

As in the proof of Theorem 1, $F(X, y)$ is a continuous function of y for every fixed X, and $G(y)$ is continuous by Theorem 1; consequently, we may integrate $F(X, y) - G(y)$ wo y within the interval $\alpha \leqslant y \leqslant \beta$ and by using §9.10, (30), we deduce that

$$\left| \int_{\alpha}^{\gamma} dy \left\{ \int_{a}^{X} f(x, y)\, dx - \int_{a}^{\infty} f(x, y)\, dx \right\} \right| < \frac{\epsilon(\gamma - \alpha)}{\beta - \alpha} \leqslant \epsilon,$$

where $\alpha \leqslant \gamma \leqslant \beta$, for all $X > X_0$.

By §9.35, the order of integration in the first repeated integral may be reversed, giving

$$\left| \int_{a}^{X} dx \int_{\alpha}^{\gamma} f(x, y)\, dy - \int_{\alpha}^{\gamma} dy \int_{a}^{\infty} f(x, y)\, dx \right| \leqslant \epsilon$$

for all $X > X_0$. It follows that the L.H.S. tends to zero when $X \to \infty$. The result is (35).

Theorem 3. (i) Let $f(x, y)$ and $(\partial/\partial y) f(x, y)$ be continuous functions of position in the rectangle $a \leqslant x < \infty$, $\alpha \leqslant y \leqslant \beta$.

(ii) Let

$$\int_{a}^{\infty} f(x, y)\, dx \quad \text{converge and let} \quad \int_{a}^{\infty} \frac{\partial}{\partial y} f(x, y)\, dx$$

converge uniformly wo y in the interval $\alpha \leqslant y \leqslant \beta$. Then, in this interval,

$$\frac{d}{dy} \int_{a}^{\infty} f(x, y)\, dx = \int_{a}^{\infty} \frac{\partial}{\partial y} f(x, y)\, dx. \tag{37}$$

Proof. Put

$$G(y) = \int_{a}^{\infty} f(x, y)\, dx, \quad H(y) = \int_{a}^{\infty} \frac{\partial f}{\partial y}\, dx.$$

By Theorem 2, we may replace $f(x, y)$ in (35) by $(\partial/\partial y)\, f(x, y)$. After doing so and integrating wo y on the R.H.S. we obtain

$$\int_{\alpha}^{\gamma} H(y)\, dy = \int_{a}^{\infty} dx\, [f(x, \gamma) - f(x, \alpha)] = G(\gamma) - G(\alpha)$$

where $\alpha \leqslant \gamma \leqslant \beta$.

By Theorem 1, $H(y)$ is a continuous function of y, and so, by §9.16, Theorem 2, the L.H.S. has the derivative $H(\gamma)$ wo γ. It follows that the R.H.S. can be differentiated and that $H(\gamma) = G'(\gamma)$, which is equivalent to (37).

Note. The conditions stated in the above theorems are *sufficient* conditions. Thus, in Theorem 1, in order that $G(y)$ may be continuous, it is *sufficient*, but not *necessary*, that $F(X, y)$ should converge uniformly to $G(y)$.

10.17   Tests for uniform convergence

The following theorems provide tests which, when they are satisfied by a given convergent integral, are *sufficient* to enable one to decide that it converges uniformly.

Theorem 1. If there exists a positive function $g(x)$, independent of y, such that $|f(x, y)| \leqslant g(x)$ in the rectangle $a \leqslant x < \infty$, $\alpha \leqslant y \leqslant \beta$, and such that

$$\int_{a}^{\infty} g(x)\, dx \quad \text{converges, then} \quad \int_{a}^{\infty} f(x, y)\, dx \quad \text{converges}$$

uniformly wo y in the interval $\alpha \leqslant y \leqslant \beta$.

Proof. Let (X_1, y) and (X_2, y) be two points with the same y in the said rectangle, and $X_1 < X_2$. Then

$$\left| \int_{X_1}^{X_2} f(x, y)\, dx \right| \leqslant \int_{X_1}^{X_2} |f(x, y)|\, dx \leqslant \int_{X_1}^{X_2} g(x)\, dx$$

from which the proof can easily be completed with the aid of §§10.9 and 10.15.

Corollary. The theorem holds good in particular when $g(x) = C/x^s$ with $s > 1$ and C independent of x and y.

Note. A similar theorem and corollary apply to infinite integrals of the second kind, *mutatis mutandis.*

Theorem 2. If $g(x)$ is continuous and if

$$\int_{a}^{\infty} g(x)\, dx \quad \text{converges, then} \quad \int_{a}^{\infty} e^{-xy} g(x)\, dx$$

converges uniformly wo y in any interval $0 \leqslant y \leqslant \beta$.

Proof. If $0 \leqslant y$ the factor e^{-xy}, as a function of x, is positive monotonic decreasing for all x. Hence, by §9.31, (80), since $g(x)$ is continuous, if $X_1 < X_2$ a number X exists $(X_1 \leqslant X \leqslant X_2)$ such that

$$\left| \int_{X_1}^{X_2} e^{-xy} g(x)\, dx \right| = \left| e^{-X_1 y} \int_{X_1}^{X} g(x)\, dx \right| \leqslant \left| \int_{X_1}^{X} g(x)\, dx \right|$$

from which the proof can easily be completed with the aid of §§ 10.9 and 10.15.

Note. The factor e^{-xy} may be replaced by any positive function $f(x,y)$ which decreases monotonically as x increases and is such that $f(a,y) < k$, where k is independent of y. Also, $g(x)$ may be replaced by $g(x,y)$ provided that

$$\int_a^\infty g(x,y)\, dx \text{ converges uniformly.}$$

Theorem 3. If $g(x)$ is continuous and if, as $X \to \infty$,

$$\int_a^X g(x)\, dx \quad \text{is bounded, then} \quad \int_a^\infty e^{-xy} g(x)\, dx$$

converges uniformly wo y in any interval $0 < \alpha \leqslant y \leqslant \beta$.

Proof. Let

$$\int_a^X g(x)\, dx = G(X) \quad \text{and suppose} \quad |G(X)| < K,$$

where K is independent of X.

Since $g(x)$ is continuous, by §9.31, (80), if $X_1 < X_2$ a number X exists $(X_1 \leqslant X \leqslant X_2)$ such that

$$\int_{X_1}^{X_2} e^{-xy} g(x)\, dx = e^{-X_1 y} \int_{X_1}^{X} g(x)\, dx = e^{-X_1 y} \{ G(X) - G(X_1) \}.$$

Now, given $\epsilon > 0$, let X_0 be chosen so that $e^{-X_0 \alpha} = \epsilon/2K$, and take $X_1 > X_0$. Then, since $e^{-X_1 y} < e^{-X_0 \alpha}$,

$$\left| \int_{X_1}^{X_2} e^{-xy} g(x)\, dx \right| < e^{-X_0 \alpha} \{ |G(X)| + |G(X_1)| \} \leqslant \frac{\epsilon}{2K} (K + K) = \epsilon.$$

Since the choice of X_0 was independent of y, the theorem follows.

Note. The factor e^{-xy} may be replaced by any function $f(x,y)$ decreasing monotonically to zero as x increases, uniformly wo y. Also, $g(x)$ may be replaced by $g(x,y)$ provided that $\int_a^X g(x,y)\, dx$, as $X \to \infty$, lies within bounds which are independent of y.

Examples

(1) Evaluate the integral

$$G(y) = \int_0^\infty e^{-xy} \frac{\sin x}{x} dx \quad (0 \leqslant y).$$

Suppose $0 < y$; then (§10.16, Theorem 3) we can differentiate wo y under the integral sign, since the resulting integral is uniformly convergent (§10.17, Theorem 3); thus

$$G'(y) = \int_0^\infty -e^{-xy} \sin x \, dx = -\frac{1}{1+y^2}$$

and hence, by integration, $G(y) = A - \tan^{-1} y$, where A is independent of y. Now let $y \to \infty$, then $\tan^{-1} y \to \frac{1}{2}\pi$ and, since $|(\sin x)/x| \leqslant 1$,

$$|G(y)| < \int_0^\infty e^{-xy} dx = \frac{1}{y} \to 0.$$

It follows that $A = \frac{1}{2}\pi$ and that

$$G(y) = \int_0^\infty e^{-xy} \frac{\sin x}{x} dx = \frac{1}{2}\pi - \tan^{-1} y \quad (0 < y). \tag{38}$$

Further, by Theorem 2 and the Example of §10.8 with $s = 1$, the integral $G(y)$ converges uniformly in an interval $0 \leqslant y \leqslant \beta$. It is therefore continuous at $y = 0$ (§10.16, Theorem 1). Also, $\tan^{-1} y$ is continuous at $y = 0$. Hence, by letting $y \to 0$ in (38),

$$G(0) = \int_0^\infty \frac{\sin x}{x} dx = \frac{1}{2}\pi. \tag{39}$$

It easily follows by substituting ax for x that

$$\int_0^\infty \frac{\sin ax}{x} dx = \frac{1}{2}\pi \quad (a > 0),$$

$$= 0 \quad (a = 0),$$

$$= -\frac{1}{2}\pi \quad (a < 0). \tag{40}$$

(2) Consider the three related integrals

$$I = \int_0^\infty \frac{\sin xy}{x(1+x^2)} dx, \quad J = \int_0^\infty \frac{\cos xy}{1+x^2} dx, \quad K = \int_0^\infty \frac{x \sin xy}{1+x^2} dx.$$

The first two converge uniformly in any interval of values of y, by Theorem 1. For, in the second integral $|\cos xy| \leqslant 1$ and we can take $g(x) = 1/(1+x^2)$. In the first integral, suppose $-K < y < K$; then, since $|(\sin xy)/(xy)| \leqslant 1$, it follows that $|(\sin xy)/x| < K$ and we can take $g(x) = K/(1+x^2)$.

The third integral converges uniformly in any interval from which $y = 0$ is excluded, by Theorem 3 and *Note*. For, if $0 < \alpha \leqslant y$, then

$$0 \leqslant \int_0^X \sin xy \, dx = \frac{1 - \cos Xy}{y} \leqslant \frac{1+1}{\alpha} = \frac{2}{\alpha}$$

while the factor $x/(1+x^2)$ decreases monotonically to zero as x increases from $x = 1$ onwards.

Note next that

$$I = 0, \quad J = \tfrac{1}{2}\pi, \quad K = 0, \quad \text{when} \quad y = 0, \tag{i}$$

and that, by (40), $\quad I + K = \tfrac{1}{2}\pi \quad$ when $\quad y > 0.$ (ii)

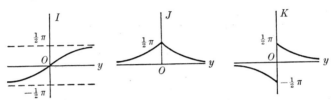

Fig. 9.

Now suppose $y \geqslant 0$. Then since J and K are uniformly convergent wo y, we can differentiate I and J wo y under the integral sign, obtaining

$$dI/dy = J \quad (0 \leqslant y), \qquad dJ/dy = -K \quad (0 < y). \tag{iii}$$

From (iii) and (ii),

$$d^2J/dy^2 = -dK/dy = dI/dy = J \quad (0 < y)$$

and hence $J = A e^y + B e^{-y}$. Now J remains finite when $y \to +\infty$; consequently $A = 0$, $J = B e^{-y} \ (0 < y)$.

To find B, let $y \to 0$; then, by (i) and since J and e^{-y} are continuous at $y = 0$, we get $\tfrac{1}{2}\pi = B$ and therefore

$$J = \tfrac{1}{2}\pi e^{-y} \quad (0 \leqslant y). \tag{41}$$

Next, from (iii) and (41), $dI/dy = \tfrac{1}{2}\pi e^{-y} \ (0 \leqslant y)$, and hence, since I is continuous at $y = 0$ and $I = 0$ at $y = 0$,

$$I = \tfrac{1}{2}\pi(1 - e^{-y}) \quad (0 \leqslant y). \tag{42}$$

Finally, from (i), (iii) and (41),

$$K = 0 \quad (y = 0), \quad K = \tfrac{1}{2}\pi e^{-y} \quad (0 < y). \tag{43}$$

We note that I and K are odd functions of y, and that J is even. Their graphs are indicated in Fig. 9.

(3) Prove that
$$I = \int_0^\infty e^{-x^2}\,dx = \frac{\sqrt{\pi}}{2}. \tag{44}$$

Proof. By first changing the variable of integration from x to xy, and then multiplying both sides by e^{-y^2}, we see that, if $0 < y$,

$$I e^{-y^2} = e^{-y^2} \int_0^\infty y e^{-x^2 y^2}\,dx = \int_0^\infty y e^{-(x^2+1)y^2}\,dx.$$

The last integral converges uniformly in any interval $0 < \alpha \leqslant y \leqslant \beta$, for we may put $g(x) = \beta e^{-(x^2+1)\alpha^2}$ and apply the test of §10.17, Theorem 1, since $0 < y e^{-(x^2+1)y^2} < \beta e^{-(x^2+1)\alpha^2}$.

By §10.16, Theorem 2, we may therefore integrate wo y from $y = \alpha$ to $y = \beta$ under the integral sign, obtaining

$$I \int_\alpha^\beta e^{-y^2}\,dy = \int_0^\infty dx \int_\alpha^\beta y e^{-(x^2+1)y^2}\,dy$$

$$= \int_0^\infty \frac{e^{-(x^2+1)\alpha^2} - e^{-(x^2+1)\beta^2}}{2(x^2+1)}\,dx.$$

The first integral on the right is a continuous function of α at $\alpha = 0$; for it converges, uniformly wo α, since the integrand is not greater than $\frac{1}{2}/(x^2+1)$, and so we may apply §10.17, Theorem 1. Also, the integral on the left is a continuous function of α, by §9.16. We may therefore let $\alpha \to 0$, obtaining

$$I \int_0^\beta e^{-y^2}\,dy = \int_0^\infty \frac{1 - e^{-(x^2+1)\beta^2}}{2(x^2+1)}\,dx = \tfrac{1}{4}\pi - G(\beta)$$

where $0 < G(\beta) < e^{-\beta^2}(\tfrac{1}{4}\pi)$ since $e^{-(x^2+1)\beta^2} \leqslant e^{-\beta^2}$.

Now let $\beta \to \infty$; then $G(\beta) \to 0$, and since the integral on the left converges to I we find $I^2 = \tfrac{1}{4}\pi$ and (44) follows.

(4) If $0 < y$ and $-1 < n$, and if

$$I_n = \int_0^\infty x^n e^{-x^2 y^2}\,dx, \quad \text{prove that} \quad I_{n+2} = -\frac{1}{2y}\frac{dI_n}{dy}.$$

Proof. By §10.17, Theorem 1, the integral

$$\int_0^\infty y x^{n+2} e^{-x^2 y^2}\,dx$$

is uniformly convergent wo y in any interval $0 < \alpha \leqslant y \leqslant \beta$, since

$$0 < \int_0^\infty y x^{n+2} e^{-x^2 y^2}\,dx < \int_0^\infty \beta x^{n+2} e^{-x^2 \alpha^2}\,dx$$

which is convergent and independent of y. The required result follows by differentiating I_n wo y under the integral sign. The condition $-1 < n$ ensures convergence at the lower limit.

Corollary. By changing the variable of integration from x to t by means of the substitution $xy = t$ in I_0 and I_1, we find $I_0 = \sqrt{\pi}/(2y)$ and $I_1 = 1/(2y^2)$. Hence follow, if n is a positive integer,

$$I_{2n} = \left(-\frac{1}{2y}\frac{d}{dy}\right)^n \left(\frac{\sqrt{\pi}}{2y}\right), \quad I_{2n+1} = \left(-\frac{1}{2y}\frac{d}{dy}\right)^n \left(\frac{1}{2y^2}\right).$$

10.18   The error function

The function erf (x), defined by

$$\text{erf}(x) = \frac{2}{\sqrt{\pi}} \int_0^x e^{-t^2}\, dt \tag{45}$$

is called the *error* function. The factor $2/\sqrt{\pi}$ is inserted so that erf (x) may approach the limiting value 1 when $x \to \infty$, as we see from (44). The following few values of x and erf (x) show that erf (x) quickly approaches its limiting value, being within $\frac{1}{2}$ per cent. of it when $x = 2$:

x	0·5	1	1·5	2
erf (x)	0·5205	0·8427	0·9661	0·9953

Examples 10 A

Determine which of the integrals in Examples 1–7 exist. Evaluate those that exist.

(1) (i) $\displaystyle\int_0^\infty e^{-pt} dt,$ (ii) $\displaystyle\int_0^\infty t e^{-pt} dt$ $(p > 0).$

(2) (i) $\displaystyle\int_0^\infty e^{-pt} \cos \omega t\, dt,$ (ii) $\displaystyle\int_0^\infty e^{-pt} \sin \omega t\, dt$ $(p > 0).$

(3) (i) $\displaystyle\int_0^\infty \frac{dx}{1+x^2},$ (ii) $\displaystyle\int_0^\infty \frac{dx}{(1+x)^2},$ (iii) $\displaystyle\int_0^\infty \frac{x\, dx}{(1+x)^2},$ (iv) $\displaystyle\int_0^\infty \frac{dx}{(1-x)^2}.$

(4) (i) $\displaystyle\int_0^1 \log x\, dx,$ (ii) $\displaystyle\int_0^1 x^{-\frac{1}{3}}\, dx,$ (iii) $\displaystyle\int_{-1}^3 (x-1)^{-\frac{2}{3}}\, dx.$

(5) (i) $\displaystyle\int_1^\infty \frac{dx}{\sqrt{x}},$ (ii) $\displaystyle\int_0^a \frac{dx}{\sqrt{(a^2-x^2)}},$ (iii) $\displaystyle\int_0^{\frac{1}{2}\pi} \frac{d\theta}{\sin\theta},$ (iv) $\displaystyle\int_0^{\frac{1}{2}\pi} \frac{d\theta}{\sqrt{(\tan\theta)}}.$

(6) (i) $\displaystyle\int_0^{\frac{1}{2}\pi} \tan\theta\, d\theta,$ (ii) $\displaystyle\int_0^{\frac{1}{2}\pi} \sec\theta\, d\theta,$ (iii) $\displaystyle\int_0^{\frac{1}{2}\pi} (\sec\theta - \tan\theta)\, d\theta.$

(7) (i) $\displaystyle\int_1^\infty \frac{\log x}{x^2}\, dx,$ (ii) $\displaystyle\int_1^\infty \left(\frac{\log x}{x}\right)^2 dx,$ (iii) $\displaystyle\int_0^1 \left(\frac{\log x}{x}\right)^2 dx.$

(8) If $0 < a < 1$ and $1 < b$, prove that the integrals

$$\int_a^1 \frac{dx}{x\log x}, \quad \int_0^a \frac{dx}{x\log x}, \quad \int_1^b \frac{dx}{x\log x}, \quad \int_b^\infty \frac{dx}{x\log x}$$

all diverge.

(9) If n is a positive integer and $p > 0$, prove that

$$\int_0^\infty t^n e^{-pt} dt = \frac{n}{p} \int_0^\infty t^{n-1} e^{-pt} dt = \frac{n!}{p^{n+1}}.$$

(10) Prove that, if $s > 0$, $\displaystyle \int_0^1 x^{s-1} \log x \, dx = -\frac{1}{s^2}$.

Prove also that, if $s \leqslant 0$,

$$\int_\delta^1 x^{s-1} \log x \, dx \to -\infty \quad \text{as} \quad \delta \to 0.$$

(11) If $f(x) = ax^3 + bx^2 + cx + d$, prove that

$$\int_0^\infty x^2 f(x) e^{-x} dx = \tfrac{3}{2} f(2) + \tfrac{1}{2} f(6).$$

(12) If $y = e^{-x}\{2 - \theta + (\theta - 1)x\}$, show that $y \geqslant 0$ for all $x \geqslant 0$ and $1 \leqslant \theta \leqslant 2$. Show that the centroid of the area under the graph of y from $x = 0$ to $+\infty$ is the point

$$x = \theta, \quad y = \tfrac{1}{8}(\theta^2 - 4\theta + 5).$$

(13) Prove that, when $n \to \infty$,

(i) $\displaystyle \lim \int_0^1 \frac{dx}{\sqrt[n]{(1-x^n)}} = 1$,   (ii) $\displaystyle \lim \int_0^1 \sqrt[n]{(1-x^n)} \, dx = 1$.

(14) By putting $x = 1/t$, prove that

$$\int_0^\infty \frac{1}{1+x^\lambda} \frac{dx}{1+x^2} = \int_0^\infty \frac{x^\lambda}{1+x^\lambda} \frac{dx}{1+x^2} = \frac{\pi}{4}.$$

(15) By putting $x = -t$, prove that

$$\int_{-\infty}^\infty \frac{\operatorname{sech} x \, dx}{1+e^{\lambda x}} = \int_{-\infty}^\infty \frac{e^{\lambda x} \operatorname{sech} x \, dx}{1+e^{\lambda x}} = \frac{\pi}{2}.$$

(16) If $a > 0$ and $n > 1$ show that

$$\int_a^\infty \{x - \sqrt{(x^2 - a^2)}\}^n dx = \frac{a^{n+1}}{n^2 - 1}.$$

(17) Prove that $\displaystyle \int_0^1 x^x e^{-x} \log x \, dx = -\frac{e-1}{e}$.

(18) Prove that $\displaystyle \int_0^{\frac{1}{2}\pi} \sin x \log(\sin x) \, dx = -1 + \log 2$.

(19) Prove that $\displaystyle \int_{-\infty}^\infty \frac{dx}{(e^{x-a}+1)(e^{-x}+1)} = \frac{ae^a}{e^a - 1}$.

(20) Prove that

$$\int_{-2}^2 \frac{dx}{\sqrt{(2+x)} + \sqrt{(2-x)}} = 4 - 2\sqrt{2}\log(1 + \sqrt{2}).$$

Examples 10 B

(1) Prove that
$$\int_0^\infty \log\left(x+\frac{1}{x}\right)\frac{dx}{x^2+1} = \pi\log 2.$$
[Put $x = \tan\theta$.]

(2) Prove that

(i) $\int_0^{\frac{1}{2}\pi} \log(2\sin\theta)\, d\theta = 0$, (ii) $\int_0^\pi \theta\log(2\sin\theta)\, d\theta = 0$.

(3) If $a > 0$ prove that
$$\frac{1}{a^2+1^2}+\frac{1}{a^2+2^2}+\ldots+\frac{1}{a^2+n^2} < \frac{\pi}{2a}.$$

(4) Prove that
$$\tfrac{1}{1}+\tfrac{1}{3}+\tfrac{1}{5}+\ldots+\frac{1}{2n-1} = \tfrac{1}{2}\gamma+\log 2+\tfrac{1}{2}\log n+\eta_n$$
where γ denotes Euler's constant and $\eta_n \to 0$ when $n \to \infty$.

(5) By using (14) prove that
$$1-\tfrac{1}{2}+\tfrac{1}{3}-\tfrac{1}{4}+\ldots\text{ to inf.} = \log 2.$$

(6) Prove that the series
$$\frac{1}{2(\log 2)^s}+\frac{1}{3(\log 3)^s}+\frac{1}{4(\log 4)^s}+\ldots$$
converges if $s > 1$, but diverges if $s \leqslant 1$.

(7) Show that
$$\operatorname{cosech} 2 + \operatorname{cosech} 3 + \operatorname{cosech} 4 + \ldots < \log\coth\tfrac{1}{2} < 0.772.$$

(8) If $f''(x) > 0$ in $[a, b]$, $a < b$, prove that
$$f\left(\frac{a+b}{2}\right) < \frac{1}{b-a}\int_a^b f(x)\, dx < \tfrac{1}{2}\{f(a)+f(b)\}.$$

(9) By applying Example 8 to the function $f(x) = 1/x$, show that
$$\tfrac{2}{3}+\tfrac{2}{5}+\ldots+\frac{2}{2n-1} < \log n < \tfrac{1}{2}+\left(\tfrac{1}{2}+\tfrac{1}{3}+\ldots+\frac{1}{n-1}\right)+\frac{1}{2n}.$$
Deduce that $\tfrac{1}{2} < \gamma < 2(1-\log 2)$, where γ denotes Euler's constant.

(10) By applying Example 8 to the function $f(x) = 1/x^2$, show that
$$\frac{3}{2} < \frac{1}{1^2}+\frac{1}{2^2}+\frac{1}{3^2}+\ldots\text{ to inf.} < \frac{5}{3}.$$

Examples 10 C

(1) Find the values of the constants in (i)...(x) for which the given integrals converge:

(i) $\int_0^\infty x^{n-1} e^{-x} dx,$

(ii) $\int_0^\infty e^{-ax} \cos bx \, dx,$

(iii) $\int_0^\infty \frac{x^{a-1}}{1+x} dx,$

(iv) $\int_0^\infty \frac{x^{a-1}}{1+x^n} dx,$

(v) $\int_0^\infty \frac{x^{a-1} dx}{(1+x)^{a+b}},$

(vi) $\int_0^1 \frac{dx}{(1-x^n)^{1/s}} \quad (0 < n),$

(vii) $\int_0^1 x^{m-1} (1-x)^{n-1} dx,$

(viii) $\int_0^{\frac{1}{2}\pi} \sin^{m-1}\theta \cos^{n-1}\theta \, d\theta,$

(ix) $\int_0^\infty \frac{x^{a-1} - x^{b-1}}{1-x} dx,$

(x) $\int_0^\pi \sin^m x \, (1 - \cos x)^n \, dx.$

(2) Show that
$$\int_\delta^X e^{\sin x} \frac{dx}{x} \quad \text{diverges when } \delta \to 0 \quad \text{and when } X \to \infty.$$

(3) Show that $\displaystyle\int_0^\infty \frac{\cos mx}{a^2 + x^2} dx$ converges absolutely.

(4) Show that $\displaystyle\int_a^\infty \frac{\sin x \, dx}{\sqrt{(x^3 - a^3)}}$ converges absolutely.

(5) Show that
$$\int_\pi^X \log(4\cos^2 x) \, dx \quad \text{is bounded as} \quad X \to \infty.$$

Deduce that $\displaystyle\int_\pi^\infty \frac{\log(4\cos^2 x) \, dx}{(\log x)^s}$ converges if $0 < s.$

(6) Show by integration by parts that
$$\int_\pi^\infty \frac{\cos x}{\log x} dx = \int_\pi^\infty \frac{\sin x \, dx}{x(\log x)^2}$$
and hence that the integral on the left converges.

(7) Prove that $\displaystyle\int_0^\infty \cos(x^2) \, dx$ converges.

[Note that in this example the integrand $\cos(x^2)$ does not tend to zero as $x \to \infty$; whereas a necessary condition for the convergence of an infinite series is that the nth term should tend to zero as $n \to \infty$.]

(8) Prove that $\displaystyle\int_0^\infty \frac{\sin x \tanh x}{x^s} dx$ exists if $0 < s < 3.$

(9) Prove that $\displaystyle\int_0^1 \frac{1}{\sqrt{x}} \sin\frac{1}{x} dx$ converges absolutely.

(10) By integration by parts show that, if $1 < n$ and $y \neq 0$,

$$\int_0^\infty x^n e^{-x^2 y^2}\, dx = \frac{n-1}{2y^2} \int_0^\infty x^{n-2} e^{-x^2 y^2}\, dx.$$

If $I_n = \displaystyle\int_0^\infty x^n e^{-x^2}\, dx$, and n is a positive integer, show that

$$I_{2n+1} = \tfrac{1}{2}n!, \quad I_{2n} = \frac{(2n-1)!}{(n-1)!}\frac{\sqrt{\pi}}{2^{2n}}.$$

Examples 10 D

(1) Prove that $xy^2/(x-y)$ converges to y^2 when $x \to \infty$, uniformly wo y over any interval of values of y.

(2) If $f(x,y) = xy^2/(xy-1)$ and $g(y) = y$, prove that, over any interval $-K \leqslant y \leqslant K$, the function $f(x,y)$ converges to $g(y)$ when $x \to \infty$, but that the convergence is not uniform wo y.

(3) If $f(x,y) = xy/(xy+1)$ prove that, over any interval $-K \leqslant y \leqslant K$, the function $f(x,y)$ converges when $x \to \infty$, but not uniformly wo y.

(4) Prove that $\tan^{-1}(xy)$ is convergent when $x \to \infty$, over any interval of values of y, but not uniformly wo y over any interval that includes $y = 0$.

(5) Prove that y^x is convergent over the interval $0 \leqslant y \leqslant 1$ when $x \to \infty$, but not uniformly wo y.

(6) Prove that $f(x,y)$, if uniformly convergent wo y, remains uniformly convergent when multiplied by a bounded function of y.

Examples 10 E

(1) Consider the two integrals

$$\int_0^X \frac{\sin xy}{x}\, dx, \quad \int_0^\infty \frac{\sin xy}{x}\, dx.$$

Show that the first is a continuous function of y in any interval $-K \leqslant y \leqslant K$ for any fixed X, but that the second is not continuous at $y = 0$.

(2) Show that if $0 \leqslant y$

$$\left| \int_0^X e^{-xy} \sin x\, dx - \frac{1}{1+y^2} \right| \leqslant \frac{(1+y)\, e^{-Xy}}{1+y^2}.$$

Deduce that the integral converges to $1/(1+y^2)$ uniformly wo y in any interval $0 < \alpha \leqslant y \leqslant \beta$.

(3) Determine the interval of values of y in which each of the following integrals converges uniformly wo y:

(i) $\displaystyle\int_1^\infty \frac{dx}{x^2+y^2}$,   (ii) $\displaystyle\int_0^\infty \frac{dx}{x^2+y^2}$,   (iii) $\displaystyle\int_0^\infty e^{-xy}\, dx$,

(iv) $\displaystyle\int_0^\infty y e^{-xy}\, dx$,   (v) $\displaystyle\int_0^\infty e^{-x^2 y^2}\, dx$,   (vi) $\displaystyle\int_0^\infty e^{-x^2+2xy}\, dx$.

(4) If $p > 0$ and n is a positive integer, prove, by applying § 10.16, Theorem 3, that

$$\int_0^\infty e^{-pt}\,dt = \frac{1}{p}, \qquad \int_0^\infty t^n e^{-pt}\,dt = \frac{n!}{p^{n+1}}.$$

(5) If $-1 < s$ and n is a positive integer, prove that

$$\int_0^1 x^s\,dx = \frac{1}{s+1}, \qquad \int_0^1 x^s (\log x)^n\,dx = \frac{(-)^n n!}{(s+1)^{n+1}}.$$

(6) If $a > 0$ and n is a positive integer, prove that

$$\int_0^\infty \frac{dx}{a^2+x^2} = \frac{\pi}{2a}, \qquad \int_0^\infty \frac{dx}{(a^2+x^2)^n} = \frac{1.3...(2n-3)}{2.4...(2n-2)}\frac{\pi}{2a^{2n-1}},$$

obtaining the second integral in two ways.

(7) If a and b are unequal and positive, prove that

$$\int_0^\infty \frac{dx}{(x+a)(x+b)} = \frac{1}{b-a}\log\frac{b}{a}.$$

Deduce the values of

$$\int_0^\infty \frac{dx}{(x+a)^2(x+b)}, \qquad \int_0^\infty \frac{dx}{(x+a)^2(x+b)^2}.$$

(8) If $p > 0$, $\omega > 0$, and $\tan\theta = \omega/p$, $0 < \theta < \tfrac{1}{2}\pi$, prove that

$$\int_0^\infty e^{-pt}\,t^n \cos\omega t\,dt = \frac{n!\cos(n+1)\theta}{(p^2+\omega^2)^{\frac{1}{2}(n+1)}}.$$

(9) Prove that $\displaystyle\int_0^1 \frac{\tan^{-1}(ax)}{x\sqrt{(1-x^2)}}\,dx = \tfrac{1}{2}\pi\log\{a+\sqrt{(1+a^2)}\}.$

(10) Prove that, if a and b are positive,

$$\int_0^\infty \frac{\log(1+a^2x^2)}{1+b^2x^2}\,dx = \frac{\pi}{b}\log\frac{a+b}{b}.$$

(11) Prove that $\displaystyle\int_0^\infty \frac{1-\cos xy}{xe^x}\,dx = \tfrac{1}{2}\log(1+y^2).$

(12) If $\displaystyle G = \int_0^\infty \exp\left(-x^2-\frac{a^2}{x^2}\right)dx$, prove that $\dfrac{dG}{da} = -2G.$

Deduce that $\qquad G = \tfrac{1}{2}\sqrt{\pi}e^{-2a} \quad (0 \leqslant a).$

(13) Given that I, J, K denote respectively the integrals:

$$\int_0^\infty e^{-x^2}\frac{\sin ax}{x}\,dx, \qquad \int_0^\infty e^{-x^2}\cos ax\,dx, \qquad \int_0^\infty e^{-x^2}x\sin ax\,dx,$$

prove that $\qquad dI/da = J, \quad dJ/da = -K, \quad 2K = aJ$

and hence that

$$I = \tfrac{1}{2}\pi\,\mathrm{erf}\,(\tfrac{1}{2}a), \quad J = \tfrac{1}{2}\sqrt{\pi}\exp(-\tfrac{1}{4}a^2), \quad K = \tfrac{1}{4}a\sqrt{\pi}\exp(-\tfrac{1}{4}a^2).$$

(14) Prove that
$$y = \int_0^\infty \exp(-t^2) \sinh 2xt \, dt$$

satisfies the equation $dy/dx - 2xy = 1$ and hence that

$$y = \exp(x^2) \int_0^x \exp(-t^2) \, dt.$$

(15) Obtain the results of § 10.17, Example 1, by considering as a function of y the integral
$$\int_0^\infty e^{-x} \frac{\sin xy}{x} \, dx.$$

(16) Verify that

$$\int_0^1 dy \int_0^\infty \frac{\sin xy}{x} \, dx = \int_0^\infty dx \int_0^1 \frac{\sin xy}{x} \, dy.$$

[The infinite integral on the left is not uniformly convergent, so this example shows that uniform convergence is not a *necessary* condition for integration under the integral sign.]

(17) Prove that, uniformly wo a $(0 \leqslant a)$,

$$\int_0^\infty \frac{1 - \cos a\theta}{\theta^2} \, d\theta = \frac{\pi a}{2}.$$

By integrating from 0 to a under the integral sign, deduce that

$$\int_0^\infty \frac{a\theta - \sin a\theta}{\theta^3} \, d\theta = \frac{\pi}{2} \frac{a^2}{2!},$$

$$\int_0^\infty \frac{\cos a\theta - 1 + \frac{1}{2} a^2\theta^2}{\theta^4} \, d\theta = \frac{\pi}{2} \frac{a^3}{3!}.$$

(18) Show that, if $0 \leqslant a$,

$$\int_0^\infty \left(\sin \frac{a}{x} \right)^2 dx = \int_0^\infty \left(\frac{\sin at}{t} \right)^2 dt = \frac{\pi a}{2}.$$

(19) If n is a positive integer, prove that

$$\int_0^\infty \frac{\sin^{2n+1} x}{x} \, dx = \frac{1 . 3 ... (2n-1)}{2 . 4 ... 2n} \frac{\pi}{2}.$$

(20) If $0 \leqslant a$, prove that:

(i) $\displaystyle\int_0^\infty \frac{\sin^3 ax}{x^3} \, dx = \frac{3\pi a^2}{8}$, (ii) $\displaystyle\int_0^\infty \frac{\sin^2 ax \cos ax}{x^2} \, dx = \frac{\pi a}{4}$,

(iii) $\displaystyle\int_0^\infty \frac{\sin^2 2ax \sin^2 ax}{x^2} \, dx = \frac{\pi a}{4}$.

(21) Evaluate:

(i) $\displaystyle\int_0^\infty \frac{\sin ax \cos bx}{x} \, dx$, (ii) $\displaystyle\int_0^\infty \frac{\sin ax \sin bx}{x^2} \, dx$,

(iii) $\displaystyle\int_0^\infty \frac{\cos ax - \cos bx}{x^2} \, dx$, (iv) $\displaystyle\int_0^\infty \frac{\sin ax \sin bx \sin cx}{x^3} \, dx$.

(22) Prove that, if $0 < t$,

$$\int_0^\infty (e^{-x} - e^{-tx}) \frac{dx}{x} = \log t.$$

(23) Prove that, if $ab > 0$,

$$\int_0^\infty (\cos ax - \cos bx) \frac{dx}{x} = \log \frac{b}{a}.$$

(24) *Frullani's integral.* Let $f(x)$ be continuous for $0 \leqslant x$, and let

$$\int_\delta^X f(x)\, dx, \quad 0 < \delta, \quad \text{be bounded as} \quad X \to +\infty.$$

Then, by §10.8, the integral $\phi(X) = \int_\delta^X f(x)\, dx/x$ converges when $X \to +\infty$.

Theorem. If a and b are positive,

$$\int_0^\infty \frac{f(ax) - f(bx)}{x}\, dx = f(0) \log \frac{b}{a}.$$

[The integral on the left is called Frullani's integral. Examples 22, 23 are particular cases.]

Proof. By putting $ax = t$ in the first part of the integral, with the range of integration from δ to X, and $bx = t$ in the second part, we find

$$\int_\delta^X \frac{f(ax) - f(bx)}{x}\, dx = \int_{a\delta}^{aX} \frac{f(t)}{t}\, dt - \int_{b\delta}^{bX} \frac{f(t)}{t}\, dt$$

$$= \int_{a\delta}^{b\delta} \frac{f(t)}{t}\, dt - \int_{aX}^{bX} \frac{f(t)}{t}\, dt. \tag{i}$$

From Cauchy's principle of convergence (§10.9) it follows that, since a and b are positive and $\phi(X)$ converges, the second term in (i) tends to zero when $X \to +\infty$.

Also, by the first mean value theorem for integrals (§9.30), since $f(t)$ is continuous there exists a number ξ between $a\delta$ and $b\delta$ such that

$$\int_{a\delta}^{b\delta} \frac{f(t)}{t}\, dt = f(\xi) \int_{a\delta}^{b\delta} \frac{dt}{t} = f(\xi) \log \frac{b}{a}.$$

The theorem now follows when $\delta \to 0$ and $X \to +\infty$ in (i), since $\xi \to 0$ when $\delta \to 0$.

(25) Evaluate the integral

$$\int_0^\infty \frac{\sin ax \sin bx}{x}\, dx, \quad b^2 \neq a^2;$$ and show that the integral diverges when $b^2 = a^2 \neq 0$.

(26) Prove that if a and b are positive

$$\int_0^1 \frac{x^{a-1} - x^{b-1}}{(1 + x^a)(1 + x^b)} \frac{dx}{\log (1/x)} = \tfrac{1}{2} \log \frac{b}{a}.$$

(27) Prove that, if $a > 0, b > 0$,

$$\int_0^\infty \frac{\tan^{-1} ax \tan^{-1} bx}{x^2}\, dx = \tfrac{1}{2}\pi \left\{ a \log \left(1 + \frac{b}{a} \right) + b \log \left(1 + \frac{a}{b} \right) \right\}.$$

(28) If $0 \leqslant \lambda < \mu$, prove that

$$\int_{-\infty}^{\infty} \frac{\sin \lambda(x-a)}{x-a} \cdot \frac{\sin \mu(x-b)}{x-b} \, dx = \frac{\pi \sin \lambda(a-b)}{a-b}.$$

(29) Show that the minimum value of the integral

(i) $\displaystyle\int_0^\infty e^{-x^2}(1-ax)^2 \, dx$ is $\dfrac{\pi-2}{2\sqrt{\pi}}$ when $a = \dfrac{2}{\sqrt{\pi}}$;

(ii) $\displaystyle\int_0^\infty e^{-x^2}(1-ax-bx^2)^2 \, dx$ is $\dfrac{(\pi-3)\sqrt{\pi}}{3\pi-8}$.

(30) Show that the minimum value of the integral

$$\int_0^\infty e^{-x}(1+a_1 x + a_2 x^2 + \dots + a_n x^n)^2 \, dx,$$

considered as a function of $a_1, a_2, \dots, a_n$, is $1/(n+1)$.

11

INDEFINITE INTEGRALS

11.1

In the last two chapters it was taken for granted that the reader was familiar with the standard indefinite integrals, and with the methods of reducing other indefinite integrals to standard forms, when this was possible. In the present chapter, indefinite integrals are considered more generally. For a comprehensive account, reference should be made to, e.g. G. H. Hardy: *The Integration of Functions of a Single Variable*, Cambridge Tracts in Mathematics, No. 2.

11.2

The leading general theorem concerning indefinite integrals is that the indefinite integral of every rational function, i.e. any function reducible to the form $P(x)/Q(x)$, where $P(x)$ and $Q(x)$ are polynomials, can be expressed as the sum of a finite number of terms of the form Ax^r, $A/(x-a)^s$ or $A \log |x-a|$, where r and s are positive integers, provided that such constants as A and a may have complex values. If all such constants are to be real, and if the coefficients in $P(x)$ and $Q(x)$ are real, then the integral can be expressed as the sum of a finite number of terms of the form

$$Ax^r, A/(x-a)^s, A \log |x-a|, A \log (ax^2+bx+c)$$

and
$$A \tan^{-1}\{(x-a)/b\},$$

where r and s are positive integers and A, a, b, c are real constants.

We first recapitulate some properties of polynomials.

11.3 One polynomial: general properties

Definition. A function of the form

$$f(x) \equiv c_0 x^n + c_1 x^{n-1} + c_2 x^{n-2} + \ldots + c_{n-1} x + c_n \tag{1}$$

in which the coefficients $c_0, c_1, c_2, \ldots$ are independent of x, $c_0 \neq 0$, and n is a positive integer, is called a *polynomial* in x of the nth degree. An equation of the form

$$c_0 x^n + c_1 x^{n-1} + c_2 x^{n-2} + \ldots + c_{n-1} x + c_n = 0 \tag{2}$$

is called a polynomial equation of the nth degree.

We say that $x = \alpha$ *satisfies* equation (2), or that α is a *root* of equation (2), if $c_0 \alpha^n + c_1 \alpha^{n-1} + \ldots + c_n \equiv 0$; for example, 3 is a root of the equation $x^2 - 2x - 3 = 0$ because $3^2 - 2.3 - 3 \equiv 0$.

A root of the equation $f(x) = 0$ is also called a *zero* of the polynomial $f(x)$.

Theorem 1. Every polynomial equation has at least one root.

This is called *the fundamental theorem of algebra*. It is not true in the field of real numbers, since an equation (e.g. $x^2 = -1$) may not have any real root. Evidently, therefore, the theorem cannot be proved as long as we restrict ourselves to real numbers. It can be proved at an early stage in the theory of functions of a complex variable.

Theorem 2. When the polynomial $f(x)$ is divided by $x - \alpha$, the remainder is $f(\alpha)$.

This is called *the remainder theorem*.

Proof. The steps of the process of long division can be continued until the remainder is independent of x. If at this stage the remainder is R and the quotient $Q(x)$, then identically

$$f(x) \equiv (x - \alpha) Q(x) + R. \tag{3}$$

Putting $x = \alpha$, we see that $f(\alpha) = 0 + R$, or $R = f(\alpha)$.

Corollary.
$$f(x) \equiv (x - \alpha) Q(x) + f(\alpha), \tag{4}$$

where $Q(x)$ is a polynomial of degree $n - 1$, in which the term of highest degree is $c_0 x^{n-1}$.

Theorem 3. If α is a root of the equation $f(x) = 0$, then $x - \alpha$ is a factor of $f(x)$.

This is called *the factor theorem*.

Proof. By definition of a root, $f(\alpha) = 0$, and so (4) becomes

$$f(x) \equiv (x - \alpha) Q(x) \tag{5}$$

showing that $x - \alpha$ is a factor of $f(x)$.

Corollary. Every polynomial has at least one linear factor.

This follows at once from Theorems 1 and 3.

Theorem 4. Every polynomial of degree n can be expressed as the product of n linear factors, not necessarily distinct.

Proof. By the last Corollary, the polynomial $Q(x)$ has at least one linear factor if $n - 1 > 0$. Let $x - \alpha_2$ be such a factor; then we can put $f(x) \equiv (x - \alpha)(x - \alpha_2) Q_2(x)$, the term of highest degree in $Q_2(x)$ being $c_0 x^{n-2}$. Continuing in this way, we finally obtain $f(x)$ expressed as the product of n linear factors and the non-zero constant c_0, thus

$$f(x) \equiv (x - \alpha)(x - \alpha_2)(x - \alpha_3)\ldots(x - \alpha_n) c_0. \tag{6}$$

Theorem 5. A polynomial equation of degree n cannot have more than n distinct roots.

Proof. Suppose that $f(x)$ has been expressed in the form (6) and that $\alpha, \alpha_2, \alpha_3 \ldots \alpha_n$ are all distinct. Let β be different from any of them; then, by (6),

$$f(\beta) = (\beta - \alpha)\,(\beta - \alpha_2)\,(\beta - \alpha_3)\ldots(\beta - \alpha_n)\,c_0.$$

It follows that β cannot be a root, for $f(\beta) = 0$ is impossible, since every factor on the R.H.S. differs from zero.

Corollary 1. If an equation of the form (2) is satisfied by more than n values of x, then all the $n + 1$ coefficients $c_0, c_1, \ldots$ must be zero, i.e.

$$c_0 = 0, \quad c_1 = 0, \quad c_2 = 0, \ldots, \quad c_n = 0.$$

Corollary 2. If two polynomials of the form (1) are equal for more than n values of x, then coefficients of like powers of x must be equal.

For if, for more than n values of x,

$$b_0 x^n + b_1 x^{n-1} + \ldots + b_n = c_0 x^n + c_1 x^{n-1} + \ldots + c_n$$

then the equation

$$(b_0 - c_0)\,x^n + (b_1 - c_1)\,x^{n-1} + \ldots + b_n - c_n = 0$$

is satisfied by more than n values of x and hence, by Corollary 1,

$$b_0 - c_0 = 0, \quad b_1 - c_1 = 0, \ldots, \quad b_n - c_n = 0$$

and therefore $\quad b_0 = c_0, \quad b_1 = c_1, \ldots, \quad b_n = c_n.$

Theorem 6. If the coefficients $c_0, c_1, \ldots, c_n$ are all real, then complex roots of the equation $f(x) = 0$ occur in conjugate pairs; i.e. if $\lambda + i\mu$ is one root, with λ, μ both real and $\mu \neq 0$, then $\lambda - i\mu$ is another root.

Proof. First, let λ, μ be any real numbers.

Let $f(\lambda + i\mu) = L + iM$, where L, M are real.

Then $f(\lambda - i\mu) = L - iM$, since the coefficients $c_0, c_1, \ldots$ are real.

Next, let $\lambda + i\mu$ be a root, i.e. let $f(\lambda + i\mu) = 0$; then $L + iM = 0$ and therefore $L = 0$, $M = 0$, since L, M are real.

It follows that $f(\lambda - i\mu) = 0$ and hence, if $\mu \neq 0$, that $\lambda - i\mu$ is another root.

Corollary. Every polynomial with real coefficients can be expressed as the product of real factors each of which is linear or quadratic.

For if $\lambda + i\mu$ is a root, then $\lambda - i\mu$ is also a root, and it follows from the factor theorem and (6) that, if $\mu \neq 0$, $f(x)$ will have among its factors the two distinct linear factors $x - \lambda - i\mu$ and $x - \lambda + i\mu$ and consequently will have the real quadratic factor

$$(x - \lambda - iu)\,(x - \lambda + i\mu) = (x - \lambda)^2 + \mu^2 = x^2 - 2\lambda x + \lambda^2 + \mu^2. \quad (7)$$

11.4   Two polynomials. Highest Common Factor

Let N_1, N_2 be two polynomials of degrees n_1, n_2 respectively, with $n_1 \geqslant n_2 \geqslant 1$.

Divide N_1 by N_2; let the quotient be Q_1 of degree $n_1 - n_2$ and the remainder N_3 of degree $n_3 < n_2$.

Then divide N_2 by N_3, letting the quotient be Q_2 of degree $n_2 - n_3$ and the remainder N_4 of degree $n_4 < n_3$.

This process can be continued until we reach a stage at which the remainder is independent of x, because at every step the degree of the remainder is reduced and must therefore be reduced to zero after at most n_1 steps. We thus obtain a succession of identities in x of the form

$$N_1 \equiv Q_1 N_2 + N_3 \qquad (n_1 \geqslant n_2 > n_3),$$

$$N_2 \equiv Q_2 N_3 + N_4 \qquad (n_2 > n_3 > n_4), \qquad (8)$$

$$\cdots\cdots\cdots\cdots\cdots\cdots\cdots\cdots\cdots\cdots\cdots\cdots$$

$$N_{s-2} \equiv Q_{s-2} N_{s-1} + N_s \quad (n_{s-2} > n_{s-1} > n_s = 0),$$

where N_s is a constant with respect to x.

It follows that, if N_1 and N_2 have any common factor (except a mere constant) it must be a factor of N_3, of N_4, of N_5, ... and finally of N_{s-1}, but not of N_s since N_s does not involve x. There are now two alternatives: either (i) $N_s = 0$, or (ii) $N_s \neq 0$:

(i) If $N_s = 0$, then N_{s-1} is a factor of N_{s-2} and hence also of N_{s-3}, ... of N_3, of N_2, and of N_1; thus N_{s-1} is a common factor of N_1 and N_2. But every common factor of N_1 and N_2 is a factor of N_{s-1}. It follows that N_{s-1} is the highest common factor (H.C.F.) of N_1 and N_2, that is, the common factor of highest degree.

(ii) If $N_s \neq 0$, then N_1 and N_2 have no common factor involving x. We say that they are *prime* to one another, or are mutually prime, and that their H.C.F. is 1.

Consequently, if H is the highest common factor, then

$$H = N_{s-1} \quad \text{if} \quad N_s = 0, \quad H = 1 \quad \text{if} \quad N_s \neq 0. \qquad (9)$$

11.5

Theorem. Let H be the H.C.F. of two polynomials N_1, N_2 of degrees n_1, n_2 respectively.

Mutually prime polynomials A, B exist such that, identically in x,

$$H \equiv A N_1 + B N_2. \qquad (10)$$

Proof. Suppose $n_1 \geqslant n_2$. From equations (8) we may express $N_3, N_4, \ldots$ each in terms of N_1, N_2. Thus, we get in succession

$$\left. \begin{aligned} N_3 &\equiv N_1 - Q_1 N_2 \\ N_4 &\equiv -Q_2 N_1 + (1 + Q_1 Q_2) N_2 \end{aligned} \right\} \tag{11}$$

$$\cdots\cdots\cdots\cdots\cdots\cdots\cdots\cdots\cdots$$

(i) In case $N_s = 0$, the last step in this succession will give an identity of the form $\qquad N_{s-1} \equiv AN_1 + BN_2 \tag{12}$

i.e. of the form (10), since in this case $H = N_{s-1}$

(ii) In case $N_s \neq 0$, the last step in the succession will give an identity of the form $N_s \equiv A_1 N_1 + B_1 N_2$, i.e. after dividing by N_s (a constant) of the form $\qquad 1 \equiv AN_1 + BN_2 \tag{13}$

i.e. of the form (10), since in this case $H = 1$.

That A and B are mutually prime follows because H is a factor of N_1 and of N_2, so that when (10) is divided by H, we get the identity $1 \equiv AF_1 + BF_2$ where $F_1 = N_1/H$, $F_2 = N_2/H$, and since now the L.H.S. has no factor, therefore A and B cannot have a common factor.

11.6

Let N_1, N_2 be mutually prime polynomials.

Theorem 1. Any polynomial P can be expressed in the form

$$P \equiv P_2 N_1 + P_1 N_2 \tag{14}$$

in an infinity of ways.

Proof. Since N_1, N_2 have no common factor, an identity of the form $1 \equiv AN_1 + BN_2$ holds good, by (13). After multiplying by P we get $P \equiv PAN_1 + PBN_2$.

This is one way of expressing P in the form (14). We may now replace PA by $PA + FN_2$ and PB by $PB - FN_1$, where F denotes any polynomial. The number of ways is therefore infinite.

Theorem 2. Any polynomial P can be expressed in the form

$$P \equiv P_2 N_1 + P_1 N_2 \quad (p_2 < n_2), \tag{15}$$

where p_2, n_2 denote the degrees of P_2, N_2 respectively.

There is only one such way of expressing P.

Proof. Let $P \equiv A_2 N_1 + A_1 N_2$ be one expression of P in the form (14), and suppose that the degree of A_2 exceeds that of N_2. Let A_2 be divided by N_2 and let the result be $A_2 \equiv KN_2 + P_2$ where $p_2 < n_2$. We thus obtain $\qquad P \equiv (KN_2 + P_2) N_1 + A_1 N_2 \equiv P_2 N_1 + P_1 N_2$

where $P_1 \equiv KN_1 + A_1$. This is of the form (15).

To show that there is only one pair of polynomials P_1, P_2 that satisfies (15), suppose that there could be another pair Q_1, Q_2, such that

$$P \equiv Q_2 N_1 + Q_1 N_2 \quad (q_2 < n_2), \tag{16}$$

where q_2 is the degree of Q_2. Then, from (15) and (16), after subtraction,

$$(Q_2 - P_2)N_1 \equiv (P_1 - Q_1) N_2.$$

Now N_1, N_2 have no common factor. Consequently, N_2 must be a factor of $Q_2 - P_2$. But this is impossible, since the degree of $Q_2 - P_2$ is less than n_2. Hence Q_1, Q_2 cannot differ from P_1, P_2.

Corollary. If the degree of P is less than $n_1 + n_2$, then P can be expressed uniquely in the form

$$P \equiv P_2 N_1 + P_1 N_2 \quad (p_1 < n_1,\ p_2 < n_2). \tag{17}$$

11.7 Rational fractions

Let $F(x)$, $N(x)$ be two polynomials. The fraction $F(x)/N(x)$, or briefly F/N, is called a *rational fraction*. We shall assume that F and N have no common factor.

Any function reducible to a rational fraction by the four simple operations of algebra is called a *rational function*. Such a function is defined for all values of x except the roots of the equation $N(x) = 0$.

If the degree of the numerator F is less than that of the denominator N, the fraction F/N is called a *proper fraction*. If the degree of F is not less than that of N, the fraction F/N can be expressed as the sum of a polynomial and a proper fraction; for we can then divide F by N, obtaining an identity of the form $F \equiv QN + P$ in which Q is a polynomial and the degree of P is less than that of N, and hence follows

$$\frac{F}{N} \equiv Q + \frac{P}{N} \tag{18}$$

where P/N is a proper fraction; this being an identity in the sense that it holds good for all values of x except the roots of the equation $N(x) = 0$.

11.8 Integration of rational fractions

By (18), the integral of an improper fraction can be expressed as a sum of integrals of a polynomial and a proper fraction. As the integral of a polynomial is elementary, we need only consider integrals of proper fractions. We begin by showing how a given proper fraction may be expressed as a sum of proper fractions of the simplest types.

11.9 Partial fractions

Theorem. Let P/N be a proper fraction. If the denominator N is the product of two factors N_1, N_2 prime to one another, then the proper fraction P/N can be expressed as the sum of two simpler proper fractions.

Proof. Let p, n, n_1, n_2 be the degrees of P, N, N_1, N_2 respectively, where $p < n = n_1 + n_2$. Then P can be uniquely expressed in the form (17) and after dividing by N, $= N_1 N_2$, we obtain identically (see end of §11.7)

$$\frac{P}{N} \equiv \frac{P_1}{N_1} + \frac{P_2}{N_2}. \tag{19}$$

We say that P/N is thus expressed as the sum of two *partial fractions*.

Corollary 1. If $N \equiv N_1 N_2 ... N_s$ is the product of s factors each prime to the others, then the proper fraction P/N can be uniquely expressed as the sum of s proper fractions: thus

$$\frac{P}{N} \equiv \frac{P_1}{N_1} + \frac{P_2}{N_2} + ... + \frac{P_s}{N_s}. \tag{20}$$

For, by repeated applications of (19), we get

$$P/N \equiv P_1 N_1 + Q_1/(N_2 N_3 ... N_s)$$
$$\equiv P_1/N_1 + P_2/N_2 + Q_2/(N_3 ... N_s)...$$

until the form (20) is reached.

Corollary 2. If $N \equiv N_1 N_2^2 N_3^3 ... N_s^s$, the factors N_1, N_2, ...being prime to one another, then the proper fraction P/N can be uniquely expressed in the form

$$\frac{P}{N} \equiv \frac{P_1}{N_1} + \frac{P_2}{N_2^2} + ... + \frac{P_s}{N_s^s} \tag{21}$$

where $p_1 < n_1, p_2 < 2n_2, ..., p_s < sn_s$.

11.10

We now recall that, by §11.3, Theorem 4, a polynomial N of degree n is theoretically expressible as a product of n linear factors, not necessarily distinct. Suppose that N has been so expressed and put

$$N_1 \equiv (x - \alpha_1)(x - \alpha_2)..., \quad N_2^2 \equiv (x - \beta_1)^2 (x - \beta_2)^2 ...,$$
$$N_3^3 \equiv (x - \gamma_1)^3 (x - \gamma_2)^3 ...,$$

so that N_1 is the product of n_1 factors each of which occurs once only, N_2 is the product of n_2 factors each of which occurs twice only, and so on. Then, identically, taking the coefficient of x^n in N to be unity,

$$N \equiv N_1 N_2^2 N_3^3 \dots N_s^s \tag{22}$$

where $n = n_1 + 2n_2 + 3n_3 + \dots + sn_s$ and $N_1, N_2, \dots$ are all prime to each other.

It follows from §11.9, Corollary 2, that every proper fraction P/N can be uniquely expressed in the form (21), with $N_1, N_2, \dots$ each the product of distinct linear factors.

11.11

Further, it follows from §11.3, Theorem 6, Corollary, that, provided the coefficients of the polynomial N are real, it can be expressed in the form (22), where each of the factors $N_1, N_2, \dots$ is the product of real factors, linear or quadratic. Consequently, in this case, the proper fraction P/N can be expressed in the form (21), where each of $N_1, N_2, \dots$ is a product of distinct real linear or quadratic factors.

11.12

We next observe that we can find the factors $N_1, N_2, \dots$ by the elementary process of finding the H.C.F. of two polynomials (§11.4), without having to face the problem of expressing N as a product of its linear factors $x - \alpha_1, \dots x - \beta_1, \dots$.

It will sufficiently illustrate the general case if we take $s = 4$ and put

$$N \equiv N_1 N_2^2 N_3^3 N_4^4, \ H_1 \equiv N_2 N_3^2 N_4^3, \ H_2 \equiv N_3 N_4^2, \ H_3 \equiv N_4. \tag{23}$$

Consider N and its derivative dN/dx. Evidently in dN/dx no linear factor of N_1 will occur, each linear factor of N_2 will occur once, each linear factor of N_3 twice, and each linear factor of N_4 three times. Hence, and similarly, apart from numerical factors,

$$H_1 \text{ will be the H.C.F. of } N \text{ and } dN/dx,$$

$$H_2 \text{ will be the H.C.F. of } H_1 \text{ and } dH_1/dx,$$

$$H_3 \text{ will be the H.C.F. of } H_2 \text{ and } dH_2/dx.$$

Consequently N_1, N_2, N_3, N_4 can be found by elementary algebra: we first find H_1, H_2, H_3; then

$$N_4 \equiv H_3, \ N_3 \equiv H_2/N_4^2, \ N_2 \equiv H_1/N_3^2 N_4^3, \ N_1 \equiv N/N_2^2 N_3^3 N_4^4. \tag{24}$$

11.13   Hermite's method of integrating rational fractions

From (21) and §11.12 it follows that the integration of any rational fraction P/N can, by the elementary operations of addition, subtraction, multiplication and division, be made to depend upon the integration of fractions of the type A/M^r where A, M are polynomials; and although the linear factors of M are unknown, it is known that they are distinct.

We therefore consider the integration of A/M^r. It is not necessary here to suppose that this is a proper fraction.

Since M has no repeated factor, M and its derivative M', $= dM/dx$, have no common factor and therefore, by §11.6, Theorem 1, we can find polynomials A_1, A_2 so that $A \equiv A_2 M + A_1 M'$. Then, if $r > 1$,

$$\int \frac{A\,dx}{M^r} = \int \frac{A_1 M'}{M^r}\,dx + \int \frac{A_2\,dx}{M^{r-1}}. \tag{25}$$

After integrating the first term on the right by parts, we find

$$\int \frac{A\,dx}{M^r} = -\frac{A_1}{(r-1)\,M^{r-1}} + \int \frac{B\,dx}{M^{r-1}} \tag{26}$$

where $B = A_2 + A_1'/(r-1)$. We can now find polynomials B_1, B_2 such that $B \equiv B_2 M + B_1 M'$ and thence obtain, if $r > 2$,

$$\int \frac{A\,dx}{M^r} = -\frac{A_1}{(r-1)\,M^{r-1}} - \frac{B_1}{(r-2)\,M^{r-2}} + \int \frac{C\,dx}{M^{r-2}} \tag{27}$$

where $C = B_2 + B_1'/(r-2)$. After at most $r-1$ steps of this kind, we reach a stage at which the integrand of the integral on the right is of the form S/M. If this is not a proper fraction we can find Q and T such that $S/M = Q + T/M$ where T/M is a proper fraction. We shall then have

$$\int \frac{A\,dx}{M^r} = R(x) + \int \frac{T\,dx}{M} \tag{28}$$

where $R(x)$ is rational and the degree of T is less than that of M. There will now be two alternatives: either (i) $T \equiv 0$ or (ii) $T \not\equiv 0$:

(i) If $T \equiv 0$, we see that the integral is rational and will have been found without a knowledge of the linear factors of $M(x)$, i.e. without solving the equation $M(x) = 0$.

(ii) If $T \not\equiv 0$, to complete the integration it will be necessary (except in special cases, e.g. when T is proportional to dM/dx) to solve the equation $M(x) = 0$, if possible. When this has been done, the fraction

T/M can be expressed as a sum of real simple partial fractions of the form

$$\Sigma A/(x-\alpha) + \Sigma(Bx+C)/(ax^2+2bx+c), \tag{29}$$

where $ac > b^2$ and A, B, C now denote constants. The integrals of these fractions are not rational: thus

$$\int \frac{A\,dx}{x-\alpha} = A\log|x-\alpha|, \tag{30}$$

$$\int \frac{(Bx+C)\,dx}{ax^2+2bx+c} = \frac{B}{2a}\log(ax^2+2bx+c) + \frac{aC-bB}{a\sqrt{(ac-b^2)}}\tan^{-1}\frac{ax+b}{\sqrt{(ac-b^2)}}. \tag{31}$$

Note 1. There will be cases in which it will be preferable to express M entirely as a product of its linear factors, whether real or complex. In such cases, T/M will consist solely of terms of the form $A/(x-\alpha)$, and among them complex terms will occur in conjugate pairs, assuming that the coefficients in M are real.

Note 2. Note that the rational part of the integral can always be found without solving the equation $M(x) = 0$.

Examples

(1) Show that the integral

$$\int \frac{ax^2+2bx+c}{(Ax^2+2Bx+C)^2}\,dx \quad (A \neq 0),$$

is rational (i) if $AC - B^2 = 0$, or (ii) if $Ac + Ca = 2Bb$.

(i) If $AC - B^2 = 0$ the denominator is of the form $A^2(x-\alpha)^4$ and by putting $\xi = x - \alpha$ the integral can be expressed as

$$\int \left(\frac{\lambda}{\xi^2} + \frac{\mu}{\xi^3} + \frac{\nu}{\xi^4}\right) d\xi = -\frac{\lambda}{\xi} - \frac{\mu}{2\xi^2} - \frac{\nu}{3\xi^3}$$

where λ, μ, ν are constants; this is rational.

(ii) If $AC - B^2 \neq 0$, then $Ax^2 + 2Bx + C$ and its derivative $2Ax + 2B$ are prime to one another and, by (17), constants λ, μ, ν can be found such that

$$ax^2+2bx+c \equiv \lambda(Ax^2+2Bx+C) + (\mu x+\nu)(2Ax+2B). \tag{32}$$

Putting $M = Ax^2 + 2Bx + C$, the integral then reduces to

$$\lambda\int\frac{dx}{M} - \int(\mu x+\nu)\,d\left(\frac{1}{M}\right) = (\lambda+\mu)\int\frac{dx}{M} - \frac{\mu x+\nu}{M} \tag{33}$$

by integrating the second integral on the left by parts.

By equating coefficients of like powers of x in (32), we see that the constants λ, μ, ν satisfy the equations

$$A(\lambda + 2\mu) = a, \quad B(\lambda + \mu) + A\nu = b, \quad C\lambda + 2B\nu = c.$$

These equations have a unique solution which may be written as

$$\frac{\lambda}{2AH - 2a\Delta} = \frac{\mu}{2a\Delta - AH} = \frac{\nu}{2b\Delta - BH} = \frac{1}{2A\Delta}$$

where $H = Ac + Ca - 2Bb$, $\Delta = AC - B^2 \neq 0$.

If $H = 0$, the solution reduces to

$$-\lambda/a = \mu/a = \nu/b = 1/A$$

and the integral to $-(ax + b)/\{A(Ax^2 + 2Bx + C)\}$ which is rational.

(2) Find a reduction formula for the integral

$$I_n = \int \frac{dx}{(x^2 + 2px + q)^n} \quad (q - p^2 \neq 0).$$

(i) *Using the method of the text*: Put $M = x^2 + 2px + q$. Since $q \neq p^2$, M cannot be a perfect square; hence M and M' are mutually prime and, by (17), we can find constants λ, μ, ν such that

$$1 \equiv \lambda M + (\mu x + \nu) M' \equiv \lambda(x^2 + 2px + q) + 2(\mu x + \nu)(x + p)$$

and unity on the L.H.S. could be replaced by any more convenient constant, in this case by $q - p^2$. In fact, we find at once

$$q - p^2 \equiv x^2 + 2px + q - (x + p)^2$$

and hence, if $n > 1$,

$$(q - p^2) I_n = \int \frac{dx}{M^{n-1}} + \frac{1}{2(n-1)} \int (x + p)\, d\left(\frac{1}{M^{n-1}}\right)$$

$$= I_{n-1} + \frac{1}{2(n-1)} \left(\frac{x + p}{M^{n-1}} - I_{n-1}\right),$$

which gives

$$(2n - 2)(q - p^2) I_n = (x + p)/M^{n-1} + (2n - 3) I_{n-1}. \qquad (34)$$

(ii) *Second method: beginning with an integration by parts.*
Replace dx by $d(x + p)$: then

$$\int \frac{dx}{M^n} = \int \frac{1}{M^n} d(x + p) = \frac{x + p}{M^n} + \int \frac{2n(x + p)^2}{M^{n+1}} dx$$

and hence, after putting $(x + p)^2 = M - (q - p^2)$,

$$I_n = (x + p)/M^n + 2n I_n - 2n(q - p^2) I_{n+1}$$

which reduces to (34) when we replace n by $n - 1$.

(iii) *Third method: beginning with a differentiation.*
Differentiate the function $(x+p)/M^n$: we obtain

$$\frac{d}{dx}\frac{x+p}{M^n} = \frac{1}{M^n} - \frac{2n(x+p)^2}{M^{n+1}}.$$

As in (ii), now put $(x+p)^2 = M - (q-p^2)$; then integrate and replace n by $n-1$ to get (34).

Note. In (ii), instead of replacing dx by $d(x+p)$, we could begin:

$$\int \frac{dx}{M^n} = \frac{x}{M^n} + \int \frac{2nx(x+p)\,dx}{M^{n+1}}.$$

In (iii), we could begin by differentiating x/M^n instead of $(x+p)/M^n$. In either case a little algebra would lead to (34).

11.14   A useful rule

Let $x-\alpha$ be a non-repeated factor of the polynomial $N(x)$, and put $N(x) \equiv (x-\alpha)M(x)$. Then $x-\alpha$ is prime to $M(x)$ and, by §11.6, Theorem 2, any polynomial $P(x)$ can be expressed uniquely in the form

$$P(x) \equiv AM(x) + (x-\alpha)Q(x), \tag{35}$$

where A is a constant, and the fraction $P(x)/N(x)$ in the form

$$\frac{P(x)}{N(x)} = \frac{A}{x-\alpha} + \frac{Q(x)}{M(x)}. \tag{36}$$

Rule. The constant A is given by the rule

$$A = P(\alpha)/N'(\alpha). \tag{37}$$

Proof. If we differentiate the identity $N(x) \equiv (x-\alpha)M(x)$ and then put $x = \alpha$, we get $N'(\alpha) = M(\alpha)$. Hence and by putting $x = \alpha$ in (35), $A = P(\alpha)/M(\alpha) = P(\alpha)/N'(\alpha)$.

Example

Evaluate the integral $\quad I = \displaystyle\int_{-\infty}^{\infty} \frac{x^{2m}}{1+x^{2n}}\,dx,$

where m, n are positive integers, and $m < n$ for convergence.
The roots of the equation $1 + x^{2n} = 0$ are

$$\exp(\pm s\pi i/2n) \quad (s = 1, 3, ..., 2n-1).$$

Let α be a typical root, so that $\alpha^{2n} = -1$. Then, using (37),

$$\frac{x^{2m}}{1+x^{2n}} = \Sigma \frac{\alpha^{2m}}{2n\alpha^{2n-1}(x-\alpha)} = \Sigma - \frac{\alpha^{2m+1}}{2n(x-\alpha)}.$$

Put $\alpha = \exp(s\pi i/2n) = \lambda + i\mu$, $-\alpha^{2m+1}/2n = L + iM$, where λ, μ, L, M are real; then a conjugate pair of partial fractions is

$$\frac{L+iM}{x-\lambda-i\mu} + \frac{L-iM}{x-\lambda+i\mu} = \frac{2L(x-\lambda)-2M\mu}{(x-\lambda)^2+\mu^2}$$

and by integration, with $\xi = x - \lambda$, and since $\mu > 0$,

$$\int_{-\infty}^{\infty} \frac{2L(x-\lambda)-2M\mu}{(x-\lambda)^2+\mu^2}\,dx = \int_{-\infty}^{\infty} \frac{2L\xi-2M\mu}{\xi^2+\mu^2}\,d\xi = -2\pi M.$$

Now $L + iM = (-1/2n)(\cos s\theta + i \sin s\theta)$, where $\theta = (2m+1)\pi/2n$; therefore $M = (-1/2n)\sin s\theta$, and by summing wo s,

$$I = (-2\pi)(-1/2n)\{\sin\theta + \sin 3\theta + \ldots + \sin(2n-1)\theta\}.$$

Multiplying both sides by $2n\sin\theta/\pi$, we find

$$(2n\sin\theta/\pi)I = 1 - \cos 2\theta + \cos 2\theta - \cos 4\theta + \ldots - \cos 2n\theta$$

$$= 1 - \cos 2n\theta = 1 - \cos(2m+1)\pi = 2$$

and hence $I = \pi/(n\sin\theta)$. Further, since the integrand is even,

$$\int_0^{\infty} \frac{x^{2m}\,dx}{1+x^{2n}} = \tfrac{1}{2}I = \frac{\pi}{2n\sin\theta} = \frac{\pi}{2n\sin(2m+1)\pi/2n}.$$

By putting $u = x^{2n}$, $a = (2m+1)/2n$, we deduce that

$$\int_0^{\infty} \frac{u^{a-1}}{1+u}\,du = \frac{\pi}{\sin a\pi} \quad (0 < a < 1).$$

11.15

The operation of differentiating under the integral sign with respect to a parameter in the integrand can be useful in finding indefinite integrals.

We know that the partial derivatives $\partial^2 F/\partial x\,\partial y$ and $\partial^2 F/\partial y\,\partial x$ are equal in a region in which they are both continuous functions of position, by §7.2. Supposing this condition to be satisfied and $F(x,y)$ to be defined by

$$F(x,y) = \int f(x,y)\,dx \tag{38}$$

we have

$$\frac{\partial}{\partial x}\frac{\partial F}{\partial y} = \frac{\partial}{\partial y}\frac{\partial F}{\partial x} = \frac{\partial f}{\partial y}$$

and hence

$$\frac{\partial F}{\partial y} = \int \frac{\partial f}{\partial y}\,dx. \tag{39}$$

This formula enables us, from a known indefinite integral (38), to deduce another by differentiating partially wo y under the integral sign.

Example

From
$$F(x,a) = \int \sin ax \, dx = -\frac{\cos ax}{a} \quad (a \neq 0),$$

follows
$$\int x \cos ax \, dx = \frac{x \sin ax}{a} + \frac{\cos ax}{a^2}$$

by differentiating on the L.H.S. partially wo a under the integral sign. The result holds good at all points of a plane in which x, a are Cartesian coordinates, except points on the axis $a = 0$. For, if $a \neq 0$ both $\partial^2 F/\partial x \partial a$ and $\partial^2 F/\partial a \partial x$ exist and are continuous, but if $a = 0$ the function $F(x,a)$ itself is not defined and so its partial derivatives do not exist.

11.16 Rational exponential or trigonometric integrands

We next note the general theorems that integrals of the form, or reducible to the form

$$\text{(i)} \ \int R(e^x) \, dx, \quad \text{or} \quad \text{(ii)} \ \int R(\sin \theta, \cos \theta) \, d\theta,$$

where (i) $R(e^x)$ denotes a rational function of e^x, i.e. a rational function of u when we put $u = e^x$, and (ii) $R(\sin \theta, \cos \theta)$ denotes a rational function of $\sin \theta$ and $\cos \theta$, can be integrated by simple substitutions which reduce them to integrals with rational integrands.

In (i) put $u = e^x$, $dx = du/u$; then

$$\int R(e^x) \, dx = \int R(u) \frac{du}{u} = \int R_1(u) \, du, \tag{40}$$

where $R_1(u)$ denotes a rational function of u. In (ii) put

$$t = \tan\frac{\theta}{2}, \quad \sin \theta = \frac{2t}{1+t^2}, \quad \cos \theta = \frac{1-t^2}{1+t^2}, \quad d\theta = \frac{2dt}{1+t^2}; \tag{41}$$

then the integral becomes

$$\int R\left(\frac{2t}{1+t^2}, \frac{1-t^2}{1+t^2}\right) \frac{2dt}{1+t^2} = \int R_2(t) \, dt, \tag{42}$$

where $R_2(t)$ denotes a rational function of t.

Note. The value of these theorems lies in their generality: they show that integrals of the forms (i) and (ii) can always be found in theory, on the assumption that (apart from special cases, e.g. when the integrand is of the form f'/f) we can find the linear factors of the denominators of $R_1(u)$ and $R_2(t)$. In practice, the standard rules of elementary integration would often be more expeditious than these general theorems. Thus, we should not think of putting $u = e^x$ in order to integrate $\cosh x$, nor $t = \tan \tfrac{1}{2}\theta$ in order to integrate $\sin 2\theta$.

11.17   Irrational integrands

When square root, cube root, ... signs occur in an integrand we try, by means of a substitution, to transform the integral into one in which the integrand is a rational function, a rational exponential function or a rational trigonometric function of the new variable of integration.

The commonest kind of irrationality is the square root of one linear or quadratic function, in which case the general form of the integrand is $R(x, y)$ where R denotes a rational function of x and of y, and y is of the form $\sqrt{(ax+b)}$ or $\sqrt{(ax^2+2bx+c)}$, the variable of integration being x.

11.18   Square root of one linear function in the integrand

When $\sqrt{(ax+b)}$ occurs in the integrand, we make the substitution

$$ax+b = u^2, \quad x = (u^2-b)/a, \quad dx = 2u\,du/a.$$

If there is no other irrationality in the integrand, this substitution will convert the integral into one in which the integrand is a rational function of u, the new variable of integration.

11.19   Square root of a quadratic in the integrand

When the square root of a quadratic occurs in the integrand, an algebraic substitution can be found that will transform the integral into one with a rational integrand. But, in practice, trigonometric or hyperbolic substitutions may be easier to apply. The following table lists a few substitutions by which integrals of the types indicated in the left-hand column by their differentials can be converted into integrals with rational trigonometric or hyperbolic integrands.

(i)	$R\{x, \sqrt{(a^2-x^2)}\}\,dx$	$x = a\sin\theta,\ x = a\tanh u.$
(ii)	$R\{x, \sqrt{(x^2+a^2)}\}\,dx$	$x = a\tan\theta,\ x = a\sinh u.$
(iii)	$R\{x, \sqrt{(x^2-a^2)}\}\,dx$	$x = a\sec\theta,\ x = a\cosh u.$
(iv)	$R\{x, \sqrt{(q+2px-x^2)}\}\,dx$	$x-p = \xi$ leads to type (i).
(v)	$R\{x, \sqrt{(x^2+2px+q)}\}\,dx$	$x+p = \xi$ leads to (ii) or (iii).
(vi)	$R\{x, \sqrt{(x^2+2px+q)}\}\,dx$	$u = x+\sqrt{(x^2+2px+q)}.$
(vii)	$R\{x, \sqrt{(ax-x^2)}\}\,dx$	$x = a\sin^2\theta.$
(viii)	$R\{x, \sqrt{(x^2+ax)}\}\,dx$	$x = a\sinh^2 u.$

(ix) $R\{x, \sqrt{(x^2 - ax)}\}\, dx$ $x = a \cosh^2 u.$

(x) $R\{x, (x-a)^{\frac{1}{2}} (b-x)^{\frac{1}{2}}\}\, dx$ $x = a \cos^2 \theta + b \sin^2 \theta,$

or $R\left\{x, \left(\dfrac{x-a}{b-x}\right)^{\frac{1}{2}}\right\} dx$ or $\dfrac{x-a}{b-x} = u^2 \quad (a < x < b).$

(xi) $R\{x, (x-a)^{\frac{1}{2}} (x-b)^{\frac{1}{2}}\}\, dx$ $x = b \cosh^2 u - a \sinh^2 u,$

or $R\left\{x, \left(\dfrac{x-b}{x-a}\right)^{\frac{1}{2}}\right\} dx$ or $\dfrac{x-b}{x-a} = v^2 \quad (a < b < x).$

(xii) $R\{x, (a-x)^{\frac{1}{2}} (b-x)^{\frac{1}{2}}\}\, dx$ $x = a \cosh^2 u - b \sinh^2 u,$

or $R\left\{x, \left(\dfrac{a-x}{b-x}\right)^{\frac{1}{2}}\right\} dx$ or $\dfrac{a-x}{b-x} = v^2 \quad (x < a < b).$

11.20 Reduction to a standard form

Let $y = \sqrt{(ax^2 + 2bx + c)}$.

Theorem. Any rational function $R(x, y)$ of x and y can be reduced to the form indicated by

$$R(x, y) = S(x) + T(x)/y, \tag{43}$$

where $S(x)$, $T(x)$ denote rational functions of x only.

Proof. Since

$$y^2 = ax^2 + 2bx + c \text{ and } y^3 = (y^2)y, \ y^4 = (y^2)^2, \ y^5 = (y^4)y, \dots$$

we can begin by writing $R(x, y)$ in the form

$$R(x, y) = (L + My)/(P + Qy),$$

where L, M, P, Q are polynomials in x. Then, after multiplying numerator and denominator by $P - Qy$, we get

$$R(x, y) = \frac{LP - MQy^2}{P^2 - Q^2 y^2} + \frac{(MP - LQ)\, y^2}{(P^2 - Q^2 y^2)\, y}$$

which, when we put $y^2 = ax^2 + 2bx + c$, is of the form (43).

Note. The theorem would apply if $\sqrt{(ax^2 + 2bx + c)}$ were replaced by the square root of any rational function of x.

11.21

It follows from (43) that the integral of $R(x, y)\, dx$ can be reduced to the standard form expressed by

$$\int R(x, y)\, dx = \int S(x)\, dx + \int \frac{T(x)}{y}\, dx. \tag{44}$$

The integration of rational fractions has been discussed, so it remains to consider the second integral on the right of (44). The rational fraction $T(x)$ can first be expressed as the sum of powers of x and proper partial fractions; it will then be necessary to evaluate integrals of the three types:

$$\text{(i)} \int \frac{x^n \, dx}{y}, \quad \text{(ii)} \int \frac{dx}{(x-\alpha)^n \, y}, \quad \text{(iii)} \int \frac{(hx+k) \, dx}{(Ax^2+2Bx+C)^n \, y} \quad (45)$$

where n is a positive integer, $y = \sqrt{(ax^2+2bx+c)}$ and $AC > B^2$. We consider these types in turn.

$$\text{(i)} \ I_n = \int \frac{x^n \, dx}{y}, \quad y = \sqrt{(ax^2+2bx+c)}. \quad (46)$$

We could begin by making a trigonometric or hyperbolic substitution, as in §11.19.

Alternatively, we can obtain a reduction formula, as follows: Begin by differentiating $x^m y$. Since $dy/dx = (ax+b)/y$, we find

$$(d/dx)(x^m y) = mx^{m-1}y + x^m(ax+b)/y$$

$$= \{mx^{m-1}(ax^2+2bx+c) + x^m(ax+b)\}/y$$

$$= \{(m+1)\,ax^{m+1} + (2m+1)\,bx^m + mcx^{m-1}\}/y$$

and hence, by integration,

$$x^m y = (m+1)\,aI_{m+1} + (2m+1)\,bI_m + mcI_{m-1}$$

from which follows, on putting $m+1 = n$,

$$naI_n = x^{n-1}y - (2n-1)\,bI_{n-1} - (n-1)\,cI_{n-2}. \quad (47)$$

This is the reduction formula required; by applying it repeatedly, I_n is seen to depend upon I_1 and I_0, both of which are elementary integrals.

$$\text{(ii)} \ I_n = \int \frac{dx}{(x-\alpha)^n \, y}. \quad (48)$$

We could make a substitution, as in §11.19; or we could make the substitution $x-\alpha = 1/v$, which would reduce this integral to one of type (i); or we could work out a reduction formula by the method suggested in Examples 11 D, 1, or by differentiating wo α under the integral sign.

$$\text{(iii)} \ I_n = \int \frac{(hx+k) \, dx}{(Ax^2+2Bx+C)^n \, y} \quad (AC > B^2). \quad (49)$$

*We could at once make a substitution, as in §11.19.

* See also, Hardy: *Course of Pure Mathematics*, §142 (Cambridge University Press).

Alternatively, we can work out reduction formulae (Examples 11, D, 2), by which I_n is shown to depend upon I_1 and I_0. It remains to consider I_1, since I_0 is elementary. Put

$$I_1 = \int \frac{(hx + k)\,dx}{(Ax^2 + 2Bx + C)\,\sqrt{(ax^2 + 2bx + c)}} . \tag{50}$$

We have $Ax^2 + 2Bx + C \equiv (1/A)\{(Ax + B)^2 + D^2\},$

where $D^2 = AC - B^2 > 0$. Hence, when we put

$$Ax + B = D\xi, \quad x = (D\xi - B)/A, \quad dx = (D/A)\,d\xi,$$

the integral takes the form

$$\int \frac{(h'\xi + k')\,d\xi}{(\xi^2 + 1)\,\sqrt{(a'\xi^2 + 2b'\xi + c')}} .$$

Now make the substitution

$$\xi = (\lambda t + 1)/(t - \lambda), \quad d\xi = -(\lambda^2 + 1)\,dt/(t - \lambda)^2.$$

The integral remains of the same form

$$\int \frac{(lt + m)\,dt}{(t^2 + 1)\,\sqrt{(\alpha t^2 + 2\beta t + \gamma)}},$$

where $\beta = -b'\lambda^2 - (c' - a')\lambda + b'$ and the other coefficients also depend on λ. We can now make $\beta = 0$ by choosing λ to be either of the roots (evidently real) of the equation

$$b'\lambda^2 + (c' - a')\lambda - b' = 0. \tag{51}$$

We shall then have a result of the form

$$I_1 = \int \frac{lt\,dt}{(t^2 + 1)\,\sqrt{(\alpha t^2 + \gamma)}} + \int \frac{m\,dt}{(t^2 + 1)\,\sqrt{(\alpha t^2 + \gamma)}} .$$

In the last integral, put $t = 1/s$; then, taking $t > 0$, $s > 0$ (see *Note* below),

$$I_1 = \int \frac{lt\,dt}{(t^2 + 1)\,\sqrt{(\alpha t^2 + \gamma)}} - \int \frac{ms\,ds}{(s^2 + 1)\,\sqrt{(\gamma s^2 + \alpha)}} .$$

These two integrals are of the same form. To reduce them to standard forms, put $\alpha t^2 + \gamma = u^2$, $\gamma s^2 + \alpha = v^2$.

Examples

(1) $\displaystyle \int \frac{dx}{x(x + 1)\,\sqrt{(x^2 + 1)}} = \int \left(\frac{1}{x} - \frac{1}{x + 1}\right) \frac{dx}{\sqrt{(x^2 + 1)}}$

$$= \int \frac{dx}{x\,\sqrt{(x^2 + 1)}} - \int \frac{dx}{(x + 1)\,\sqrt{(x^2 + 1)}} .$$

In the first of these integrals put $x = 1/u$, in the second $x+1 = 1/v$; we get

$$\int \frac{-du}{u\sqrt{(1+1/u^2)}} + \int \frac{dv}{v\sqrt{\{1+(1-1/v)^2\}}}$$

$$= -\int \frac{du}{\sqrt{(u^2+1)}} + \int \frac{dv}{\sqrt{(2v^2-2v+1)}} \quad (u>0, v>0)$$

$$= -\sinh^{-1} u + (1/\sqrt{2})\sinh^{-1}(2v-1)$$

$$= -\sinh^{-1}(1/x) + (1/\sqrt{2})\sinh^{-1}\{(1-x)/(1+x)\} \quad (x>0).$$

Note. When we put u and v above under the respective square root signs, it is assumed that $u>0$ and $v>0$; that is, $x>0$ and $x+1>0$. Hence the result holds good for $x>0$.

(2) $$I = \int_{-1}^{1} \frac{(x+2)\,dx}{(2x^2+2x+1)\sqrt{(3x^2+4x+2)}}.$$

Put $2x+1 = \xi$, $x = \frac{1}{2}(\xi-1)$, $dx = \frac{1}{2}d\xi$. Then we find

$$I = \int_{-1}^{3} \frac{(\xi+3)\,d\xi}{(\xi^2+1)\sqrt{(3\xi^2+2\xi+3)}}.$$

Here, equation (51) is $\lambda^2 + (3-3)\lambda - 1 = 0$, i.e. $\lambda^2 = 1$; hence $\lambda = \pm 1$. Either $\lambda = 1$ or $\lambda = -1$ will do. If we take $\lambda = 1$, we put $\xi = (t+1)/(t-1)$, $t = (\xi+1)/(\xi-1)$, and as ξ increases from -1 to 3, t first decreases from 0 to $-\infty$, and then decreases from $+\infty$ to 2, so the resulting integral will have to be evaluated in two parts.

If we take $\lambda = -1$, we put $\xi = (1-t)/(1+t)$, $t = (1-\xi)/(1+\xi)$ and now t decreases steadily from $+\infty$ to $-\frac{1}{2}$. Of the two values, $\lambda = -1$ is therefore the more convenient. So taking $\lambda = -1$, we find

$$I = \int_{-\frac{1}{2}}^{\infty} \frac{(2+t)\,dt}{(t^2+1)\sqrt{(2+t^2)}}.$$

Next, considering that part of the integrand is an even function of t, and the other part odd, we see that

$$I = \left(\int_0^{\frac{1}{2}} + \int_0^{\infty}\right) \frac{2dt}{(t^2+1)\sqrt{(2+t^2)}} + \int_{\frac{1}{2}}^{\infty} \frac{t\,dt}{(t^2+1)\sqrt{(2+t^2)}}$$

or, with $t = 1/s$ in the first of these,

$$I = \left(\int_2^{\infty} + \int_0^{\infty}\right) \frac{2s\,ds}{(s^2+1)\sqrt{(2s^2+1)}} + \int_{\frac{1}{2}}^{\infty} \frac{t\,dt}{(t^2+1)\sqrt{(2+t^2)}}.$$

Now put $2s^2+1 = u^2$, $2+t^2 = v^2$. Then the integrals reduce to standard forms, and we find

$$I = 2(\pi - \tan^{-1} 3 - \tan^{-1} 1) + \frac{1}{2}\log 5 \doteq 3.019.$$

11.22   Other types of irrational integrands

A few other types of integrals with irrational integrands, and sub-stitutions that will convert them into integrals with rational integrands, are indicated below:

(i) *Square roots of two linear functions in the integrand.*

$$\int R(x, y, z)\, dx, \quad y = \sqrt{(\alpha x + \beta)}, \quad z = \sqrt{(\gamma x + \delta)}, \tag{52}$$

where R denotes a rational function of x, y and z.

The substitution

$$\alpha x + \beta = t^2, \quad x = (t^2 - \beta)/\alpha, \quad dx = 2t\, dt/\alpha \tag{53}$$

will convert the integral (52) into one of the form

$$\int R_1\{t, \sqrt{(at^2 + c)}\}\, dt \tag{54}$$

in which the only irrationality is the square root of a quadratic func-tion of the variable of integration t.

(ii) *Fractional powers of one linear function.*

$$\int R(x, y)\, dx, \quad y = (ax + b)^{1/n} \tag{55}$$

where R denotes a rational function of x and y, and n is the lowest common denominator of the indices of the fractional powers of $ax + b$ that occur in the integrand.

The substitution

$$ax + b = t^n, \quad x = (t^n - b)/a, \quad dx = nt^{n-1}\, dt/a \tag{56}$$

will convert the integral into one in which the integrand is a rational function of t.

(iii) *Fractional powers of the quotient of two linear functions.*

$$\int R(x, y)\, dx, \quad y = \left(\frac{\alpha x + \beta}{\gamma x + \delta}\right)^{1/n}. \tag{57}$$

The substitution

$$\frac{\alpha x + \beta}{\gamma x + \delta} = t^n, \quad x = -\frac{\delta t^n - \beta}{\gamma t^n - \alpha}, \quad dx = \frac{n(\alpha\delta - \beta\gamma)\, t^{n-1}\, dt}{(\gamma t^n - \alpha)^2} \tag{58}$$

will convert the integral into one with a rational integrand.

11.23   Elliptic integrals

An integral of the form $\int R(x,y)\,dx$, where y denotes the square root of a cubic or quartic polynomial in x, is called an *elliptic* integral, because the rectification of a conic (in particular, of an ellipse) leads to an integral of this kind. There is an extensive literature on elliptic integrals and the 'elliptic functions' associated with them.

Examples 11 A

(1) Show that the following integrals are rational functions of x and find them, assuming that n is any integer except unity in (i):

(i) $\displaystyle\int \frac{x\,dx}{(x^2+c)^n}$,

(ii) $\displaystyle\int \frac{3x^4+2x^3-6x}{(x^3+x+1)^2}\,dx$,

(iii) $\displaystyle\int \frac{x^3-3x^2+1}{x^2(x^2-1)^2}\,dx$,

(iv) $\displaystyle\int \frac{5x^3+x^2+3x}{(x^3+3x+1)^3}\,dx$.

(2) Find the rational parts of the integrals:

(i) $\displaystyle\int \frac{15x^2+7x}{(x^3+x+1)^2}\,dx$,

(ii) $\displaystyle\int \frac{dx}{(x^4-6x^2+1)^2}$.

(3) Find the integrals:

(i) $\displaystyle\int \frac{x^5\,dx}{x^3+1}$,

(ii) $\displaystyle\int \frac{x^3+x^2}{(x^3-1)(x^4+1)}\,dx$,

(iii) $\displaystyle\int \frac{dx}{x-x^n}$,

(iv) $\displaystyle\int \frac{5x^9-2x^7+3x^2}{(x^5+1)^2}\,dx$.

(4) Show that
$$\int_{-\infty}^{0} \frac{dx}{(x^3-1)^2} = \frac{4\pi}{9\sqrt{3}}.$$

(5) Obtain reduction formulae for the integrals:

(i) $\displaystyle\int \frac{dx}{(x^2+c)^n}$,   (ii) $\displaystyle\int \frac{dx}{(a^2-x^2)^n}$,   (iii) $\displaystyle\int \frac{dx}{(x-a)^n(b-x)^n}$.

Also, show how to find the integral (iii) by means of the substitution
$$\xi = (x-a)/(b-x).$$

(6) If $-\pi < \alpha < \pi$, prove that

(i) $\displaystyle\int_0^\infty \frac{dx}{x^2+2x\cos\alpha+1} = \frac{\alpha}{\sin\alpha}$

and evaluate the integrals:

(ii) $\displaystyle\int_0^\infty \frac{x\,dx}{(x^2+2x\cos\alpha+1)^2}$,

(iii) $\displaystyle\int_0^\infty \frac{dx}{(x^2+2x\cos\alpha+1)^2}$.

Sketch the graphs of these three integrals as functions of α for all values of α.

(7) Prove that if $\alpha \neq n\pi$:

(i) $\displaystyle\int_{-1}^{1} \frac{dx}{1 - 2x\cos\alpha + x^2} = \frac{\frac{1}{2}\pi}{|\sin\alpha|}$,

(ii) $\displaystyle\int_{0}^{\infty} \frac{x^2\,dx}{x^4 - 2x^2\cos 2\alpha + 1} = \frac{\frac{1}{4}\pi}{|\sin\alpha|}$.

(8) Prove that if $a > 0$:

(i) $\displaystyle\int_{0}^{\infty} \frac{dx}{x^2 + 2x\cosh a + 1} = \frac{a}{\sinh a}$,

(ii) $\displaystyle\int_{0}^{1} \frac{dx}{x^2 - 2x\sinh a - 1} = -\frac{1}{2\cosh a}\log(e^a \coth \tfrac{1}{2}a)$.

(9) Determine constants A, B, C, D such that

$$\frac{x^4 + 1}{(x^2 + 1)^4} = \frac{d}{dx}\frac{Ax^5 + Bx^3 + Cx}{(x^2 + 1)^3} + \frac{D}{x^2 + 1}.$$

Hence prove that $\displaystyle\qquad 0.502 < \int_{0}^{1}\frac{x^4 + 1}{(x^2 + 1)^4}\,dx < 0.503.$

(10) Let α be a double root of the equation $N(x) = 0$, so that we can put $N(x) = (x - \alpha)^2 M(x)$, where $M(\alpha) \neq 0$.

If the rational fraction $P(x)/N(x)$ is expressed in the form

$$\frac{P(x)}{N(x)} = \frac{Q(x)}{M(x)} + \frac{A}{(x - \alpha)^2} + \frac{B}{x - \alpha},$$

where $Q(x)$ is a polynomial and A, B are constants, show that

$$A = P(\alpha)/M(\alpha) = 2P(\alpha)/N''(\alpha),$$

$$B = \left[\frac{d}{dx}\frac{P(x)}{M(x)}\right]_{x=\alpha} = \frac{2}{N''(\alpha)}\left[P'(\alpha) - \frac{P(\alpha)N'''(\alpha)}{3N''(\alpha)}\right].$$

(11) The polynomial $N(x)$ has only simple zeros α, β, γ,.... If

$$\frac{1}{\{N(x)\}^2} = \Sigma\left(\frac{A}{(x - \alpha)^2} + \frac{B}{x - \alpha}\right),$$

where A, B are constants, show that

$$A = 1/\{N'(\alpha)\}^2, \quad B = -N''(\alpha)/\{N'(\alpha)\}^3.$$

(12) If $N(x) = (x - \alpha)^s M(x)$, where $M(\alpha) \neq 0$, show that the coefficient of $1/(x - \alpha)^{s-r}$ in the expression of $P(x)/N(x)$ in partial fractions is

$$\frac{1}{r!}\left[\frac{d^r}{dx^r}\frac{P(x)}{M(x)}\right]_{x=\alpha} \quad (1 \leqslant r \leqslant s - 1).$$

(13) The polynomial $N(x)$ has no simple zeros. Show that necessary and sufficient conditions that the integral of $P(x)/N(x)$ should be rational are that the relation

$$\left[\frac{d^{r-1}}{dx^{r-1}}(x - a)^r\frac{P(x)}{N(x)}\right]_{x=a} = 0$$

should hold good for every r-fold zero a of $N(x)$.

(14) Show that $\displaystyle\int_0^\infty \frac{dx}{1+x^n} = \frac{\pi}{n}\operatorname{cosec}\frac{\pi}{n}$ $(n > 1)$.

If $\displaystyle I_s = \int_0^\infty \frac{dx}{(1+x^n)^s}$ show that $\displaystyle I_{s+1} = \left(1 - \frac{1}{ns}\right)I_s$ $(s > 1/n > 0)$.

(15) Show that

$$\int_0^1 \frac{dx}{1+x^4} = \frac{\pi + 2\log(1+\sqrt2)}{4\sqrt2}, \qquad \int_0^\infty \frac{dx}{1+x^4} = \frac{\pi}{2\sqrt2}.$$

Examples 11 B

Evaluate the following integrals:

(1) $\displaystyle\int \frac{x\,dx}{1+\sqrt x}.$ (2) $\displaystyle\int \frac{dx}{2-x-\sqrt x}.$ (3) $\displaystyle\int \frac{dx}{x+\sqrt{(x-1)}}.$

(4) $\displaystyle\int_0^1 \frac{1-\sqrt x}{1+\sqrt x}\,dx.$ (5) $\displaystyle\int_3^8 \frac{\sqrt{(x+1)}}{x}\,dx.$ (6) $\displaystyle\int_2^5 \frac{dx}{1+\sqrt{(x-1)}}.$

(7) $\displaystyle\int_1^\infty \frac{\sqrt{(2x-1)}}{x^2}\,dx.$ (8) $\displaystyle\int_0^\infty \frac{dx}{(1+x)(1+\sqrt x)}.$

Examples 11 C

Evaluate the following integrals by using the given substitutions, or otherwise.

(1) (i) $\displaystyle\int_0^a \frac{dx}{\sqrt{(2ax-x^2)}},$ (ii) $\displaystyle\int_0^{2a} \sqrt{(2ax-x^2)}\,dx;$ $x = 2a\sin^2\theta.$

(2) (i) $\displaystyle\int_0^a \frac{dx}{\sqrt{(x^2+2ax)}},$ (ii) $\displaystyle\int_0^{2a} \sqrt{(x^2+2ax)}\,dx;$ $x = 2a\sinh^2 u.$

(3) (i) $\displaystyle\int_{2a}^{4a} \frac{dx}{\sqrt{(x^2-2ax)}},$ (ii) $\displaystyle\int_{2a}^{3a} \sqrt{(x^2-2ax)}\,dx;$ $x = 2a\cosh^2 u.$

(4) (i) $\displaystyle\int_a^b \frac{dx}{x\sqrt{\{(x-a)(b-x)\}}},$ (ii) $\displaystyle\int_a^b \frac{dx}{(a+b-x)^2\sqrt{\{(x-a)(b-x)\}}};$

$x = a\cos^2\theta + b\sin^2\theta$ $(0 < a < b).$

(5) $\displaystyle\int_b^R \frac{dx}{\sqrt{\{(x-a)(x-b)\}}};$ $x = b\cosh^2 u - a\sinh^2 u,$

or $v = \sqrt{(x-a)} + \sqrt{(x-b)}$ $(a < b < R).$

(6) $\displaystyle\int_{-R}^a \frac{dx}{\sqrt{\{(a-x)(b-x)\}}} + \int_b^R \frac{dx}{\sqrt{\{(x-a)(x-b)\}}}$ $(-R < a < b < R).$

(7) Show that $\displaystyle\int_0^a \frac{dx}{\{x+\sqrt{(a^2-x^2)}\}^2} = \frac{1}{a\sqrt2}\log(1+\sqrt2).$

Examples 11D

(1) Obtain reduction formulae for the integrals:

(i) $\displaystyle\int \frac{x^n\,dx}{\sqrt{(ax^3+bx^2+cx+d)}}\,.$

(ii) $\displaystyle\int \frac{dx}{(x-\alpha)^n\,y}$ where $y=\sqrt{(ax^2+2bx+c)}.$

[Begin by differentiating $y/(x-\alpha)^m$.]

(iii) $\displaystyle\int \frac{x\,dx}{(x^2+1)^n\,\sqrt{(ax^2+c)}}\,.$ $\left[\text{Differentiate }\dfrac{\sqrt{(ax^2+c)}}{(x^2+1)^m}\,.\right]$

(iv) $\displaystyle\int \frac{dx}{(x^2+1)^n\,\sqrt{(ax^2+c)}}\,.$ $\left[\text{Differentiate }\dfrac{x\sqrt{(ax^2+c)}}{(x^2+1)^m}\,.\right]$

(2) If $y=\sqrt{(ax^2+2bx+c)}$ and if

$$I_n=\int \frac{x\,dx}{(Ax^2+2Bx+C)^n\,y}, \qquad J_n=\int \frac{dx}{(Ax^2+2Bx+C)^n\,y},$$

show how to obtain reduction formulae of the type

$$I_n=\alpha I_{n-1}+\beta J_{n-1}+\gamma J_{n-2}+(\lambda x+\mu)\,y/X^{n-1},$$
$$J_n=\alpha' I_{n-1}+\beta' J_{n-1}+\gamma' J_{n-2}+(\lambda'x+\mu')\,y/X^{n-1},$$

where $X=Ax^2+2Bx+C$.

[*Hint*: Differentiate $y/(Ax^2+2Bx+C)^m$ and $xy/(Ax^2+2Bx+C)^m$ and combine the results.]

Evaluate the following integrals:

(3) $\displaystyle\int \frac{x^2+x+1}{x\sqrt{(x^2-1)}}\,dx.$

(4) $\displaystyle\int \frac{dx}{(x^2-1)\sqrt{(x^2+2x+2)}}\,.$

(5) $\displaystyle\int \frac{x\,dx}{(x+1)^2\,\sqrt{(x^2-x+1)}}\,.$

(6) $\displaystyle\int \frac{dx}{x(x-1)\,\sqrt{(3x^2-2x+1)}}\,.$

(7) $\displaystyle\int \frac{(x+1)\,dx}{(x^2+4)\,\sqrt{(5-x^2)}}\,.$

(8) $\displaystyle\int \frac{(x+1)\,dx}{(x^2+9)\,\sqrt{(x^2+5)}}\,.$

(9) $\displaystyle\int_1^\infty \frac{dx}{x(x^2+1)\sqrt{(x^2+2)}}\,.$

(10) $\displaystyle\int_{\sqrt 2}^\infty \frac{dx}{(x^2-1)^3\,\sqrt{(x^2-2)}}\,.$

(11) $\displaystyle\int_0^\infty \frac{dx}{(x^2+1)\,\sqrt{(2x^2+x+2)}}\,.$

(12) $\displaystyle\int \frac{dx}{(x^2-2x+2)\,\sqrt{(2x^2-x-3)}}\,.$

(13) $\displaystyle\int \frac{x\,dx}{(x^2+1)^2\,\sqrt{(2x^2+1)}}\,.$

(14) $\displaystyle\int_3^\infty \frac{(x-1)\,dx}{(2x^2-6x+5)\,\sqrt{(7x^2-22x+19)}}\,.$

Examples 11 E

Find the given integrals:

(1) $\displaystyle\int \frac{dx}{\sqrt{(x^2-1)}}$,   by putting (i) $x-1 = u^2$,   (ii) $x+1 = v^2$.

(2) $\displaystyle\int \frac{dx}{\sqrt{x}+\sqrt[3]{x}}$.

(3) $\displaystyle\int \frac{x\,dx}{(x-1)^{\frac{1}{2}}+(x-1)^{\frac{5}{6}}}$.

(4) $\displaystyle\int \frac{x\,dx}{\sqrt[5]{(x+1)}}$.

(5) $\displaystyle\int_1^3 \frac{(x-1)^{\frac{1}{2}}}{(x+1)^{\frac{3}{2}}}\,dx$.

(6) $\displaystyle\int_0^1 \frac{dx}{\sqrt[n]{(1-x^n)}}$.

(7) $\displaystyle\int_0^1 \frac{dx}{x^{\frac{1}{2}}+x^{\frac{2}{3}}}$.

(8) $\displaystyle\int_0^1 \frac{x\,dx}{(1-x)^{\frac{1}{3}}}$.

(9) $\displaystyle\int_0^1 \frac{1-x^{\frac{1}{3}}}{1+x^{\frac{2}{3}}}\,dx$.

(10) $\displaystyle\int_5^8 \frac{dx}{\sqrt{(x+4)}-\sqrt{(x-4)}}$.

(11) $\displaystyle\int_0^{\frac{1}{2}\pi} \frac{d\theta}{\sqrt{(\cot\theta)}+\sqrt{(\tan\theta)}}$.

(12) $\displaystyle\int_0^1 \frac{dx}{2\sqrt{(1+x)}+\sqrt{x}}$.

(13) $\displaystyle\int_2^3 \frac{dx}{\{1+\sqrt{(x-1)}\}\sqrt{(x-2)}}$.

(14) Show that of the two integrals

$$ \text{(i)} \quad \int \sqrt{(ae^x+b)}\,dx, \quad \text{(ii)} \quad \int \sqrt{(a\sin\theta+b)}\,d\theta, $$

the first is elementary, the second is elliptic.

(15) If n is a positive integer and $R(x)$ a rational function of x, show that

$$ \int \{\sqrt{(x-a)}+\sqrt{(x-b)}\}^{1/n}\,R(x)\,dx \quad (a < b < x) $$

can be found by putting $x = b\cosh^2 nu - a\sinh^2 nu$.

(16) Show that the integral

$$ \int R\{x, \sqrt{(\alpha x+\beta)}, \sqrt{(\gamma x+\delta)}\}\,dx, $$

where $R(x,y,z)$ denotes a rational function of x, y, z, is converted into the integral of a rational function of t by the substitution

$$ 4x = \frac{\beta}{\alpha}\left(t-\frac{1}{t}\right)^2 - \frac{\delta}{\gamma}\left(t+\frac{1}{t}\right)^2. $$

(17) Prove that the integral

$$ \int_1^\infty \{x-\sqrt{(x^2-1)}\}^n\,dx $$

converges if $n > 1$ and has the value $1/(n^2-1)$.

12

DOUBLE INTEGRALS

12.1 Dissection of a rectangle

In the xy plane, consider the rectangle bounded by the lines $x = a$, $x = b, y = \alpha, y = \beta, a < b, \alpha < \beta$. Let the region within this rectangle, including its boundary, be denoted by $[a,b;\ \alpha,\ \beta]$ or by R, as convenient.

Given $\delta > 0$, let the interval $[a,b]$ on the axis of x be divided into m subintervals each of length not exceeding δ, so that $m\delta \geqslant b - a$. Let the points of division be $x_1, x_2, \ldots$ so that if we put $x_0 = a, x_m = b$, then (cf. §9.1)
$$a = x_0 < x_1 < x_2 \ldots < x_{m-1} < x_m = b. \tag{1}$$

Also, given $\delta' > 0$, let the interval $[\alpha, \beta]$ on the axis of y be divided into n subintervals each of length not exceeding δ', so that $n\delta' \geqslant \beta - \alpha$. Let the points of division be $y_1, y_2, \ldots$ so that, with $y_0 = \alpha, y_n = \beta$,
$$\alpha = y_0 < y_1 < y_2 \ldots < y_{n-1} < y_n = \beta. \tag{2}$$

Now, through the points of division indicated by (1) and (2), let lines be drawn parallel to the axes of y and x respectively. The resulting division of R into mn subrectangles will be called a dissection $D(\delta, \delta')$ of R.

The subrectangles can be grouped in rows or in columns. Let R_{ij} be the subrectangle in the ith column and jth row (i increasing from 1 to m, and j increasing from 1 to n, in the directions of x and y increasing, respectively). Let a_{ij} be the area of R_{ij}. Put
$$\delta_i = x_i - x_{i-1} \leqslant \delta, \quad \delta_j = y_j - y_{j-1} \leqslant \delta'; \tag{3}$$
then we note that $\qquad a_{ij} = \delta_i \delta_j \leqslant \delta\delta'. \tag{4}$

In particular, for the dissection $D(\delta, \delta)$ in which $\delta' = \delta$,
$$a_{ij} \leqslant \delta^2. \tag{5}$$

12.2 Integration over a rectangle

Let $f(x,y)$ be a bounded function over the rectangle $R[a,b;\ \alpha,\beta]$. Let $f_{ij} = f(\xi_{ij}, \eta_{ij})$ denote the value of $f(x,y)$ at an arbitrary point (ξ_{ij}, η_{ij}) in the rectangle R_{ij} in the dissection $D(\delta, \delta)$. For brevity, write D for $D(\delta, \delta)$ and put
$$\sigma(D) = \sum_{j=1}^{n} \sum_{i=1}^{m} f_{ij} \delta_i \delta_j. \tag{6}$$

Definition. The double integral of the bounded function $f(x, y)$ over the rectangle R will be defined by

$$\iint_R f(x, y) \, dx \, dy = \lim_{\delta \to 0} \sigma(D) \tag{7}$$

provided that this limit exists.

When $f(x, y) = 1$, $\sigma(D) =$ sum of the areas of all the subrectangles $R_{ij} =$ area of the rectangle R, so that in this case the integral certainly exists and

$$\iint_R dx \, dy = (b - a)(\beta - \alpha). \tag{8}$$

Theorems corresponding to Theorems 1, 2, 3 of §9.2 follow at once from the definition and are proved in a similar way. In particular, if m and M are the closest bounds of $f(x, y)$ over the rectangle R, that is, if m is the g.l.b. and M the l.u.b., then

$$m(b - a)(\beta - \alpha) \leqslant \iint_R f(x, y) \, dx \, dy \leqslant M(b - a)(\beta - \alpha) \tag{9}$$

and hence, for some number μ between m and M,

$$\iint_R f(x, y) \, dx \, dy = \mu(b - a)(\beta - \alpha). \tag{10}$$

The number μ may be called the *average* or *mean* value of the function $f(x, y)$ over the rectangle. It is given explicitly by

$$\mu = \frac{\displaystyle\iint_R f(x, y) \, dx \, dy}{\text{area of } R}.$$

12.3   A necessary and sufficient condition of integrability

Let m_{ij}, M_{ij} be the closest bounds of $f(x, y)$ over the subrectangle R_{ij} in the dissection $D(\delta, \delta)$ of the rectangle $R[a, b; \alpha, \beta]$. The dissection will again be denoted briefly by D.

The *lower sum* $s(D)$ and the *upper sum* $S(D)$, for the function $f(x, y)$ over the rectangle R, are defined by

$$s(D) = \Sigma\Sigma m_{ij} \delta_i \delta_j, \quad S(D) = \Sigma\Sigma M_{ij} \delta_i \delta_j \tag{11}$$

the summation in each case extending over the values

$$i = 1, 2, \ldots, m; \quad j = 1, 2, \ldots, n.$$

Since $m_{ij} \leqslant f_{ij} \leqslant M_{ij}$ and $\delta_i \delta_j > 0$, therefore

$$m_{ij} \delta_i \delta_j \leqslant f_{ij} \delta_i \delta_j \leqslant M_{ij} \delta_i \delta_j$$

and by summing over all the subrectangles, by (11) and (6),

$$s(D) \leqslant \sigma(D) \leqslant S(D). \tag{12}$$

Theorem. A necessary and sufficient condition that the function $f(x,y)$ should be integrable over the rectangle R, in other words that the limit of $\sigma(D)$ when $\delta \to 0$ should exist, is

$$S(D) - s(D) \to 0 \quad \text{when} \quad \delta \to 0. \tag{13}$$

The truth of this theorem will be taken for granted. The proof follows the same steps as the proof of §9.4, (14): it first shows that (13) is a necessary and sufficient condition that $s(D)$ and $S(D)$ should approach the same limit when $\delta \to 0$, and then (12) shows that the limit of $\sigma(D)$ exists and is equal to the common limit of $s(D)$ and $S(D)$.

12.4   A continuous function is integrable

From the necessary and sufficient condition (13) it easily follows that a function which is a continuous function of position (§5.7, V) over a closed rectangle R is integrable over R.

Proof. Let $f(x,y)$ be a continuous function of position over the closed rectangle $R[a, b: \alpha, \beta]$.

Let $A = (b-a)(\beta - \alpha)$ be the area of R.

Given any $\epsilon > 0$, put $\epsilon_1 = \epsilon/A$.

Since $f(x,y)$ is a continuous function of position over R, by §5.5, II a number δ_0 exists such that, for every dissection $D(\delta, \delta)$ in which $\delta < \delta_0$, the spread $M_{ij} - m_{ij}$ of $f(x,y)$ in every subrectangle R_{ij} will be less than ϵ_1. Thus δ_0 exists such that, for all $\delta < \delta_0$, by (11),

$$S(D) - s(D) = \Sigma \Sigma (M_{ij} - m_{ij}) \delta_i \delta_j < \epsilon_1 \Sigma \Sigma \delta_i \delta_j = \epsilon_1 A = \epsilon.$$

It follows that $S(D) - s(D) \to 0$ when $\delta_0 \to 0$, and hence, by §12.3, that $f(x,y)$ is integrable.

12.5   Connection with volume

Let $f(x,y)$ be a positive continuous function of position over the rectangle $\qquad R[a, b; \alpha, \beta], \quad a < b, \quad \alpha < \beta.$

Consider the rectangular prism bounded by the planes

$$x = a, \quad x = b, \quad y = \alpha, \quad y = \beta.$$

Let V be the volume of the piece of this prism between the plane $z = 0$ and the surface $z = f(x,y)$: i.e. bounded above by a portion S of the surface $z = f(x,y)$ and standing on the rectangle R as base.

Also, let V_{ij} denote the volume of the piece of the rectangular prism bounded above by an element S_{ij} of the surface $z = f(x, y)$ and standing on the subrectangle R_{ij}. Evidently $V = \Sigma V_{ij}$.

We recall that a continuous function attains its closest upper and lower bounds. Consequently, there is a point p_{ij} on S_{ij} at which $z = m_{ij}$ and a point P_{ij} at which $z = M_{ij}$. Let planes be drawn through p_{ij} and P_{ij} parallel to $z = 0$. They will include between them and the plane $z = 0$ two rectangular prisms, both standing on the base R_{ij}, one of height m_{ij} and volume $m_{ij} \delta_i \delta_j$, the other of height M_{ij} and volume $M_{ij} \delta_i \delta_j$, and evidently V_{ij} will lie between these two volumes, being possibly equal to either or both, that is

$$m_{ij} \delta_i \delta_j \leqslant V_{ij} \leqslant M_{ij} \delta_i \delta_j. \tag{14}$$

By summing over all the subrectangles R_{ij}, we see that

$$s(D) \leqslant V \leqslant S(D). \tag{15}$$

But, as we have seen, $f(x, y)$ being continuous, $s(D)$ and $S(D)$ approach a common limit when $\delta \to 0$, and from (15) it follows that this common limit is V. The common limit is also, by definition, the integral of $f(x, y)$ over the rectangle R. Thus the integral and volume are equal.

If the function $f(x, y)$ is negative over part of R, the surface S will lie partly below the plane $z = 0$. In this case the corresponding part of the integral will be negative and the integral itself will be numerically equal to the part of the volume above $z = 0$, minus the part below.

Note. In the course of the above discussion, the physical idea of volume has been taken for granted. In a strictly analytical approach, the volume would be *defined* to be the integral when it exists.

12.6 Evaluation of a double integral by repeated integration

(i) Note first that, if $g(x)$ is a continuous function of x in the interval $[a, b]$ and if $D(\delta)$ is a dissection of the interval (§9.1), then numbers $\xi_1, \xi_2, \ldots$ exist such that

$$\int_a^b g(x)\, dx = g(\xi_1)\, \delta x_1 + g(\xi_2)\, \delta x_2 + \ldots + g(\xi_m)\, \delta x_m \tag{16}$$

where $x_{i-1} < \xi_i < x_i$, $\delta x_i = x_i - x_{i-1}$ $(i = 1, 2, \ldots, m)$. For we can put

$$\int_a^b g(x)\, dx = \int_{x_0}^{x_1} g(x)\, dx + \int_{x_1}^{x_2} g(x)\, dx + \ldots + \int_{x_{m-1}}^{x_m} g(x)\, dx \tag{17}$$

and (16) follows by applying the simple mean value theorem for integrals (§9.11) to each of the integrals on the R.H.S.

(ii) Secondly, we recall that by §9.26, if $f(x, y)$ is a continuous function of position over the rectangle $R[a, b; \alpha, \beta]$, then the integral

$$G(y) = \int_a^b f(x, y)\, dx \tag{18}$$

is a continuous and therefore integrable function of y in the interval $\alpha \leqslant y \leqslant \beta$.

12.7

Theorem. If $z = f(x, y)$ is a continuous function of position over the rectangle R, then

$$\iint_R z\, dx\, dy = \int_a^b dx \int_\alpha^\beta z\, dy = \int_\alpha^\beta dy \int_a^b z\, dx. \tag{19}$$

Proof. We first convert the double summation in (6) into a single summation. This we are able to do by a convenient choice of the point (ξ_{ij}, η_{ij}) in R_{ij}.

Since m, n are finite integers, we can put (6) in the form

$$\sigma(D) = \sum_{j=1}^n \delta_j \left\{ \sum_{i=1}^m f(\xi_{ij}, \eta_{ij})\, \delta_i \right\} \tag{20}$$

where D denotes $D(\delta, \delta)$.

(i) Now first replace η_{ij} by η_j, where $y_{j-1} \leqslant \eta_j \leqslant y_j$ and η_j is independent of i.

(ii) Secondly, observe that, by (16), since $f(x, \eta_j)$ is a continuous function of x, numbers ξ_{ij} exist such that

$$\sum_{i=1}^m f(\xi_{ij}, \eta_j)\, \delta_i = \int_a^b f(x, \eta_j)\, dx = G(\eta_j) \tag{21}$$

where $G(\eta_j)$ is given by putting $y = \eta_j$ in (18). Consequently,

$$\sigma(D) = \sum_{j=1}^n G(\eta_j)\, \delta_j. \tag{22}$$

Now in the dissection $D(\delta, \delta)$ let $\delta \to 0$. Then both sides tend to limits, which are therefore equal. The limit of the L.H.S. is the double integral on the left of (19) which is known to exist, by §12.4. The limit of the R.H.S. is, by the definition in §9.1, the simple integral over the interval $[\alpha, \beta]$ of $G(y)$, which is continuous by §12.6 (ii).

Thus in (19) the double integral is equal to the second of the repeated integrals. By §9.35 it is also equal to the first.

Examples

(1) Since $\cos(x+2y)$ is a continuous function of position over the rectangle $[0, \pi; \, 0, \frac{1}{2}\pi]$, we may put $z = \cos(x+2y)$ in (19), with $a = 0$, $b = \pi$, $\alpha = 0$, $\beta = \frac{1}{2}\pi$. Then one of the repeated integrals is

$$\int_0^\pi dx \int_0^{\frac{1}{2}\pi} \cos(x+2y)\, dy = \int_0^\pi \frac{dx}{2} [\sin(x+\pi) - \sin x] = -2.$$

It may be verified that the value of the other repeated integral is also -2, and this is the value of the double integral of $\cos(x+2y)$ over the rectangle.

(2) Note that, if X is a function of x only, and Y a function of y only, and both are continuous, then

$$\iint_R X Y\, dx\, dy = \int_a^b X\, dx \times \int_\alpha^\beta Y\, dy \qquad (23)$$

the product of two independent simple integrals.

(3) Note also that the double integral may sometimes be evaluated by means of one of the repeated integrals, though the other repeated integral may not be easily expressible in a finite form. Thus, consider the integral

$$I = \int_0^1 \int_0^{\frac{1}{2}\pi} \sqrt{(1 - x\sin^2\theta)}\, dx\, d\theta.$$

By integrating first with respect to x, we find

$$I = \int_0^{\frac{1}{2}\pi} \frac{2(1 - \cos^3\theta)\, d\theta}{3\sin^2\theta} = \frac{2}{3} \int_0^{\frac{1}{2}\pi} \left(\cos\theta + \frac{1}{1+\cos\theta}\right) d\theta = \tfrac{4}{3}.$$

But that $\frac{4}{3}$ is also the value of the other repeated integral, viz.

$$\int_0^1 E(x)\, dx, \quad \text{where} \quad E(x) = \int_0^{\frac{1}{2}\pi} \sqrt{(1 - x\sin^2\theta)}\, d\theta,$$

cannot be so easily verified.

12.8   Triple integral over a rectangular volume

Let $f(x, y, z)$ be a bounded function of x, y, z throughout the volume V enclosed by the planes $x = a$, $x = b$, $y = \alpha$, $y = \beta$, $z = A$, $z = B$.

Let $D(\delta, \delta, \delta)$, or briefly D, denote a dissection of the volume V in which the interval $[a, b]$ is divided into l subintervals δ_i, the interval

$[\alpha, \beta]$ is divided into m subintervals δ_i, and the interval $[A, B]$ into n subintervals δ_k, no subinterval exceeding δ. Put

$$\sigma(D) = \Sigma\Sigma\Sigma f_{ijk}\delta_i\delta_j\delta_k \tag{24}$$

where f_{ijk} denotes the value of $f(x, y, z)$ at an arbitrary point in the rectangular element of volume $\delta_i\delta_j\delta_k$.

Definition. The triple integral of $f(x, y, z)$ throughout the volume V is defined by

$$\iiint_V f(x, y, z)\, dx\, dy\, dz = \lim_{\delta \to 0} \sigma(D) \tag{25}$$

provided that this limit exists.

We can prove, in particular, that the limit exists when $f(x, y, z)$ is a continuous function of position throughout V, and that the value of the triple integral is then given by

$$\iiint_V f(x, y, z)\, dx\, dy\, dz = \int_A^B dz \int_\alpha^\beta dy \int_a^b f(x, y, z)\, dx \tag{26}$$

or by any of the other five repeated integrals obtained by changing the order of the integrations wo x, y, z on the R.H.S.

12.9   Triangular region

Next suppose the opposite corners (a, α) and (b, β) of the rectangle $R[a, b;\ \alpha, \beta]$ to be joined by a curve C along which x and y increase steadily, the equation of C being expressible as $y = \phi(x)$ or $x = \psi(y)$, where ϕ and ψ are continuous increasing monotonic functions.

The curve C divides R into two 'triangular' pieces. Let T be the piece, of area A, bounded by the lines $x = a$, $y = \beta$ and the curve C; let T' be the other piece, of area A'. Let A° be the area of R, so that $A + A' = A^\circ$.

In any dissection D of R, let $A(D)$ be the sum of the areas of the subrectangles in T that have no point in common with the curve C, and let $A'(D)$ be the sum of the areas of the subrectangles in T' that have no point in common with C. Also, let $A''(D)$ be the sum of the areas of the remaining subrectangles each of which has at least one point in common with C. Then $A(D) + A'(D) + A''(D) = A^\circ$.

Now consider a sequence of dissections D, $D_1, ..., D_r ...$ each of which, after the first, is a refinement (§ 9.5) of the one before it. Evidently, as r increases, $A(D)$ and $A'(D)$ increase monotonically to the limits A and A' respectively, and therefore $A''(D)$ decreases monotonically to zero.

12.10

Let $f(x,y)$ be a continuous function of position over T, and let $f_1(x,y)$ be defined over the rectangle R by

$$f_1(x,y) = f(x,y) \text{ over } T \text{ and its boundary}, \qquad (27)$$
$$= 0 \text{ over } T',$$

so that $f_1(x,y)$ will in general be discontinuous across the curve C but no discontinuity can exceed K, where K is such that

$$-K < f(x,y) < K$$

over T.

Theorem. The double integral of $f_1(x,y)$ over R exists.

Proof. In any dissection $D(\delta, \delta)$ of R let m_{ij}, M_{ij} be the closest bounds of $f_1(x,y)$ over the subrectangle R_{ij}. Then the difference between the upper and lower sums for $f_1(x,y)$ over the rectangle R, arising from the dissection $D, = D(\delta, \delta)$, is given by

$$S(D) - s(D) = \Sigma\Sigma(M_{ij} - m_{ij})\,\delta_i\,\delta_j. \qquad (28)$$

By §12.3 it is necessary and sufficient to prove that this difference tends to zero as $\delta \to 0$.

We can replace the R.H.S. of (28) by three summations, the first Σ over all the subrectangles belonging entirely to T, the second Σ' over those belonging entirely to T', and the third Σ'' over those having at least one point in common with the curve C.

Now Σ tends to zero when $\delta \to 0$, for the same kind of reason as in §12.4; Σ' is identically zero because $f_1(x,y) = 0$ over T'; while Σ'' tends to zero because it cannot exceed $K \cdot A''(D)$, which tends to zero since K is finite and $A''(D) \to 0$ by §12.9.

Thus $S(D) - s(D) \to 0$ and the theorem is proved.

12.11

Definition. The double integral of $f(x,y)$ over T is defined to be equal to the double integral of $f_1(x,y)$ over R, which has just been proved to exist; that is,

$$\iint_T f(x,y)\,dx\,dy = \iint_R f_1(x,y)\,dx\,dy. \qquad (29)$$

Theorem. The double integral of $f(x,y)$ over T is equal to either of two repeated integrals: viz., if $z = f(x,y)$, then

$$\iint_T z\,dx\,dy = \int_\alpha^\beta dy \int_a^{\psi(y)} z\,dx = \int_a^b dx \int_{\phi(x)}^\beta z\,dy \qquad (30)$$

where $\phi(x)$ and $\psi(y)$ are as defined in §12.9.

Proof. By (29), we have to prove that the double integral of $f_1(x, y)$ over R is equal to each of the repeated integrals in (30).

Now the double integral of $f_1(x, y)$ over R is, by definition, given by

$$\iint_R f_1(x, y) \, dx \, dy = \lim_{\delta \to 0} \sigma(D), \tag{31}$$

where
$$\sigma(D) = \sum_j \sum_i (f_1)_{ij} \delta_i \delta_j \tag{32}$$

and $(f_1)_{ij}$ denotes the value of $f_1(x, y)$ at an arbitrary point (ξ_{ij}, η_{ij}) in the subrectangle R_{ij}.

As in §12.7 we again put $\eta_{ij} = \eta_j$, where $y_{j-1} \leqslant \eta_j \leqslant y_j$ so that η_{ij} is independent of i: that is, η_{ij} is the same for every subrectangle between $y = y_{j-1}$ and $y = y_j$.

We next refine the dissection $D(\delta, \delta)$, if necessary, by adding parallels to the axis of y through the points where the lines $y = \eta_j$ meet the curve C, i.e. the points $x = \psi(\eta_j)$, $y = \eta_j$. Let D_1 denote the refined dissection and $\sigma(D_1)$ the corresponding double summation. Let m and n be the number of x- and y-subintervals respectively in D_1, and put $\delta_i = x_i - x_{i-1}$, $\delta_j = y_j - y_{j-1}$. Then, m and n being positive integers, we can put

$$\sigma(D_1) = \sum_{j=1}^{n} \delta_j \left\{ \sum_{i=1}^{m} f_1(\xi_{ij}, \eta_j) \delta_i \right\}. \tag{33}$$

Now, since $f_1(x, \eta_j)$ is a continuous function of x from $x = a$ to $x = \psi(\eta_j)$, and $f_1(x, \eta_j) = 0$ from $x = \psi(\eta_j)$ to $x = b$, it follows, as in §12.7, that the ξ_{ij} can be chosen so that the summation in brackets in (33) is equal to the integral of $f(x, \eta_j)$ from $x = a$ to $x = \psi(\eta_j)$, and hence that

$$\sigma(D_1) = \sum_{j=1}^{n} \delta_j \int_a^{\psi(\eta_j)} f(x, \eta_j) \, dx. \tag{34}$$

Since the integral $\int_a^{\psi(y)} f(x, y) \, dx$

is a continuous function of y, it follows from the definition of a simple integral that, when $\delta \to 0$,

$$\lim_{\delta \to 0} \sigma(D_1) = \int_\alpha^\beta dy \int_a^{\psi(y)} f(x, y) \, dx.$$

Thus the double integral of $f_1(x, y)$ over R, or the double integral of $f(x, y)$ over T, is equal to one of the repeated integrals in (30). That it is also equal to the other can be proved in the same kind of way.

12.12

Similar considerations apply to double integrals over the 'triangular' region T' bounded by the curve C and the lines $y = \alpha$, $x = b$.

Let $f(x, y)$ be continuous over T' and its boundary; then the double integral of $f(x, y)$ over T' may be defined, in the first place as a double integral over the rectangle R(cf. §12.11), and proved to be equal to either of two repeated integrals: thus, if $z = f(x, y)$,

$$\iint_{T'} z\,dx\,dy = \int_\alpha^\beta dy \int_{\psi(y)}^b z\,dx = \int_a^b dx \int_\alpha^{\phi(x)} z\,dy. \qquad (35)$$

Further, we may consider the 'triangular' regions into which the rectangle R is divided by a curve C_1 which joins the corners (a, β) and (b, α) and which has an equation of the form $y = \phi_1(x)$ or $x = \psi_1(y)$, where ϕ_1 and ψ_1 denote continuous decreasing functions. The double integral of a continuous function over either of these triangular regions may be similarly defined and proved to be equal to each of the corresponding repeated integrals.

12.13   Double integral over a simple closed region

By a 'simple' closed region we shall mean any closed region that can be divided into a finite number of rectangular or 'triangular' subregions of the types just considered.

The double integral of a function $f(x, y)$ over a simple closed region can then be defined as the sum of the double integrals over the rectangular or triangular parts into which it can be divided.

Often the sum of the double integrals over two adjacent such parts can be combined into one double integral. For example, if the closed region is the ellipse $x^2/a^2 + y^2/b^2 = 1$, it is divided into four 'triangular' quarters by the coordinate axes. If these four quarters are denoted by T_1, T_2, T_3 and T_4, belonging respectively to the first, second, third and fourth quadrants, and if $z = f(x, y)$ is continuous over the ellipse (E), we can define the double integral of $f(x, y)$ over E as the sum of four double integrals, as indicated by

$$\iint_E = \iint_{T_1} + \iint_{T_2} + \iint_{T_3} + \iint_{T_4}. \qquad (36)$$

Evidently the four integrals on the R.H.S. can be combined into one repeated integral in two ways: thus,

$$\iint_E z\,dx\,dy = \int_{-a}^a dx \int_{-\phi(x)}^{+\phi(x)} z\,dy = \int_{-b}^b dy \int_{-\psi(y)}^{+\psi(y)} z\,dx,$$

where $\phi(x) = b\sqrt{(1 - x^2/a^2)}$, $\psi(y) = a\sqrt{(1 - y^2/b^2)}$.

Examples

(1) Find the volume common to the cylinder $x^2 + y^2 = b^2$ and the ellipsoid $x^2/a^2 + y^2/b^2 + z^2/c^2 = 1$, when $a > b$.

If $z = c\sqrt{(1 - x^2/a^2 - y^2/b^2)}$, and V is the volume required, then $\frac{1}{2}V$ is the double integral of z over the circular region $x^2 + y^2 = b^2$ in the xy plane and by symmetry $\frac{1}{8}V$ is given by

$$\frac{V}{8} = \int_0^b \int_0^{\sqrt{(b^2 - y^2)}} c\left(1 - \frac{x^2}{a^2} - \frac{y^2}{b^2}\right)^{\frac{1}{2}} dx\, dy.$$

A slight simplification is effected by putting $x = a\xi$, $dx = a\,d\xi$, $y = b\eta$, $dy = b\,d\eta$; then

$$\frac{V}{8} = abc \int_0^1 d\eta \int_0^{(b/a)\sqrt{(1 - \eta^2)}} (1 - \xi^2 - \eta^2)^{\frac{1}{2}} d\xi$$

which reduces to

$$\frac{V}{8} = abc \int_0^1 \left\{\sin^{-1}\frac{b}{a} + \frac{b\sqrt{(a^2 - b^2)}}{a^2}\right\} \frac{1 - \eta^2}{2} d\eta,$$

whence follows

$$V = \tfrac{8}{3}abc\{\sin^{-1}(b/a) + (b/a)\sqrt{(1 - b^2/a^2)}\}.$$

(2) It may be that one of the repeated integrals can be easily evaluated directly, but not the other. Thus, consider the double integral

$$I = \iint_T \sqrt{\{x(y - x)\}} \cos y\, dx\, dy,$$

where T denotes the triangle bounded by $x = 0$, $y = \pi$, $y = x$.

Here it is easy to integrate first wo x, and then wo y; thus by putting $x = y\sin^2\theta$, $dx = 2y\sin\theta\cos\theta\, d\theta$, $y - x = y\cos^2\theta$, we find

$$I = \int_0^\pi \cos y\, dy \int_0^y \sqrt{\{x(y - x)\}}\, dx = \tfrac{1}{8}\pi \int_0^\pi y^2\cos y\, dy = -\tfrac{1}{4}\pi^2.$$

It follows, by integrating first wo y, that

$$\int_0^\pi \sqrt{x}\, dx \int_x^\pi \sqrt{(y - x)}\cos y\, dy = -\tfrac{1}{4}\pi^2$$

but this cannot easily be verified directly.

12.14 Double integral over a circle in polar coordinates

Let $f(r, \theta)$ be a bounded function of position over the region Γ enclosed by the circle $r = 1$, where r, θ are polar coordinates.

Given $\delta > 0$, let the interval from $r = 0$ to $r = 1$ be divided into

m subintervals each not exceeding δ, so that $m\delta \geqslant 1$: let the points of division be indicated by

$$0 = r_0 < r_1 < r_2 \ldots < r_{m-1} < r_m = 1. \tag{37}$$

Let the interval from $\theta = 0$ to $\theta = 2\pi$ be divided into n subintervals each not exceeding δ, so that $n\delta \geqslant 2\pi$: let the points of division be indicated by

$$0 = \theta_0 < \theta_1 < \theta_2 \ldots < \theta_{n-1} < \theta_n = 2\pi. \tag{38}$$

Let $D(\delta, \delta)$ or briefly D denote the resulting dissection of the region Γ into elements of area bounded by the circles $r = r_0, r = r_1, r = r_2, \ldots$ and the radii $\theta = 0, \theta = \theta_1, \theta = \theta_2, \ldots$.

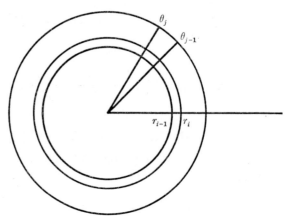

Fig. 10.

Let a_{ij} be the element of area enclosed by the circles $r = r_{i-1}$, $r = r_i$ and the radii $\theta = \theta_{j-1}, \theta = \theta_j$ (see Fig. 10).

Let f_{ij} denote the value of $f(r, \theta)$ at any point (r_{ij}, θ_{ij}) in the element of area a_{ij}. Put

$$\sigma(D) = \sum_{j=1}^{n} \sum_{i=1}^{m} f_{ij} a_{ij}. \tag{39}$$

Definition. The double integral I of $f(r, \theta)$ over the region Γ is defined by

$$I = \lim_{\delta \to 0} \sigma(D) \tag{40}$$

provided that this limit exists uniquely, whatever the points (r_{ij}, θ_{ij}) in the elements of area a_{ij}.

Now put $\delta r_i = r_i - r_{i-1}$, $\delta\theta_j = \theta_j - \theta_{j-1}$: then, since $r_i = r_{i-1} + \delta r_i$,

$$a_{ij} = (\pi r_i^2 - \pi r_{i-1}^2)(\delta\theta_j / 2\pi) = (r_{i-1} + \tfrac{1}{2}\delta r_i)\,\delta r_i\,\delta\theta_j.$$

Hence, assuming that the double integral I exists, and taking $r_{ij} = r_{i-1}$, $\theta_{ij} = \theta_{j-1}$, and therefore $f_{ij} = f(r_{i-1}, \theta_{j-1})$,

$$I = \lim_{\delta \to 0} \Sigma\Sigma f_{ij}(r_{i-1} \delta r_i \delta\theta_j + \tfrac{1}{2}\delta r_i \delta r_i \delta\theta_j).$$

Now the first of the double summations on the right is, by definition, the double integral of $rf(r, \theta)$ over the rectangle $[0, 1; 0, 2\pi]$ in a plane in which r, θ are rectangular Cartesian coordinates.

The second double summation tends to zero as $\delta \to 0$. For, put

$$J = \Sigma\Sigma f_{ij} \delta r_i \delta r_i \delta\theta_j$$

and suppose $-K < f(r, \theta) < K$: then, since $0 < \delta r_i \leqslant \delta$,

$$|J| < K\delta\Sigma\Sigma \, \delta r_i \delta\theta_j = K\delta\Sigma \, \delta r_i \Sigma \delta\theta_j = K\delta \, . \, 1 \, . \, 2\pi$$

which tends to zero with δ. It follows that

$$I = \int_0^{2\pi} \int_0^1 f(r, \theta) \, r \, dr \, d\theta. \tag{41}$$

The differential expression $r \, dr \, d\theta$ is called *the element of area in polar coordinates*.

Example

If $f(r, \theta) = 1$, the double integral given by (39) and (40) is plainly equal to the area of the circle Γ, i.e. $\pi \, . \, 1^2$. We can at once verify that this is the value given by (41).

12.15

Polar coordinates may be convenient in some cases when the field of integration is not circular. For the sake of illustration, let us find, using polar coordinates, the area of the triangle bounded by the lines $y = 0$, $x = a$, $y = x$.

If we integrate first wo r and then wo θ, we find (see Fig. 11 (i))

$$A = \int_0^{\frac{1}{4}\pi} d\theta \int_0^{a \sec\theta} r \, dr = \int_0^{\frac{1}{4}\pi} \tfrac{1}{2}a^2 \sec^2\theta \, d\theta = \tfrac{1}{2}a^2.$$

If the order of integration is reversed, it is necessary to express A as the sum of two integrals: thus (see Fig. 11 (ii))

$$A = \int_0^a r \, dr \int_0^{\frac{1}{4}\pi} d\theta + \int_a^{a\sqrt{2}} r \, dr \int_{\cos^{-1}(a/r)}^{\frac{1}{4}\pi} d\theta$$

which may be verified also to give $A = \tfrac{1}{2}a^2$.

12.16

A double integral expressed in polar coordinates may be easier to work out than the same integral expressed in Cartesian coordinates: in such a case, if the integral is given in terms of Cartesian coordinates, it will be worth while to restate it in polar coordinates.

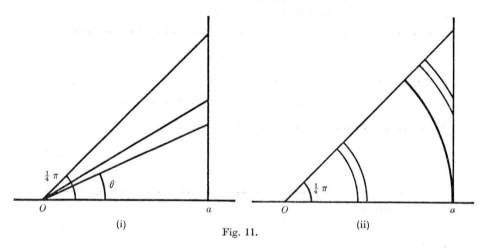

(i)   (ii)

Fig. 11.

Example

Evaluate the double integral

$$I = \iint_{\Gamma} \exp\left(-\frac{x^2 + y^2}{c^2}\right) dx\, dy,$$

where Γ denotes the circle $x^2 + y^2 = a^2$.

Restated in polar coordinates,

$$I = \int_0^{2\pi} \int_0^a \exp\left(-\frac{r^2}{c^2}\right) r\, dr\, d\theta = \pi c^2 \left\{1 - \exp\left(-\frac{a^2}{c^2}\right)\right\}$$

as may be easily verified.

12.17   Spherical polar coordinates (r, θ, ϕ)

It is known from the geometry of the sphere that the area of the zone included between two parallel planes distant h apart is equal to $2\pi a h$, where a is the radius of the sphere. Consequently, in the usual spherical polar coordinates (r, θ, ϕ) $(0 \leqslant \theta \leqslant \pi, 0 \leqslant \phi \leqslant 2\pi)$, the area of the zone between θ and $\theta + \delta\theta$ on a sphere of radius r is

$$2\pi r\{r\cos\theta - r\cos(\theta + \delta\theta)\}.$$

Hence, if δS is the area of the part of this zone between the meridians ϕ and $\phi + \delta\phi$ (see Fig. 12 (i)),

$$\delta S = 2\pi r^2 \{\cos\theta - \cos(\theta + \delta\theta)\} (\delta\phi/2\pi). \qquad (42)$$

Now, by Taylor's formula,

$$\cos(\theta + \delta\theta) - \cos\theta = -\sin\theta\,\delta\theta - \tfrac{1}{2}\cos(\theta + \lambda\delta\theta)\,\delta\theta^2,$$

where λ denotes some number between 0 and 1. It follows, since $-1 \leqslant \cos(\theta + \lambda\delta\theta) \leqslant 1$, that

$$\delta S = r^2 (\sin\theta + \alpha\delta\theta)\,\delta\theta\,\delta\phi, \qquad (43)$$

where $-\tfrac{1}{2} \leqslant \alpha \leqslant \tfrac{1}{2}$.

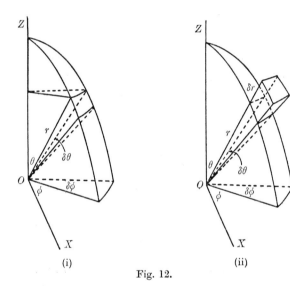

Fig. 12.

It follows that, if $f(\theta, \phi)$ is a bounded function of position on the surface of the sphere, then the double integral of $f(\theta, \phi)$ over any portion S of the spherical surface can be defined by

$$\iint_S f(r, \theta)\,dS = \iint_S f(\theta, \phi)\,r^2 \sin\theta\,d\theta\,d\phi = r^2 \iint_S f(\theta, \phi)\sin\theta\,d\theta\,d\phi \qquad (44)$$

provided that the integral exists. For the contribution to the integral on the L.H.S. of (44) arising from terms like the one involving α in (43) is zero in the limit, by the same kind of reasoning as that by which it was shown that $J \to 0$ in §12.14.

The differential expression $r^2 \sin\theta\,d\theta\,d\phi$ is called *the element of area* in spherical polar coordinates on the surface of the sphere.

12.18

Let a cone of any shape have its apex at the centre of a sphere of radius r. Let the cone intercept a portion a of the area of the surface of the sphere and a portion v of the volume of the sphere. Then the ratio of a to the whole surface of the sphere is the same as the ratio of v to the whole volume, that is, $a/(4\pi r^2) = v/(\frac{4}{3}\pi r^3)$ and therefore $v = \frac{1}{3}ra$.

Consequently, the volume of the sphere intercepted by the elementary cone which intercepts the surface in the element of area δS given by (42) or (43) can be written in the form

$$\tfrac{1}{3}r\delta S = \tfrac{1}{3}r^3(\sin\theta + \alpha\delta\theta)\,\delta\theta\,\delta\phi.$$

The element of volume δV included between r, $r+\delta r$; $\theta, \theta+\delta\theta$; and $\phi, \phi+\delta\phi$ is therefore expressible by

$$\begin{aligned}
\delta V &= \tfrac{1}{3}\{(r+\delta r)^3 - r^3\}\,(\sin\theta + \alpha\,\delta\theta)\,\delta\theta\,\delta\phi \\
&= (r^2\,\delta r + r\,\delta r^2 + \tfrac{1}{3}\delta r^3)\,(\sin\theta + \alpha\,\delta\theta)\,\delta\theta\,\delta\phi \\
&= (r^2 + \beta\,\delta r)\,(\sin\theta + \alpha\,\delta\phi)\,\delta r\,\delta\theta\,\delta\phi, \quad\quad (45)
\end{aligned}$$

where $\beta = r + \tfrac{1}{3}\delta r$ (see Fig. 12 (ii)).

Suppose $0 < r < 1$. Then this δV is accurate for certain values of α, β such that $-\tfrac{1}{2} \leqslant \alpha \leqslant \tfrac{1}{2}$, $0 < \beta = r + \tfrac{1}{3}\delta r < r + \delta r \leqslant 1$. It follows, by the same kind of reasoning as in §12.14, that if $f(r, \theta, \phi)$ is a bounded function of position throughout the volume of the sphere of radius unity, then the triple integral of $f(r, \theta, \phi)$ throughout any portion V of the volume of the sphere can be defined by

$$\iiint_V f(r, \theta, \phi)\,dV = \iiint_V f(r, \theta, \phi)\,r^2\sin\theta\,dr\,d\theta\,d\phi \quad\quad (46)$$

provided that the triple integral exists.

The differential expression $r^2\sin\theta\,dr\,d\theta\,d\phi$ is called *the element of volume* in spherical polar coordinates.

Example

Verify the formula $V = \frac{4}{3}\pi a^3$ for the volume V of a sphere of radius a.

(i) Using spherical polar coordinates, with the pole at the centre of the sphere, we have

$$\begin{aligned}
V &= \int_0^{2\pi}\int_0^{\pi}\int_0^{a} r^2\sin\theta\,dr\,d\theta\,d\phi \\
&= \int_0^{2\pi} d\phi \int_0^{\pi} \sin\theta\,d\theta \int_0^{a} r^2\,dr = 4\pi a^3/3.
\end{aligned}$$

(ii) With the pole at the lowest point of the sphere and the axis of z passing through the centre,

$$V = \int_0^{2\pi} \int_0^{\frac{1}{2}\pi} \int_0^{2a\cos\theta} r^2 \sin\theta \, dr \, d\theta \, d\phi.$$

Integrating first wo r, we find

$$V = \frac{8a^3}{3} \int_0^{2\pi} d\phi \int_0^{\frac{1}{2}\pi} \cos^3\theta \sin\theta \, d\theta = 4\pi a^3/3.$$

12.19 Change of variables of integration

Let $f(u,v)$, $g(u,v)$ be continuous differentiable functions over a rectangle $R[u_0, U; v_0, V]$ in a plane in which u, v are Cartesian co-ordinates. Put

$$x = f(u,v), \quad y = g(u,v) \tag{47}$$

and suppose that the Jacobian $\partial(x,y)/\partial(u,v)$ does not vanish at any point in R. Then (§6.12) equations (47) define u and v inversely as continuous differentiable functions of x and y over a region Ω in the xy plane corresponding to R, such that the correspondence is one-one. To the line $u = u_0$ will correspond a curve given in terms of the parameter v by $x = f(u_0, v), y = g(u_0, v), v_0 \leqslant v \leqslant V$. To the lines $u = U$, $v = v_0$, $v = V$ will correspond similar curves. These four curves will form the boundary of the region Ω which, in ordinary cases, will be a curvilinear quadrilateral.

Let $D(\delta, \delta)$ denote a dissection of R into subrectangles R_{ij} by means of the lines $u = u_0$, $u = u_1, \ldots,$ $u = u_m = U$ parallel to the axis of v, and $v = v_0$, $v = v_1, \ldots,$ $v = v_n = V$ parallel to the axis of u. Let $\delta u_i = u_i - u_{i-1}$, $\delta v_j = v_j - v_{j-1}$.

Let ω_{ij} be the area of the elementary curvilinear quadrilateral in the xy plane that corresponds to the subrectangle of area $\delta u_i \delta v_j$ in the uv plane. Let $F(x,y)$ be a continuous function of position over Ω and let F_{ij} denote the value of $F(x,y)$ at an arbitrary point in ω_{ij}. Put

$$\sigma(D) = \sum_{j=1}^{n} \sum_{i=1}^{m} F_{ij} \omega_{ij}. \tag{48}$$

Definition. The double integral of $F(x,y)$ over Ω is defined by

$$I = \lim_{\delta \to 0} \sigma(D). \tag{49}$$

Let P be the point $x = f(u,v)$, $y = g(u,v)$, and let $PQ_1P'Q_2$ denote the elementary curvilinear quadrilateral of area ω_{ij}. Also let P_1, P_2 be the points whose coordinates are respectively

$$(x + x_1 \delta u, y + y_1 \delta u), \quad (x + x_2 \delta v, y + y_2 \delta v),$$

where $x_1 = \partial x/\partial u, x_2 = \partial x/\partial v, \dots$, so that P_1, P_2 are points on the tangents at P to the curves $v =$ const., $u =$ const. passing through P (Fig. 13).

If the tangents PP_1, PP_2 have the same orientation as the coordinate axes Ox, Oy respectively, then (§ 7.25) the area of the parallelogram of

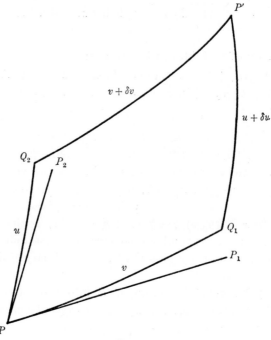

Fig. 13

which PP_1, PP_2 are two sides is $\{\partial(x, y)/\partial(u, v)\}\, du\, dv$. We shall assume here that ω_{ij} is expressible in the form

$$\omega_{ij} = \left\{\frac{\partial(x, y)}{\partial(u, v)} + \eta\right\}\delta u\, \delta v \tag{50}$$

and that $\partial(x, y)/\partial(u, v)$ and η are continuous functions of position in the rectangle R and that $\eta \to 0$ when $\delta \to 0$.

On this assumption it follows, in the same kind of way as in § 12.14, that the double integral I given by (48) and (49) reduces to the double integral over the rectangle R expressed by

$$I = \int_{v_0}^{V}\int_{u_0}^{U} F(x, y)\frac{\partial(x, y)}{\partial(u, v)}\, du\, dv, \tag{51}$$

where $x = f(u, v), y = g(u, v)$.

The differential expression $\{\partial(x,y)/\partial(u,v)\}\,du\,dv$ is called *the element of area* in the coordinates u, v.

A particular case is the change of variables from Cartesian to polar coordinates already discussed. In this case, equations (47) are $x = r\cos\theta$, $y = r\sin\theta$, and we find $\partial(x,y)/\partial(r,\theta) = r$, the circles $r = \text{const.}$ and the radii $\theta = \text{const.}$ having the same orientation as the lines $x = \text{const.}$, $y = \text{const.}$ Hence, over any ordinary field of integration,

$$\iint F(x,y)\,dx\,dy = \iint F(r\cos\theta, r\sin\theta)\,r\,dr\,d\theta. \tag{52}$$

Examples

(1) If $x = u/(u^2+v^2)$, $y = -v/(u^2+v^2)$, verify that the curves $u = \text{const.}$, $v = \text{const.}$ are circles touching the coordinate axes at the origin. Find the area enclosed by the circles $u = a$, $u = b$; $v = \alpha$, $v = \beta$; $0 < a < b$, $0 < \alpha < \beta$.

The circles are

$$x^2 + y^2 - x/u = 0, \quad x^2 + y^2 + y/v = 0.$$

We find $\quad\quad\quad \partial(x,y)/\partial(u,v) = 1/(u^2+v^2)^2.$

The area required is the area indicated by the curvilinear quadrilateral $ABCD$ in the xy plane (Fig. 14 (ii)), and this is equal to

$$\int_\alpha^\beta \int_a^b \frac{du\,dv}{(u^2+v^2)^2}$$

over the rectangle $ABCD$ in the uv plane (Fig. 14 (i)). The evaluation of this double integral is easy but lengthy.

(2) Find the area of the parallelogram enclosed by the straight lines
$$2x + 3y = 3, \quad 2x + 3y = 8, \quad 2y - 7x = -4, \quad 2y - 7x = 6.$$

Put $u = 2x + 3y$, $v = 2y - 7x$, so that the lines $u = \text{const.}$, $v = \text{const.}$ have the same orientation as the lines $x = \text{const.}$, $y = \text{const.}$ It is not necessary to solve for x and y in terms of u and v, since

$$\partial(x,y)/\partial(u,v) = 1/\{\partial(u,v)/\partial(x,y)\} = \tfrac{1}{25}.$$

Hence the required area is

$$\int_3^8 \int_{-4}^6 \frac{du\,dv}{25} = 2.$$

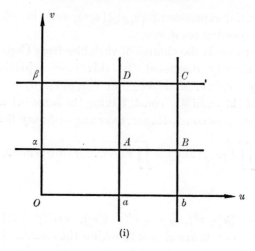

(i)

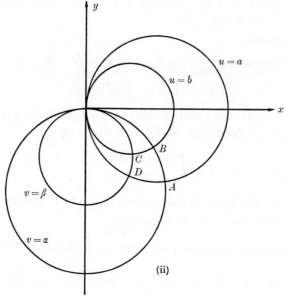

(ii)

Fig. 14.

12.20   Three dimensions

In three dimensions the formula corresponding to (51) is

$$I = \int_{w_0}^{W} \int_{v_0}^{V} \int_{u_0}^{U} F(x,y,z)\, \frac{\partial(x,y,z)}{\partial(u,v,w)}\, du\,dv\,dw,$$

where $x = f(u,v,w)$, $y = g(u,v,w)$, $z = h(u,v,w)$, and the normals to the surfaces $u = \text{const.}$, $v = \text{const.}$, $w = \text{const.}$ have the same orientation as the coordinate axes Ox, Oy, Oz, the normals being drawn in the directions of u, v, w increasing.

12.21   Area of a curved surface

The method of §12.17 will not apply to a non-spherical surface.

In the notation of §12.5 we now consider the area, on the surface $z = f(x, y)$, of the portion S intercepted by the rectangular prism standing on the rectangle R as base. We suppose the function $f(x, y)$ to be differentiable in the rectangle R, so that (§5.21) the surface $z = f(x, y)$ will have a tangent plane at every point of S.

Let $Q(x_{ij}, y_{ij}, z_{ij})$ be any point on the element S_{ij} intercepted on S by the elementary rectangular prism standing on the subrectangle R_{ij}, and let A_{ij} be the element of area intercepted by this elementary prism on the tangent plane at Q. Further, let the normal to S at Q make an acute angle γ with the axis of z. Then γ will also be the acute angle between the tangent plane at Q and the coordinate plane $z = 0$. Since the area of the subrectangle R_{ij} is the orthogonal projection of A_{ij} on the plane $z = 0$, it follows that $\delta_i \delta_j = A_{ij} \cos \gamma$ and hence that $A_{ij} = \delta_i \delta_j \sec \gamma$. We are thus led to the following definition:

Definition. The area A of the portion S of the surface $z = f(x, y)$ intercepted by the rectangular prism standing on the rectangle R as base is defined by

$$A = \lim_{\delta \to 0} \Sigma\Sigma A_{ij} = \iint_R \sec \gamma \, dx \, dy \tag{53}$$

provided that the limit exists.

Having regard to §12.13, this definition may be extended so that R may signify any simple closed region in the xy plane and A the area of the portion of surface intercepted by the right cylinder standing on R as base.

Example

Find the area of the part of the surface $9az^2 = 2(x+y)^3$ intercepted by the triangular prism standing on the triangle whose vertices are the points $(0, 0)$, $(a, 0)$, $(0, a)$ in the xy plane.

Putting $F(x, y, z) \equiv 9az^2 - 2(x+y)^3$, we know from the coordinate geometry of three dimensions that the direction cosines of the normal to the given surface are proportional to F_x, F_y, F_z, that is, $-6(x+y)^2$, $-6(x+y)^2$, $18az$, and are therefore proportional to

$$-(x+y)^{\frac{1}{2}}, \quad -(x+y)^{\frac{1}{2}}, \quad (2a)^{\frac{1}{2}}.$$

Hence follows $\cos\gamma = \sqrt{a}/\sqrt{(a+x+y)}$ and, by (53), if A is the required area,

$$A = \frac{1}{\sqrt{a}}\int_0^a\int_0^{a-x} \sqrt{(a+x+y)}\,dx\,dy = \frac{4(1+\sqrt{2})\,a^2}{15}$$

as may be verified.

12.22

Surface given parametrically. In (53) let R in the xy plane be the region Ω corresponding to the rectangle $[u_0, U; v_0, V]$ in the transformation (47): $x = f(u, v)$, $y = g(u, v)$. Also, suppose the curves $u = $ const., $v = $ const. in the xy plane to have the same orientation as the lines $x = $ const., $y = $ const. Then the Jacobian $\partial(x, y)/\partial(u, v)$ will be positive and (53) will give

$$A = \int_{v_0}^V\int_{u_0}^U \sec\gamma\,\frac{\partial(x, y)}{\partial(u, v)}\,du\,dv.$$

If we now suppose $z = \phi(x, y)$ to be the equation of the surface of which A is the area, by substituting $x = f(u, v)$, $y = g(u, v)$ we can express z also in terms of u and v, and we then have three equations of the form

$$x = f(u, v), \quad y = g(u, v), \quad z = h(u, v)$$

which express the coordinates of any point on the surface as functions of the parameters u and v.

We can now express $\cos\gamma$ in terms of u and v. For the tangent at any point P to the curve $v = $ const. passing through P will have direction ratios x_1, y_1, z_1 and the tangent at P to the curve $u = $ const. passing through P will have direction ratios x_2, y_2, z_2 where $x_1 = \partial x/\partial u$, $x_2 = \partial x/\partial v$, ... The normal to the surface at P, being perpendicular to both these tangents, will have direction ratios $y_1 z_2 - y_2 z_1$, $z_1 x_2 - z_2 x_1$, $x_1 y_2 - x_2 y_1$.

If we now put $E = \Sigma x_1^2$, $F = \Sigma x_1 x_2$, $G = \Sigma x_2^2$, then, by Lagrange's identity,

$$\Sigma(y_1 z_2 - y_2 z_1)^2 = EG - F^2$$

and hence, since $x_1 y_2 - x_2 y_1 > 0$, the direction cosine $\cos\gamma$ is given by

$$\cos\gamma = (x_1 y_2 - x_2 y_1)/\sqrt{(EG - F^2)}$$

and the formula for A becomes

$$A = \int_{v_0}^V\int_{u_0}^U \sqrt{(EG - F^2)}\,du\,dv.$$

12.23 Solid angle

Definition. The solid angle subtended by a surface S at a point O is defined to be the area intercepted on a sphere of unit radius, centre O, by the cone with apex at O whose generators pass through the periphery of S.

It is assumed that any ray drawn from O meets S in not more than one point and that S has a tangent plane at every point.

Let P be any point belonging to an element δS of S. Let χ be the acute angle between the normal to S at P and the ray OP. The elementary cone, apex O, which intercepts the element δS on S, will intercept an element of area $\delta S \cos \chi$, approximately, on the sphere, centre O, radius OP. The same elementary cone will therefore intercept on the sphere of unit radius, centre O, an element of area approximately equal to $\delta S \cos \chi / r^2$, where $r = OP$. Consequently, the solid angle ω subtended at O by S will be given by the double integral

$$\omega = \iint_S \frac{\cos \chi \, dS}{r^2}.$$

Example

Find the solid angle subtended at the point $(0, 0, c)$ by the rectangle $[0, a;\ 0, b]$ in the xy plane.

Let $P(x, y, 0)$ be the coordinates of any point in the given rectangle, and let C be the point $(0, 0, c)$.

The angle χ between the normal to the rectangle and the ray CP is $\cos^{-1} CO/CP$, and hence $\cos \chi = c/\sqrt{(x^2 + y^2 + c^2)}$. Also

$$CP^2 = x^2 + y^2 + c^2.$$

The required solid angle is therefore given by

$$\omega = \int_0^b \int_0^a \frac{c \, dx \, dy}{(x^2 + y^2 + c^2)^{\frac{3}{2}}}.$$

Integrating first wo x, we find

$$\omega = ac \int_0^b \frac{dy}{(c^2 + y^2)\sqrt{(a^2 + c^2 + y^2)}}$$

and hence, after putting $y = 1/\eta$ and then $1 + (a^2 + c^2)\eta^2 = \zeta^2$, or otherwise,

$$\omega = \tan^{-1} \frac{ab}{c\sqrt{(a^2 + b^2 + c^2)}}.$$

12.24 Mean values

Let $f = f(x, y, z)$ be a continuous function of position throughout a volume V. The *average* or *mean* value $\bar{f}$ of f throughout the volume V is defined by

$$\bar{f} = \frac{1}{V} \iiint_V f \, dV.$$

For instance, if s is the distance of a fixed point C from the element δV at P, then the mean value of s is given by

$$\bar{s} = \frac{1}{V} \iiint_V s \, dV.$$

This is called the *average* or *mean distance* of the point C from the volume V.

Similar definitions apply to average or mean values over surfaces.

Example

Find the mean distance of the point $(0, 0, c)$ from the surface of the sphere $x^2 + y^2 + z^2 = a$ when (i) $c > a$, (ii) $c < a$.

(i) Using spherical polar coordinates, the distance s of the point $(0, 0, c)$ from the point (a, θ, ϕ) is given by $s^2 = c^2 + a^2 - 2ca \cos \theta$. The mean distance $\bar{s}$ is given by

$$4\pi a^2 . \bar{s} = \int_0^{2\pi} \int_0^{\pi} s . a^2 \sin \theta \, d\theta \, d\phi.$$

Since $s \, ds = ca \sin \theta \, d\theta$, and s varies from $c - a$ to $c + a$ while θ varies from 0 to π, therefore

$$\begin{aligned}
4\pi a^2 \bar{s} &= \int_0^{2\pi} a^2 \, d\phi \int_{c-a}^{c+a} \frac{s^2 \, ds}{ac} \\
&= 2\pi a^2 \frac{1}{3ac} \{(c+a)^3 - (c-a)^3\} \\
&= 4\pi a^2 (3c^2 + a^2)/3c
\end{aligned}$$

and hence $\bar{s} = c + a^2/3c$, if $c > a$.

(ii) If $c < a$, then s varies from $a - c$ to $a + c$ while θ varies from 0 to π, and therefore

$$4\pi a^2 . \bar{s} = \int_0^{2\pi} a^2 \, d\phi \int_{a-c}^{a+c} \frac{s^2 \, ds}{ac}$$

and in this case we find $\bar{s} = a + c^2/3a$.

From the results of (i) and (ii) we infer that, if A and B are two concentric spheres, then the mean distance of a point of A from the surface of B is the same as the mean distance of a point of B from the surface of A.

12.25   Discontinuities in the integrand

When the integrand is not a continuous function of position over the field of integration, the integral may or may not exist.

The simplest types of discontinuity are (i) discontinuity at a point, (ii) discontinuity along a line or curve. Without attempting a general discussion, we give a few examples.

Examples

(1)
$$I = \iint_T \frac{\tan y \, dx \, dy}{\frac{1}{4}\pi - y}$$

where T denotes the triangle bounded by the lines $x = 0$, $y = 0$, $x + y = \frac{1}{4}\pi$.

The integrand tends to infinity in the corner $x = 0$, $y = \frac{1}{4}\pi$, of T. To avoid this corner, consider the integral over the trapezoidal region L bounded by $x = 0$, $y = 0$, $y = \frac{1}{4}\pi - \delta$, $x + y = \frac{1}{4}\pi$. Over L the integrand is a continuous function of position and so the integral exists: let it be denoted by J. Then I may be defined as the limit of J when $\delta \to 0$.

The value of J is found in the usual way from either of the repeated integrals to which it is equal: thus,

$$J = \int_0^{\frac{1}{4}\pi - \delta} \frac{\tan y \, dy}{\frac{1}{4}\pi - y} \int_0^{\frac{1}{4}\pi - y} dx = \int_0^{\frac{1}{4}\pi - \delta} \tan y \, dy$$

and hence $I = \lim_{\delta \to 0} J = \lim_{\delta \to 0} [\log \sec y]_0^{\frac{1}{4}\pi - \delta} = \frac{1}{2}\log 2.$

(2)
$$I = \iint_T \frac{dx \, dy}{\sqrt{(x - y)}}$$

where T denotes the triangle bounded by $y = 0$, $x = a$, $y = x$.

Here the integrand is infinite on the line $y = x$. Let J denote the integral over the triangle L bounded by $y = 0$, $x = a$, $y = x - \delta$. Then, since the integrand is continuous over L,

$$J = \iint_L \frac{dx \, dy}{\sqrt{(x - y)}} = \int_\delta^a dx \int_0^{x - \delta} \frac{dy}{\sqrt{(x - y)}} = \frac{2}{3}(2a^{\frac{3}{2}} - 3a\delta^{\frac{1}{2}} + \delta^{\frac{3}{2}}).$$

We now define I as $\lim_{\delta \to 0} J$. Hence $I = 4a^{\frac{3}{2}}/3.$

(3)
$$I = \iint_Q \tan^{-1}\frac{y}{x} \, dx \, dy$$

where Q denotes the positive quadrant of the circle $x^2 + y^2 = a^2$.

The integrand is not continuous at $(0,0)$. Let this point be excluded from the field of integration by a small quadrant of a circle of radius δ, and change to polar coordinates. Then it is natural to define I by

$$I = \lim_{\delta \to 0} \int_\delta^a \int_0^{\frac{1}{2}\pi} \theta \,.\, r \, dr \, d\theta. \quad \text{Hence} \quad I = \frac{\pi^2 a^2}{16}.$$

Note. If the field of integration were to extend into the second quadrant, the integrand $\tan^{-1}(y/x)$ would have to be defined in the extended part of the field.

(4)
$$I = \iint_T \frac{e^y \, dx \, dy}{\sqrt{(xy - x^2)}}$$

where T denotes the triangle bounded by $x = 0$, $y = 1$, $y = x$.

The integrand tends to infinity along the lines $x = 0$, $y = x$. Let J denote the integral over the triangle L bounded by $x = \delta$, $y = 1$, $y = x + \delta$. Over L the integrand is continuous; we can therefore define I as the limit of J when $\delta \to 0$, provided that the limit exists. Now, putting $x = y \sin^2 \theta$, we find

$$J = \int_{2\delta}^1 e^y \, dy \int_\alpha^\beta 2 \, d\theta,$$

where $\alpha = \sin^{-1} \sqrt{(\delta/y)}$, $\beta = \sin^{-1} \sqrt{(1 - \delta/y)}$; and evidently

$$I = \lim_{\delta \to 0} J = \int_0^1 e^y \, dy \int_0^{\frac{1}{2}\pi} 2 \, d\theta = (e - 1)\pi.$$

12.26   Infinite field of integration

Taking a simple case, suppose that $f(x, y)$ is continuous and that $f(x, y) \geqslant 0$ at every point of the xy plane, and consider the possibility of defining the double integral of $f(x, y)$ over the whole plane.

Let $R_1, R_2, \ldots, R_n, \ldots$ be a sequence of simple closed regions (§ 12.13) such that (i) each region R_n is entirely contained in the next R_{n+1}, (ii) the distance from any fixed point in R_1 to the nearest point on the boundary of R_n tends to infinity with n.

Let I_n denote the double integral of $f(x, y)$ over the region R_n. Then, since $f(x, y) \geqslant 0$, I_n is a positive increasing monotonic function of n and therefore tends to a limit or to infinity as $n \to \infty$.

Let $R_1', R_2', \ldots, R_n', \ldots$ be any other sequence of simple closed regions also possessing the properties (i), (ii). Let I_n' be the double integral of $f(x, y)$ over the region R_n'.

Theorem. If I_n tends to a limit I, then I_n' tends to the same limit I.

Proof. Let n be fixed. Then, because of (ii), a number N exists such that R_n' is entirely contained in R_N. Because $f(x, y) \geqslant 0$ and I_N tends to the limit I when $N \to \infty$, it follows that $I'_n < I_N < I$. Since I'_n is monotonic increasing, it follows further that $\lim I'_n$ exists and that, if $I' = \lim I'_n$, then $I' \leqslant I$.

By the same kind of reasoning, since I' exists, we see that also $I \leqslant I'$.

From $I' \leqslant I$ and $I \leqslant I'$ must follow $I' = I$.

12.27

A consequence of the last theorem is that if $f(x, y)$ is continuous and if $f(x, y) \geqslant 0$ at every point of the xy plane, we can construct the following definition:

Definition. Let $R_1, R_2, ..., R_n, ...$ be any sequence of simple closed regions having the properties (i), (ii) above. Then

$$\iint_W f(x, y) \, dx \, dy = \lim I_n \quad \text{when} \quad n \to \infty, \tag{54}$$

where W on the L.H.S. denotes the whole xy plane, and the R.H.S. denotes the limit when $n \to \infty$ of the double integral over the region R_n, provided that the limit exists.

12.28

The definition is satisfactory because, having regard to the above theorem, if $\lim I_n$ exists for any one sequence of simple closed regions $R_1, R_2, ...$ having the properties (i), (ii), then it exists for every such sequence and is the same for all of them.

The same definition would apply if $f(x, y)$ were everywhere negative. But if $f(x, y)$ were not one-signed, the definition would begin with the same words, but the proviso at the end would be: *provided that the limit exists for every such sequence and is the same for all of them.*

Example

Consider the integral $\quad \iint e^{-x^2-y^2} \, dx \, dy.$

Let I_R denote the integral over the circle $x^2 + y^2 = R^2$. Then, changing to polar coordinates,

$$I_R = \int_0^{2\pi} \int_0^R e^{-r^2} r \, dr \, d\theta = \pi(1 - e^{-R^2}).$$

Hence, when $R \to \infty$, $\lim I_R$ exists and equals π. We can therefore put

$$\iint_W e^{-x^2-y^2} \, dx \, dy = \pi,$$

where W denotes the whole xy plane. Since this integral exists and the integrand is positive, it follows, in particular, that

$$\lim_{R \to \infty} \int_{-R}^{R} \int_{-R}^{R} e^{-x^2-y^2}\,dx\,dy = \lim_{R \to \infty} \left(\int_{-R}^{R} e^{-x^2}\,dx \int_{-R}^{R} e^{-y^2}\,dy \right) = \pi,$$

where the field of integration, before proceeding to the limit, is now the square bounded by the lines $x = \pm R$, $y = \pm R$. Hence follows further (cf. §10.17, (44))

$$\left(\lim_{R \to \infty} \int_{-R}^{R} e^{-x^2}\,dx \right)^2 = \pi, \quad \text{or} \quad \int_{-\infty}^{\infty} e^{-x^2}\,dx = \sqrt{\pi}. \tag{55}$$

12.29   The Gamma function

Definition. For positive values of n, the gamma function, denoted by $\Gamma(n)$, may be defined by

$$\Gamma(n) = \int_{0}^{\infty} e^{-x} x^{n-1}\,dx \quad (0 < n). \tag{56}$$

The condition $0 < n$ ensures the convergence of the integral at the lower limit, by §10.7, Theorem 2. It is convergent at the upper limit for all real values of n, by §10.7, Theorem 1.

By integrating by parts, and then replacing n by $n+1$, we find

$$\Gamma(n+1) = n\Gamma(n) \quad (0 < n). \tag{57}$$

By repeated application of this formula follows, if n is a positive integer,

$$\Gamma(n+1) = n(n-1)\ldots3.2.1\,\Gamma(1) = n! \tag{58}$$

since $\Gamma(1) = 1$, from (56) by direct integration.

Again, if $n = \tfrac{1}{2}$, by (56) and (55),

$$\Gamma(\tfrac{1}{2}) = \int_{0}^{\infty} e^{-x} x^{-\frac{1}{2}}\,dx = 2\int_{0}^{\infty} e^{-t^2}\,dt = \sqrt{\pi}. \tag{59}$$

Using (57) and (59), we find further

$$\Gamma(\tfrac{3}{2}) = \tfrac{1}{2}\sqrt{\pi}, \quad \Gamma(\tfrac{5}{2}) = \tfrac{1}{2}\tfrac{3}{2}\sqrt{\pi}, \quad \Gamma(\tfrac{7}{2}) = \tfrac{1}{2}\tfrac{3}{2}\tfrac{5}{2}\sqrt{\pi}, \ldots. \tag{60}$$

Next, from (57) we have

$$\Gamma(n) = \Gamma(n+1)/n \tag{61}$$

from which we see that $\Gamma(n) \to +\infty$ when $n \to 0+$.

From this last property and by using (58) and (60), a sketch of the graph of $\Gamma(n)$ can be drawn for positive values of n(see Fig. 15).

Definition. For real negative non-integral values of n, let $\Gamma(n)$ be defined by (61), first for values of n between -1 and 0, then for values between -2 and -1, then for values between -3 and -2, and so on.

In particular, from this definition follows

$$\Gamma(-\tfrac{1}{2}) = -2\sqrt{\pi}, \quad \Gamma(-\tfrac{3}{2}) = \tfrac{2}{1}\tfrac{2}{3}\sqrt{\pi}, \quad \Gamma(-\tfrac{5}{2}) = -\tfrac{2}{1}\tfrac{2}{3}\tfrac{2}{5}\sqrt{\pi}, \dots \quad (62)$$

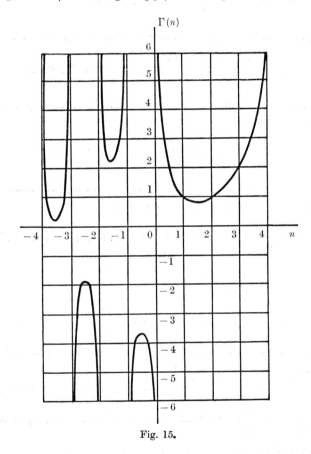

Fig. 15.

With the aid of these values and (61), the sketch of the graph can be continued for negative values of n.

Example

Prove that, if $0 < n < 1$, $0 < b$,

$$\int_0^\infty \frac{\cos bx}{x^n}\, dx = \frac{b^{n-1}}{2\Gamma(n)}\frac{\pi}{\cos \tfrac{1}{2}n\pi} \qquad (63)$$

and that, if $0 < n < 2,\ 0 < b$,

$$\int_0^\infty \frac{\sin bx}{x^n}\,dx = \frac{b^{n-1}}{2\Gamma(n)}\frac{\pi}{\sin \frac{1}{2}n\pi}. \tag{64}$$

Proof. In (56) replace x by xy, $0 < y$, and divide by y^n: then

$$\frac{\Gamma(n)}{y^n} = \int_0^\infty e^{-xy}\,x^{n-1}\,dx.$$

Multiply by $\cos by$ and integrate wo y from α to β, where $0 < \alpha < \beta$; we get

$$\Gamma(n)\int_\alpha^\beta \frac{\cos by}{y^n}\,dy = \int_\alpha^\beta dy \int_0^\infty \cos by\; e^{-xy}\,x^{n-1}\,dx.$$

By §10.17, Theorem 1, the last integral is uniformly convergent wo y if $\alpha \leqslant y$, since

$$\left|\int_0^\infty \cos by\; e^{-xy}\,x^{n-1}\,dx\right| \leqslant \int_0^\infty e^{-x\alpha}\,x^{n-1}\,dx$$

which converges and is independent of y. We may therefore (§10.16) reverse the order of integration, obtaining

$$\Gamma(n)\int_\alpha^\beta \frac{\cos by}{y^n}\,dy = \int_0^\infty x^{n-1}\,dx \int_\alpha^\beta e^{-xy}\cos by\; dy = G(\beta) - G(\alpha),$$

where

$$G(y) = \int_0^\infty x^{n-1} e^{-xy}\frac{b\sin by - x\cos by}{b^2 + x^2}\,dx.$$

Now

$$|G(y)| < \int_0^\infty x^{n-1} e^{-xy}\frac{b+x}{b^2+x^2}\,dx \leqslant \int_0^\infty x^{n-1}\frac{b+x}{b^2+x^2}\,dx,$$

which converges at both limits, since $0 < n < 1$. Hence $G(y)$ converges uniformly wo y in any interval $0 \leqslant y$.

Again, since $(b+x)/(b^2+x^2)$ is evidently bounded, we can put

$$|G(y)| < K\int_0^\infty x^{n-1} e^{-xy}\,dx = K\Gamma(n)/y^n \to 0 \quad \text{when } y \to \infty.$$

We can therefore let $\alpha \to 0$ and $\beta \to 0$, and get

$$\Gamma(n)\int_0^\infty \frac{\cos by}{y^n}\,dy = 0 - G(0) = \int_0^\infty \frac{x^n\,dx}{b^2+x^2}.$$

Putting $t = x^2/b^2$ in the last integral and using the integral worked out at the end of §11.14, since $0 < \frac{1}{2}(n+1) < 1$ if $0 < n < 1$, we get

$$\int_0^\infty \frac{x^n\,dx}{b^2+x^2} = \frac{1}{2}b^{n-1}\int_0^\infty \frac{t^{\frac{1}{2}(n+1)-1}}{1+t}\,dt = \frac{1}{2}b^{n-1}\frac{\pi}{\sin\{\frac{1}{2}(n+1)\pi\}}.$$

Hence follows (63). In the same kind of way we may prove (64), in which $n < 2$ is a sufficient condition for convergence at the lower limit of integration.

After putting $b = 1$, $n = 1/2$, $x = t^2$ in (63) and (64), we find

$$\int_0^\infty \cos(x^2)\,dx = \frac{\sqrt{\pi}}{2\sqrt{2}} = \int_0^\infty \sin(x^2)\,dx. \tag{65}$$

By using the first relation given in Examples 12F, 8, and putting $n = 1 - m$, results (63) and (64) can be expressed in the form

$$\int_0^\infty x^{m-1} \cos bx\,dx = \Gamma(m)\cos \tfrac{1}{2}m\pi/b^m \quad (0 < m < 1); \tag{66}$$

$$\int_0^\infty x^{m-1} \sin bx\,dx = \Gamma(m)\sin \tfrac{1}{2}m\pi/b^m \quad (-1 < m < 1). \tag{67}$$

12.30 The Beta function

Definition. For positive values of m and n, the beta function $B(m,n)$ may be defined by

$$B(m,n) = \int_0^1 x^{m-1}(1-x)^{n-1}\,dx. \tag{68}$$

The conditions $0 < m$, $0 < n$ ensure convergence at the lower and upper limits respectively.

By putting $x = \cos^2\theta$, we see that $B(m,n)$ can also be expressed by

$$B(m,n) = 2\int_0^{\frac{1}{2}\pi} \cos^{2m-1}\theta \sin^{2n-1}\theta\,d\theta. \tag{69}$$

12.31 Relation between the Gamma and Beta functions

Consider the double integral

$$I = \int_0^\infty \int_0^\infty e^{-x^2-y^2} x^{2m-1} y^{2n-1}\,dx\,dy \quad (0 < m,\ 0 < n). \tag{70}$$

In polar coordinates,

$$I = \int_0^\infty e^{-r^2} r^{2m+2n-1}\,dr \int_0^{\frac{1}{2}\pi} \cos^{2m-1}\theta \sin^{2n-1}\theta\,d\theta$$

and after putting $r^2 = t$, $2r\,dr = dt$, we find, by (56) and (69),

$$I = \tfrac{1}{4}\Gamma(m+n)\,B(m,n). \tag{71}$$

Now, since the integrand in (70) is everywhere positive in the first quadrant, by §12.28 we have also

$$I = \int_0^\infty e^{-x^2} x^{2m-1}\,dx \int_0^\infty e^{-y^2} y^{2n-1}\,dy$$

from which, by putting $x^2 = \xi, y^2 = \eta$, we find

$$I = \tfrac{1}{4}\Gamma(m)\,\Gamma(n). \tag{72}$$

From (71) and (72) follows

$$B(m,n) = \frac{\Gamma(m)\,\Gamma(n)}{\Gamma(m+n)}. \tag{73}$$

12.32 Dirichlet's multiple integral

Consider the double integral

$$I = \iint_T x^{l-1} y^{m-1}\, dx\, dy, \tag{74}$$

where T denotes the triangle bounded by $x = 0$, $y = 0$, $x+y = 1$.

Formally we have, if $m > 0$,

$$I = \int_0^1 x^{l-1}\, dx \int_0^{1-x} y^{m-1}\, dy = \int_0^1 x^{l-1}\frac{(1-x)^m}{m}\, dx$$

and hence, using (73) and (57), provided that $l > 0$, $m > 0$,

$$I = \frac{\Gamma(l)\,\Gamma(m+1)}{m\Gamma(l+m+1)} = \frac{\Gamma(l)\,\Gamma(m)}{\Gamma(l+m+1)}. \tag{75}$$

12.33

Note. If $l \geqslant 1$, $m \geqslant 1$, the integrand in (74) is a continuous function of position over the triangle T. But if $0 < l < 1$, the integrand becomes infinite on the axis $x = 0$; and if $0 < m < 1$, it becomes infinite on the axis $y = 0$. In such cases, the existence of the integral should first be established.

Suppose, for example, that both $0 < l < 1$ and $0 < m < 1$. Then, since the integrand is positive and continuous at all points in the triangle except those on the axes $x = 0$ and $y = 0$, we can define the integral over T by

$$\iint_T x^{l-1} y^{m-1}\, dx\, dy = \lim_{\delta \to 0} \iint_L x^{l-1} y^{m-1}\, dx\, dy,$$

where L denotes the triangle bounded by $x = \delta$, $y = \delta$, $x+y = 1$, provided that the limit exists. Now, since the integrand is continuous over L, the integral can be found by repeated integration; thus, if we integrate wo y first,

$$\iint_L = \int_\delta^{1-\delta} x^{l-1}\, dx \int_\delta^{1-x} y^{m-1}\, dy$$

$$= \frac{1}{m}\int_\delta^{1-\delta} x^{l-1}(1-x)^m\, dx - \frac{\delta^m}{m}\int_\delta^{1-\delta} x^{l-1}\, dx.$$

The second term on the R.H.S. $= (-\delta^m/m)\{(1-\delta)^l/l - \delta^l/l\}$, and when $\delta \to 0$, $\delta^m \to 0$ since $m > 0$, $\delta^l \to 0$ since $l > 0$, and $(1-\delta)^l \to 1$; consequently, this term tends to zero.

The first term converges when $\delta \to 0$ (§10.7, Theorem 2) and hence

$$\iint_T = \lim_{\delta \to 0} \iint_L = \frac{1}{m} \int_0^1 x^{l-1} (1-x)^m \, dx = \frac{\Gamma(l)\,\Gamma(m)}{\Gamma(l+m+1)},$$

as in (75).

12.34

Another method. Make the substitution

$$x+y = u, \quad y = uv \tag{76}$$

that is, $\qquad\qquad x = u(1-v), \quad y = uv$

or, inversely, $\qquad u = x+y, \quad v = y/(x+y).$

The curves $u = $ const. in the xy plane are the straight lines $x+y = u$, where u varies from 0 to 1. The curves $v = $ const. are the lines $y/x = v/(1-v)$ radiating from the origin, where v varies from 0 to 1. Also,

$$\frac{\partial(x,y)}{\partial(u,v)} = \begin{vmatrix} 1-v & -u \\ v & u \end{vmatrix} = u.$$

The integral (74) is transformed by this substitution into

$$I = \int_0^1 \int_0^1 \{u(1-v)\}^{l-1} . (uv)^{m-1} . u \, du \, dv,$$

a double integral over the square $[0, 1; 0,1]$ in the uv plane. Hence

$$I = \int_0^1 u^{l+m-1} \, du \int_0^1 v^{m-1} (1-v)^{l-1} \, dv$$

which reduces at once to (75), by using (68), (73) and (57).

12.35

A more general multiple integral, due to Dirichlet, of which (74) is a simple case, is exemplified in three dimensions by

$$I = \iiint_T f(x+y+z)\, x^{l-1} y^{m-1} z^{n-1} \, dx \, dy \, dz$$

$$= \frac{\Gamma(l)\,\Gamma(m)\,\Gamma(n)}{\Gamma(l+m+n)} \int_0^1 f(u)\, u^{l+m+n-1} \, du, \tag{77}$$

where $l > 0$, $m > 0$, $n > 0$, and T denotes the tetrahedron bounded by the planes $x = 0$, $y = 0$, $z = 0$, $x+y+z = 1$.

B & G

Using the method of §12.34, make the substitution

$$x+y+z = u, \quad y+z = uv, \quad z = uvw \qquad (78)$$

that is,

$$x = u(1-v), \quad y = uv(1-w), \quad z = uvw$$

or, inversely,

$$u = x+y+z, \quad v = \frac{y+z}{x+y+z}, \quad w = \frac{z}{y+z}.$$

As x, y, z are all positive and $x+y+z \leqslant 1$, each of the new variables u, v, w varies from 0 to 1, and the substitution converts the tetrahedron T into a cube. Also, we find that the Jacobian

$$\partial(x, y, z)/\partial(u, v, w) = u^2 v.$$

After making the substitution, using §12.20, we see that the variables can be separated, and that the integral equals

$$\int_0^1 f(u)\, u^{l+m+n-1}\, du \int_0^1 (1-v)^{l-1}\, v^{m+n-1}\, dv \int_0^1 (1-w)^{m-1}\, w^{n-1}\, dw$$

$$= \int_0^1 f(u)\, u^{l+m+n-1}\, du \,.\, B(l, m+n) \,.\, B(m, n)$$

from which and (73) follows (77).

The formula (77) certainly holds good if $l \geqslant 1$, $m \geqslant 1$, $n \geqslant 1$ and $f(u)$ is continuous. These conditions are sufficient but not necessary: for example, the formula holds good when $f(u)$ is continuous and $0 < l < 1$, $0 < m < 1$, $0 < n < 1$ (cf. §12.33). Here we shall not go further into the question of setting down fewer sufficient conditions, but shall assume that the evaluation by formal methods of any integral given in Examples 12 F can be justified.

Examples 12 A

(1) Find the volume under the surface $xyz = c^3$, standing on the square $[a, b; a, b]$, $0 < a < b$.

(2) Evaluate $\displaystyle\iint \frac{dx\, dy}{x+y}$ over the rectangle $[0, 2; 1, 2]$.

(3) Evaluate $\displaystyle\int_0^a \int_0^{2\pi} \frac{r\, dr\, d\theta}{c + r\cos\theta}$ $(c > a)$.

(4) Evaluate $\displaystyle\int_0^3 \int_0^4 \sqrt{(x^2 + y^2)}\, dx\, dy$.

(5) Evaluate $\displaystyle\int_0^1 \int_0^1 x^5 y \exp(-x^2 y)\, dx\, dy$.

(6) Evaluate
$$\int_0^a \int_0^b \frac{dx\,dy}{(c^2+x^2+y^2)^{\frac{3}{2}}}.$$

(7) Evaluate
$$\int_0^1 \int_2^3 \log(x+y)\,dx\,dy.$$

(8) Show that
$$\int_\gamma^\delta \int_\alpha^\beta \cos(x+y)\,dx\,dy = 0$$
if $\beta = \alpha + 2n\pi$ or if $\delta = \gamma + 2n\pi$ or if $\alpha + \beta + \gamma + \delta = (2n+1)\pi$, where n is an integer.

(9) Show that
$$\int_0^1 \int_0^1 e^{-xy}\,dx\,dy \doteqdot 0.79659.$$

(10) Show that
$$\int_0^1 \int_0^{\frac{1}{2}\pi} \frac{dx\,d\theta}{\sqrt{(1-x\sin^2\theta)}} = 2.$$

(11) Show that
$$\int_0^1 \int_0^{\frac{1}{2}\pi} \frac{dk\,d\theta}{\sqrt{(1-k^2\sin^2\theta)}} = \int_0^{\frac{1}{2}\pi} \frac{\theta\,d\theta}{\sin\theta} = 2\int_0^1 \frac{\tan^{-1}x}{x}\,dx$$
$$= 2\left(\frac{1}{1^2}-\frac{1}{3^2}+\frac{1}{5^2}-\frac{1}{7^2}+\ldots\right) = 2G.$$

$[G = 0.915965594\ldots$ (Catalan's constant)].

(12) By considering the reflection of the integrand in the line $y = x$, show that
$$\int_1^a \int_1^a \frac{x-y}{(x+y)^3}\,dx\,dy = 0 \quad (0 < a).$$
Verify by repeated integration.

(13) Prove that
$$\int_0^1 dx \int_0^1 \frac{x^2-y^2}{(x^2+y^2)^2}\,dy = \frac{\pi}{4}, \quad \int_0^1 dy \int_0^1 \frac{x^2-y^2}{(x^2+y^2)^2}\,dx = -\frac{\pi}{4}.$$
Deduce that the integrand is not continuous over the square $[0, 1; 0, 1]$. Verify this.

(14) The thickness of a square plate $ABCD$, of side a, is tr^2/a^2, where t is a constant, small compared with a, and r is distance from A. Find (i) the average thickness of the plate, (ii) the position of its centroid.

(15) Find the average value over the square $[0, \pi; 0, \pi]$ of each of the functions:

    (i) $\sin x + \sin y$,   (ii) $\sin(x+y)$,   (iii) $\cos(x+y)$,

    (iv) $(\sin x + \sin y)^2$,   (v) $(\cos x + \cos y)^2$,   (vi) $\sin^2(x+y)$.

(16) Find the average value over the cube $[0, \pi; 0, \pi; 0, \pi]$ of

    (i) $(\sin x + \sin y + \sin z)^2$,   (ii) $(\cos x + \cos y + \cos z)^2$.

(17) If
$$z = a_0 + a_1 x + b_1 y + ax^2 + bxy + cy^2 + Ax^3 + Bx^2y + Cxy^2 + Dy^3,$$
prove that the average value of z over any rectangle is
$$\tfrac{1}{12}(8z_0 + z_1 + z_2 + z_3 + z_4)$$
where z_0 is the value of z at the centre, and z_1, z_2, z_3, z_4 are the values at the corners of the rectangle (cf. Simpson's rule).

(18) Show that, if $0 \leqslant a < 1$,

$$\int_0^{\frac{1}{2}\pi} \int_0^{\frac{1}{2}\pi} \frac{\sin(x+y)\,dx\,dy}{(1-a^2\sin^2 x \sin^2 y)^{\frac{3}{2}}} = \frac{1}{a}\log\frac{1+a}{1-a}.$$

Examples 12 B

(1) Evaluate

  (i) $\displaystyle\int_0^1 \int_0^y (2xy - x^2)\,dx\,dy;$ (ii) $\displaystyle\int_1^2 \int_0^x \frac{dx\,dy}{x^2 + 3y^2};$

  (iii) $\displaystyle\int_0^1 \int_0^{2+x} (x^2 + xy + 3y^2)\,dx\,dy;$

  (iv) $\iint e^{2x+3y}\,dx\,dy$ over the triangle in which $x \geqslant 0,\ y \geqslant 0,\ x+y \leqslant a;$

  (v) $\iint \sin\pi\,(x/a+y/b)\,dx\,dy$ over the triangle of which the vertices are $(0,0)$, $(a,0),(0,b);$

  (vi) $\iint (3x+y)^2\,dx\,dy$ over the area bounded by the curves $xy = 1,\ xy = 4$, and the lines $x = 1,\ x = 2$.

(2) Show that $\iint \sin\lambda(x+y)\,dx\,dy$ (λ constant), over the triangle in which $x \geqslant 0,\ y \geqslant 0,\ x+y \leqslant 1$, vanishes if λ is a root of the equation $\theta = \tan\theta$.

(3) Find the volume bounded by the coordinate planes and the plane

$$x/a + y/b + z/c = 1, \quad \text{if} \quad abc > 0.$$

(4) Find the volume bounded by the coordinate planes, the plane

$$x + y + z = c,$$

and the cylinder $x^2 + y^2 = a^2,\ c > a\sqrt{2}$.

(5) Find the volume bounded above by the surface $bz = a^2\sin^2(\pi x/l) - y^2$ and standing on the base between the curves $y = \pm a\sin(\pi x/l)$ $(0 < x < l)$.

(6) Find the volume intercepted between the sphere $x^2+y^2+z^2 = 4$ and the paraboloid $y^2+z^2 = 4(x+1)$.

(7) Find the volume common to the cylinders $x^2+y^2 = 2ax,\ z^2 = 2bx$.

(8) Find the volume of the part of the oblate spheroid $x^2+4y^2+z^2 = 4a^2$ intercepted by the cylinder $x^2+y^2 = a^2$.

(9) Find the volume of the part of the sphere $x^2+y^2+z^2 = a^2$ in the positive octant, standing on the triangle whose vertices are the points $(0,0,0)$, $(a,0,0)$, $(0,a,0)$.

(10) Show that the volume of the part of the sphere $x^2+y^2+z^2 = a^2$ between the planes $x = \pm h,\ y = \pm k$ can be found by elementary integration,

$$(h^2 + k^2 \leqslant a^2).$$

(11) Show that the volume common to the sphere $x^2+y^2+z^2 = a^2$ and the elliptic cylinder $x^2+y^2/(1-e^2) = a^2$ is

$$\tfrac{8}{3}a^3\{e\sqrt{(1-e^2)} + \cos^{-1}e\}.$$

(12) Show that the volume contained between the paraboloid of revolution $cz = x^2 + y^2$, the plane $z = 0$, and the elliptic cylinder

$$(x - x_0)^2/a^2 + (y - y_0)^2/b^2 = 1$$

is

$$(\pi ab/c)\,(x_0^2 + y_0^2 + \tfrac{1}{4}a^2 + \tfrac{1}{4}b^2).$$

(13) A metal plate has the shape of a quadrant of a circle of radius a, and its thickness at any point P is proportional to the product of the distances of P from the bounding radii. Prove that the centroid of the plate is distant $8a/15$ from each of these radii.

(14) Find the coordinates of the centroid of the positive octant of the uniform solid sphere the equation of the surface of which is $x^2 + y^2 + z^2 = a^2$.

Deduce the coordinates of the centroid of the positive octant of the uniform solid ellipsoid the equation of the surface of which is $x^2/a^2 + y^2/b^2 + z^2/c^2 = 1$.

(15) Two rectangular axes Ox, Oy are drawn in the same plane as, but not intersecting, a closed area. The centroid of the area is equidistant from both axes. Prove that the centroids of the solids formed by rotating the area about each axis in turn are at equal distances from O.

(16) If $z = a_0 + a_1 x + b_1 y + ax^2 + bxy + cy^2$, show that the average value of z over any triangle is $\tfrac{1}{3}(z_1 + z_2 + z_3)$, where z_1, z_2, z_3 are the values of z at the midpoints of the sides.

(17) Show that $$\int_0^1 \int_0^x x^3 y\, e^{-xy}\, dx\, dy = (3 - e)/(2e).$$

(18) Show that $$\int_0^1 \int_y^1 e^{-x^2}\, dx\, dy = (e - 1)/(2e).$$

(19) Show that $$\int_0^1 \int_x^1 e^{-x/y}\, dx\, dy = (e - 1)/(2e).$$

(20) Show that

$$\int_0^1 \int_0^x xy\, e^{-xy}\, dx\, dy = \frac{1}{2} \int_0^1 \int_0^1 xy\, e^{-xy}\, dx\, dy = \frac{1}{2}\left(\frac{1}{2^2} - \frac{1}{3^2 \cdot 1!} + \frac{1}{4^2 \cdot 2!} - \frac{1}{5^2 \cdot 3!} + \cdots\right).$$

Examples 12C

(1) Find the area between the circles $r = 2a \cos \theta$, $r = c$ ($c < 2a$).

(2) Evaluate $$\iint (x^2 + y^2)\, \sqrt{(a^2 - x^2 - y^2)}\, dx\, dy$$

over the circle $x^2 + y^2 = a^2$.

(3) Evaluate $$\int_0^a \int_{x^2/a}^{\sqrt{(2ax - x^2)}} \frac{dx\, dy}{\sqrt{(x^2 + y^2)}}.$$

(4) The axes of two equal circular cylinders intersect at right angles. Show that the common volume is $16a^3/3$, where a is the radius of each. Also show that the area of the surface of the common volume is $8a^2$.

If the axes intersect at an angle β, show that the common volume is then $16a^3/(3 \sin \beta)$, and that the area of its surface is $8a^2/\sin \beta$.

(5) The axes of three equal circular cylinders, each of radius a, intersect at right angles. Show that the common volume is $8(2-\sqrt{2})\,a^3$.

(6) Find the average value of $x/(a^2+x^2+y^2)$ over the positive quadrant of the circle $x^2+y^2 = a^2$.

(7) Find the average value of $\sqrt{(4x^2+4y^2+a^2)}$ over the circle $x^2+y^2 = a^2$.

(8) Find the average value (i) of r, (ii) of θ, over the semicircle $r = a\cos\theta$, $0 < \theta < \frac{1}{2}\pi$.

(9) ABC is an isosceles triangle, right-angled at C, and P is any point inside the triangle. Prove that the mean value of AP is $0\cdot765\,AC$ and that the mean value of the angle PAC is $25°\,9'$, approximately.

(10) The mass per unit area at any point P of a square lamina of side a is mr/a, where m is a constant and r is the distance of P from one corner. Show that the mass of the lamina is $\frac{1}{3}ma^2\{\sqrt{2}+\log{(1+\sqrt{2})}\}$.

(11) Evaluate
$$\iint \frac{dx\,dy}{\sqrt{(x^2+y^2)}}$$
(i) over the circle $x^2+y^2 = a^2$, (ii) over a square with its centre at the origin and having the same area as the circle. Deduce that $\log{(3+2\sqrt{2})} < \sqrt{\pi}$.

(12) Using spherical polar coordinates, find the radius of gyration, about a diameter, of (i) a uniform solid sphere of radius a, (ii) a uniform spherical shell of radius a.

(13) Using spherical polar coordinates, find (i) the position of the centroid of a uniform solid circular cone of height h and base-radius r, (ii) its radius of gyration about its own axis.

(14) A cone of semi-vertical angle α has its apex on the surface of a sphere of radius a and its axis passes through the centre of the sphere. Show that the common volume is $(4\pi a^3/3)(1-\cos^4\alpha)$.

(15) Referring to the common volume defined in the last example, find (i) the distance of its centroid from the apex of the cone, (ii) its radius of gyration about its own axis.

(16) Show that, if V denotes the region enclosed by the sphere $x^2+y^2+z^2 = a^2$, then
$$\iiint_V \exp{[-(x^2+y^2+z^2)/c^2]}\,dx\,dy\,dz = \pi c^3\left\{\sqrt{\pi}\operatorname{erf}\left(\frac{a}{c}\right) - \frac{2a}{c}\exp{(-a^2/c^2)}\right\}.$$

(17) Prove that
$$\int_0^a \int_{-\pi}^{\pi} \frac{r\,dr\,d\theta}{\sqrt{(a^2-r^2)}\,\sqrt{(r^2+h^2-2hr\cos\theta)}} = \pi^2 \quad (h < a).$$

[This provides a verification of the fact that a distribution of electricity over a circular disc of radius a, with surface density proportional to $1/\sqrt{(a^2-r^2)}$, makes the potential constant over the disc.]

(18) Show that the area of the region enclosed by the parabolas

$$ay = x^2, \quad by = x^2, \quad \alpha x = y^2, \quad \beta x = y^2 \quad (a < b, \alpha < \beta)$$

is $\frac{1}{3}(b-a)(\beta-\alpha)$.

Also show that, over the same region,

$$\iint xy\, dx\, dy = \tfrac{1}{12}(b^2-a^2)(\beta^2-\alpha^2).$$

(19) If $u = xy, v = y/x$, find the area enclosed by the curves $u = 1, u = 3$, $v = \frac{1}{2}, v = 2$.

(20) By means of the substitution $x+y = u, y = uv$, or otherwise, prove that

$$\iint_T \{xy(1-x-y)\}^{\frac{1}{2}}\, dx\, dy = \frac{2\pi}{105}$$

where T denotes the triangle bounded by $x = 0, y = 0, x+y = 1$.

(21) If $x+iy$ is an analytic function of $u+iv$, and if a region of the uv plane is dissected into elementary squares, show that the corresponding elements of area in the xy plane approximate to squares.

(22) If $x+iy = (u+iv)^2$, show that the curves $u = $ const., $v = $ const. are two systems of confocal parabolas.

If O be the common focus, and the parabolas $u = u_1 > 0, v = v_1 > 0$ intersect at P, prove that $\iint dx\, dy/\sqrt{(x^2+y^2)}$ taken over the area bounded by $u = u_1$ and the straight lines OX and OP is $2u_1 v_1$.

(23) Show that

$$\int_0^3 dx \int_0^{\sqrt{(5-5x^2/9)}} dy = \frac{3\pi\sqrt{5}}{4}.$$

Verify this result by expressing the integral in terms of variables λ, μ defined by

$$\lambda+\mu = \sqrt{\{(x+2)^2+y^2\}}, \quad \lambda-\mu = \sqrt{\{(x-2)^2+y^2\}}.$$

(24) If $x = c \cosh u \cos v, y = c \sinh u \sin v$, show that the curves $u = $ const., $v = $ const. are orthogonal confocal ellipses and hyperbolas. Also show that the area enclosed by the curves $u = u_0, u = U, v = v_0, v = V$ is

$$\tfrac{1}{4}c^2\{(V-v_0)(\sinh 2U - \sinh 2u_0) - (U-u_0)(\sin 2V - \sin 2v_0)\},$$

where $0 \leqslant u_0 < U < \infty, 0 \leqslant v_0 < V \leqslant \frac{1}{2}\pi$.

(25) By making the substitution $x+y = X\sqrt{2}, -x+y = Y\sqrt{2}$, prove that

$$\iint_T e^{-(x+y)^2}\, dx\, dy = \tfrac{1}{2}(1-e^{-a^2}),$$

where T denotes the triangle of sides $x = 0, y = 0, x+y = a$.

(26) If r_1 and r_2 are the distances of the point (x, y) from the foci $(\pm c, 0)$ of the ellipse $x^2/a^2 + y^2/b^2 = 1$, show, by means of the transformation

$$x+iy = c \cosh(u+iv)$$

or otherwise, that, over the interior of the ellipse,

$$\iint \left(\frac{1}{r_1}+\frac{1}{r_2}\right) dx\, dy = 4\pi b.$$

Examples 12 D

(1) Show that the area of the part of the cylinder $z^2 = 4ax$ intercepted by the cylinder $x^2 + y^2 = a^2$ is $5\pi a^2/2$.

(2) Calculate the area of the part of the surface $xy = az$ intercepted by the cylinder $(x^2 + y^2)^2 = 2a^2xy$.

(3) Find the position of the centroid of the hemispherical surface

$$x^2 + y^2 + z^2 = a^2 \quad (0 < z).$$

Deduce the coordinates of the centroid (i) of the part of the surface in the positive octant, (ii) of the part of the volume in the positive octant.

(4) The coordinates of a point on the paraboloid $x^2/a + y^2/b = 2z$ are expressed in terms of θ and ϕ by means of the equations $x = a \tan \theta \cos \phi, y = b \tan \theta \sin \phi$. Show that θ is the angle which the normal at the point makes with the axis of z, and that the area of the cap of the paraboloid cut off by the curve $\theta = \gamma$ is

$$\tfrac{2}{3} \pi ab (\sec^3 \gamma - 1).$$

(5) By means of the substitution $x = at \cos \phi, y = bt \sin \phi$, show that the area of the curved surface of the part of the cone $x^2/a^2 + y^2/b^2 = z^2/c^2$ cut off by the plane $z = \lambda c$ is $\tfrac{1}{2}\lambda^2 s \sqrt{(ab)}$, where s is the perimeter of an ellipse of which the squares of the semiaxes are $b(c^2 + a^2)/a$ and $a(c^2 + b^2)/b$.

(6) Let V denote the part above the xy plane of the volume of the sphere $x^2 + y^2 + z^2 = a^2$ intercepted by the cylinder $x^2 + y^2 = ax$. Let S denote the part above the xy plane of the surface of the sphere intercepted by the cylinder.

(i) Find the volume V, the coordinates of its centroid, and its radius of gyration about the axis of the cylinder.

(ii) Find the area S, the coordinates of its centroid, and its radius of gyration about the axis of the cylinder.

(7) Prove that the area of the surface of the ellipsoid of revolution whose parametric equations are

$$x = a \sin \theta \cos \phi, \quad y = a \sin \theta \sin \phi, \quad z = c \cos \theta$$

is given by $\quad S = 2\pi a^2 \{1 + (\sin^{-1} e)/(ee')\} \quad (c > a);$

$$= 2\pi a^2 \{1 + (e'^2/e) \sinh^{-1} (e/e')\} \quad (c < a);$$

where in each case e is the eccentricity of a meridional section and $e' = \sqrt{(1 - e^2)}$.

(8) A point O lies on the axis of a circular disc of radius a. If c is the distance of O from the centre of the disc, find the solid angle subtended at O by the disc.

(9) A point O lies on the axis of a square of side $2a$. If a is the distance of O from the centre of the square, find the solid angle subtended at O by the square.

(10) From the result of the example in §12.23, deduce the solid angle subtended at the point $(0, 0, c)$ by the rectangle $[a, b; \alpha, \beta]$ in the xy plane.

(11) Let P be any point on the surface of a cube of edge a, of which O is one corner and C is the centre. Find the mean value of (i) OP^2, (ii) CP^2.

Also find the mean value of OP^2 and of CP^2 when P is any point in the volume of the cube.

(12) Find the mean distance of the point $(0, 0, c)$ from the volume of the sphere $x^2 + y^2 + z^2 = a^2$ when (i) $c > a$, (ii) $c < a$.

(13) Find the average value of the function $x^2 y^2$ over the surface of the sphere $x^2 + y^2 + z^2 = a^2$. Also find the average value of $x^2 y^2$ throughout the volume of the sphere.

(14) Find the average value of $x^2 y^2 z^2$ taken throughout the volume of the ellipsoid $x^2/a^2 + y^2/b^2 + z^2/c^2 = 1$.

(15) Prove that the mean value of $(\alpha x + \beta y + \gamma z)^{2n}$ over the surface of the sphere $x^2 + y^2 + z^2 = 1$ is $(\alpha^2 + \beta^2 + \gamma^2)^n/(2n+1)$, and find the mean value taken throughout the volume of the sphere.

(16) The region in the xy plane enclosed by the ellipse $x^2/a^2 + y^2/b^2 = 1$ subtends the solid angle ω at the point $(0, 0, c)$ on the axis of z. Show that

$$\omega = \frac{4bc}{a\sqrt{(a^2+c^2)}} \int_0^1 \frac{\lambda^2 u^2}{1 - \lambda^2 u^2} \frac{du}{\sqrt{(1-u^2)}\sqrt{(1-k^2 u^2)}}$$

where $\lambda^2 = a^2/(a^2+c^2)$, $k^2 = (a^2-b^2)/(a^2+c^2)$, and $a > b$.

Examples 12 E

(1) Evaluate the following integrals:

(i) $\displaystyle\int_1^\infty \int_1^\infty \frac{dx\, dy}{(x^2+y^2)^2}$,

(ii) $\displaystyle\int_0^1 \int_0^1 \log(x+y)\, dx\, dy$,

(iii) $\displaystyle\int_0^a \int_0^a \frac{dx\, dy}{\sqrt{(x^2+y^2)}}$,

(iv) $\displaystyle\int_0^{\frac{1}{2}\pi} \int_0^{\sin^2\phi} \frac{dx\, d\phi}{\sin^2\phi \sqrt{(1-x)}}$,

(v) $\displaystyle\int_0^a \int_0^a \frac{xy\, dx\, dy}{(x^2+y^2)^{\frac{3}{2}}}$,

(vi) $\displaystyle\int_0^a \int_0^{a-x} \sin\frac{\pi y}{a} \frac{dx\, dy}{\sqrt{x}\sqrt{(a-y)}}$,

(vii) $\displaystyle\int_0^4 \int_x^4 \frac{\log y\, dx\, dy}{\sqrt{(y-x)}}$,

(viii) $\displaystyle\int_0^1 \int_y^1 \frac{f'(x)\, dx\, dy}{\sqrt{(xy-y^2)}}$,

(ix) $\displaystyle\int_0^\infty \int_0^\infty e^{-y^2(1+x^2)} y\, dx\, dy$,

(x) $\displaystyle\int_0^a dy \int_y^a \cos\left(\frac{\pi y}{2x}\right) dx$.

(2) Evaluate

(i) $\displaystyle\int_0^1 \int_x^1 xy \log(1+y^4)\, dx\, dy$,

(ii) $\displaystyle\int_0^1 \int_y^2 x^3 y\, e^{xy}\, dx\, dy$.

(3) Show that $\displaystyle\int_0^\infty \int_0^\infty x^2 e^{-x^2 - y^2}\, dx\, dy = \tfrac{1}{8}\pi$.

(4) Show that $\displaystyle\int_0^\infty \int_0^\infty e^{-x^2-y^2} \cos(x^2+y^2)\, dx\, dy = \tfrac{1}{8}\pi$.

(5) Show that

$$\int_{-\infty}^\infty \int_{-\infty}^\infty \frac{dx\, dy}{(x^2+y^2+a^2)^{n+1}} = \frac{\pi}{na^{2n}} \quad (0 < n).$$

(6) Show that, over the region between the circles $r = 2a \cos \theta$, $r = 2b \cos \theta$, $(0 < a < b)$

$$\iint \frac{\log (x^2 + y^2)}{x^2 + y^2} \, dx \, dy = \pi \log (ab) \log \frac{b}{a}.$$

(7) Show that

$$\int_0^1 \int_0^{\frac{1}{2}\pi} \int_0^{\frac{1}{2}\pi} \frac{dx \, d\theta \, d\phi}{1 - x \sin^2 \theta \sin^2 \phi} = \pi.$$

(8) Show that

$$\int_0^\infty \int_0^\infty \int_0^\infty x e^{-x^2 - y^2 - z^2} \, dx \, dy \, dz = \tfrac{1}{8}\pi.$$

(9) Show that

$$\int_0^\infty \int_0^\infty \int_0^\infty \frac{xyz \, dx \, dy \, dz}{(x + y + z + a)^7} = \frac{1}{720a}.$$

(10) By making the substitution

$$x \cos \alpha + y \sin \alpha = X, \quad -x \sin \alpha + y \cos \alpha = Y,$$

show that
$$\iint_C \frac{f(x \cos \alpha + y \sin \alpha)}{\sqrt{(1 - x^2 - y^2)}} \, dx \, dy = \pi \int_{-1}^1 f(X) \, dX,$$

where C denotes the area enclosed by the circle $x^2 + y^2 = 1$.

(11) By making the substitution $x + y = u$, $y - x = uv$, show that

$$\iint_T \exp \left(\frac{y - x}{x + y} \right) \, dx \, dy = \tfrac{1}{4} a^2 \frac{e^2 - 1}{e},$$

where T denotes the triangle bounded by $x = 0$, $y = 0$, $x + y = a$.

(12) Show that

(i) $\displaystyle \int_1^\infty dx \int_1^\infty \frac{x - y}{(x + y)^3} \, dy = -\tfrac{1}{2},$ (ii) $\displaystyle \int_1^\infty dy \int_1^\infty \frac{x - y}{(x + y)^3} \, dx = \tfrac{1}{2},$

(iii) $\displaystyle \lim_{R \to \infty} \int_1^{\lambda R} dy \int_1^R \frac{x - y}{(x + y)^3} \, dx = \frac{1}{2} \frac{1 - \lambda}{1 + \lambda} \quad (0 < \lambda),$

(iv) $\displaystyle \iint \frac{x - y}{(x + y)^3} \, dx \, dy = \tfrac{1}{4} \log \frac{A}{a},$

where the field of integration is that bounded by $r = a$, $r = A$, $\theta = 0$, $\theta = \tfrac{1}{4}\pi$.

(13) Show that, if $0 < \delta$,

$$\int_0^1 dy \int_\delta^1 \frac{x^2 - y^2}{(x^2 + y^2)^2} \, dx = \int_\delta^1 dx \int_0^1 \frac{x^2 - y^2}{(x^2 + y^2)^2} \, dy$$

but that this is not true if $\delta = 0$.

(14) Show that, if $f'(y)$ is continuous,

$$\int_0^a dx \int_0^x \frac{f'(y) \, dy}{\sqrt{\{(a - x)(x - y)\}}} = \pi \{ f(a) - f(0) \}.$$

(15) Express in spherical polar coordinates the triple integral

$$\int_0^\infty \int_0^\infty \int_0^\infty z^n e^{-x^2 - y^2 - z^2} \, dx \, dy \, dz.$$

Deduce the reduction formula

$$I_n = \frac{n-1}{2} I_{n-2}, \quad \text{where} \quad I_n = \int_0^\infty x^n e^{-x^2}\, dx.$$

(16) The integral I_n being defined as in the last example, express I_n^3 as a triple integral in spherical polar coordinates. Deduce that

$$I_2 = \frac{2}{\pi} I_0^3, \quad I_5 = 8I_1^3, \quad I_8 = \frac{210}{\pi} I_2^3, \quad I_{11} = 480\, I_3^3, \ldots.$$

(17) Show that (see also the first of Examples 12 F)

$$\int_{-\infty}^\infty \int_{-\infty}^\infty e^{-x^4-y^4}\, dx\, dy = \sqrt{\pi} \int_0^{\frac{1}{2}\pi} \frac{d\phi}{\sqrt{(1-\frac{1}{2}\sin^2\phi)}}.$$

(18) Show that, if $a > 0$, $ab - h^2 > 0$,

(i) $$\int_{-\infty}^\infty \int_{-\infty}^\infty \exp[-(ax^2 + 2hxy + by^2)]\, dx\, dy = \frac{\pi}{\sqrt{(ab-h^2)}};$$

(ii) $$\int_{-\infty}^\infty \int_{-\infty}^\infty (Ax^2 + 2Hxy + By^2)\exp[-(ax^2 + 2hxy + by^2)]\, dx\, dy$$
$$= \tfrac{1}{2}\pi(aB + Ab - 2hH)/(ab-h^2)^{\frac{3}{2}}.$$

Examples 12 F

(1) Show that
$$\int_{-\infty}^\infty e^{-x^4}\, dx = \tfrac{1}{2}\Gamma(\tfrac{1}{4}).$$

(2) Show that
$$\int_0^\infty x^m \exp(-x^n)\, dx = \frac{1}{n} \Gamma\left(\frac{m+1}{n}\right) \quad (-1 < m, 0 < n).$$

(3) Show that, if $-1 < m$, $-1 < n$,
$$\int_0^{\frac{1}{2}\pi} \cos^m\theta \sin^n\theta\, d\theta = \tfrac{1}{2}\Gamma\left(\frac{m+1}{2}\right)\Gamma\left(\frac{n+1}{2}\right)\Big/\Gamma\left(\frac{m+n+2}{2}\right).$$

(4) Show that, if $-1 < n$,
$$\int_0^{\frac{1}{2}\pi} \sin^n\theta\, d\theta = \int_0^{\frac{1}{2}\pi} \cos^n\theta\, d\theta = \tfrac{1}{2}\sqrt{\pi}\,\Gamma\left(\frac{n+1}{2}\right)\Big/\Gamma\left(\frac{n+2}{2}\right).$$

(5) Show that, if $-1 < n < 1$,
$$\int_0^{\frac{1}{2}\pi} \tan^n\theta\, d\theta = \tfrac{1}{2}\Gamma\left(\frac{1+n}{2}\right)\Gamma\left(\frac{1-n}{2}\right).$$

(6) Show, by evaluating the integral
$$\int_0^{\frac{1}{2}\pi} \sin^{2n-1}\theta \cos^{2n-1}\theta\, d\theta$$

in two ways, that
$$\Gamma(n)\,\Gamma(n+\tfrac{1}{2}) = 2^{1-2n}\sqrt{\pi}\,\Gamma(2n).$$

(7) If $0 < m$, $0 < n$, show that $B(m,n) = B(n,m)$, and that

$$B(m,n) = \int_1^\infty \frac{(x-1)^{n-1}}{x^{m+n}}\,dx = \int_0^\infty \frac{x^{n-1}}{(1+x)^{m+n}}\,dx.$$

(8) Show that, if $0 < n < 1$,

$$\Gamma(n)\,\Gamma(1-n) = B(n, 1-n) = \int_0^\infty \frac{x^{n-1}}{1+x}\,dx$$

and deduce (see end of the example in §11.14) that, if $0 < n < 1$,

$$\Gamma(n)\,\Gamma(1-n) = \frac{\pi}{\sin n\pi}, \qquad \Gamma(1+n)\,\Gamma(1-n) = \frac{n\pi}{\sin n\pi}.$$

Using (57) and (61), show that these hold good for all real non-integral values of n.

(9) If $p > 0$, $q > 0$, and $p+q = 1$, show that

$$\int_0^1 \frac{dx}{x^p\,(1-x)^q} = \frac{\pi}{\sin p\pi} = \frac{\pi}{\sin q\pi}.$$

(10) If H denotes the semicircle $x^2 + y^2 - x < 0 < y$, evaluate

$$\iint_H \frac{dx\,dy}{\sqrt{(xy)}}$$

   (i) by changing to polar coordinates,
   (ii) by the substitution $x = u/(1+v^2)$, $y = uv/(1+v^2)$.

(11) Show that

$$\iint_T f(x+y)\,x^{l-1}\,y^{m-1}\,dx\,dy = \frac{\Gamma(l)\,\Gamma(m)}{\Gamma(l+m)} \int_0^h f(u)\,u^{l+m-1}\,du,$$

where T denotes the triangle bounded by $x = 0$, $y = 0$, $x+y = h$.

(12) Prove that, if a, b, l, m, λ, μ are positive, then

$$\iint_T x^{l-1}\,y^{m-1}\,dx\,dy = \frac{a^l b^m}{\lambda\mu}\,\frac{\Gamma(l/\lambda)\,\Gamma(m/\mu)}{\Gamma(l/\lambda + m/\mu + 1)}$$

where T denotes the region in the first quadrant bounded by the axes $x = 0$, $y = 0$ and the curve $(x/a)^\lambda + (y/b)^\mu = 1$.
  Deduce the area of the ellipse $x^2/a^2 + y^2/b^2 = 1$.
  Under similar conditions show that

$$\iiint_T x^{l-1}\,y^{m-1}\,z^{n-1}\,dx\,dy\,dz = \frac{a^l b^m c^n}{\lambda\mu\nu}\,\frac{\Gamma(l/\lambda)\,\Gamma(m/\mu)\,\Gamma(n/\nu)}{\Gamma(l/\lambda + m/\mu + n/\nu + 1)},$$

where T denotes the region in the first octant bounded by the coordinate planes and the surface $(x/a)^\lambda + (y/b)^\mu + (z/c)^\nu = 1$.
  Deduce the volume of the ellipsoid $x^2/a^2 + y^2/b^2 + z^2/c^2 = 1$.

(13) Show that, if $n > 1$,

$$\int_0^a (a^n - x^n)^{-1/n}\,dx = \frac{\pi}{n}\,\operatorname{cosec}\frac{\pi}{n}.$$

(14) Sketch the curve $x^4 + y^4 = a^4$. Show that the area bounded by the curve is

$$\{\Gamma(\tfrac{1}{4})\}^2 \, (a^2/2\sqrt{\pi}).$$

(15) Sketch the curve $x^4 + y^4 = a^2 x^2$. Show that the whole area bounded by the curve is $\pi a^2 / \sqrt{2}$.

(16) Show that $\displaystyle\iiint_T \log(x+y+z)\,dx\,dy\,dz = -\tfrac{1}{18}$

where T denotes the tetrahedron bounded by $x = 0,\ y = 0,\ z = 0,\ x+y+z = 1$.

(17) Show that $\displaystyle\iiint_V \frac{dx\,dy\,dz}{\sqrt{(1-x^2-y^2-z^2)}} = \pi^2$

where V denotes the interior of the sphere $x^2 + y^2 + z^2 = 1$.

(18) Show that the volume in the first octant bounded by the coordinate planes and the surface $x^n + y^n + z^n = a^n\ (n > 0)$ is

$$\left\{\Gamma\left(1+\frac{1}{n}\right)\right\}^3 a^3 \Big/ \Gamma\left(1+\frac{3}{n}\right).$$

(19) Show that

$$\iiint_T x^{l-1} y^{m-1} z^{n-1} (1-x-y-z)^{p-1}\,dx\,dy\,dz = \frac{\Gamma(l)\,\Gamma(m)\,\Gamma(n)\,\Gamma(p)}{\Gamma(l+m+n+p)},$$

where $l,\ m,\ n,\ p$ are positive, and T denotes the tetrahedron bounded by the planes $x = 0,\ y = 0,\ z = 0,\ x+y+z = 1$.

(20) Evaluate the integral

$$\int_0^\infty \int_0^\infty \int_0^\infty \frac{xyz\,dx\,dy\,dz}{(x+y+z+a)^n} \qquad (a > 0,\ n > 6),$$

  (i) directly as a repeated integral,
  (ii) as an example of Dirichlet's integral over the tetrahedron

$$x \geqslant 0,\ y \geqslant 0,\ z \geqslant 0,\quad x+y+z \leqslant h,$$

followed by letting h tend to ∞.

(21) Show that $\displaystyle\int_{-\infty}^\infty \cos\frac{\pi x^2}{2}\,dx = 1.$

(22) Show that $\displaystyle\int_0^\infty \cos(x^n)\,dx = \frac{1}{n}\Gamma\left(\frac{1}{n}\right)\cos\frac{\pi}{2n} \quad (1 < n).$

(23) Show that, of the two double integrals

$$\int_0^R \int_0^R \cos(x^2+y^2)\,dx\,dy, \qquad \int_0^R \int_0^{\frac{1}{2}\pi} \cos(r^2)\,r\,dr\,d\theta,$$

when $R \to \infty$ the first tends to the limit zero, the second does not tend to a limit.

(24) Show that

$$\int_0^\infty \int_0^\infty \int_0^\infty \sin(x^2 + y^2 + z^2)\, dx\, dy\, dz = \frac{\pi\sqrt{\pi}}{8\sqrt{2}},$$

but that

$$\int_0^\infty \int_0^{\frac{1}{2}\pi} \int_0^{\frac{1}{2}\pi} \sin(r^2)\, r^2 \sin\theta\, dr\, d\theta\, d\phi$$

does not exist.

(25) (i) Show that, if $0 < m$, $0 < n$, $a < b$,

$$\int_a^b (x-a)^{m-1}(b-x)^{n-1}\, dx = (b-a)^{m+n-1}\, B(m,n).$$

(ii) If l, m, n are positive and $l + m + n = 1$, and if $f'(z)$ is continuous, show that

$$\int_0^a (a-x)^{l-1}\, dx \int_0^x (x-y)^{m-1}\, dy \int_0^y (y-z)^{n-1} f'(z)\, dz$$

$$= \Gamma(l)\, \Gamma(m)\, \Gamma(n)\, \{f(a) - f(0)\}.$$

(26) Prove that, if $0 < a$, $0 < n$,

$$\int_0^\infty e^{-ax} x^{n-1} \cos bx\, dx = \frac{\Gamma(n) \cos n\theta}{r^n},$$

$$\int_0^\infty e^{-ax} x^{n-1} \sin bx\, dx = \frac{\Gamma(n) \sin n\theta}{r^n},$$

where $r = \sqrt{(a^2 + b^2)}$, $\theta = \tan^{-1}(b/a)$.

ANSWERS TO EXAMPLES

Examples 1, p. 20

(1) (i) $(-2, 2)$; (ii) $[-1, 1]$;

(iii) $[1-\sqrt{3}, 1+\sqrt{3}]$; (iv) $(-1, 1)$;

(v) $(-\infty, 1)$, $(3, \infty)$; (vi) $(-\frac{7}{2}, \frac{1}{2})$;

(vii) $(0, 2)$; (viii) $(-1, 0)$, $(3, \infty)$;

(ix) $(1, 3)$; (x) $\{\frac{1}{2}(5-\sqrt{10}), \frac{1}{2}(5+\sqrt{10})\}$;

(xi) $(-\infty, -4)$, $(0, 1)$; (xii) $(\frac{1}{3}, 1)$, $(1, 3)$;

(xiii) $(-\infty, -2)$, $(-\frac{1}{2}, 1)$; (xiv) $(-1, 1)$, $(\frac{3}{2}, \infty)$;

(xv) $(\frac{14}{5}, 3)$, $(4, 5)$;

(xvi) $(-1, \frac{1}{2}(3-\sqrt{17}))$, $(0, \frac{1}{2}(3+\sqrt{17}))$;

(xvii) $(0, 1)$; (xviii) $(-\frac{1}{2}, 0)$, $(1, \infty)$.

(2) (iii) $f(0+) = 1$; $f(x)$ is odd; (iv) $f(0+) = \frac{1}{2}\pi$; $f(x)$ is odd;

(v) $f(x) = 1$ $(x \neq 0)$, $f(0) = 2$;

(vi) $f(x) = (\sin x)/x$ $(x \neq 0)$, $f(0) = 3$;

(vii) $f(x) = 0$, all x; (viii) $f(x) = 0$ $(x \neq n\pi)$, $f(n\pi) = n\pi$;

(ix) $f(x) = 0$ $(0 \leqslant x < 1)$, $f(1) = 1$; (x) $f(x) = 1$ $(0 < x)$, $f(0) = 0$;

(xi) $f\{\frac{1}{2}(2n+1)\pi\} = 1$, $f(x) = 0$, $x \neq \frac{1}{2}(2n+1)\pi$;

(xii) $f(x) = 1$, $x \neq n\pi$; $f(n\pi) = 0$.

(3) (i) The square of sides $x = \pm a$, $y = \pm a$.

(ii) The lines $y^2 = x^2$ $(x^2 \geqslant a^2)$, and the lines $x^2 = a^2$ $(y^2 \leqslant a^2)$.

(iii) The line $x+y = a$ $(x^2 \geqslant a^2)$, and the lines $x = a$ $(y^2 \leqslant a^2)$, $y = a$ $(x^2 \leqslant a^2)$.

(4) (i) 2 if n is odd, 1 if n is even; (ii) 1;

(iii) π if n is odd, $\frac{1}{2}\pi$ is n is even.

(5) $x|x| = x^2$ $(0 \leqslant x)$. $x/|x| = 1$ $(0 < x)$, undefined at $x = 0$.
$\frac{1}{2}(1+x/|x|) = 1$ $(0 < x)$, $= 0$ $(x < 0)$, undefined at $x = 0$.

(6) The function is equal to $\frac{1}{2} \coth \frac{1}{2}x$.

(7) $e^x = \cosh x + \sinh x$, $e^{-x} = \cosh x - \sinh x$.

(8) (i) $y = x$ $(0 \leqslant x)$, $y = -x$ $(x < 0)$;

(ii) $y = \sqrt{x}$ $(0 \leqslant x)$, $y = \sqrt{(-x)}$ $(x < 0)$;

(iii) $y = 1/\sqrt{x}$ $(0 < x)$, $y = 1/\sqrt{(-x)}$ $(x < 0)$; undefined at $x = 0$.

(9) Let y denote the given function; then:

(i) $y = \log x$ $(0 < x)$, and y is even; (ii) y is even;

(iii) y is periodic with period π, undefined at $x = n\pi$;

   (iv) y is undefined between $(2n-1)\pi$ and $2n\pi$;

   (v) y is undefined between $\frac{1}{2}(2n-1)\pi$ and $n\pi$;

   (vi) $y \doteqdot \frac{1}{2}x^2$ (x small), $y \doteqdot |x| - \log 2$ (x large);

   (vii) $y = \log(x^2-1)$ is even, undefined if $x^2 \leqslant 1$;

   (viii) y is undefined if $x \leqslant 1$;   (ix) y is odd, undefined if $x^2 \leqslant 1$;

   (x) y is undefined if $x \leqslant 1$;

   (xi) $\tan^{-1}\{2x/(1-x^2)\}$ is discontinuous at $x = \pm 1$.

(10) Let y denote the given function; then:

   (i) y is odd, $dy/dx = 0$ at $x = 0$;

   (ii) y is even, $dy/dx \to +\infty$ as $x \to 0+$;

   (iii) y is odd, $y \to +\infty$ as $x \to 0+$; (iv) y not defined if $x < 0$;

   (v) y is even, $y \to 1-$ as $x \to 0$;

   (vi) y is even, oscillating between the lines $y = \pm x$;

   (vii) y is even, oscillating between the parabola $y = x^2$ and the axis of x;

   (viii) y is undefined if $x \leqslant 0$; when $x \to 0+$, $y \to 0-$ and $dy/dx \to -\infty$;

   (ix) $y = -\log x \to +\infty$ when $x \to 0+$;

   (x) $y \to 1-$ when $x \to 0+$, $y \to 0+$ when $x \to 0-$;

   (xi) y is even;   (xii) y is undefined for $x \leqslant 0$.

(15) $\sinh x = \pm \sqrt{(\frac{1}{2}\sqrt{3})}$. ($\pm 0.8314$, ± 0.9306).

Examples 2A, p. 48

(1) $-1, 1$.

(2) $-1, +1$ at $x = 0$, ± 2, etc.; $+1, -1$ at $x = \pm 1$, ± 3, etc.

(4) $-\frac{1}{2}\pi, +\frac{1}{2}\pi$.   (6) $1/a, -1/a$.

(7) $f(a) = 1/a, f'(a) = -1/a^2; f''(a-0) = -1/a^3, f''(a+0) = 2/a^3$.

Examples 2B, p. 49

(1) $\frac{1}{2}(a+b)$.   (2) $\phi(x)$ is not continuous at $x = a$.

(3) $\cos^{-1}(2/\pi)$.

(4) $\xi = 1/\alpha$, where α is any positive root of the equation $\alpha = \tan \alpha$.

(5) $\phi'(x)$ does not exist at $x = 1$ (a point in the open interval).

(6) (i) $\frac{1}{4}$; (ii) $\frac{4}{9}$.

(19) (i) yes; (ii) yes; (iii) no;

   (iv) yes: in this case $f'(x)$ and $g'(x)$ vanish together at $x = 0$, showing that the conditions under which Cauchy's mean value theorem is proved are not necessary, though they are sufficient.

Examples 2 C, p. 51

(1) (i) a,   (ii) $\log a$,
     (iii) $(p-q)/(m-n)$,   (iv) $\frac{1}{2}a^2$,
     (v) $1/\pi$,   (vi) 0,
     (vii) $-e^{-a}$,   (viii) 0,
     (ix) 1,   (x) $\log a$,
     (xi) 2,   (xii) 0,
     (xiii) $\pi/(2\alpha \sin \alpha)$,   (xiv) $\log (a/p)$,
     (xv) $\log 2$,   (xvi) m/n,
     (xvii) $-\frac{1}{4}$,   (xviii) $\log (a/c)$.

(3) $f'(0) = 0, f''(0+) = 2, f''(0-) = 0.$

(4) $f'(0+) = 0, f'(x)$ oscillates infinitely as $x \to 0+$.

(5) $f(0+) = b/d, f(0-) = a/c; f'(0+) = 0, f'(0-) = 0.$

(7) (ii) 1;   (iii) $e, e.$

(8) (i) When $x \to 0+$, $f(x) \to 1$, $f'(x) \to -\infty$. Also, $f'(x) < 0$, $0 < x < 1/e$; $f'(1/e) = 0; f'(x) > 0, 1/e < x.$

(ii) $f'(x) = 0, x = 0, x = e^{-\frac{1}{2}}; f''(x) = 0, x = e^{-\frac{3}{2}}; f(x)$ is even, $f(x) = 2x^2 \log x$ for $0 < x.$

(iii) $f(x)$ is even ; $f(x) \to 1$ $(x \to \infty); f'(x) > 0$ $(0 < x); f'(0+) = 0.$

(iv) $f(x)$ is odd; $f(0+) = 1, f(0-) = -1; f'(x) < 0.$

(v) $f(x)$ is even; $f(x) \to 1-0 (x \to \infty)$; $f'(x) > 0$ $(0 < x)$; $f'(0+) = 0.$ Inflexions at $x = \pm\sqrt{(\frac{2}{3})}.$

(vi) $f(0+) = f(0-) = e$; $f'(0+) = f'(0-) = -\frac{1}{2}e.$

(9) (a) When $x \to 0$, $\lim f(x) = 1 = f(0)$; therefore $f(x)$ is continuous at $x = 0$. Also, if $x \neq 0$, $f'(x) = (x \cos x - \sin x)/x^2$ and hence, when $x \to 0, f'(x) \to 0$ (l'Hospital). Therefore $f'(0) = 0.$ (b) Similar reasoning.

Examples 2D, p. 53

(3) $n = \frac{2}{3}, K = (9a/4)^{\frac{2}{3}}.$

(5) (i) $y_1^2 = m^2(y^2 - A^2 + B^2)$,   (ii) $y_1^2 = n^2(y^2 - 4AB).$

(10) (i) $1/(x \log x \log \log x)$,   (ii) $\cot x \cot (\log \sin x)$,
      (iii) $2 \log x (x^{\log x})/x$,   (iv) $x^{\log \log x}(1 + \log \log x)/x$,
      (v) $(\log x)^{\log x}(1 + \log \log x)/x$,   (vi) $\log a(1 + \log x) x^x a^{x^x}$,
      (vii) $-\log a/\{x(\log x)^2\}$,   (viii) $(\log_x a)^x \{\log (\log_x a) - 1/\log x\}$,
      (ix) $2/(1 + x^2)$,   (x) $2/\sqrt{(1 - x^2)}$,
      (xi) $1/\{2x\sqrt{(x^2 - 1)}\}$,   (xii) $\sin \alpha/(1 - 2x \cos \alpha + x^2)$,
      (xiii) $\sqrt{a}/\{(a + x) \sqrt{x}\}$,   (xiv) $4/(1 - x^2)$,
      (xv) $-\cos \theta/\{\sin \theta \sqrt{(1 - 2 \sin^2 \theta)} \sqrt{(3 \sin^2 \theta - 1)}\}$,

(xvi) $a \sin \alpha/(x^2 + a^2 - 2ax \cos \alpha)$,   (xvii) $1/\{2(1+x^2)\}$,

(xviii) $\sqrt{(b^2 - a^2)}/(b + a \cos x)$,

(xix) $(2-x)/[2(x+2)\sqrt{\{x(x+1)(x+4)\}}]$,

(xx) $3(x+1)/\{(x-2)\sqrt{(1+x^3)}\}$,

(xxi) $\dfrac{1-(1+k')x}{1-(1-k')x} \dfrac{1-k'}{2\sqrt{\{x(1-x)(1-k^2 x)\}}}$.

Examples 3B, p. 69

(1)   (i) $n!/(1-x)^{n+1} (n > 1)$,   (ii) $n!(ad-bc)(-c)^{n-1}/(cx+d)^{n+1}$,

     (iii) $(2n)!(1-x)^{-n-\frac{1}{2}}/(2^{2n} n!)$,   (iv) $2^{n-1} \cos(2x + n\pi/2)$,

     (v) $\frac{1}{2}\{3^n \cos(3x + n\pi/2) + \cos(x + n\pi/2)\}$,

     (vi) $2^{n-2}\{\cos(2x + n\pi/2) + 2^n \cos(4x + n\pi/2) + 3^n \cos(6x + n\pi/2)\}$,

     (vii) $\frac{3}{4}\{\sin(x + n\pi/2) - 3^{n-1}\sin(3x + n\pi/2)\}$,

     (viii) $\frac{1}{4}\{\cos(x + n\pi/2) - 3^n \cos(3x + n\pi/2)\}$,

     (ix) $2^{\frac{1}{2}n} e^x \cos(x + n\pi/4)$,

     (x) $\sinh x$ (n odd), $\cosh x$ (n even),

     (xi) $a^n \cosh ax$ (n odd), $a^n \sinh ax$ (n even),

     (xii) $e^{3x}\{3^n + 13^{\frac{1}{2}n} \sin(2x + n \tan^{-1}\frac{2}{3})\}$,

     (xiii) $e^{ax}\{a^n x^3 + 3na^{n-1} x^2 + 3n(n-1)a^{n-2}x + n(n-1)(n-2)a^{n-3}\}$
         $(n \geqslant 3)$,

     (xiv) $n! e^x(1 + {}^nC_1 x/1! + {}^nC_2 x^2/2! + \ldots + x^n/n!)$,

     (xv) $\{x^2 - n(n-1)\}\sin(x + n\pi/2) - 2nx \cos(x + n\pi/2)$,

     (xvi) $e^x\{1/x - n/x^2 + n(n-1)/x^3 - \ldots + (-)^n n!/x^{n+1}\}$,

     (xvii) $6(n-4)!/(-x)^{n-3} (n > 3)$,

     (xviii) $(n-1)!\{p(-)^{n-1}/(x-a)^n + q/(b-x)^n\}$.

(6) $D^n(u+iv) = m(m-1)\ldots(m-n+1)X^{m-n} \exp[i\{(m-n)\phi + \frac{1}{2}n\pi\}]$,   where
$X = \sqrt{(1+x^2)}$, $\phi = \tan^{-1}x$.

(27) $P = (2x)^n - n(n-1)(n-2)(n-3)(2x)^{n-4}/2! + \ldots$;
     $Q = n(n-1)(2x)^{n-2}/1! - n(n-1)(n-2)\ldots(n-5)(2x)^{n-6}/3! + \ldots$.

Examples 3C, p. 77

(3) $\Sigma \sin(\beta - \gamma)\sin(x - \alpha) \equiv 0$.

Examples 4D, p. 121

(2) $-\frac{1}{3}$.   (3) $\frac{1}{45}$.

Examples 4F, p. 126

(10) $2\cdot0408, 2\cdot1179$.   (21) $1\cdot426, -0\cdot88; 6$.

Examples 4G, p. 130

(1) (0·86, 0·65), 0·56. (2) 1·2. (3) 4·75.

(4) $x_1 = -0.684$, $x_2 = 0.79$. (5) 2·08.

(6) 0·72. (8) 0·932. (9) 4·4934.

(10) 2·074.

Examples 5A, p. 151

(1) (i) The interior of the circle $x^2 + y^2 = a^2$, including the circumference.

(ii) The region 'within' the parabola $y^2 = 4ax$, including the parabola.

(iii) The interior of the circle $(x-a)^2 + (y-b)^2 = \delta^2$.

(iv) The half-plane above the line $3x + 4y = 5$.

(v) The interior of the square bounded by the four straight lines $\pm(x-a)\pm(y-b) = \delta$.

(vi) The half-plane below the line $3x + 4y + 5 = 0$.

(vii) The region exterior to the circle $x^2 + y^2 = 4$ and below the line $y = x$.

(viii) The region between the parabola $y^2 = 8x$ and its tangent $y = x + 2$.

(ix) The part of the interior of the circle $x^2 + y^2 = 25$ in which $xy < 12$.

(x) The interior of the rectangle of sides $x = a \pm \alpha$, $y = b \pm \beta$.

(xi) The interior of the triangle of sides $x = 1$, $y = \pm x$.

(xii) The second quadrant bounded by the coordinate axes; also, the triangle of sides $x = 0$, $y = 0$, $x - y = 1$.

(2) (i) $(1+m^2)/(1-m^2)$, (ii) $(1+m)/(1-m)$; in each case the limit depends on m.

(3) (i) 2, (ii) $\frac{1}{2}$, (iii) $(1+2m)/(2+m)$.
    (i) 2, (ii) $-\frac{1}{2}$, (iii) $\frac{1}{3}$.

(4) (i) 0, (ii) 1.

(5) (i) 1, (ii) 0.

Examples 5B, p. 152

(3) $-\frac{3}{2}$.

Examples 5C, p. 153

(8) 5.

Examples 6A, p. 179

(1) (2, 4). (3) $\frac{1}{2}$, $-\frac{1}{3}$, $-\frac{1}{6}$.

(4) $1/a$. (5) $-a/b$.

Examples 6C, p. 181

(4) $\partial r/\partial y = \sin\theta\sin\phi$, $\partial\theta/\partial y = (\cos\theta\sin\phi)/r$, $\partial\phi/\partial y = \cos\phi/(r\sin\theta)$, $\partial r/\partial z = \cos\theta$, $\partial\theta/\partial z = -\sin\theta/r$, $\partial\phi/\partial z = 0$.

Examples 6D, p. 185

(1)   (i) $v = \sin u$,   (ii) $v = \tan u$,

     (iii) $u^2 = v^2 + 1$,   (iv) $v(1 - u^2) = u$.

(2) $u^2 + v^2 = 1$.

(6) (i) $u^2 = v + 2w$,   (ii) $vw + wu + uv = b - a^2$.

Examples 6E, p. 186

(8)   (i) $2xy + C$,   (ii) $-\frac{1}{2}\log(x^2 + y^2) + C$,

    (iii) $-e^{-2xy}\cos(x^2 - y^2) + C$,   (iv) $\sqrt{\{\sqrt{(x^2 + y^2)} - x\}} + C$,

    (v) $(x + x^2y + y^3)/(x^2 + y^2) + C$,

    (vi) $(\sin x \sinh y)/(\cosh 2y + \cos 2x) + C$,

    (vii) $\tan^{-1}(\sin x/\cosh y) + C$.

Examples 7B, p. 216

(1)   (i) $x^2y - xy^2 + 3x^2 + 5y^2$,   (ii) $\frac{1}{2}\log\{(x + y)/(x - y)\}$,

    (iii) $-(2/\sqrt{3})\tan^{-1}\{(2x + y)/y\sqrt{3}\}$,

    (iv) ye^{3x},   (v) $\log\{xy/\sqrt{(x^2 - y^2)}\}$,

    (vi) $(\sin n\theta)/(nr^n)$,

    (vii) $\log x + \int dv/\{v + f(v)\}, v = y/x$.

(2) $x^3 - y^3 + 2xy^2 - 3xy + y - 2$.   (5) $\mu = y, u = x^2y^2 - 3xy^2 + 3y^3$.

(6) $\mu = x(x^2y + 1), u = \frac{1}{2}(x^2y + 1)^2(x^2 - 2). \ x/\sqrt{(x^2 - 2)}$.

(8) $y = x + Cx^2$.

(9) $e^{-2x}f(u)$, where $u = e^{-2x}(y + 2\cos x - \sin x)$.

(10) $\dfrac{1}{(y - 2x)(y + 3x)} f\left(\dfrac{(y + 3x)^7}{(y - 2x)^2}\right)$.

(11) $(1 - n)v = \log\{(1 - n)u\}$.

Examples 7D, p. 219

(6) $-(F_{33}F_1F_2 - F_{13}F_2F_3 - F_{23}F_1F_3 + F_{12}F_3^2)/F_3^3$.

(8) $\{G_tG_y^2(F_{tt}F_x^2 - 2F_{xt}F_xF_t + F_{xx}F_t^2)$

                   $- F_tF_x^2(G_{tt}G_y^2 - 2G_{yt}G_yG_t + G_{yy}G_t^2)\}/(F_tG_y)^3$.

Examples 7 *E*, p. 220

(8) $Axy(x^2 - y^2) + B$.

Examples 8 *A*, p. 246

(5) (i) Max. $+1$, min. -1;   (ii) max. $+1$, min. -1;

(iii) max. $\cosh 1 = 1\cdot543$, min. 1.

(8) Max. $\frac{14}{15}$ at $\theta = \frac{1}{6}\pi, \frac{5}{6}\pi$; max. $\frac{13}{15}$ at $\frac{1}{2}\pi$; min. $\frac{2}{5}\sqrt{3}$ at $\theta = \frac{1}{3}\pi, \frac{2}{3}\pi$.

Examples 8 *B*, p. 247

(1) (i) Saddle point,   (ii) min.,   (iii) neither,

     (iv) saddle,   (v) saddle,   (vi) min.,

     (vii) min.,   (viii) neither,   (ix) neither.

Examples 8 *C*, p. 248

(1) $(0, 0, 10)$ saddle, $(3, 0, -17)$ min.

(2) $(0, 0, 1)$ saddle, $(\frac{2}{3}, -\frac{4}{3}, \frac{31}{27})$ max.

(3) $(0, 0, 0)$ neither, $(0, \sqrt{2}, 4)$ max., $(0, -\sqrt{2}, 4)$ max.

(4) $(0, 0, 0)$ max., $(0, \pm\frac{1}{2}\sqrt{3}, -\frac{9}{8})$ min., $(\pm\frac{1}{2}, 0, -\frac{1}{8})$ neither.

(5) The axis of x and the axis of y lie entirely on the surface, so there can be no true max. or min. point on either. The point (a, a, a^6) is a max-point.

(6) $(a\lambda, b\lambda, \frac{1}{2}\lambda^{-1})$, $\lambda = \{-c \pm \sqrt{(a^2 + b^2 + c^2)}\}/(a^2 + b^2)$, one max., one min.

(7) $(0, 0, 0)$ saddle, $(4a, 4a, 0)$ saddle, $(2a, 2a, 16a^2)$ max.

(8) $(\pm 1/\sqrt{2}, \pm 1/\sqrt{2}, 2)$ both max.

(9) $(\frac{1}{2}e^{-\frac{1}{2}}, \frac{1}{2}e^{-\frac{1}{2}}, -\frac{1}{8}e^{-1})$ min., $(1, 0, 0)$ and $(0, 1, 0)$ both saddle.

(10) $(0, 0, 0)$ neither max. nor min.

(11) The line $y = x$ lies entirely on the surface, so no point on it can be a strict max. or min. $(\pm\frac{1}{2}\sqrt{3}, \mp\frac{1}{2}\sqrt{3}, -\frac{9}{2})$ both min.

(12) $(0, 0, 0)$ saddle, $(-\frac{5}{3}, -\frac{5}{3}, \frac{125}{27})$ max.

(13) $(-ac/\lambda, -bc/\lambda, c^2/\lambda)$, $\lambda = 1 + a^2 + b^2$, min.

(14) $(ac - b, -c, 3)$ min.

(15) $\{(hf - bg)/(ab - h^2), (gh - af)/(ab - h^2), \Delta/(ab - h^2)\}$, min. if $a > 0$, max. if $a < 0$.

(20) $(\frac{1}{2}\pi, 0, a)$ max., $(0, \frac{1}{2}\pi, b)$ saddle, $(\pi, \frac{1}{2}\pi, -b)$ saddle, $(\frac{1}{2}\pi, \pi, -a)$ min.

(24) $\pm a^n$.

(29) $(1, -1, \frac{1}{2})$ saddle; $\left(\dfrac{2 - \sqrt{10}}{3}, \dfrac{7 - 2\sqrt{10}}{9}, \dfrac{4 + \sqrt{10}}{6}\right)$ max.;

$\left(\dfrac{2 + \sqrt{10}}{3}, \dfrac{7 + 2\sqrt{10}}{9}, \dfrac{4 - \sqrt{10}}{6}\right)$ min.

Examples 8 D, p. 250

(5) $(ab-h^2)\,r^4-(a+b)\,r^2+1 = 0$; giving two positive values of r^2, the squares of the semi-axes of the ellipse $ax^2 + 2hxy + by^2 = 1$ $(ab-h^2 > 0)$.

If $ab-h^2 < 0$ the same equation in r^2 has one positive root, the square of the real semi-axis of the hyperbola $ax^2 + 2hxy + by^2 = 1$ $(ab-h^2 < 0)$; and one negative root, the square of the imaginary semi-axis, or the negative square of the real semi-axis of the conjugate hyperbola.

(8) $(3\sqrt{3})\,ab/4$ $(a > b)$.

(11)   (i) $(a, a, a, 3a)$ min., $(-a, -a, -a, -3a)$ max.;

      (ii) $(a, a, a, 3a)$ min.;

     (iii) $(a, a, a, 3a)$ neither, $(\frac{1}{4}a, -2a, -2a, -15a/4)$ etc., all max.

(12) $(\frac{1}{3}, \frac{2}{3}, \frac{2}{3}, 25)$ min., $(-\frac{1}{3}, -\frac{2}{3}, -\frac{2}{3}, 49)$ max.

(13) Neither max. nor min. at $(3a, 0, 0)$, etc.

(15) See, e.g. Salmon, *Solid Geometry*, for the lengths of the semi-axes of the ellipsoid given by the general equation; or other standard text on the coordinate geometry of three dimensions.

(16) The max. and min. values are the roots of the quadratic equation $\Sigma a^2 l^2/(a^2 - \lambda) = 0$.

(18) $P(0, 0)$, $Q(3a \pm 2a\sqrt{3}, 0)$; and the points where P and Q coincide. The lines $y = \pm(x-3)$ cut the curves orthogonally in points which give PQ^2 saddle values.

Examples 9 A, p. 286

(2) $-\frac{12}{7}, \frac{6}{7}, -\frac{4}{35}$.   (4) 8.

(5) $25\pi/2 - 25(\phi-\theta) + 12 + \sqrt{21}$, $25\pi/2 - 25(\phi+\theta) - \sqrt{21}$, $25(\phi-\theta) - \sqrt{21}$,
$25(\phi+\theta) - 12 + \sqrt{21}$; $\theta = \frac{1}{2}\sin^{-1}(\frac{3}{5})$, $\phi = \frac{1}{2}\sin^{-1}(\frac{4}{5})$.

(6) $h/3 + V/(\pi h^2)$.   (7) $\frac{1}{2}h$.   (9) $\frac{1}{9}$.   (10) 0.

(13) (i) $m = n = 0, \pi$; $m = n \neq 0, \frac{1}{2}\pi$; $m \neq n, 0$.

     (ii) $l+m+n$ odd, $I = 0$; l, m, n all zero, $I = \pi$; two zero, $I = 0$; one zero, $I = 0$ or $\frac{1}{2}\pi$ (see (i)); none zero, $I = 0$ unless $l = m+n$ or $m = n+l$ or $n = l+m$, in which cases $I = \frac{1}{4}\pi$.

(15) $Re\{(e^z - 1)/(1 - e^{-z/n})\}$, $z = x + iy$.

Examples 9 B, p. 287

(8) $\frac{1}{3}\pi + 2\sqrt{3}$.

Examples 9 D, p. 291

(7) $\xi = \frac{1}{2}\pi$, $X = \frac{2}{3}\pi$.

Examples 9*E*, p. 293

(1) $x/\{a^2\sqrt{(a^2-x^2)}\}$. (3) $-\pi/4a^2$. (5) $\sin y + 2\cos y$.

(6) $x\sinh x - \cosh x \log\cosh x + A\,e^x + B\,e^{-x}$.

(7) $y = \frac{1}{2} - e^{-t} + \frac{1}{2}e^{-2t}(0 < t \leqslant 1)$; $(e-1)\,e^{-t} - \frac{1}{2}(e^2-1)\,e^{-2t}(1 < t)$.

(8) $y = A + Be^x + e^{e^x}$.

(9) $y = e^t(At + B + t\log t)$.

(10) (i) $2/(3-2x)$, (ii) $20x^3/9$, (iii) $x(A - 6\log x)$,

(iv) $\sqrt{x}\{\cos(\frac{1}{2}\sqrt{3}\log x) + (5/\sqrt{3})\sin(\frac{1}{2}\sqrt{3}\log x)\}$.

(12) $f(x) = A/x$, $\phi(y) = A\log y + B$.

(13) $h^2 = a^2(2a^2 + b^2)/(3a^2 + b^2)$. (14) $\pi/(4a^3b)$, $\pi/(4ab^3)$.

(15) $(2n + \frac{1}{2})\pi/\sin\alpha$, $\{1 + (2n + \frac{1}{2})\pi\cos\alpha\}/\sin^3\alpha$, $\{(2n + \frac{1}{2})\pi + \cos\alpha\}/\sin^3\alpha$.

(16) $I_1 = \sin x/x^3 - \cos x/x^2$, $I_2 = 3\sin x/x^5 - 3\cos x/x^4 - \sin x/x^3$,
$I_3 = 15\sin x/x^7 - 15\cos x/x^6 - 6\sin x/x^5 + \cos x/x^4$.

(23) $\log(1 + y^2) - 2 + 2y\tan^{-1}(1/y)$. (26) $\frac{1}{8}\pi\log 2$.

Examples 10*A*, p. 319

(1) (i) $1/p$, (ii) $1/p^2$.

(2) (i) $p/(p^2 + \omega^2)$, (ii) $\omega/(p^2 + \omega^2)$.

(3) (i) $\frac{1}{2}\pi$, (ii) 1,
(iii) diverges to $+\infty$, (iv) diverges at $x = 1$.

(4) (i) -1, (ii) $\frac{3}{2}$, (iii) $6 \times 2^{\frac{1}{3}}$.

(5) (i) Diverges to $+\infty$, (ii) $\frac{1}{2}\pi$,
(iii) diverges to $+\infty$ at the lower limit, (iv) $\pi/\sqrt{2}$.

(6) (i) Diverges to $+\infty$, (ii) diverges to $+\infty$, (iii) $\log 2$.

(7) (i) 1, (ii) 2, (iii) diverges to $+\infty$.

(8) (i), (ii) Diverge to $-\infty$; (iii), (iv) diverge to $+\infty$.

Examples 10*C*, p. 322

(1) (i) $0 < n$; (ii) $0 < a$; (iii) $0 < a < 1$;
(iv) $0 < a < n$, or $n < a < 0$;
(v) $0 < a$, $0 < b$; (vi) $0 < n$, $1 < s$; (vii) $0 < m$, $0 < n$;
(viii) $0 < m$, $0 < n$; (ix) $0 < a < 1$, $0 < b < 1$ (unless $a = b$);
(x) $-1 < m$, $-1 < m + 2n$.

Examples 10*E*, p. 323

(3) (i) All y, (ii) $0 < \alpha \leqslant |y|$, (iii) $0 < \alpha \leqslant y$,
(iv) $0 < \alpha \leqslant y$, (v) $0 < \alpha \leqslant |y|$, (vi) all y.

(7) $\dfrac{1}{b-a}\left(\dfrac{1}{a} - \dfrac{1}{b-a}\log\dfrac{b}{a}\right)$, $\dfrac{1}{(b-a)^2}\left(\dfrac{1}{a} + \dfrac{1}{b} - \dfrac{2}{b-a}\log\dfrac{b}{a}\right)$.

(21) (i) $\frac{1}{2}\pi$ if $a > b \geqslant 0$, 0 if $b > a \geqslant 0$, $\frac{1}{4}\pi$ if $b = a > 0$;

    (ii) $\frac{1}{2}\pi b$ if $a > b \geqslant 0$,   (iii) $\frac{1}{2}\pi(b-a)$ if $0 < a, 0 < b$;

    (iv) $\frac{1}{2}\pi bc$ if $a \geqslant b \geqslant c \geqslant 0, a - b \geqslant c$;

        $\frac{1}{8}\pi(2bc + 2ca + 2ab - a^2 - b^2 - c^2)$ if $a \geqslant b \geqslant c > 0$, $a - b \leqslant c$.

(25) $\frac{1}{2}\log|(a+b)/(a-b)|$.

Examples 11A, p. 348

(1) (i) $-1/\{2(n-1)(x^2+c)^{n-1}\}$;   (ii) $-(3x^2+x+1)/(x^3+x+1)$;

    (iii) $-(x-2)/\{2x(x^2-1)\}$,   (iv) $-(x+\frac{1}{6})/(x^3+3x+1)^2$.

(2) (i) $(3x^2-x-3)/(x^3+x+1)$,   (ii) $(-3x^3+17x)/32(x^4-6x^2+1)$.

(3) (i) $\frac{1}{3}x^3 - \frac{1}{3}\log(x^3+1)$,   (ii) $\frac{1}{3}\log(x^3-1) - \frac{1}{4}\log(x^4+1)$,

    (iii) $\dfrac{1}{n-1}\log\dfrac{x^{n-1}}{1-x^{n-1}}$,   (iv) $\dfrac{x^3+1}{x^5+1} + \log(x^5+1)$.

(5) (i) $2ncI_{n+1} = (2n-1)I_n + x/(x^2+c)^n$,

    (ii) $2na^2I_{n+1} = (2n-1)I_n + x/(a^2-x^2)^n$,

    (iii) $(a-b)^2 nI_{n+1} = 2(2n-1)I_n + (2x-a-b)/\{(x-a)^n(b-x)^n\}$.

(6) (ii) $\frac{1}{2}(\sin\alpha - \alpha\cos\alpha)/\sin^3\alpha$,   (iii) $\frac{1}{2}(\alpha - \sin\alpha\cos\alpha)/\sin^3\alpha$.

(9) $A = \frac{3}{8}, B = \frac{2}{3}, C = \frac{5}{8}, D = \frac{3}{8}$.

Examples 11B, p. 350

(1) $\frac{2}{3}u^3 - u^2 + 2u - 2\log(1+u)$ ($u = \sqrt{x}$).

(2) $-\frac{2}{3}\log\{(u-1)(u+2)^2\}$ ($u = \sqrt{x}$).

(3) $\log(u^2+u+1) - (2/\sqrt{3})\tan^{-1}\{(2u+1)/\sqrt{3}\}$ ($u = \sqrt{(x-1)}$).

(4) $3 - 4\log 2$.   (5) $2 + \log 3 - \log 2$.   (6) $2 - \log 9 + \log 4$.

(7) $1 + \frac{1}{2}\pi$.   (8) $\frac{1}{2}\pi$.

Examples 11C, p. 350

(1) (i) $\frac{1}{2}\pi$,   (ii) $\frac{1}{2}\pi a^2$.

(2) (i) $\log(2+\sqrt{3})$,   (ii) $\{3\sqrt{2} - \log(1+\sqrt{2})\}a^2$.

(3) (i) $2\log(1+\sqrt{2})$,   (ii) $\{\sqrt{3} - \frac{1}{2}\log(2+\sqrt{3})\}a^2$.

(4) (i) $\pi/\sqrt{(ab)}$,   (ii) $\frac{1}{2}\pi(a+b)/(ab)^{\frac{3}{2}}$.

(5) $2\log\{\sqrt{(R-a)} + \sqrt{(R-b)}\} - \log(b-a)$.

(6) $2\log[\{\sqrt{(R+a)} + \sqrt{(R+b)}\}\{\sqrt{(R-a)} + \sqrt{(R-b)}\}/(b-a)]$.

Examples 11D, p. 351

(1) (i) $2x^{n+1}y = (2n+5)aI_{n+3} + (2n+4)bI_{n+2} + (2n+3)cI_{n+1} + (2n+2)dI_n$;

where $y = \sqrt{(ax^3 + bx^2 + cx + d)}$.

    (ii) $n(a\alpha^2 + 2b\alpha + c)I_{n+1} + (2n-1)(a\alpha+b)I_n + (n-1)aI_{n-1} = -y/(x-\alpha)^n$.

(iii) $2n(a-c)I_{n+1}-(2n-1)aI_n = \sqrt{(ax^2+c)}/(x^2+1)^n.$

(iv) $2n(a-c)I_{n+1}-(2n-1)(2a-c)I_n+2(n-1)aI_{n-1}$
$$= -x\sqrt{(ax^2+c)}/(x^2+1)^n.$$

(3) $\sqrt{(x^2-1)}+\cosh^{-1}x+\sec^{-1}x \quad (x>1).$

(4) $\frac{1}{2}\sinh^{-1}\frac{1}{x+1}-\frac{1}{2\sqrt{5}}\sinh^{-1}\frac{2x+3}{x-1} \quad (x>1).$

(5) $\dfrac{\sqrt{(x^2-x+1)}}{3(x+1)}+\dfrac{1}{2\sqrt{3}}\sinh^{-1}\dfrac{(x-1)\sqrt{3}}{x+1} \quad (x>-1).$

(6) $-\sinh^{-1}\dfrac{x-1}{x\sqrt{2}}-\dfrac{1}{\sqrt{2}}\sinh^{-1}\dfrac{x\sqrt{2}}{x-1} \quad (x>1).$

(7) $-\frac{1}{6}\left\{\log\dfrac{3+\sqrt{(5-x^2)}}{3-\sqrt{(5-x^2)}}+\tan^{-1}\dfrac{2\sqrt{(5-x^2)}}{3x}\right\}.$

(8) $\frac{1}{2}\tan^{-1}\dfrac{\sqrt{(x^2+5)}}{2}+\dfrac{1}{12}\log\dfrac{3\sqrt{(x^2+5)}+2x}{3\sqrt{(x^2+5)}-2x}.$

(9) $(1/\sqrt{2})\log(\sqrt{2}+\sqrt{3})+\frac{1}{2}\log 2-\log(1+\sqrt{3}).$

(10) $\frac{1}{8}\pi.$ 　　　　　　　　　　　　(11) $\sqrt{2}\tan^{-1}\frac{1}{2}.$

(12) $\dfrac{1}{5\sqrt{2}}\left\{\tan^{-1}\dfrac{x-4}{\sqrt{(4x^2-2x-6)}}+3\tanh^{-1}\dfrac{\sqrt{(4x^2-2x-6)}}{3x-2}\right\}.$

(13) $\sqrt{(2x^2+1)}/2(x^2+1)+\tan^{-1}\sqrt{(2x^2+1)}.$

(14) $\log(1+\sqrt{7})-\frac{1}{2}\log 10.$

Examples 11 E, p. 352

(1) $2\log\{\sqrt{(x-1)}+\sqrt{(x+1)}\}.$

(2) $2u^3-3u^2+6u-6\log(u+1) \quad (u=x^{\frac{1}{6}}).$

(3) $6u^7/7-6u^5/5+2u^3, \quad u=(x-1)^{\frac{1}{6}}.$

(4) $5u^9/9-5u^4/4, \quad u=(x+1)^{\frac{1}{5}}.$ 　　(5) $2\log(1+\sqrt{2})-\sqrt{2}.$

(6) $(\pi/n)\operatorname{cosec}(\pi/n) \quad (1<n).$ 　　(7) $3(\log 4-1).$

(8) $\frac{9}{10}.$ 　　　　　(9) $3(\frac{1}{2}+\frac{1}{2}\log 2-\frac{1}{4}\pi).$ 　　(10) $2\sqrt{3}-\frac{5}{3}.$

(11) $\frac{1}{2}\pi(\sqrt{2}-1).$ 　　　(12) $-(6-4\sqrt{2})/3+(8\sqrt{3}/9)\tan^{-1}(3\sqrt{3}-2\sqrt{6}).$

(13) $\log(3+2\sqrt{2})+2-2\sqrt{2}.$

Examples 12 A, p. 386

(1) $c^3\{\log(b/a)\}^2.$ 　　　(2) $3\log(\frac{4}{3}).$ 　　　　(3) $2\pi\{c-\sqrt{(c^2-a^2)}\}.$

(4) $20+(9/2)\log 3+(\frac{32}{3})\log 2.$ 　　(5) $(3-e)/(2e).$

(6) $(1/c)\tan^{-1}\{ab/c\sqrt{(a^2+b^2+c^2)}\}.$ 　　　(7) $9\log(\frac{4}{3})-\frac{3}{2}.$

(14) $2t/3,(5a/8,5a/8).$

(15) (i) $4/\pi$,   (ii) 0,   (iii) $-4/\pi^2$,

    (iv) $1+8/\pi^2$,   (v) 1,   (vi) $\frac{1}{2}$.

(16) (i) $\frac{3}{2}+24/\pi^2$,   (ii) $\frac{3}{2}$.

Examples 12 B, p. 388

(1) (i) $\frac{1}{6}$,   (ii) $(\pi\log 2)/(3\sqrt{3})$,   (iii) $18\frac{23}{24}$,

    (iv) $(2e^{3a}-3e^{2a}+1)/6$,   (v) ab/π,

    (vi) $45\log 2+48\frac{3}{8}$.

(3) $abc/6$.   (4) $\pi a^2 c/4 - 2a^3/3$.   (5) $16a^3 l/9\pi b$.

(6) $10\pi/3$ or $22\pi/3$.   (7) $128a^{\frac{5}{2}}b^{\frac{1}{2}}/15$.   (8) $(8a^3/9)(2\pi+3\sqrt{3})$.

(9) $(\pi a^3/24)(5\sqrt{2}-4)$.

(10) $\frac{1}{8}V = \frac{1}{3}hk\sqrt{(a^2-h^2-k^2)} - \frac{1}{3}a^3\tan^{-1}[hk/\{a\sqrt{(a^2-h^2-k^2)}\}]$
$+\frac{1}{2}h(a^2-\frac{1}{3}h^2)\sin^{-1}\{k/\sqrt{(a^2-h^2)}\} + \frac{1}{2}k(a^2-\frac{1}{3}k^2)\sin^{-1}\{h/\sqrt{(a^2-k^2)}\}$.

(14) $(3a/8, 3a/8, 3a/8)$, $(3a/8, 3b/8, 3c/8)$.

Examples 12 C, p. 389

(1) $(2a^2-c^2)\cos^{-1}(c/2a) + \frac{1}{2}c\sqrt{(4a^2-c^2)}$.   (2) $4\pi a^5/15$.

(3) a.   (6) $(4-\pi)/\pi a$.   (7) $(5\sqrt{5}-1)a/6$.

(8) (i) $16a/9\pi \doteqdot 0{\cdot}566a$,   (ii) $\frac{1}{4}\pi - 1/\pi \doteqdot 26° 45'$.

(12) (i) $a\sqrt{(\frac{2}{5})}$,   (ii) $a\sqrt{(\frac{2}{3})}$.   (13) (ii) $r\sqrt{(\frac{3}{10})}$.

(15) (i) $(1+\cos^2\alpha+\cos^4\alpha)a/(1+\cos^2\alpha)$,

    (ii) $k^2 = (2a^2/5)(1+\cos^2\alpha+\cos^4\alpha-3\cos^6\alpha)/(1+\cos^2\alpha)$.

(19) $\log 4$.

Examples 12 D, p. 392

(2) $(4-\pi)a^2/3$.

(3) $(0, 0, \frac{1}{2}a)$, $(\frac{1}{2}a, \frac{1}{2}a, \frac{1}{2}a)$, $(3a/8, 3a/8, 3a/8)$.

(6) (i) $(3\pi-4)a^3/9$; $\{12a/5(3\pi-4), 0, 45\pi a/64(3\pi-4)\}$;

    $(a/10)\sqrt{(195\pi-548)}/\sqrt{(3\pi-4)}$.

    (ii) $(\pi-2)a^2$; $\{2a/3(\pi-2), 0, \pi a/4(\pi-2)\}$; $(a/6)\sqrt{(33\pi-98)}/\sqrt{(\pi-2)}$.

(8) $2\pi\{1 - c/\sqrt{(a^2+c^2)}\}$.   (9) $\frac{2}{3}\pi$.

(10) $f(a, \alpha) + f(b, \beta) - f(a, \beta) - f(b, \alpha)$, where $f(a, b) = \tan^{-1}[ab/\{c\sqrt{(a^2+b^2+c^2)}\}]$.

(11) (i) $7a^2/6$,   (ii) $5a^2/12$. $a^2, \frac{1}{4}a^2$.

(12) (i) $c+a^2/5c$,   (ii) $3a/4 + c^2/2a - c^4/20a^3$.

(13) $a^4/15$, $a^4/35$.   (14) $a^2b^2c^2/315$.

(15) $3(\alpha^2+\beta^2+\gamma^2)^n/\{(2n+1)(2n+3)\}$.

Examples 12E, p. 393

(1)　(i)　$\frac{1}{8}(\pi - 2)$,　　　(ii)　$\log 4 - \frac{3}{2}$,　　　(iii)　$2a \log (1 + \sqrt{2})$,

　　(iv)　2,　　　　　(v)　$(2 - \sqrt{2})\,a$,　　　(vi)　$4a/\pi$,

　　(vii)　$(\frac{64}{9})\,(\log 8 - 1)$, (viii)　$\pi\{f(1) - f(0)\}$,　　(ix)　$\frac{1}{4}\pi$,

　　(x)　a^2/π.

(2)　(i)　$\frac{1}{8}(\log 4 - 1)$,　　　　　(ii)　$\frac{1}{2}(2e^2 - 3e + 6)$.

Examples 12F, p. 395

(10)　$\pi/\sqrt{2}$.　　　　(20)　$1/\{(n-1)(n-2)\ldots(n-6)\,a^{n-6}\}$.

INDEX